Solid-State Physics

Advanced Texts in Physics

This program of advanced texts covers a broad spectrum of topics which are of current and emerging interest in physics. Each book provides a comprehensive and yet accessible introduction to a field at the forefront of modern research. As such, these texts are intended for senior undergraduate and graduate students at the MS and PhD level; however, research scientists seeking an introduction to particular areas of physics will also benefit from the titles in this collection.

Springer
Berlin
Heidelberg
New York
Hong Kong
London
Milan
Paris
Tokyo

Harald Ibach   Hans Lüth

Solid-State Physics

An Introduction to Principles of Materials Science

Third Extensively Updated and Enlarged Edition

With 263 Figures, 17 Panels and 100 Problems

Springer

Professor Dr. Harald Ibach
Institut für Schichten und Grenzflächen
Forschungszentrum Jülich GmbH, 52425 Jülich and
Rheinisch-Westfälische Technische Hochschule
52062 Aachen, Germany
e-mail: h.ibach@fz-juelich.de

Professor Dr. Hans Lüth
Institut für Schichten und Grenzflächen
Forschungszentrum Jülich GmbH, 52425 Jülich and
Rheinisch-Westfälische Technische Hochschule
52062 Aachen, Germany
e-mail: h.lueth@fz-juelich.de

Title of the German original edition:
H. Ibach, H. Lüth: *Festkörperphysik. Einführung in die Grundlagen* (Sechste Auflage)

Library of Congress Cataloging-in-Publication Data: Ibach, H., 1941– [Festkörperphysik. English] Solid-state physics: an introduction to principles of materials science/ Harald Ibach, Hans Lüth. – 3rd extensively up-dated and enl. ed. p.cm. – (Advanced texts in physics, ISSN 1439-2674) Includes bibliographical references and index. ISBN 354043870X (acid-free paper) 1. Solid state physics. I. Lüth, H. (Hans) II. Title. III. Series. QC 176.I2313 2003 530.4'1–dc21 2002036466

ISSN 1439-2674

ISBN 3-540-43870-X 3rd Edition Springer-Verlag Berlin Heidelberg New York

ISBN 3-540-58573-7 2nd Edition Springer-Verlag Berlin Heidelberg New York

Springer-Verlag Berlin Heidelberg New York
a member of BertelsmannSpringer Science+Business Media GmbH

http://www.springer.de

Typesetting: K+V Fotosatz, Beerfelden
Cover design: *design & production* GmbH, Heidelberg

Printed on acid-free paper SPIN 10783587 56/3141/ba 5 4 3 2 1 0

Preface to the Third Edition

Our German textbook "Festkörperphysik" has meanwhile appeared in its 6th edition, extensively revised and extended in comparison to the latest 2nd English edition. Presently, the book has been translated into Japanese, Korean and Polish and is used as a standard text in many universities around the world. It is therefore high time to carefully revise the English text and bring it up to par with the latest 6th German edition. The sections on "High Temperature Superconductors" as well as Panel XVI on "Shubnikov-de Haas Oscillations and Quantum Hall Effect" are completely revised according to the present deeper understanding of the phenomena. This 3rd English edition has furthermore been expanded by several chapters to meet the educational requirements for recent fields of research. We let ourselves be guided by the idea that modern teaching of solid state physics emphasizes aspects of material science and its applications, in particular in solid state electronics. Accordingly, deviations from the ideal periodic solid have gained more weight in the text: we now consider phase diagrams of alloys, some basics of defect physics and amorphous solids. Because of the importance of strained layer systems in device physics, inclusion of the fundamentals of crystal elasticity theory seems (again) necessary, so a new chapter is devoted to this topic. The additional chapter on the excitation spectrum of a superconductor is intended to further the understanding of modern research on superconductor/normal conductor interfaces as well as on applications in superconductor electronics. For similar reasons, sections on the metal/semiconductor Schottky-contact and on the basic concepts of important semiconductor devices have been included in the new edition. With all of these additions we have tried to maintain the spirit of the book, namely to put the phenomena into a general frame of an atomistically founded understanding of solid state physics.

We thank Dr. Klaus Dahmen, Dr. Arno Förster, Dr. Margret Giesen, Dr. Michel Marso, Prof. Dr. Angela Rizzi and Dr. Thomas Schäpers for discussions on special topics and many suggestions for improving the presentation.

We express our thanks to Dr. H.J. Koelsch, Dr. T. Schneider and Mr. C.-D. Bachem of Springer-Verlag for the pleasant collaboration.

Jülich and Aachen, January 2003 *H. Ibach · H. Lüth*

Preface to the Second Edition

Our German textbook *"Festkörperphysik"* has become rather popular among German-speaking students, and is currently produced in its 4th edition. Its English version has already been adopted by many universities in the United States and other countries. This new 2nd edition corresponds to the 4th edition in German.

In addition to correcting some typographical errors and making small improvements in the presentation, in the present edition some chapters have been revised or extended. Panel V, for example, has been extended to include a description of angle-resolved photoemission and its importance for the study of electronic band structures. Section 10.10 on high-temperature superconductors has been completely rewritten. This active field of research continues to progress rapidly and many new results have emerged since the publication of the first edition. These results shed new light on much of the fundamental physics.

The new version of Sect. 10.10 has been developed in discussions with colleagues who are themselves engaged in superconductivity research. We thank, in particular, Professor C. Calandra from the University of Modena and Dr. R. Wördenweber of the Institute of Thin Film and Ion Technology at the Research Centre Jülich.

The revision of the problems was done with the help of Dr. W. Daum, Dr. A. Förster, A. Leuther and Ch. Ohler. We would like to thank them for their efforts. We also thank Dr. Margret Giesen for numerous improvements to the manuscript as well as Dr. Angela Lahee for the competent translation of the revised or new sections.

Jülich and Aachen, April 1995 *H. Ibach · H. Lüth*

Preface to the First Edition

In recent decades solid state physics has seen many dramatic new developments and has become one of the largest independent branches of physics. It has simultaneously expanded into many new areas, playing a vital role in fields that were once the domain of the engineering and chemical sciences. A consequence of this explosive development is that no single university lecturer can today be expected to have a detailed knowledge of all aspects of this vast subject; likewise, it is impossible to conceive of a course that could offer students a comprehensive understanding of the entire discipline and its many applications.

In view of this situation, it is particularly valuable to have a textbook that gives a concise account of the essential elements of the physics of solids. In this book the fundamental aspects of solid state physics are presented according to the scheme: Chemical bonding, structure, lattice dynamics, and electronic properties. We believe that this sequence is the optimum choice for tutorial purposes. It enables the more difficult concepts to be introduced at a point where a basic understanding of fundamental ideas has already been achieved through the study of simple models and examples. In addition to this carefully structured exposition of classical solid state theory based on the periodic solid and the one-electron approximation, the book also includes comprehensive descriptions of the most active areas in modern research: magnetism, superconductivity and semiconductor physics.

The chapter on magnetism discusses the exchange coupling of both localized and delocalized electrons, and will even guide the reader to the point when he or she can appreciate modern thin-film experiments. The standard picture of superconductivity is elucidated by means of a simplified presentation of BCS theory. A section is also devoted to the novel high-temperature superconductors. This field, however, remains in such a state of flux that it was necessary to confine the treatment to some selected experimental results and a few central ideas about this fascinating phenomenon. The chapter on semiconductors contains, in addition to a general introduction to these materials and their properties, detailed descriptions of semiconductor heterostructures, superlattices, epitaxy, and the quantum Hall effect.

In solid state physics, the interaction between theory and experiment has always played, and continues to play, a vital role. We have thus attempted throughout this book to steer a middle course in which both theory and experiment are adequately represented. Where a theoretical approach is helpful and not too cumbersome, we have not hesitated in challenging the reader with the necessary abstract concepts. Furthermore, we have tried to include theoretical methods and concepts, for example, those of group theory, that are indispensible for an understanding of contemporary original publications dealing with solid state theory.

The concise presentation of the essential theoretical aspects is complemented by the inclusion of selected experimental methods and examples, summarized in the form of self-contained panels. These offer the reader the opportunity to test and consolidate the material already studied and may prove helpful in stimulating further study in areas of particular interest.

Students will also benefit significantly from working through the extensive series of problems that relate to each chapter. These examples are not restricted to calculations based on the methods described in the text; in many cases they lead into areas that lie outside the scope of the main presentation. All of the examples have been put to the test in our own lecture courses. Nonetheless, the student may often need a helping hand or some preparatory instruction from a lecturer. The problems will be useful to both students and lecturers; they are designed to stimulate further study and to illustrate the connections between different disciplines.

This book is a translation of the third edition of the original German text. The authors consider it their immensely good fortune to have been supported by Dr. Angela Lahee in the translation and editing of this work. We are also grateful to numerous colleagues who over the years have offered valuable suggestions about the presentation of the book or have supplied the experimental material described herein. For her critical reading of parts of the manuscript and the page proofs we thank in particular Dr. Angela Rizzi. Other valuable contributions were made by Dr. W. Daum, Mr. Ch. Stuhlman, Dr. M. Wuttig and Mr. G. Bogdanyi. The figures were prepared with great care and patience by Mrs. U. Marx-Birmans and Mr. H. Mattke. The German manuscript was typed by Mrs. D. Krüger, Mrs. Jürss-Nysten and Mrs. G. Offermann. We express our thanks to Dr. H. Lotsch and Mr. C.-D. Bachem of Springer-Verlag for the pleasant collaboration.

Jülich, January 1991 *H. Ibach · H. Lüth*

Contents

1 Chemical Bonding in Solids

Solid-state physics is the physics of that state of matter in which a large number of atoms are chemically bound to produce a dense solid aggregate. The emphasis in this statement is placed on the large number of atoms involved, since that number is of the order of 10^{23} cm^{-3}. At first sight it would seem to be a hopeless task to try to apply traditional scientific methods for the description of such a vast number of atoms. However, it is exactly the large number of atoms involved that in fact makes a quantitative description possible by means of new models, applicable specifically to solids. A prerequisite, though, for the success of these models, is that the participating atoms are not simply chosen at random from the periodic table of elements; the solid must be composed of a limited number of different elements whose atoms are arranged in space with a certain order. Thus, for the solid state physicist, the showpieces are the "elementary" crystals, i.e., three-dimensional periodic arrangements of atoms of one type, or chemical compounds of two elements. An understanding of solids and of their characteristic properties thus requires that we first achieve a fundamental understanding of two particular phenomena: the first is that of the forces that hold together the atoms of a solid, in other words, the chemical bonding between atoms. The second important aspect is the structural ordering of the atoms within the solid. A consideration of these two basic questions forms the content of the first two chapters. In both cases it will only be possible to give a short introduction and for a more detailed discussion of these phenomena the reader is referred to textbooks on quantum chemistry and crystallography.

1.1 The Periodic Table of the Elements

By way of introduction to the topic of chemical bonding, we will take a brief look at the construction of the periodic table of the elements.

The electronic states of an atom are classified according to the one-electron states of the radially symmetric potential. Thus we have $1s$, $2s$, $2p$, $3s$, $3p$, $3d$, $4s$, $4p$, $4d$, $4f, \ldots$ states where the numbers give the principal quantum number, n, and the letters s, p, d, f correspond to the values of the electron's orbital angular momentum ($l = 0, 1, 2, 3, \ldots$). This classification stems from the picture in which the potential for each electron includes the effect of all other electrons by representing them as a continuous fixed charge distribu-

Table 1.1. The build-up of the periodic table by successive filling of the electronic energy shells. Indicated on the left of each column is the outer electron level that is being progressively filled, and in brackets is its maximum allowed occupation number. See also cover page

1*s* (2) H, He	4*s* (2) K, Ca	5*p* (6) In → Xe
2*s* (2) Li, Be	3*d* (10) Transition metals Sc → Zn	6*s* (2) Cs, Ba
2*p* (6) B → Ne	4*p* (6) Ga → Kr	4*f* (14) Rare earths Ce → Lu
3*s* (2) Na, Mg	5*s* (2) Rb, Sr	5*d* (10) Transition metals La → Hg
3*p* (6) Al → Ar	4*d* (10) Transition metals Y → Cd	6*p* (6) Tl → Rn

tion which, to a greater or lesser extent, screens the potential of the bare nucleus. In addition to the principal quantum number n and the orbital angular momentum quantum number l, there is also a magnetic quantum number m which can take $(2l+1)$ different values (ranging from $-l$ to $+l$). According to the Pauli exclusion principle, each state can be occupied by at most two electrons of opposite spin. As a function of increasing nuclear charge this then leads to the periodic tables whose structure is outlined in Table 1.1. From the order of the energy levels of the hydrogen atom, one would expect that after the 3*p*-states are filled, the next states to be occupied would be the 3*d*. But in fact, as can be seen from Table 1.1, this is not the case; following the 3*p*-levels those next occupied are the 4*s*. The subsequent filling of the 3*d*-states gives rise to the first series of transition metals (the 3*d*-metals). Similarly, one also finds 4*d*- and 5*d* transition metals. The same effect for the *f*-states leads to the so-called rare earths. The reason for this anomaly is that the electrons in *s*-states have a nonvanishing probability of being located at the nucleus thereby reducing for them the screening effect of the other electrons. Hence the *s*-electrons possess lower energy.

If one considers a thought experiment in which several initially isolated atoms are gradually brought closer together, their interaction with one another will lead to a splitting of each of their energy levels. If a very large number of atoms are involved, as in the case of a real solid, then the energy levels will lie on a quasicontinuous scale and one therefore speaks of energy bands (Fig. 1.1). The width of the band (i.e., the broadening) depends on the overlap of the wavefunctions concerned. Thus for the deep lying levels the broadening is small, and these "core levels" retain their atomic shell-like character even in the solid. For the highest occupied levels, on the other hand, the broadening is so large that the *s*-, *p*- and where present, *d*-levels merge into a single band. It is the electrons in this uppermost band that are responsible for the chemical bonding between atoms, and hence one speaks of the valence band. The ultimate source of the chemical bonding is the reduction in electronic energy which results from the level broadening. This, despite the increase in repulsion between the nuclei, leads to a decrease in the total energy as a function of atomic separation until the point where the equilibrium separation is reached – i.e., the point of minimum total energy.

The type of bonding in a solid is determined essentially by the degree of overlap between the electronic wavefunctions of the atoms involved. At

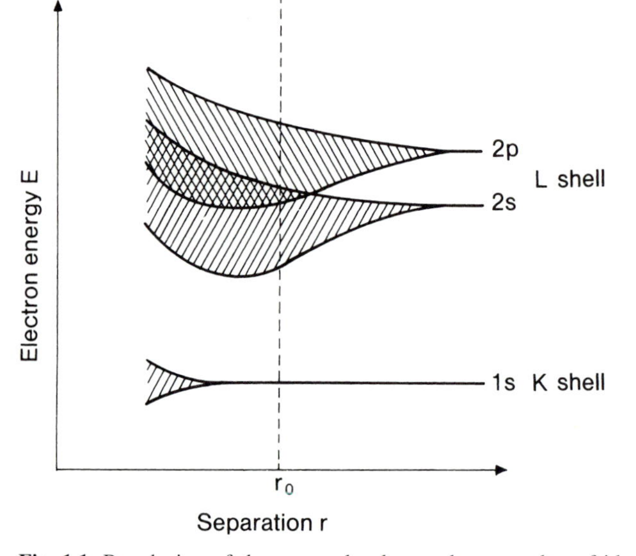

Fig. 1.1. Broadening of the energy levels as a large number of identical atoms from the first row of the periodic table approach one another (schematic). The separation r_0 corresponds to the approximate equilibrium separation of chemically bound atoms. Due to the overlap of the $2s$ and $2p$ bands, elements such as Be with two outer electrons also become metallic. Deep-lying atomic levels are only slightly broadened and thus, to a large extent, they retain their atomic character

the one extreme, this overlap may be limited to neighboring atoms; in other cases the wavefunctions may be spread over many atoms. In the former case, the degree of overlap, and thus the strength of the bonding, is dependent not only on the separation of neighboring atoms, but also on the bond angles. This is referred to as directional bonding or *covalent* bonding.

In its purest form, covalent bonding is realized between a few elements of equal "valence", i.e. elements with the same outer electronic configuration. However, an equal electronic configuration is neither a necessary nor a sufficient condition for covalent bonding. What is important is simply the relative extent of the wavefunctions in comparison to the interatomic separation. If the extent of the wavefunctions is large compared to the nearest-neighbor distance, then the exact position of the nearest neighbors plays an insignificant role in producing the greatest possible overlap with many atoms. In this case, the packing density is more important than the position of the next neighbors. Here one speaks of non-directional bonding. This regime in which the wavefunctions spread over a distance that is large in comparison to the atomic separation is characteristic of *metallic* bonding.

However, there is a further type of non-directional bonding with extremely small overlap of wavefunctions; this is the *ionic* bond. It occurs in cases where the transfer of an electron from one atom to another is sufficiently energetically favorable. A prerequisite for ionic bonding is thus a dissimilarity of the atoms involved.

In the following sections we will explore the various types of bonding in greater detail.

1.2 Covalent Bonding

We have characterized covalent bonding in solids as a bonding for which the interaction between nearest neighbor atoms is of prime importance. It is therefore possible to derive many of the essential properties of covalent solids using the quantum chemistry of molecules. For our discussion we shall refer to the simplest model for bonding, namely of a diatomic molecule with a single bonding electron.

The Hamiltonian for this molecule contains the kinetic energy of the electron and the Coulomb interaction between all partners (Fig. 1.2 a).

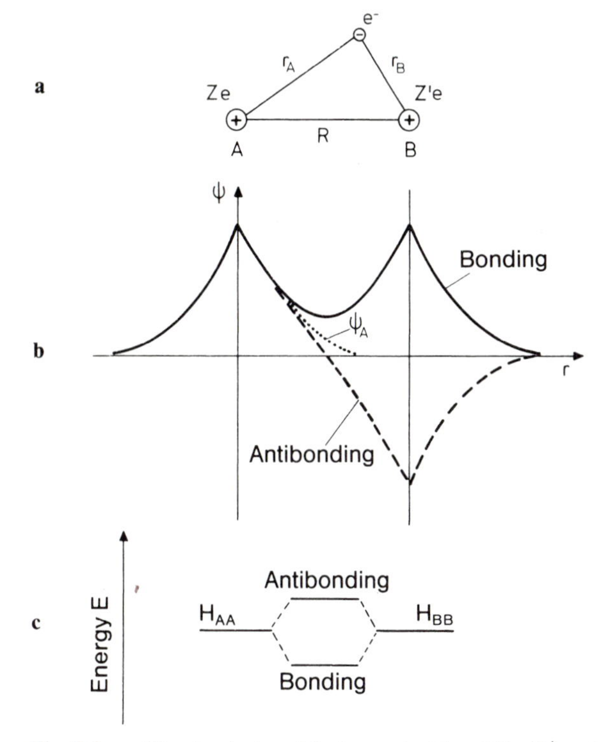

Fig. 1.2 a–c. The simplest model of a covalent bond (the H_2^+ molecule). (**a**) Definition of the symbols in (1.1). (**b**) Bonding and antibonding combinations of atomic orbitals. The bonding combination leads to an accumulation of charge between the nuclei which in turn gives rise to a reduction in the Coulomb energy. (**c**) The splitting of the atomic energy level into the bonding and antibonding states. The greatest gain in energy is achieved when the bonding state is fully occupied – i.e., contains two electrons – and the antibonding state is empty (covalent bonding)

$$\mathscr{H} = -\frac{\hbar^2}{2m}\Delta - \frac{Z e^2}{4\pi\varepsilon_0 r_A} - \frac{Z' e^2}{4\pi\varepsilon_0 r_B} + \frac{ZZ' e^2}{4\pi\varepsilon_0 R} \,. \tag{1.1}$$

The appropriate molecular orbital ψ_{mo} for the electron would be the solution to the Schrödinger equation

$$\mathscr{H}\psi_{\mathrm{mo}} = E\,\psi_{\mathrm{mo}} \,. \tag{1.2}$$

However, even in this simple case, it is necessary to rely on approximate solutions. The expectation value of the ground-state energy may be calculated using such an approximate solution, according to

$$E' = \frac{\int \psi^* \mathscr{H} \psi \, d\boldsymbol{r}}{\int \psi^* \psi \, d\boldsymbol{r}} \,. \tag{1.3}$$

The approximate solution ψ may be set equal to a linear combination of states of the two separate atoms:

$$\psi = c_A \psi_A + c_B \psi_B \,. \tag{1.4}$$

Here the wavefunctions and their coefficients are real.

It is possible to show that any trial function such as ψ always leads to an energy E' that lies above the true value E (see Problem 1.8). The best values for the coefficients c_A and c_B are those which lead to a minimum value of E'.

With the abbreviations

$$S = \int \psi_A \psi_B \, d\boldsymbol{r} \quad \text{(overlap integral)} \,, \tag{1.5a}$$

$$H_{AA} = \int \psi_A \mathscr{H} \psi_A \, d\boldsymbol{r} \,, \tag{1.5b}$$

$$H_{AB} = \int \psi_A \mathscr{H} \psi_B \, d\boldsymbol{r} \,, \tag{1.5c}$$

one obtains the following expression for E', which then has to be minimized

$$E' = \frac{c_A^2 H_{AA} + c_B^2 H_{BB} + 2 c_A c_B H_{AB}}{c_A^2 + c_B^2 + 2 c_A c_B S} \,. \tag{1.6}$$

For the minimum with respect to c_A and c_B we require

$$\frac{\partial E'}{\partial c_A} = \frac{\partial E'}{\partial c_B} = 0 \,, \tag{1.7}$$

which leads to the following secular equations

$$c_A (H_{AA} - E') + c_B (H_{AB} - E' S) = 0 \,, \tag{1.8a}$$

$$c_A (H_{AB} - E' S) + c_B (H_{BB} - E') = 0 \,. \tag{1.8b}$$

Their solution is given by the condition that the determinant vanishes, i.e.,

$$(H_{AA} - E')(H_{BB} - E') - (H_{AB} - E' S)^2 = 0 \,. \tag{1.9}$$

For simplicity, we consider a molecule with two identical nuclei (for example, H_2^+) for which we have $H_{AA}=H_{BB}$. With the single atomic eigenvalue $H_{AA}=H_{BB}$ of the individual free atoms, we then obtain two new molecular orbitals with the energies

$$E_\pm \lesssim E'_\pm = \frac{H_{AA} \pm H_{AB}}{1 \pm S} \ . \tag{1.10}$$

When the two nuclei are infinitely far apart we have $S=0$ on account of (1.5 a), whereas when the nuclear positions coincide, we have $S=1$. From (1.10) it follows that the spatial overlap of the wavefunctions ψ_A and ψ_B leads to a splitting of the original energy level $H_{AA}=H_{BB}$ into a higher and a lower molecular energy level (Fig. 1.2 c). The molecular orbital corresponding to the higher energy level is known as *antibonding*, and the other is *bonding*. In the molecule the electron occupies the lower-lying bonding orbital thereby giving rise to a reduction in the total energy. This reduction corresponds to the binding energy of the covalent bond.

From the foregoing discussion one sees that only partially occupied single-atomic orbitals, i.e., those containing less than two electrons, can participate in covalent bonding: Since the bonding molecular orbital can contain only two electrons (the Pauli principle allows two opposite spin states), any further electrons would have to occupy the higher-lying antibonding orbital, which would counteract the original gain in energy (cf. Problem 1.7).

For the diatomic molecules considered here the bonding molecular orbital consists of an additive combination of ψ_A and ψ_B, i.e. $\psi_{mo}=\psi_A+\psi_B$ [in (1.4) we have $c_A=c_B$ for identical nuclei]. As shown in Fig. 1.2, this leads to an increase in the electronic charge density between the nuclei. The antibonding combination $\psi_{mo}=\psi_A-\psi_B$, on the other hand, results in a decrease of this charge density.

One sees that covalent bonding is associated with a "piling-up" of charge between the atoms that form the molecule or solid concerned. It is the spatial overlap of the wavefunctions that is responsible for this bonding and which also determines the energy gain of the bonding orbitals of the molecule or solid, and thereby the binding energy. As shown in Fig. 1.3, for particular atomic orbitals (*s, p, d*, etc.) there are some orientations that favor the overlap and others that are unfavorable. This is the underlying reason for this strongly directional character of the covalent bonding, which is particularly evident in the covalently bound crystals of diamond (C), Si and Ge with their tetrahedral coordination (Fig. 1.4).

Let us take a closer look at this tetrahedral bonding for the example of diamond. From a consideration of its electronic configuration $1s^2$, $2s^2$, $2p^2$, one would expect a carbon atom to be able to participate in only two covalent bonds (corresponding to the two $2p$-orbitals each occupied by one electron). However, when these atoms form part of a crystal it is clear that a larger reduction in total energy is produced if four bonding orbitals can overlap. In a one-electron picture this may be understood in terms of the following simplified description: One of the electrons from the $2s$-orbital is

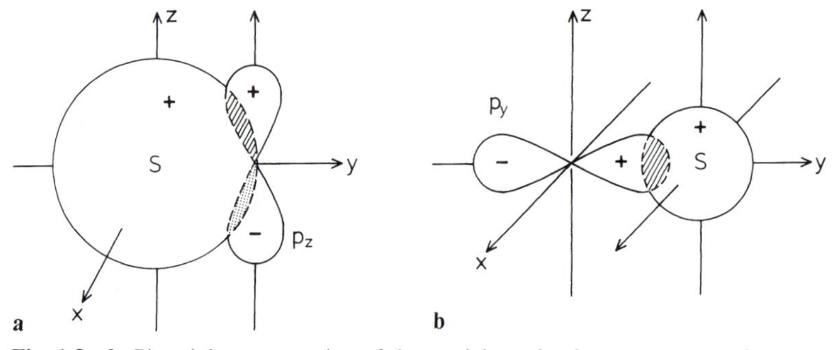

Fig. 1.3 a, b. Pictorial representation of the spatial overlap between an s- and a p-wavefunction of hydrogen. The "size" of the wavefunction is illustrated by a surface of constant wavefunction amplitude. (**a**) A situation in which the overlap cancels due to sign reversal of the p_z-wavefunction. (**b**) Nonvanishing overlap between s- and p_x-wavefunctions

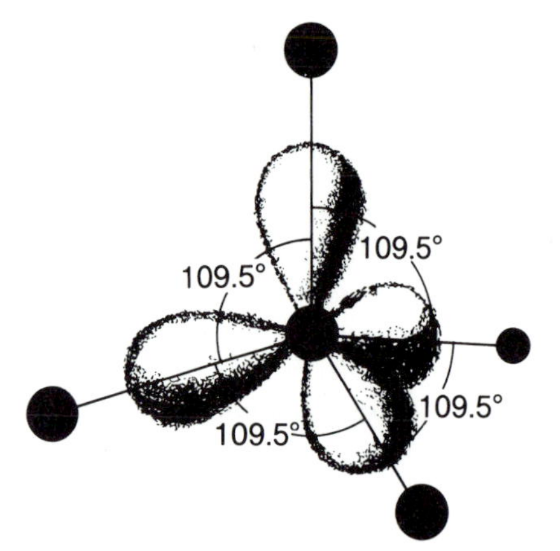

Fig. 1.4. The tetrahedral configuration of nearest neighbors in the lattice of C, Si, Ge and α-Sn. This structure is favored because its periodic repetition fills three-dimensional space, and because it enables the formation of sp^3 hybrid orbitals from the s, $p_x\,p_y$ and p_z states. The figure displays the orbitals of diamond (C). The orbitals of Si, Ge, and α-Sn possess additional nodes

excited into the empty $2p$-orbital. Each of the $2p$-orbitals and the single $2s$-orbital now contains one electron and thus each of these can participate in a covalent bond. The overlap with the wavefunctions of the nearest neighbors is maximized if four new wavefunctions are formed by a linear combination of the original $2s$, $2p_x$, $2p_y$, and $2p_z$-orbitals. These new molecular orbitals are known as sp^3 hybrids and their creation is called hybridization (Problem 1.9). The gain in energy that is produced by the overlap in the tetrahedral configuration is more than enough to compensate for the energy needed to promote the $2s$-electron into the $2p$-orbital.

If one now considers carbon atoms arranged in the diamond structure, in which each atom is surrounded by four other atoms positioned at the corners of a tetrahedron (Fig. 2.12), one finds that in the sp^3 hybridized state all the available electrons can be shared with the neighbors such that only the bonding orbitals are occupied. This leads to a fully occupied valence band that is separated from the next highest (antibonding) band by an energy gap. Energy can only be supplied in large quanta that are sufficient to promote an electron across this band gap. At low temperature such covalently bonded solids are therefore nonconducting. If the bandgap is not too large, however, electrons can be promoted by thermal excitation leading to a measurable conductivity. In this case one speaks of semiconductors. A more precise definition of these will be given in Chaps. 9 and 12.

Instead of the sp^3 hybrid orbital carbon can also form a planar hybrid orbital from one $2s$ and two $2p$ functions (Problem 1.10). These orbitals yield a planar 120° star called sp^2. An additional p_z orbital containing one electron lies perpendicular to the plane of the star. The overlap between p_z orbitals of neighboring C atoms leads to a further bonding, the so-called π bonding. This type of bonding is found within the layers of the graphite structure of carbon. The bonding between the covalently bound layers of graphite is of the van der Waals type (Sect. 1.6) and is thus relatively weak.

An interesting spatial structure involving sp^2 orbitals is that of the fullerenes, whose most prominent member is C_{60} (Fig. 1.5). Their structure includes pentagons. For topological reasons, 12 pentagons are necessary to produce a closed structure. In addition to the 12 pentagons, the C_{60} cluster also contains 20 hexagons. Even larger molecules can be produced by the inclusion of more hexagons. C_{60} clusters can also form the basis of three-dimensional crystal structures, for example when other atoms such as alkali or alkaline earth metals are included.

A complete saturation of the covalent bonding is possible for the group IV elements C, Si, Ge and α-Sn in the three-dimensional space-filling tetrahedral configuration. The group V elements P, As, Sb, demand a three-fold coordination for saturation; they form planar layer structures. Correspondingly, the group IV elements Te and Se occur in a chain-like structure with two-fold coordination.

Covalently bonded solids can, of course, also be produced from combinations of different elements. As an example, we consider boron nitride. The two elements involved have the electronic structure: $B(2s^2,2p^1)$;

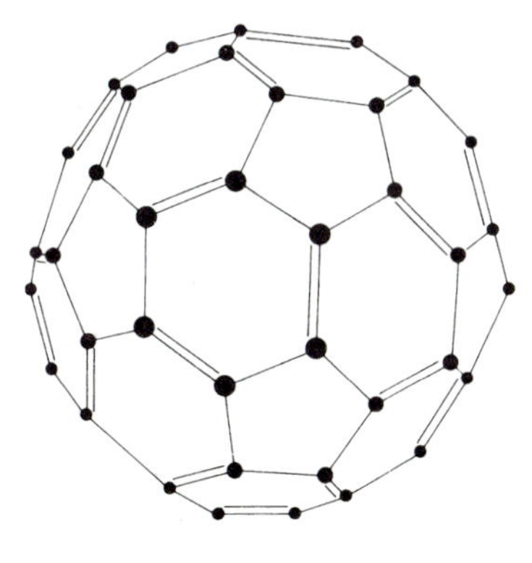

Fig. 1.5. The structure of C_{60}

N$(2s^2,2p^3)$. These elements can also bind in the diamond structure with tetrahedral coordination. Each boron atom is thereby surrounded by four nitrogen atoms and vice versa. The shared bonding electrons are comprised of five electrons from the nitrogen atom and three from boron. The total number of electrons per atom is thus the same as in the case of the carbon diamond structure. However, because two different elements are involved, the bonding has an ionic character. This will be discussed in the next section.

Typical examples of binding energy for purely covalently bonded crystals are:

C (diamond):	7.30 eV per atom (712 kJ/mol);
Si:	4.64 eV per atom (448 kJ/mol);
Ge:	3.87 eV per atom (374 kJ/mol).

1.3 Ionic Bonding

To understand ionic bonding one needs to consider the ionization energies and electron affinities of atoms. The *ionization energy* I is defined as the energy that must be supplied in order to remove an electron from a neutral atom. The *electron affinity* A is the energy that is gained when an additional electron is added to a neutral atom. Ionic bonding is produced whenever an element with a relatively low ionization energy is combined with an element with a high electron affinity. As an example we consider the combination of sodium and chlorine. The ionization energy of sodium is 5.14 eV, and the electron affinity of chlorine 3.71 eV. Thus in order to transfer one electron from a sodium atom to a chlorine atom requires an energy expenditure of 1.43 eV. The electrostatic attraction between the two ions leads to an energy gain that increases as they approach closer to one another, with a minimum separation that is determined by the sum of their two ionic radii. This electrostatic attraction contributes an energy gain of 4.51 eV thus giving an overall gain in energy of 3.08 eV. It is therefore possible for sodium and chlorine to form a diatomic molecule with a strongly ionic character. Three-dimensional crystals can also be produced in this way. In the structure so formed, each chlorine atom is surrounded by sodium neighbors, and vice versa. The exact structure is determined by the optimal use of space for the given ionic radii, and by the condition that the Coulomb attraction between oppositely charged ions should be greater than the repulsion between ions of the same sign. Figure 1.6 shows two structures that are typical for two-ion crystals; the sodium-chloride and the cesium-chloride structures.

The ionic radii are the essential factor in determining the minimum separation because, if the ions were to approach still closer, a strong overlap between the ionic electron clouds would occur. For fully occupied electron shells such as these (Sect. 1.2), the Pauli principle would then require

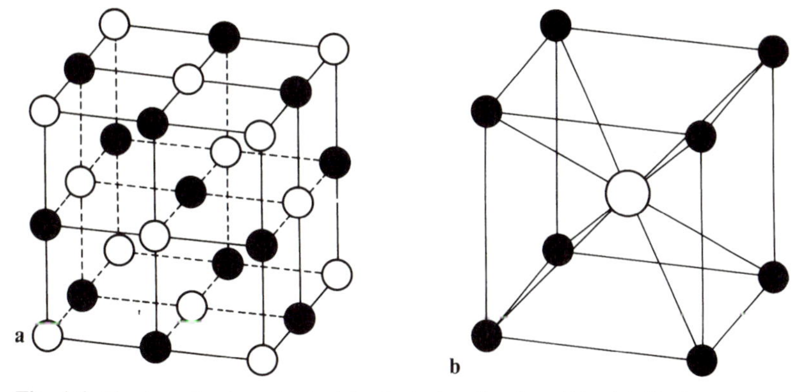

Fig. 1.6. The two structures typical for ionic bonding in solids: (**a**) NaCl structure; (**b**) CsCl structure

that higher lying antibonding orbitals should become occupied. This should lead to a steep increase in the energy and therefore a strong repulsion.

Whereas this repulsive contribution to the total energy, like the covalent bond, can only be derived from a quantum mechanical calculation, the attractive Coulomb contribution to the ionic bond can be described simply as a sum over the Coulomb potentials ion sites. For the potential energy between two singly charged ions i and j at a separation r_{ij} one writes

$$\varphi_{ij} = \pm \frac{e^2}{4\pi\varepsilon_0 r_{ij}} + \frac{B}{r_{ij}^n} ; \tag{1.11}$$

the second term describes the repulsion between the two electron clouds. It is an heuristic term containing two free parameters n and B. It should, of course, be possible to determine these parameters from an exact quantum mechanical treatment of the problem. However, a commonly used approach is to derive the values of these parameters from a fit to measured experimental quantities (ionic separation, compressibility, etc.); this yields values of n mainly in the range 6 to 10.

The form of a typical potential energy curve is depicted in Fig. 1.7. The total potential energy due to all other ions j at the site of ion i is given by the summation

$$\varphi_i = \sum_{i \neq j} \varphi_{ij} . \tag{1.12}$$

If r is the separation of nearest neighbors one can write

$$r_{ij} = r p_{ij} , \tag{1.13}$$

where the p_{ij} depend on the particular crystal structure. If the crystal contains N ion pairs, then its total potential energy is given by

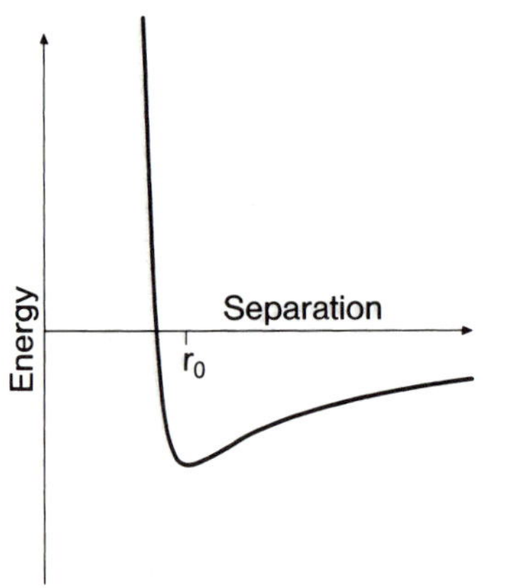

Fig. 1.7. Potential energy as a function of the separation of two ions

$$\Phi = N\,\varphi_i = N\left(-\frac{e^2}{4\pi\varepsilon_0 r}\sum_{i\neq j}\frac{\pm 1}{p_{ij}} + \frac{B}{r^n}\sum_{i\neq j}\frac{1}{p_{ij}^n}\right) . \tag{1.14}$$

For each possible structure one has the characteristic quantity

$$A = \sum_{i\neq j}\frac{\pm 1}{p_{ij}} \tag{1.15}$$

which is known as the *Madelung constant.* For the sodium-chloride structure $A = 1.748$ and for the cesium-chloride structure $A = 1.763$.

Some typical binding energies are

NaCl: 7.95 eV per ion pair (764 kJ/mol);
NaI: 7.10 eV per ion pair (683 kJ/mol);
KrBr: 6.92 eV per ion pair (663 kJ/mol).

In ionic crystals, it is not possible for the electrons to move about freely between ions unless a large amount of energy (~ 10 eV) is supplied. Solids with ionic bonding are therefore nonconducting. However, the pres-

Table 1.2. The electronegativity of selected elements [1.1]

H						
2.1						
Li	Be	B	C	N	O	F
1.0	1.5	2.0	2.5	3.0	3.5	4.0
Na	Mg	Al	Si	P	S	Cl
0.9	1.2	1.5	1.8	2.1	2.5	3.0
K	Ca	Sc	Ge	As	Se	Br
0.8	1.0	1.3	1.8	2.0	2.4	2.8
Rb	Sr	Y	Sn	Sb	Te	I
0.8	1.0	1.3	1.8	1.9	2.1	2.5

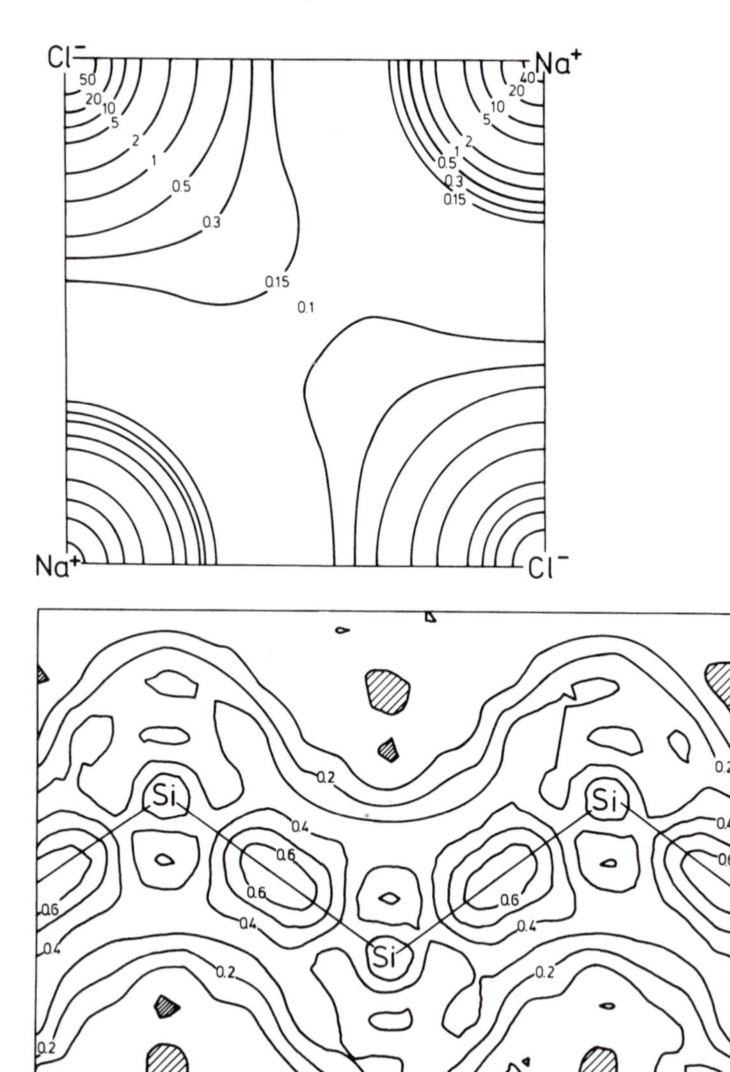

Fig. 1.8. Density of valence electrons in a typical ionic crystal (NaCl) and in a typical covalently bound crystal (Si) [1.2, 3]. One clearly sees the concentration of charge along the bond between Si atoms, whereas in the ionic bonding, the electrons are almost spherically distributed around the ions

ence of defects means that at high temperatures the ions themselves can move around, giving rise to ionic conduction.

Ionic bonding and covalent bonding are two limiting cases, of which only the latter can exist in solids composed of a single type of atom. In the majority of cases the bonding is of an intermediate nature representing a mixture of the two extremes. A qualitative measure of the ionicity of a bond is made possible by the electronegativity scale. This scale was first developed by Pauling as a result of observations of bond energies. Subsequently Millikan made use of the physical quantities ionization energy, I, and electron affinity, A, to derive the following definition of the electronegativity of an element:

$$X = 0.184\,(I + A)\ . \tag{1.16}$$

If the ionization energy and electron affinity are expressed in electron volts one obtains the Pauling electronegativity scale of Table 1.2. The higher the ionization energy and electron affinity of an atom, the greater is its tendency to draw the electrons of a bond towards itself. In a bond between two atoms it is therefore always the atom with the higher electronegativity that is the anion. The difference in the electronegativity of the two atoms is a measure of the ionicity of the bond.

The difference between ionic bonding and covalent bonding, in particular the different electron density distributions, is illustrated in Fig. 1.8. This shows contours of constant electron density which may be deduced for example from X-ray diffraction studies. For ionic bonding (NaCl) the electrons are concentrated around the atoms, whereas for covalent bonding (Si) there is an accumulation of electrons between the atoms.

1.4 Metallic Bonding

It is possible to regard metallic bonding as an extreme case of bonding in which the electrons are accumulated between the ion cores. However, in contrast to covalent bonding, the electrons now have wavefunctions that are very extended in comparison to the separation between atoms. As an example, Fig. 1.9 shows the radial part of the $3d$ and $4s$ wavefunction of nickel in the metallic state. The $4s$ wavefunction still has significant amplitude even at a distance half way to the third nearest neighbors; thus many neighbors are involved in the bonding. This leads to a strong screening of the positive ion core, and to a bonding that has certain similarities to covalent bonding. However, due to the strong "smearing out" of the valence electrons over the whole crystal, the bonding is not directional as in the case of covalent crystals. The crystal structure of metals is thus determined, to a large extent, by the desire for optimum filling of space (Sect. 2.5).

Unlike the s-electrons, the d-electrons of transition metals are localized, and the overlap is correspondingly smaller. The d-electrons produce a kind of covalent framework in the transition metals and produce the main contribution to the binding energy.

The valence band of metals, comprising the outer s-, p- and sometimes d-electrons, is not fully occupied (Table 1.1). As a consequence of the quasi-continuous distribution of states on the energy scale for a large number of atoms, one can supply energy to the electrons in infinitesimally small portions; in particular, they can be accelerated by an externally applied

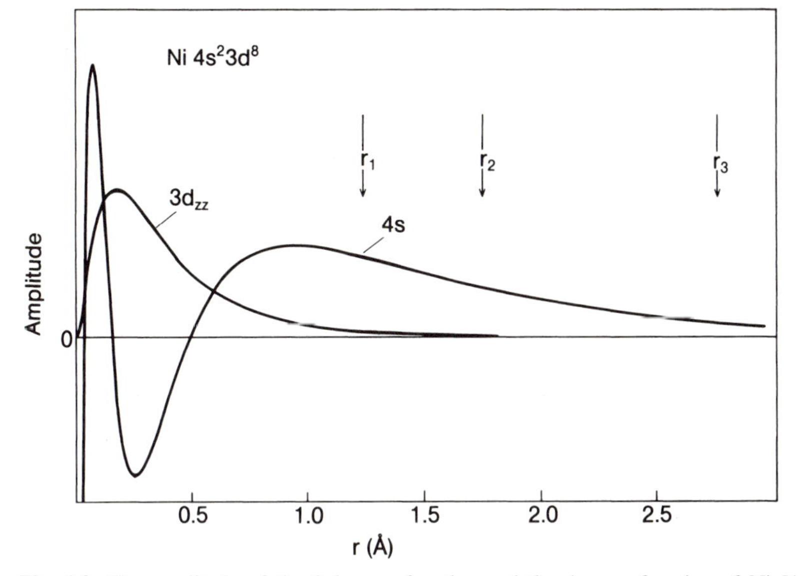

Fig. 1.9. The amplitude of the $3d_{zz}$-wavefunction and the $4s$-wavefunction of Ni [1.4]. The half-distances to the first, second and third nearest neighbors (r_1, r_2 and r_3) are shown for comparison

electric field. A particular feature of metals is thus their high electrical conductivity, which is related to their similarly high thermal conductivity. In this sense, the metallic bond is a feature that is peculiar to solids, i.e., to aggregates of many atoms.

A glance at Table 1.1 indicates that the partially filled valence band of metals can arise in various ways. The alkali metals (Li, Na, K, Rb, Cs) clearly have the necessary prerequisite since the outer atomic *s*-state is only singly occupied. For the alkaline earth metals (Be, Mg, Ca, Sr, Ba) one might initially expect a full valence band formed from the doubly occupied atomic *s*-states. However, because of the overlap in energy with the bands stemming from the (empty) atomic *p*-states of the same shell, the net result is a partially occupied joint *sp*-band. The transition metals represent a special case. Here the *s*- and *p*-states again form a broad common band. As mentioned previously, the *d*-electrons have a relatively small spatial extent (Fig. 1.9) and, due to the correspondingly small overlap with the neighboring atoms, the *d*-band of the transition metals has a smaller energy width than the *sp*-band.

The large spreading of the wavefunction of valence electrons in metals makes it particularly difficult to theoretically predict their binding energy. On the other hand, the valence electrons are free to move between the atoms. This simplifies the description of the electrical conductivity and the specific heat of the electrons, as will become clear in Chap. 6.

1.5 The Hydrogen Bond

One speaks of a hydrogen bond when a hydrogen atom is bound to two other atoms. At first sight, it is surprising that such a bond can exist since hydrogen has only one electron. However, one can imagine the hydrogen bond as follows: when hydrogen takes part in a covalent bond with a strongly electronegative atom, for example, oxygen, its single electron is almost completely transferred to the partner atom. The proton which remains can then exert an attractive force on a second negatively charged atom. Because of the extended electron cloud of the electronegative atom and the extremely small size of the proton with its strongly reduced electron screening, it is not possible for a third atom to be bound. The hydrogen atom is thus always doubly coordinated in a hydrogen bond. Such bonds are most common between strongly electronegative atoms, but are not limited to this case alone. They can be of a symmetric A–H–A type or of an asymmetric A–H...B type. A criterion for the existence of a hydrogen bond is that the observed separation of the atoms A and B is smaller than it would be if only van der Waals bonding (Sect. 1.6) were present. A further indication of the presence of hydrogen bonds may be obtained from infrared spectroscopy, in which the band corresponding to hydrogen vibrations shows a strong shift, and is often broadened, too. Generally, speaking, the phenomena associated with hydrogen bonding are quite diverse, and this type of bonding is harder to characterize than most other types. The binding energies of hydrogen bonds are of the order of 0.1 eV per bond.

It is hydrogen bonds that are responsible for linking the two chains of the double helix in the DNA molecule and, as such, these bonds play a crucial role in the mechanism of genetic reproduction. The best-known example from the realm of inorganic chemistry is water, in particular when it is in the form of ice. Each oxygen atom in ice is surrounded by four further oxygen atoms in a tetrahedral configuration and the bonding is provided by hydrogen atoms. Hydrogen bonds are also present in the liquid form of water, which leads, for example, to the anomaly in the expansion coefficient: water has its highest density at 4 °C. The reason for this is that the liquid form of water also contains complexes of H_2O molecules that are held together by hydrogen bonds. In comparison to H_2O molecules without hydrogen bonds, the former occupy a larger volume. As the temperature increases the hydrogen-bonded aggregates melt, leading to an increase in the density. Above 4 °C one finds the usual thermal expansion, i.e., the density decreases with further increasing temperature.

1.6 The van der Waals Bond

This is an additional type of bond that, in principle, is always present. However, the van der Waals bond is only significant in cases where other types of bonding are not possible, for example, between atoms with closed electron shells, or between saturated molecules. The physical source of this bonding is charge fluctuations in the atoms due to zero-point motion. The dipole moments which thereby arise cause an additional attractive force. Van der Waals forces are responsible for the bonding in molcular crystals. The bonding energy is dependent on the polarizability of the atoms involved and is mostly of the order of 0.1 eV. Typical atomic bonding radii for van der Waals bonding are considerably larger than for chemical bonding. The attractive part of the potential between atoms that interact only via van der Waals forces varies as r^{-6}, where r is the separation of the atoms (or molecules). This can be readily understood as a consequence of the dipole interaction. A dipole moment p_1 resulting form a momentary charge fluctuation gives rise to an electric field at distance r of strength $\mathscr{E} \sim p_1/r^3$. A second atom of polarizability α situated at distance r is polarized by this electric field and acquires an induced dipole moment of $p_2 \sim \alpha p_1/r^3$. Since the potential of this dipole in the field is proportional to $\mathscr{E}$ and to p_2, it follows that the attractive part of the van der Waals interaction varies as r^{-6} (Problem 1.11).

Problems

1.1 a) Calculate the Madelung constant A for a linear ionic chain.

b) Make approximate numerical calculations of the Madelung constant ($A = 1.7476$) for the NaCl lattice. Use two different approaches: First, a cubic geometry where $2ma$ is the length of a side of the cube and a the separation of nearest neighbors, and, second, a spherical geometry where ma is the radius of the sphere. In both cases the reference ion is located at the center. Carry out the calculation for m-values of $m = 97$, 98, and 99 and compare the results. What is the cause of the discrepancy?

1.2 Determine the isothermal bulk modulus

$$\kappa = V\left(\frac{\partial p}{\partial V}\right)_T$$

and the lattice energy per ion pair for NaCl using the expression for the lattice energy of N ion pairs:

$$U(r) = N\left(-\frac{e^2}{4\pi\varepsilon_0 r}A + \frac{B}{r^n}\sum_{i\neq j}\frac{1}{p_{ij}^n}\right).$$

Using the value $n = 9$, calculate B from the condition that $U(\boldsymbol{r})$ is a minimum at the equilibrium separation.

1.3 It is well known that common salt (NaCl) is easily dissolved in water. In the solution Na and Cl atoms are present as positive and negative ("solvated") ions. Show that, due to the high dielectric constant of water and the resulting screening of the Coulomb potential, the binding energy of a NaCl crystal in water is smaller than the mean thermal energy of the free ions. Calculate the equilibrium separation of the ions of a hypothetical NaCl crystal in water and show that this separation is larger than the van der Waals radius of a water molecule, thereby justifying the approximate approach used here to discuss the solubility of NaCl.

1.4 Discuss the ionicity of alkali halides with the help of Table 1.2.

1.5 Take the CsCl structure and assume that the cation radius becomes smaller while the anion radius remains constant. What happens to the binding energy? Show that for small cations the NaCl lattice becomes the preferred structure. For an even smaller cation radius the ZnS lattice has the largest binding energy (Madelung constant $A = 1.638$). Give examples.

1.6 Calculate approximately the zero-point entropy of ice. In the ice structure the oxygen atoms form a wurtzite lattice which is stabilized by hydrogen bonds between the nearest neighbor oxygen atoms. The zero-point entropy results from the possible ways of distributing the two hydrogen atoms per oxygen atom over the four bonds to the nearest neighbors.

1.7 Discuss the electron configuration of the oxygen molecule. Why is the bond strength equivalent to a chemical double bond? Why is O_2 paramagnetic? Give an argument to explain why O_2^+ is a stable ion.

1.8 Prove that the estimate of the ground-state energy obtained with the help of an approximate wavefunction using the Ritz procedure,

$$E = \frac{(\psi, \mathcal{H}\psi)}{(\psi, \psi)} ;$$

is always greater than or equal to the exact eigenvalue E_0. To do this, expand the approximate function in terms of the exact eigenfunctions ψ_i (exact eigenvalue E_i).

1.9 The carbon atom in its tetrahedral bonding configuration in diamond can be approximately represented by four $2sp^3$ wavefunctions, which are linear superpositions of the four hydrogenic $2s$, $2p_x$, $2p_y$, $2p_z$ atomic wavefunctions (ϕ_j):

$$\psi_i = \sum_j a_{ij}\,\phi_j , \quad \text{with} \quad i,j = 1,2,3,4$$

and where the possible functions ϕ_j appear in spherical coordinates as follows:

$$\phi_1 = \phi\,(2\,s) = c\,\mathrm{e}^{-\varrho}\,(1-\varrho)\;,$$
$$\phi_2 = \phi\,(2p_z) = c\,\mathrm{e}^{-\varrho}\,\varrho\cos\theta\;,$$
$$\phi_3 = \phi\,(2p_x) = c\,\mathrm{e}^{-\varrho}\varrho\sin\theta\cos\varphi\;,$$
$$\phi_4 = \phi\,(2\,p_y) = c\,\mathrm{e}^{-\varrho}\,\varrho\sin\theta\sin\varphi\;, \quad \text{and}$$
$$\varrho = \frac{Zr}{2a_0} \quad (a_0 \text{ Bohr radius, Z=nuclear charge, here that of carbon})\;.$$

One demands of the ψ_i that, like the ϕ_j, they are orthonormal, i.e.

$$\int \psi_i\,\psi_k^*\,d\boldsymbol{r} = \delta_{ik}\;.$$

a) Plot the contours of ϕ (ϱ=const, θ, φ) for the $2s$ and one of the $2p$ states on a polar diagram.
b) Prove that the orthonormality requirement for the $\psi_i(2sp^3)$ wavefunctions leads to the condition

$$\sum_j a_{ij}\,a_{kj} = \delta_{ik} \quad \text{with} \quad a_{ij} = a_{ji}^*\;.$$

c) Determine four possible ψ_i functions which fulfill the orthonormality requirement with $a_{ij}=\frac{1}{2}$ or $a_{ij}=-\frac{1}{2}$.
d) Show that the maxima of $|\psi_i|^2$ are located in tetrahedral directions and draw these by means of vectors pointing to the corners of a cube whose edges are parallel to the x, y and z axes.
e) Show that the electron density $\sum_{i=1}^{4}|\psi_i|^2$ has spherical symmetry.
f) Discuss possible reasons why the real valence electron density in a diamond crystal is nevertheless not spherical around a carbon atom but is concentrated in tetrahedral bonding directions.

1.10 a) In analogy to the sp^3 hybridization of carbon in the diamond lattice, discuss the possibility of the carbon atom forming sp^2 hybrid orbitals with three planar bonding directions. Plot a qualitative picture of the three bonding sp^2 orbitals and the remaining unpaired p_z orbital and give their electron occupation.
b) Assuming sp^2 hybridization for the carbon atoms explain the chemical bonding in the benzene molecule C_6H_6. What is the origin of the π bonding system parallel to the hexagonal ring skeleton of the six carbon atoms.
c) Assuming sp^2 hybridization explain the highly anisotropic properties of the planar hexagonal graphite lattice of carbon (quasi-metallic properties parallel to the carbon planes). How are these anisotropic properties exploited in every-day life?

1.11 As a simple quantum mechanical model for the van der Waals interaction consider two identical harmonic oscillators (oscillating dipoles) at a separation R. Each dipole consists of a pair of opposite charges whose separations are x_1 and x_2, respectively, for the two dipoles. A restoring force f acts between each pair of charges.

a) Write down the Hamiltonian $\mathscr{H}_0$ for the two oscillators without taking into account electrostatic interaction between the charges.
b) Determine the interaction energy $\mathscr{H}_1$ of the four charges.
c) Assuming $|x_1| \leq R$, $|x_2| \leq R$ approximate $\mathscr{H}_1$ as follows

$$\mathscr{H}_1 \approx -\frac{2\,e^2\,x_1\,x_2}{R^3} \ .$$

d) Show that transformation to normal coordinates

$$x_\mathrm{s} = \frac{1}{\sqrt{2}}\,(x_1 + x_2) \ ,$$

$$x_\mathrm{a} = \frac{1}{\sqrt{2}}\,(x_1 - x_2)$$

decouples the total energy $\mathscr{H} = \mathscr{H}_0 + \mathscr{H}_1$ into a symmetric and an antisymmetric contribution.
e) Calculate the frequencies ω_s and ω_a of the symmetric and antisymmetric normal vibration modes. Evaluate the frequencies ω_s and ω_a as Taylor series in $2e^2/(fR^3)$ and truncate the expansions after second order terms.
f) The energy of the complete system of two interacting oscillators can be expressed as $U = -\frac{1}{2}\hbar\,(\omega_\mathrm{s} + \omega_\mathrm{a})$. Derive an expression for the energy of the isolated oscillators and show that this is decreased by an amount c/R^6 when mutual interaction (bonding) occurs.

1.12 Calculate how the van der Waals bonding of a molecule depends on its distance, d, from a solid surface. For simplicity, choose a simple cubic lattice. Show that the result does not depend on the structure of the crystal.

2 Structure of Solid Matter

When atoms are chemically bound to one another they have well-defined equilibrium separations that are determined by the condition that the total energy is minimized. Therefore, in a solid composed of many identical atoms, the minimum energy is obtained only when every atom is in an identical environment. This leads to a three-dimensional periodic arrangement that is known as the crystalline state. The same is true for solids that are composed of more than one type of element. In this case, certain "building blocks" comprising a few atoms are the periodically repeated units. Periodicity gives rise to a number of typical properties of solids. Periodicity also simplifies the theoretical understanding and the formal theory of solids enormously. Although a real solid never possesses exact three-dimensional periodicity, one assumes perfect periodicity as a model and deals with the defects in terms of a perturbation (Sect. 2.7). Three-dimensional periodic arrangements of atoms or "building blocks" are realized in many different ways. Basic elements of the resulting crystal structures are described in Sects. 2.1–2.5.

The counterpart to the crystalline state of solids is the amorphous state. This is a state in which no long-range order exists; however, a degree of short-range order remains. Examples of amorphous solids are glasses, ceramics, gels, polymers, rapidly quenched melts and thin-film systems deposited on a substrate at low temperatures. The investigation of amorphous materials is a very active area of research. Despite enormous progress in recent years, our understanding of amorphous materials still remains far from complete. The reason is the absence of the simplifications associated with periodicity. Nonetheless, from comparison of the properties of materials in a crystalline and an amorphous state we have learned the essential features of the electronic structure, and thereby also macroscopic properties, are determined by short-range order. Thus these properties are similar for solids in the amorphous and crystalline state. In the context of this book we focus on a few structural and electronic properties of amorphous solid in Sects. 3.1, 7.6 and 9.8.

Materials of practical use are nearly always composites. Examples, some of them known to mankind since early ages, are the bronzes (alloys of copper and tin), brass (copper and zinc) or steel (in its simplest form iron with a few per cent of carbon). Modern material science has supplemented these classical, mostly binary alloys with many multi-component systems including composite materials. In the framework of this textbook, particular attention will be paid to semiconductor alloys (Chap. 12). Alloys typically do not exist as a homogenous crystalline phase; they rather consist of microcrystallites, whose composition depends on temperature, pressure, and the percentage of the constituting elements. Phase diagrams of simple binary alloys are discussed in Sect. 2.6.

2.1 The Crystal Lattice

A two-dimensional lattice is spanned by two vectors $\boldsymbol{a}$ and $\boldsymbol{b}$. Every point on the lattice can be reached by a lattice vector of the form

$$r_n = n_1\,\boldsymbol{a} + n_2\,\boldsymbol{b} \tag{2.1}$$

where n_1 and n_2 are integers. Depending on the ratio of the lengths of the vectors $\boldsymbol{a}$ and $\boldsymbol{b}$, and on the angle γ between them, lattices of various geometries can be constructed. The most general lattice, with no additional symmetry, is obtained when $a \neq b$ and $\gamma \neq 90°$ (Fig. 2.1). A planar crystal structure would be produced if every point of this "parallelogram lattice" were occupied by one atom. In this case, each elementary (or unit) cell with sides $\boldsymbol{a}$ and $\boldsymbol{b}$ would contain one atom. Such elementary cells are termed primitive. It is also possible for crystal structure to contain more than one atom per unit cell. In such cases the lattice points of Fig. 2.1 correspond to points in the crystal which all have identical environments. Such points may, but need not necessarily, lie at the center of an atom.

Other planar lattices of higher symmetry are obtained when γ, a and b take on certain special values. The rectangular lattice is obtained when $\gamma = 90°$ (Fig. 2.2). When additionally $a = b$, this becomes a square lattice. The two-dimensional plane can also be filled with a regular array of hexagons. The unit cell is then given by $a = b$ and $\gamma = 60°$. A hexagonal close packing of spheres has this unit cell. The condition $a = b$ with γ arbitrary also produces a new type of lattice. However, this lattice is more conveniently described as a "centered" rectangular lattice with $a \neq b$ and $\gamma = 90°$ (Fig. 2.2). One thereby obtains the advantage of an orthogonal coordinate system; the lattice, however, is no longer primitive.

It is easy to see that the description in terms of a centered unit cell is only useful for a rectangular lattice. If centering is introduced into a square, parallelogram or hexagonal lattice, it is always possible to describe

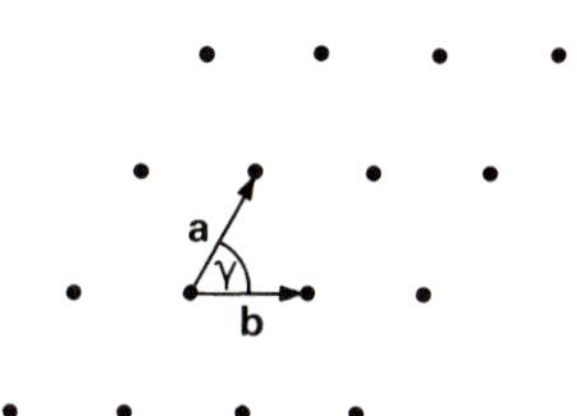

Fig. 2.1. A plane oblique lattice

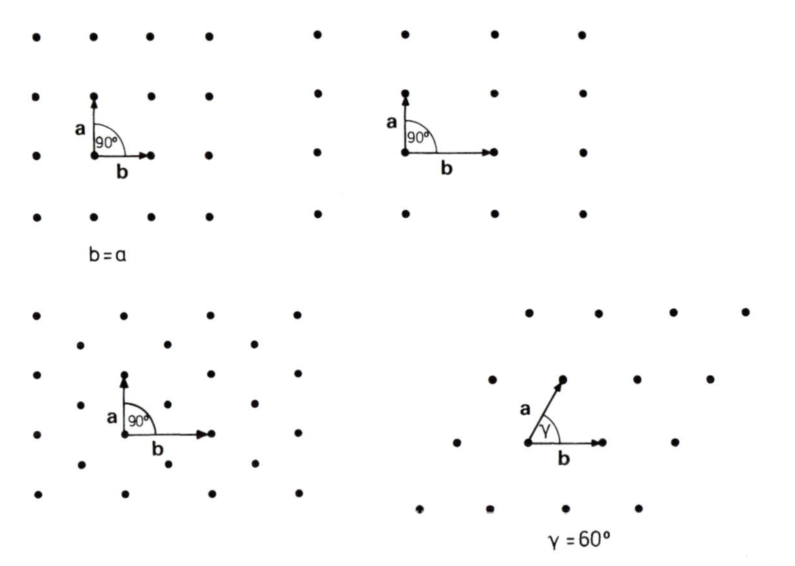

Fig. 2.2. Further two-dimensional lattices: square, rectangular, centered rectangular and hexagonal ($\gamma = 60°$, $a = b$)

the new lattice in terms of an alternative set of smaller basis vectors. In other words, the only result is that one has obtained a larger than necessary unit cell.

The discussion so far has concerned two-dimensional lattices, but the principles can be extended to lattices in three dimensions. Instead of the five possible systems of basis vectors in the plane, one now has seven possibilities (Table 2.1). These correspond to the seven distinct crystal systems of crystallography. With the addition of centering, one can construct from these basis vector systems all the possible lattices of three-dimensional space. As well as face centering, one now has the additional possibility of body centering (Fig. 2.3). As was the case in two dimensions, one can readily convince oneself that only certain centerings are useful. For exam-

Table 2.1. The seven different basis-vector systems or crystal systems. Most elements crystallize in a cubic or hexagonal structure. For this reason, and also because of their high symmetry, the cubic and hexagonal coordinate systems are particularly important

Basis vectors/crystal axes	Angles	Crystal system
$a \neq b \neq c$	$\alpha \neq \beta \neq \gamma \neq 90°$	triclinic
$a \neq b \neq c$	$\alpha = \gamma = 90°\ \beta \neq 90°$	monoclinic
$a \neq b \neq c$	$\alpha = \beta = \gamma = 90°$	orthorhombic
$a = b \neq c$	$\alpha = \beta = \gamma = 90°$	tetragonal
$a = b \neq c$	$\alpha = \beta = 90°\ \gamma = 120°$	hexagonal
$a = b = c$	$\alpha = \beta = \gamma \neq 90°$	rhombohedral
$a = b = c$	$\alpha = \beta = \gamma = 90°$	cubic

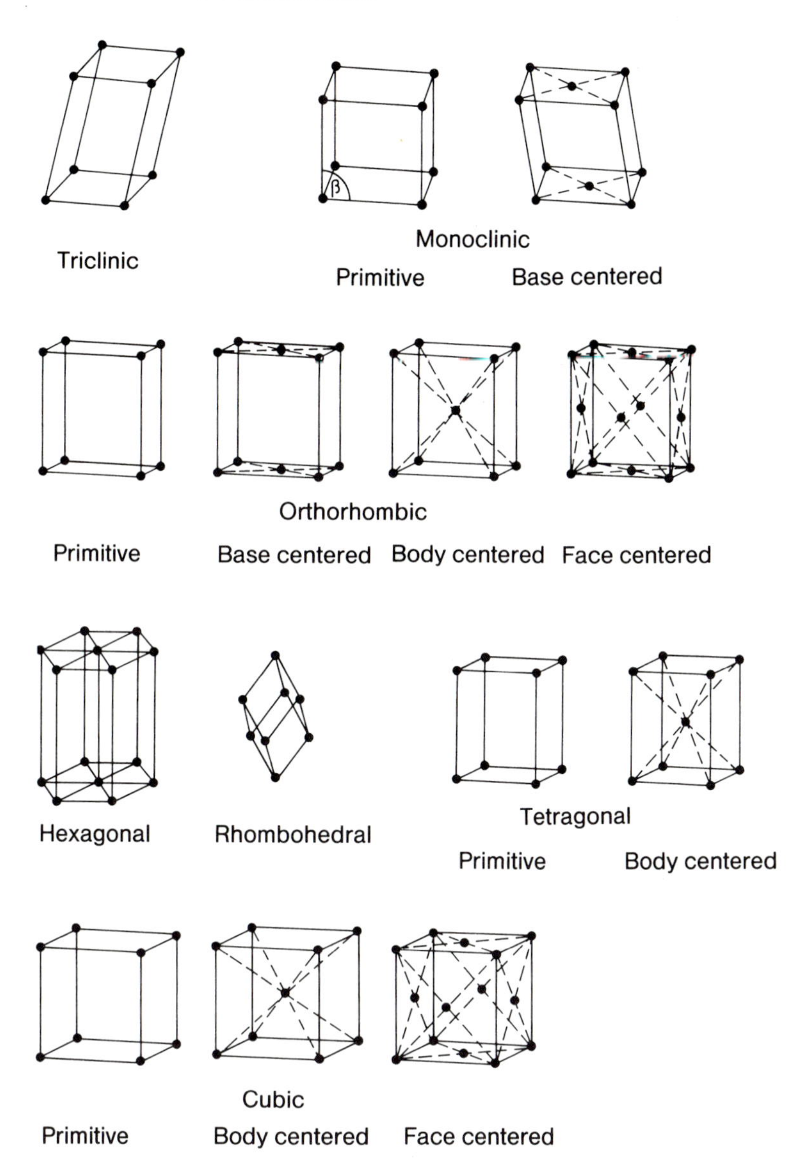

Fig. 2.3. The 14 three-dimensional Bravais lattices. The hexagonal lattice and the two centered cubic lattices are particularly important in solid state physics

ple, it would not make sense to employ a tetragonal base-centered lattice because this is equivalent to a primitive tetragonal lattice with a smaller unit cell.

2.2 Point Symmetry

Every point of the lattices just discussed represents an atom, or a more-or-less complicated group of atoms, each of which has particular symmetry properties. The symmetries and the corresponding notation will be presented in the following.

Reflection in a Plane

This symmetry is expressed mathematically by a coordinate transformation. For example, mirror symmetry about the yz-plane is represented by the transformation $y'=y$, $z'=z$, $x'=-x$. The presence of a mirror plane in a crystal structure is indicated by the symbol m. An example of a molecule possessing two perpendicular mirror planes is the water molecule (Fig. 2.7). One of the mirror planes is the plane defined by the atoms of the molecule; the other is perpendicular to this and passes through the oxygen atom, bisecting the molecule.

Inversion

Inversion symmetry is described by the coordinate transformation $x'=-x$, $y'=-y$, $z'=-z$. Thus in a sense this might be described as reflection in a point. The symbol representing inversion symmetry is $\bar{1}$. An example of a molecule possessing inversion symmetry is cyclohexane (Fig. 2.4). Homonuclear diatomic molecules also have inversion symmetry and, of course, mirror planes.

Rotation Axes

Rotational symmetry is present if a rotation through a particular angle about a certain axis, leads to an identical structure. The trivial case is, of course, rotation by 360° which inevitably leads to the same structure. The number of intermediate rotations that also result in indistinguishable ar-

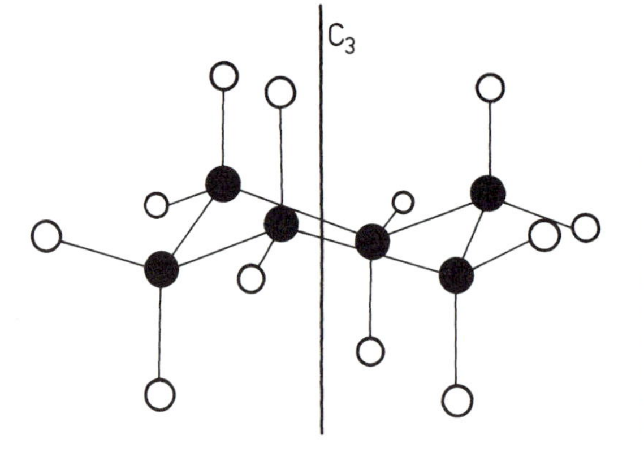

Fig. 2.4. The cyclohexane molecule (C_6H_{12}). The principal symmetry element is the 3-fold axis of rotation C_3. The molecule also has a center of inversion, three mirror planes and, perpendicular to the main axis, three 2-fold rotation axes at 120° to one another. The point group is denoted by D_{3d} (Table 2.2)

rangements is called "the order" of the rotation axis. Thus one can have 2-, 3-, 4- and 6-fold rotation axes, corresponding to invariance under rotations of 180°, 120°, 90°, and 60°. For single molecules it is also possible to have 5-fold, 7-fold, etc., axes. Small solid aggregates (clusters) may also display 5-fold rotational symmetry. An example is the particularly stable icosohedron composed of 13 atoms. Icosohedra are formed in the rapid quenching of melts. The solid which is thus formed has a quasi crystalline structure and produces sharp X-ray diffraction spots that reflect the local 5-fold symmetry [2.1]. For strictly periodic crystals, however, only 2-, 3-, 4- and 6-fold rotation axes are possible. All other orders of rotation are incompatible with the required translational symmetry. The notation describing the possible axes of rotation is given simply by the numbers 2, 3, 4 and 6.

The cyclohexane molecule shown in Fig. 2.4 has a 3-fold rotation axis. A molecule with a 6-fold axis is benzene (C_6H_6), whose carbon skeleton consists of a planar regular hexagon.

Rotation-Inversion Axes

Rotation with simultaneous inversion can be combined to give a new symmetry element – the rotation-inversion axis. It is symbolized by $\bar{2}$, $\bar{3}$, $\bar{4}$, or $\bar{6}$. Figure 2.5 illustrates a 3-fold rotation-inversion axis. From this it is evident that a 3-fold rotation-inversion axis is equivalent to a 3-fold rotation together with inversion. The 6-fold rotation-inversion axis may alternatively be represented as a 3-fold rotation axis plus a mirror plane.

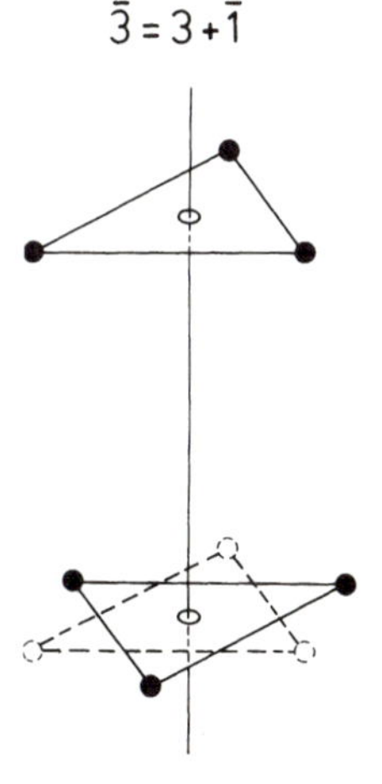

Fig. 2.5. Representation of a 3-fold rotation-inversion axis. The effect can also be described by the combination of other symmetry elements

2.3 The 32 Crystal Classes (Point Groups)

The symmetry elements discussed in the previous section may be combined with one another in various ways. Conversely, every crystal may be described by a particular combination of point-symmetry elements. To be complete, the description must satisfy a number of conditions. For example, if two successive symmetry operations are applied, the result must be a further symmetry element: $A \otimes B = C$. Three (or more successive symmetry operations must obey the so-called associativity rule: $(A \otimes B) \otimes C = A \otimes B \otimes C)$. There is an identity element E, corresponding to no operation or to a rotation about 360°, such that $A \otimes E = A$. Furthermore, every symmetry element A possesses an inverse A^{-1}, which corresponds to the reverse operation, so that $A^{-1} \otimes A = E$. These properties are the mathematical definition of a group. There are 32 distinct crystallographic point groups. If the translational symmetry is also taken into account then one obtains the 230 possible space groups. Although not necessarily true in general, we should note that for translations $A \otimes B = B \otimes A$ (the property of Abelian groups).

The 32 crystallographic point groups are most commonly represented by so-called stereographic projections. These projections were developed by crystallographers in order to obtain a systematic classification of the exposed surfaces of naturally grown crystals. The point at which each surface normal intersects a sphere is marked and then projected onto the plane perpendicular to the highest order symmetry axis. Intersection points above this plane are marked by a full circle, and those on the lower half sphere by an open circle or a cross. Hence, in the systematic representation of the point groups, the highest order axis lies in the center. Stereographic projections of two point groups are shown in Fig. 2.6. A particular point group may be denoted in three different ways:

1) by specifying a system of generating symmetry operations;
2) by specifying the international point group symbol;
3) by the Schönflies symbol.

The notation based on the generating symmetry operators is common in crystallography, whereas the Schönflies symbols have largely been adopted in group theory and spectroscopy. The Schönflies notation consists of a

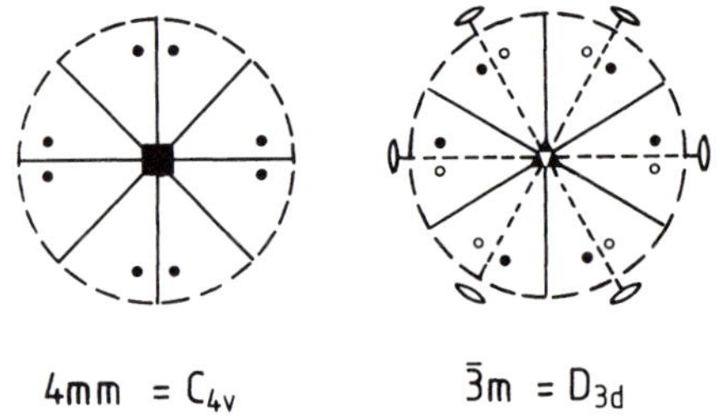

$4mm = C_{4v}$ $\bar{3}m = D_{3d}$

Fig. 2.6. Representation in stereographic projection of the symmetry elements of two point groups. The symbols (), ▲ and ■ denote 2-, 3-, and 4-fold rotation axes. The full lines are mirror planes. When the outer circle is also drawn as a full line, this indicates that the plane of the drawing is also a mirror plane

Table 2.2. The Schönflies point group symbols

	Symbol	Meaning
Classification according to rotation axes and principal mirror planes	C_j	(j=2, 3, 4, 6) j-fold rotation axis
	S_j	j-fold rotation-inversion axis
	D_j	j two-fold rotation axes $\perp$ to a (j-fold) principal rotation axis
	T	4 three- and 3 two-fold rotation axes as in a tetrahedron
	O	4 three- and 3 four-fold rotation axes as in an octahedron
	C_i	a center of inversion
	C_s	a mirror plane
Additional symbols for mirror planes	h	horizontal = perpendicular to the rotation axis
	v	vertical = parallel to the main rotation axis
	d	diagonal = parallel to the main rotation axis in the plane bisecting the 2-fold rotation axes

main symbol that characterizes the axes of rotation (if any) of the system, and a subsidiary symbol that gives the positions of the symmetry planes. The symbols and their meanings are summarized in Table 2.2. As an example, we consider the water molecule, for which the highest order rotation axis is a 2-fold axis. The symmetry planes are vertical, i.e. they lie parallel to the main axis of rotation. The corresponding Schönflies symbol is $C_{2\mathrm{v}}$. A cube has three 4-fold rotation axes, four 3-fold rotation axes and symmetry planes perpendicular to the 4-fold axes. Its Schönflies symbol is O_{h}.

2.4 The Significance of Symmetry

To the uninitiated, the correct assignment and symbolization of symmetry often seems complicated and confusing. It will thus be useful to give a brief overview of the importance of symmetry in the description of a solid. For this purpose, we must base our discussion on quantum mechanics. As we have seen, the water molecule, for example, has two mirror planes. The presence of these two mirror planes must somehow be reflected in all the physical properties of the molecule. When the electronic or vibrational properties of the molecule are described by a Hamiltonian, then this has 2-fold mirror symmetry, i.e., it remains invariant under the corresponding coordinate transformation. This invariance can also be expressed in other ways. An operator σ is assigned to the reflection. When σ operates on the Hamiltonian $\mathcal{H}$, on an eigenstate ψ or on $\boldsymbol{R}$, the result should describe $\mathcal{H}$, ψ or $\boldsymbol{R}$ in the transformed (mirror image) coordinates.

Such operators are represented as matrices. Thus the reflection of coordinates in the yz-plane is represented by the matrix operation

$$\begin{pmatrix} -1 & 0 & 0 \\ 0 & 1 & 0 \\ 0 & 0 & 1 \end{pmatrix} \begin{pmatrix} x \\ y \\ z \end{pmatrix} = \begin{pmatrix} -x \\ y \\ z \end{pmatrix} . \qquad (2.2)$$

This is a three-dimensional representation. The same operation can also be expressed in terms of three one-dimensional matrices,

$$[(-1)x; (1)y; (1)z] = (-x; y; z) ,$$

each of which acts on only one coordinate component. In this case the three-dimensional representation is known as "reducible", whereas the corresponding one-dimensional representation is called "irreducible" since it cannot be further simplified. It is easy to see that the irreducible representation of a rotation through $180°$ (a 2-fold rotation axis) is also one-dimensional: for a suitable choice of coordinates, it can be expressed simply by a sign reversal. However, for 3-, 4- and 6-fold rotation axes, except for the case of a $360°$ rotation, the operations always involve two coordinate changes. The irreducible representation is therefore two-dimensional.

If the Hamiltonian operator possesses a particular symmetry, for example mirror symmetry, then it makes no difference whether the reflection operation appears before or after the Hamiltonian operator, i.e., the two operators commute. As is well known from quantum mechanics, such operators have a common set of eigenstates. Thus the possible eigenstates of $\mathscr{H}$ can be classified according to their eigenvalues with respect to the symmetry operators. In the case of mirror symmetry and that of a 2-fold rotation axis C_2, one has $\sigma^2=1$ and $C_2^2=1$ and thus the eigenvalues can only be ± 1:

$$\begin{aligned} \sigma \Psi_+ &= +\Psi_+ , \qquad & C_2 \Psi_+ &= +\Psi_+ , \\ \sigma \Psi_- &= -\Psi_- , \qquad & C_2 \Psi_- &= -\Psi_- . \end{aligned} \qquad (2.3)$$

The eigenstates of $\mathscr{H}$ may therefore be either symmetric or antisymmetric with respect to these operators. This is often expressed by saying that the eigenstates have even or odd "parity". We have already met an example of even and odd parity in our discussion of the chemical bonding between hydrogen atoms (Sect. 1.2). The bonding state was a symmetric combination of atomic wavefunctions and therefore a state of even parity. As seen in this example, the eigenstates Ψ_+ and Ψ_- belong to distinct eigenvalues of $\mathscr{H}$. The corresponding energy levels are thus nondegenerate. From this we may conclude, for example, that the water molecule can possess only nondegenerate energy states (we ignore here any accidental degeneracy of energy levels or normal mode vibrations).

To illustrate the above discussion we will apply these ideas to the normal mode vibrations of the water molecule. Accordingly, the atoms can move symmetrically or antisymmetrically with respect to the two mirror planes of the molecule. For atoms that lie in a mirror plane, antisymmetric motion with respect to the plane implies motion perpendicular to the plane, since only then can reflection reverse the motion. The corresponding symmetric vibrations of such atoms must involve motion in the mirror plane. One of the two mirror planes of the H_2O molecule is the plane containing the three atoms (Fig. 2.7). The motions that are antisymmetric with respect to this plane are two rotations of the molecule and its translation perpendicular to the plane. The six symmetric modes with motion in the plane of the molecule are two translations, a rotation about an axis perpendicular to the plane, and the three normal mode vibrations (Fig. 2.7). Of these vibrations two are symmetric and one is antisymmetric with respect to the mirror plane *perpendicular* to the molecular plane.

For more complex molecules too, it is possible to perform such a classification of the vibrational modes and/or electronic states. However, the process becomes rather more involved for operators that have two-dimensional irreducible representations, such as C_3. If C_3 commutes with $\mathscr{H}$, then together with the state Ψ, the state $C_3\Psi$ is also an eigenstate of $\mathscr{H}$. There are now two possibilities:

1) Apart from a numerical factor, which, for suitable normalization can be made equal to 1, the state $C_3\Psi$ is identical to Ψ. Thus, in this case Ψ is totally symmetric with respect to C_3 and the operation C_3 has a one-dimensional (numerical) representation. The state Ψ is then – at least with respect to the operation – nondegenerate.
2) $C_3\Psi$ produces a new, linearly independent state Ψ', which however, due to the commutivity of C_3 and $\mathscr{H}$, must also be an eigenstate of $\mathscr{H}$ with the same eigenvalue E. The states Ψ and Ψ' are thus degenerate. Since the rotation C_3 always affects two coordinates, its irreducible representation is a two-dimensional matrix. Every eigenstate of C_3 may then be constructed as a linear combination of two functions that can

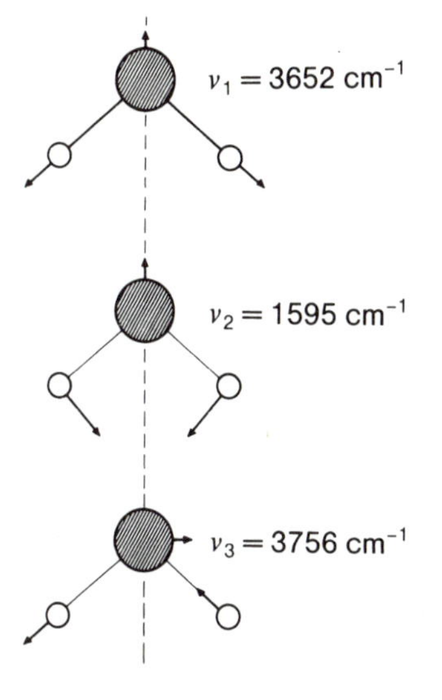

Fig. 2.7. The two symmetric and the antisymmetric vibrations of the water molecule. Together with the three rotations and three translations these give the nine normal modes corresponding to the nine degrees of freedom

be selected to be orthonormal. The energy levels are therefore 2-fold degenerate. Such degenerate levels are found for all point groups possessing more than a 2-fold rotation axis.

Particularly important in solid state physics are the diamond lattice and the face-centered and body-centered cubic lattices. These belong to the point groups T_d and O_h, respectively, the former displaying tetrahedral and the latter octahedral symmetry (Figs. 2.8, 10, 12). The representations of such symmetries affect three coordinates and thus T_d and O_h have three-dimensional irreducible representations. They are associated accordingly with 3-fold degeneracy. We will meet such states when we deal with the normal modes of these lattices (Sect. 4.5) and their electron states (Sect. 7.4).

Besides symmetry-determined degeneracy, one also finds degeneracy that results from the specific form of $\mathscr{H}$. The degeneracy associated with the angular momentum quantum number l of the hydrogen atom is well known to be a result of the $1/r$ potential, whereas the degeneracy with respect to the magnetic quantum number m stems from symmetry. The crystal symmetry also determines the number of independent components of the tensors describing macroscopic material properties. We note here for later use that second-rank tensors, such as the thermal expansion or susceptibility, have only one independent component in cubic crystals, and two in hexagonal crystals.

2.5 Simple Crystal Structures

The Face-Centered Cubic Structure

The simplest crystal structures are those in which each point of the lattice is occupied by one atom. In this case, the face-centered cubic lattice produces a face-centered cubic crystal. Each atom in this structure is surrounded by 12 nearest neighbors. The number of nearest neighbors in a particular lattice type is known as the coordination number.

The coordination number 12 represents the highest possible packing density for spheres. In a single plane of close-packed spheres the number of nearest neighbors is 6. In three dimensions there are an additional 3 nearest neighbors in each of the planes below and above. If the lattice constant of the face-centered structure is denoted by a, then the separation of nearest neighbors is given by $a/\sqrt{2}$ as is readily seen from Fig. 2.8. The closest-packed planes are sketched in Fig. 2.8 b. They lie perpendicular to the main diagonal of the cube. If one moves away from a particular close-packed plane along the main diagonal, another identical plane is reached only after passing through two differently positioned close-packed planes. This packing sequence is illustrated more clearly by Fig. 2.9. A close-packed layer has two sorts of hollow sites (visible in layer A). The second layer is obtained by placing further spheres on one of the

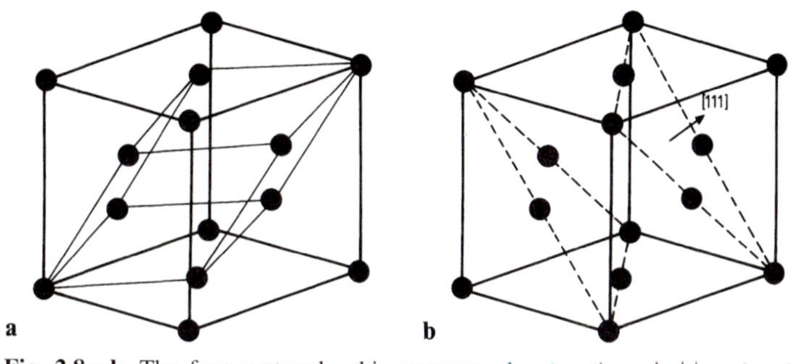

Fig. 2.8 a, b. The face-centered cubic structure showing the primitive rhombohedral unit cell (**a**). The close-packed planes are illustrated by dotted lines in (**b**). The number of nearest neighbors (the coordination number) is 12

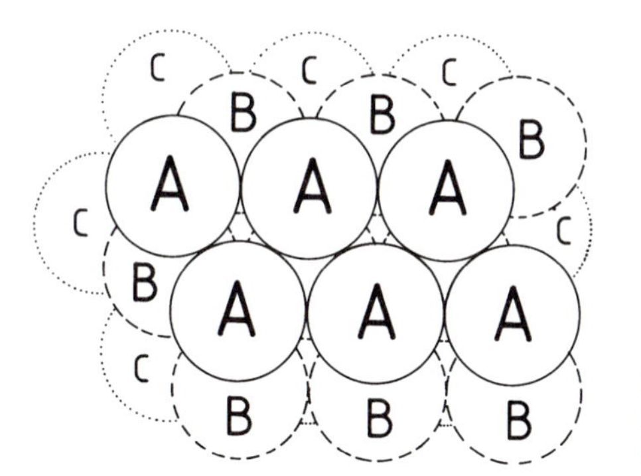

Fig. 2.9. The close-packed layers of the fcc lattice with the stacking sequence AB-CABC...

two possible sites, and the third-layer spheres lie above the other type of site. Thus the face-centered cubic structure is composed of close-packed layers in the stacking sequence ABCABC... . Each of these layers alone has hexagonal (6-fold) symmetry; however, stacked above one another in this manner the symmetry is reduced to a 3-fold rotation axis (Fig. 2.9). The face-centered cubic structure therefore has four 3-fold rotation axes as well as a mirror plane perpendicular to the 4-fold axis. It therefore belongs to the point group O_h. The face-centered cubic structure is usually denoted by the abbreviation fcc. Examples of fcc crystals are the metals Cu, Ag, Au, Ni, Pd, Pt and Al. Despite their relatively high melting points these metals are fairly soft. The reason for this is that the close-packed layers are able to slide over one another. This sliding motion occurs in the plastic deformation of these crystals. However, it does not involve an entire layer; it is limited to regions associated with so-called dislocations.

Hexagonal Close Packing

The hexagonal close-packed (hcp) structure results when close-packed planes are stacked in the sequence ABAB... . In contrast to the fcc structure, the smallest possible unit cell now contains two atoms. Thus the main axis of rotation is again 3-fold rather than 6-fold. As can be seen by considering only the layers A and B of Fig. 2.9, there exist three 2-fold rotation axes perpendicular to the 3-fold axis. Furthermore, the close-packed layer also lies in a mirror plane. The point group corresponding to these symmetry elements is $D_{3\mathrm{h}}$. As with the fcc structure, the coordination number is 12. Important metals that crystallize in the hcp structure include Zn, Cd, Be, Mg, Re, Ru and Os.

The Body-Centered Cubic Structure

The body-centered cubic (bcc) structure is shown in Fig. 2.10. For this structure the coordination number is only 8. Thus for nondirectional bonding the bcc structure would appear to be less favorable. Nonetheless, all alkali metals as well as Ba, V, Nb, Ta, W, and Mo, are found to crystallize in this structure, and Cr and Fe also possess bcc phases. At first sight this is hard to understand. However, it is important to note that in the bcc structure the 6 next-nearest neighbors are only slightly farther away than the 8 nearest neighbors. Thus, depending on the range and the nature of the wavefunctions contributing to the bonding, the effective coordination number of the bcc structure can be higher than that of the fcc structure. Figure 2.11 shows the probability functions for the positions of the lithium electrons relative to the atomic nucleus. Also shown are the half distances to the nearest (r_1), the next-nearest (r_2) and the third-nearest (r_3) neighbors for the actually occurring bcc structure and for a hypothetical fcc structure with the same nearest-neighbor separation. When the next-nearest neighbors are taken into account, it is easy to see that the bcc structure leads to a greater overlap of wavefunctions and thus an increase in the strength of the chemical bonding. This effect is enhanced by the fact that the p-orbitals in a cubic structure are oriented along the edges of the cube, thereby contributing significantly to the bonding with the next-nearest neighbors. The

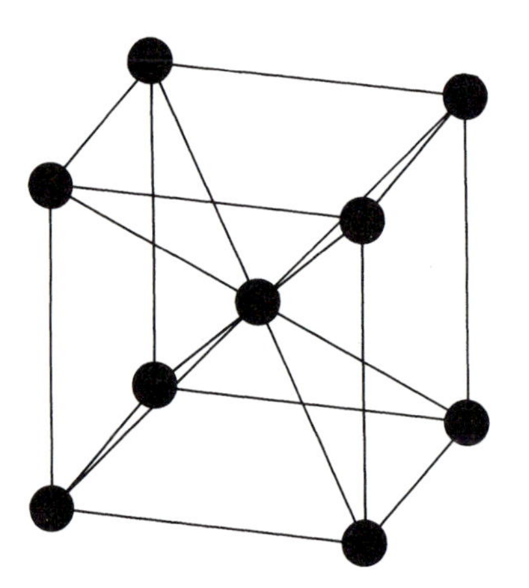

Fig. 2.10. The body-centered cubic structure with coordination number 8

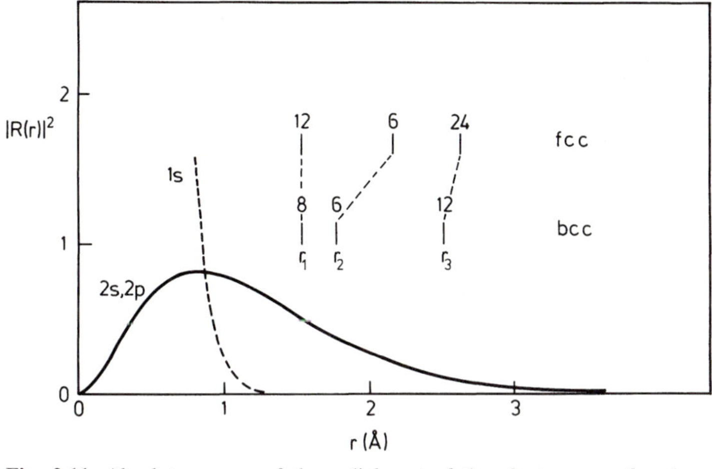

Fig. 2.11. Absolute square of the radial part of the electron wavefunctions of lithium as a function of distance from the nucleus. For the bcc structure, both the 8 nearest and the 6 next-nearest neighbors lie in a region of relatively high density. Hence for non-directional metallic bond it is possible for the bcc structure to be energetically favorable with respect to fcc. Examples of bcc metals are the alkali metals Li, Na, K, Rb, Cs and Fr. This curve may be compared with Fig. 1.9 which shows the wavefunction amplitude rather than the probability density. The decay at large distances then becomes weaker

picture changes, however, when *d*-electrons become involved in the bonding. The *d*-orbitals are directed both along the cube edges and along the diagonals of the faces. Since the *d*-orbitals are localized relatively strongly on the atoms (Fig. 1.9), they can only contribute to the bonding when they are directed towards the nearest neighbors. The fcc structure enables exactly this, which is the reason why metals with a large number of *d*-electrons often crystallize in the fcc structure.

The Diamond Structure

The diamond structure belongs to the crystal class T_d. It allows three-dimensional covalent bonding (Sect. 1.2) in which every atom is surrounded by four nearest neighbors in a tetrahedral configuration (Fig. 2.12). Thus the coordination number is 4. The diamond structure takes its name from the structure of the carbon atoms in diamond. Other elements that crystallize with the same structure are Si, Ge and α-Sn. The diamond structure can be described as two interpenetrating fcc structures that are displaced relative to one another along the main diagonal. The position of the origin of the second fcc structure, expressed in terms of the basis vectors, is $(\frac{1}{4}, \frac{1}{4}, \frac{1}{4})$. This leads to a nearest-neighbor distance of $\sqrt{3}a/4$. Since the separation of close-packed layers in the fcc structure is $\sqrt{3}a/3$, the distance of a central atom from the base of its tedrahedron of neighbors is $\frac{1}{4}$ of the total height of the tetrahedron.

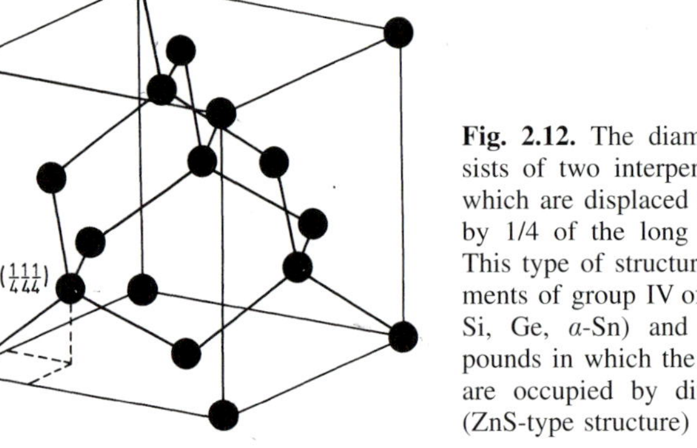

Fig. 2.12. The diamond structure. It consists of two interpenetrating fcc structures which are displaced relative to one another by 1/4 of the long diagonal of the cube. This type of structure is typical of the elements of group IV of the periodic table (C, Si, Ge, α-Sn) and also for III–V compounds in which the sites (000) and $(\frac{1}{4}\,\frac{1}{4}\,\frac{1}{4})$ are occupied by different types of atom (ZnS-type structure)

The Zinc Blende Structure

The zinc blende (ZnS) structure is closely related to the diamond structure, but now the two interpenetrating fcc structures contain different atoms. The ZnS structure is found in the most important of the compounds of group III with group V elements. Examples are GaAs, GaP and InSb. The compound ZnS which gives its name to this structure is of course also of the "zinc blende structure". The choice of name for this structure is actually slightly unfortunate, since the compound ZnS also crystallizes in a hexagonal phase with the so-called wurtzite structure. This structure, in common with the ZnS type, has tetrahedral coordination; however, the stacking sequence of the close-packed (111) planes is no longer ABCABC..., but ABAB..., thereby giving it a hexagonal structure. The wurtzite structure is also adopted by other compounds between group II and group VI elements (ZnO, ZnSe, ZnTe, CdS, CdSe). As well as the ordered packing sequences ABAB... and ABCABC... it is also possible to find mixed forms with random stacking or very long period repeats. The best-known example of these so-called "polytypes" is SiC.

Ionic Structures

Typical ionic structures, exemplified by the CsCl and NaCl structures, have already been introduced in Sect. 1.3 (Fig. 1.6). The CsCl structure is derived from the bcc structure by replacing the ion at the center of the cube by an ion of a different element. The NaCl structure is the result of placing one fcc structure within another. The coordination numbers are 8 for the CsCl structure and only 6 for the NaCl structure. As we have seen in Sect. 1.3, for equal ionic separations, the Madelung constant and thus also the ionic energy are greater for the CsCl structure. Although the differences are relatively small, it is perhaps surprising that the majority of ionic crystals prefer the NaCl structure. The explanation for this is as follows: In most cases, the radius of the cations is much smaller than that of the anion. For example,

$$r_{\mathrm{Na}} = 0.98\ \mathrm{Å}, \quad r_{\mathrm{Cl}} = 1.81\ \mathrm{Å}.$$

The cesium ion however is a large cation:

$$r_{\mathrm{Cs}} = 1.65\ \mathrm{Å}.$$

As the cation becomes smaller, the anions of a CsCl-type structure approach one another and eventually overlap. This occurs at a radius of $r^+/r^-=0.732$. For still smaller cations the lattice constant could not be further reduced and the Coulomb energy would remain constant. In this case the NaCl-type structure is favored since anion contact does not occur until a radius ratio of $r^+/r^-=0.414$ is reached. Yet smaller radius ratios are possible for the ZnS-type structure. Indeed, for ZnS itself, the ratio is $r^+/r^-=0.40$. This can be regarded as the reason why ZnS does not crystallize in the NaCl structure. This is a somewhat simplified view as it neglects the strong covalent contribution to the bonding (see also Problem 1.2).

2.6 Phase Diagrams of Alloys

Modern functional materials consist of many elements in different *phases*. The term *phase* denotes a domain of homogeneous concentration and structure on a length scale that is large compared to atomic dimensions. Separate phases can be observed even with simple binary alloys. Figure 2.13 displays the scanning electron microscope image of the polished surface of an Ag/Cu alloy. The dark and light sections represent copper- and silver-rich fcc-phases, respectively. As a rule, the various constituting phases of modern composite materials are not in thermal equilibrium. Nonetheless,

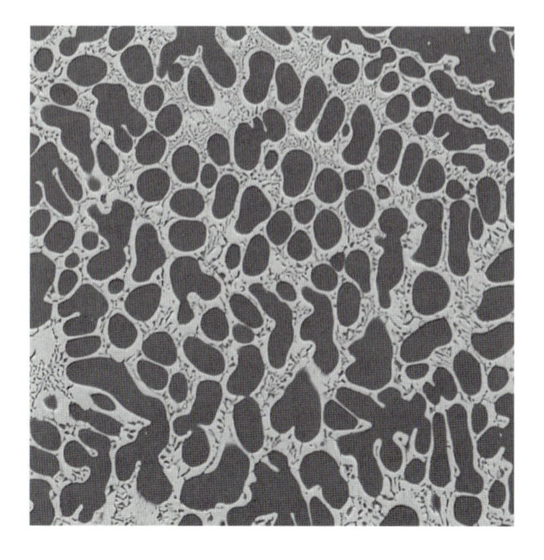

Fig. 2.13. Scanning electron microscope image of a polished specimen of an Ag/Cu alloy with 30% Ag- and 70% Cu-atoms. The dark and light areas consist of fcc-phases with about 95% Cu-atoms and about 86% Ag-atoms, respectively

the equilibrium properties are basic to the understanding. Equilibrium properties are described with the help of *phase diagrams*. In a phase diagram, the temperature is plotted vs. the concentration of one component (at the expense of another). For a particular temperature and composition, the material possesses a particular equilibrium structure. The boundaries between different structures are marked by lines. In the simplest case the lines describe the boundary between the solid and the liquid state. Phase diagrams for all important alloys have been determined experimentally by thermodynamic measurements [2.2]. In the following, we consider only the simple phase diagrams of substitutional binary alloys. A substitutional alloy consists of two types of atoms A and B, which as pure materials crystallize in the same structure. If in addition the chemical bonding is similar and the lattice constants of the pure phases are not too different, atoms A and B will assume the same lattice sites in the composite system. A number of different states exist even in this simple case: the liquid state with a complete mixture of the two components, a mixture of liquid phase and a solid phase in which atoms of either type A or B are enriched, a solid phase with micro-crystals in which either A or B are enriched, or a continuously miscible solid phase. Alloys that are continuously miscible in the solid phase have the simplest phase diagrams. The SiGe alloy is an example. Its phase diagram is displayed in Fig. 2.14. Depending on the temperature and the relative Si-concentration, the alloy exists either as a homogeneous liquid (ℓ for liquidus), as a homogeneous solid (s for solidus) or as a two-

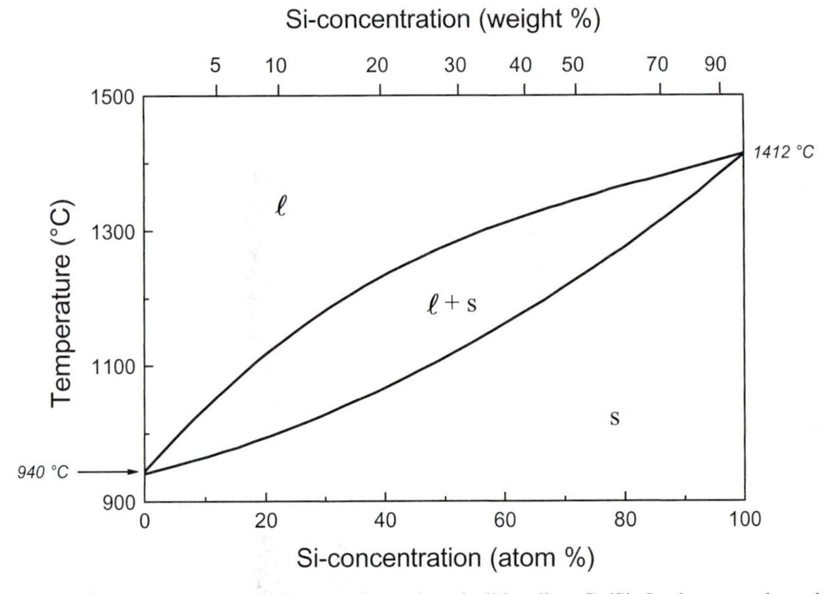

Fig. 2.14. Phase diagram for the continuously miscible alloy Ge/Si. In the range bounded by the liquidus and solidus curves a Ge-rich liquid phase coexists with a Si-rich solid phase

phase system with liquid and solid parts (ℓ+s). The realms of existence are marked by the so-called *liquidus* and *solidus* lines. Another substitutional alloy is AgCu. Here, the phase diagram is much more complex (Fig. 2.15). The reason is that the solid phase of AgCu is not completely miscible. Cu is soluble in Ag only up to a particular percentage that depends on the temperature (α-phase, left side of Fig. 2.15). Likewise is Ag in Cu only soluble up to a temperature-dependent percentage (β-phase, right side of Fig. 2.15). In the intermediate range in the so-called *miscibility gap* the solid phase consists of microcrystalline domains of an Ag-rich α-phase and a Cu-rich β-phase (see Fig. 2.13). In real systems, the size and shape of the microcrystals are nearly always determined by kinetics rather than by thermodynamic equilibrium. A defined equilibrium size of the crystallites nevertheless exists. It is determined by the minimum of the interfacial energy between the different crystallites and the elastic strain energy due to the mismatch of the lattice constants between the crystals of different composition. The strain energy decreases as the crystal size becomes smaller. On the other hand the interfacial area and thus the interfacial energy increases so that one has a minimum of the total energy for a particular crystal size.

In the course of this section we want to come to an understanding of the origin of the different phase diagrams of the two substitutional alloys SiGe and AgCu based on thermodynamic reasoning. As seen above, the crucial difference between the two systems is the possibility, respectively impossibility of a continuous mixture in the solid phase. Our thermody-

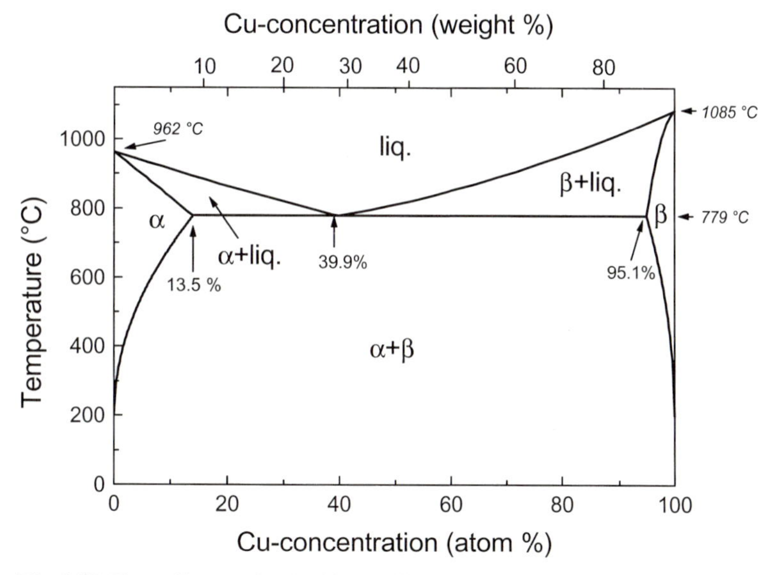

Fig. 2.15. Phase diagram for the binary alloy Ag/Cu. The system is not continuously miscible in the solid phase. Rather the alloy has a wide miscibility gap in which a Ag-rich fcc-phase (α-phase) co-exists with a Cu-rich fcc-phase (β-phase) (see Fig. 2.13)

namic considerations therefore focus on the free enthalpy associated with a mixture. The free enthalpy of a system G is a linear combination comprising of the internal energy U, the entropy S and mechanical (or other) work. In the simple case of a homogeneous, liquid or gaseous matter G is

$$G = U - TS + pV \,, \tag{2.4}$$

in which T is the temperature, p the pressure and V the volume. In addition to the work against the external pressure one needs to consider the mechanical work associated with the internal degrees of freedom of the system (see discussion above). This part is rather difficult to deal with (see also Sect. 4.5). Fortunately, the contributions of energy and entropy of mixture prevail. We therefore consider only those two contributions in the following and neglect contributions of mechanical, electrical or magnetic work. In a first step we likewise neglect thermal contributions. One further approximation is needed for the purpose of calculating the energy and entropy of mixture: this is that we admit only homogeneously mixed phases, that is we exclude spatial and temporal fluctuations. This corresponds to a *mean field approximation*, an approximation that we encounter quite frequently in the theory of solid matter. In the present context, the approximation is also known as the *Bragg–Williams approximation*. The variation in the enthalpy associated with mixing G_{mix} consists of a variation of the internal energy U_{mix}, which is the heat of solution, and the variation of the entropy due to mixing S_{mix}

$$G_{\text{mix}} = U_{\text{mix}} - T\,S_{\text{mix}} \,. \tag{2.5}$$

In a first step we calculate the variation of the internal energy under the assumption that the binding energies of the involved atoms can be represented by nearest neighbor pair interactions. This simple ansatz does not conform with reality but must suffice here. We denote the number of atoms of type A and B as N_A and N_B, and the number of nearest-neighbor bonds between atoms of type A, of type B, and between atoms A and B as N_{AA}, N_{BB}, and N_{AB}, respectively. Correspondingly, the binding energies of the atom pairs AA, BB and AB are denoted as V_{AA}, V_{BB} and V_{AB}. The coordination numbers (number of nearest-neighbors) in the two structures of interest here, the diamond structure (Si, Ge) and the fcc-structure of Ag and Cu, are $z=4$ and $z=12$, respectively. The variation of the energy due to mixing U_{mix} is

$$\begin{aligned} U_{\text{mix}} = &-(N_{AA}\,V_{AA} + N_{BB}\,V_{BB} + N_{AB}\,V_{AB}) \\ &+ \frac{1}{2}\,(z N_A\,V_{AA} + z N_B\,V_{BB}) \,. \end{aligned} \tag{2.6}$$

The sum of the first three terms is the energy after mixing and the sum of the last two terms is the energy before mixing. The factor 1/2 occurs because there are $z/2$ bonds per atom. Note that the definitions of binding energy and internal energy imply an opposite sign: a higher binding energy corresponds to a lower internal energy! The numbers of atom pairs N_{AA}, N_{BB} and N_{AB} can be expressed in terms of the concentrations

$$\begin{aligned} x_A &= N_A/N \\ x_B &= N_B/N \quad \text{with} \\ N &= N_A + N_B \end{aligned} \tag{2.7}$$

and one obtains

$$\begin{aligned} N_{AA} &= N_A\, x_A\, z/2 = N\, x_A^2\, z/2 \\ N_{BB} &= N_B\, x_B\, z/2 = N\, x_B^2\, z/2 \\ N_{AB} &= N_A\, x_B\, z/2 = N\, x_A x_B\, z/2\,. \end{aligned} \tag{2.8}$$

With that the energy of mixing U_{mix} becomes

$$U_{\text{mix}} = N\, z\, x_A\, x_B\, W_{AB} \tag{2.9}$$

with

$$W_{AB} = \frac{1}{2}(V_{AA} + V_{BB}) - V_{AB}\,. \tag{2.10}$$

If $W_{AB}<0$ then the alloy has a higher binding energy and the internal energy of the alloy is lower than the sum of the internal energy of the constituents. Since $x_A+x_B=1$ the energy of mixing has the form of a parabola.

$$U_{\text{mix}} = N\, z\, x_A(1 - x_A) W_{AB}\,. \tag{2.11}$$

Associated with a mixing of two components is always an enlargement of the entropy. This is because microscopically a mixture can be realized in many different ways. The different microscopic realizations originate from the exchange of atoms A with atoms B. The number of possibilities for an exchange of all atoms $N=N_A+N_B$ is $N!$. The exchange of atoms A and B among themselves, however, does not constitute a discernibly different microscopic state. The number of discernible states is thus $N!/(N_A!\; N_B!)$. With that the entropy of mixing becomes

$$S_{\text{mix}} = k \ln \frac{N!}{N_A!\, N_B!} = k \ln \frac{N!}{(N\, x_A)!\, (N\, x_B)!}\,. \tag{2.12}$$

With Stirling's approximation for large numbers $N \ln N! \approx N \ln N - N$, $N \gg 1$ one obtains

$$S_{\text{mix}} = -N k\, [x_B \ln x_B + (1 - x_B) \ln (1 - x_B)]\,. \tag{2.13}$$

The entropy S_{mix} is zero for $x_B=0$ and $x_B=1$ and positive for intermediate x_B. The free enthalpy of mixing is

$$\begin{aligned} G_{\text{mix}} &= U_{\text{mix}} - T\, S_{\text{mix}} \\ &= N\{z\, x_B(1-x_B) W_{AB} + k T\, [x_B \ln x_B + (1-x_B) \ln (1-x_B)]\}\,. \end{aligned} \tag{2.14}$$

The function has an extreme at x_B=0.5. Whether it is a minimum or a maximum depends on the ratio of W_{AB} and $k T$. One has

$$\text{a minimum for } z\,W_{AB}/k T < 0.5\,, \tag{2.15}$$

$$\text{a maximum for } z\,W_{AB}/k T > 0.5\,. \tag{2.16}$$

Examples for $zW_{AB}/k T$=0.5, 2.5 and 3 are displayed in Fig. 2.16. For negative values of W_{AB} the free enthalpy of mixing is always smaller than the sum of the free enthalpies of the separate components. In that case the alloy is continuously miscible. The same holds for any system, only if the temperature is high enough (2.14). The system becomes unstable if the free enthalpy has a maximum at x_B=0.5. It decomposes into two separate phases, one with a concentration of the B-component $x'_B < 1/2$ and the other one with $x''_B > 1/2$. This is independent of the sign of the entropy of mixing. For the special case when $G_{\text{mix}}(x_B)$ is symmetric around x_B=1/2 (Fig. 2.16) it is easy to see that the instability exists not only at x_B=1/2 but in the entire range between the two minima of $G_{\text{mix}}(x_B)$. Figure 2.17 displays the more general case of an asymmetric shape of $G_{\text{mix}}(x_B)$. The two concentrations x'_B and x''_B of the two phases into which the system separates are determined by the condition that the system as a hole must remain in equilibrium, that is, the chemical potentials of the two phases must be equal. The chemical potentials of the atoms of type B in the two phases are

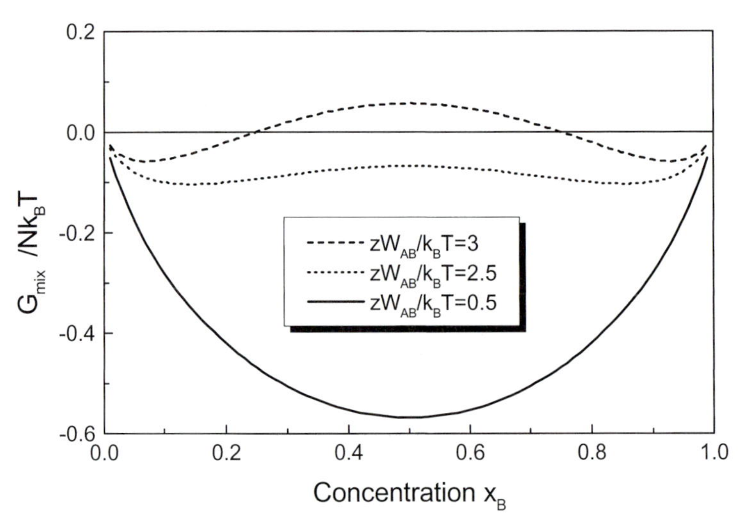

Fig. 2.16. The free enthalpy of a solid mixture according to a simple model (2.14). The solid line corresponds to the limiting case in which an intermediate maximum is not yet present. If an intermediate maximum occurs, the system is unstable against phase separation into two phases, one depleted of N_B atoms and one enriched with N_B atoms. Phase separation occurs independent of whether the enthalpy of mixing is gained in the alloying process (*dotted line*) or not (*dashed line* for concentrations around 0.5)

$$\mu'_B = \frac{\partial G'}{\partial N_B} = \frac{1}{N}\frac{\partial G'}{\partial x_B}$$
$$\mu''_B = \frac{\partial G''}{\partial N_B} = \frac{1}{N}\frac{\partial G''}{\partial x_B} \,. \tag{2.17}$$

Hence the system decomposes into two phases for which the slope of the free enthalpy (Fig. 2.17) is equal. Of all (the infinite number of) points on the curve $G(x_B)$ that fulfil the condition of an equal slope the total free enthalpy is minimal for the particular pair of concentrations x'_B and x''_B through which a common tangent to $G(x_B)$ can be drawn. This common tangent condition uniquely determines the two concentrations and x'_B and x''_B. A few simple geometrical considerations demonstrate that the system indeed separates into two phases with concentrations x'_B and x''_B, and that an alloy is unstable in the entire concentration range between x'_B and x''_B. Consider an arbitrary intermediate concentration x_{B0}. Conservation of mass during the process of phase separation requires that the numbers of B-atoms in the two phases $N'_B = N'x'_B$ and $N''_B = N''x''_B$ obey the relation

$$N'x'_B + N''x''_B = (N' + N'')x_{B0} \,. \tag{2.18}$$

The ratio of the total number of atoms in the two phases becomes therefore equal to the ratio of the concentration differences

$$\frac{N'}{N''} = \frac{x''_B - x_{B0}}{x_{B0} - x'_B} \,. \tag{2.19}$$

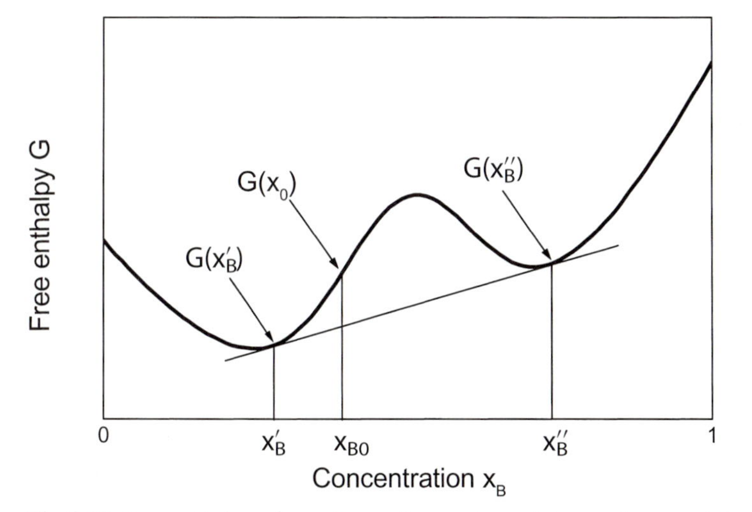

Fig. 2.17. Free enthalpy of an alloy with an intermediate maximum and two minima of a different depth. The system is unstable in the concentration range between the two points of tangency at x'_B and x''_B, and the system separates into two phases with the concentrations x'_B and x''_B. The free enthalpy of the two-phase system is described by the common tangent.

A corresponding relation holds for the component A. This simple relation resembles the lever principle in mechanics and is therefore known as the lever rule of phase diagrams. It holds for any phase separation, regardless of its nature. After phase separation the free enthalpy of the system becomes

$$G(x'_B, x''_B) = G(x'_B)\,N'/N + G(x''_B)\,N''/N\,. \tag{2.20}$$

By substituting (2.20) into (2.19) and after a suitable rearrangement the enthalpy can be written as

$$G(x'_B, x''_B) = G(x'_B) + [G(x''_B) - G(x'_B)]\frac{x_{B0} - x'_B}{x''_B - x'_B}\,, \tag{2.21}$$

which is precisely the equation describing the y-coordinate in x_0 of the common tangent. Thus, the common tangent is simply the free enthalpy after phase separation. Between x'_B and x''_B all values of the common tangent are below the free enthalpy of the system before phase separation (Fig. 2.17). The system is therefore unstable with respect to phase separation in the entire gap between x'_B and x''_B. In the context of binary alloys this gap is known as the *miscibility gap*. As can be seen from the phase diagram (Fig. 2.15) a miscibility gap does exist for the Ag/Cu alloy. If a melt of 40 atom% silver and 60 atom% copper is cooled below the temperature of solidification, the solid crystallizes in the β-phase containing about 95% Cu. The remaining melt becomes enriched with Ag, until the Cu-rich α-phase (95.1% Cu) and the silver-rich β-phase (85.9% Ag) solidify together at the so-called eutectic point, which corresponds to 39.9% Cu in the melt at a temperature of 779 °C. Upon further cooling the concentrations of the two solid phases should continue to vary in principle (Fig. 2.15). However, the equilibrium state can be achieved only for an extremely low cooling rate because of the low diffusivity of atoms in the solid phase. The concentrations in the darker and lighter areas in Fig. 2.15 therefore practically correspond to the equilibrium concentrations of the α- and β-phase at the eutectic point.

Thermodynamic reasoning can also be applied to the phase transition between the liquid and the solid phase in the case of a completely miscible alloy. Consider the free enthalpy of such a system in the liquid and solid phases (Fig. 2.18). The system is completely miscible since the free enthalpy is now represented by a function with a positive curvature in the entire concentration range. Liquid and solid phases coexist if the minima of the free enthalpies for the liquid and the solid state occur at different concentrations which is generally the case. The curves in Fig. 2.19 approximately represent the system Ge/Si. We consider first a high temperature $T > 1412$ °C. There, the free enthalpy of the melt is below that of the solid for all concentrations. This is because of the entropic contribution to the free enthalpy. At high temperatures, the entropy of the liquid phase is higher than that of the solid phase. The reason is that a liquid has more states

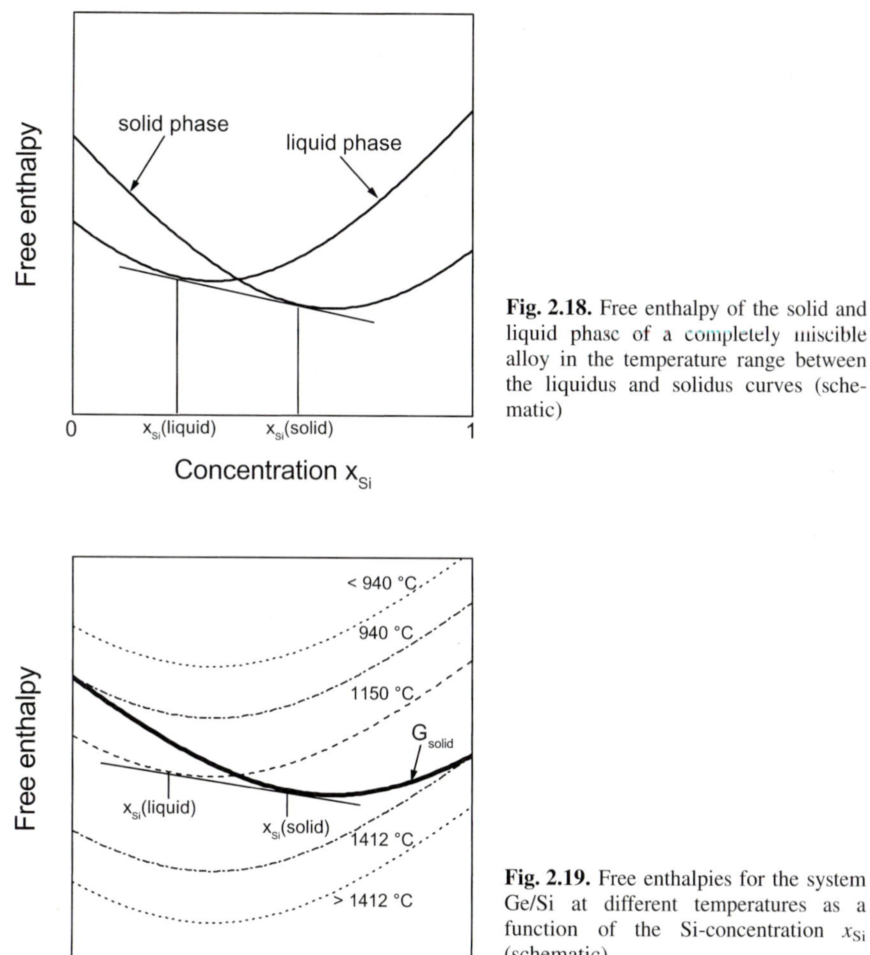

Fig. 2.18. Free enthalpy of the solid and liquid phase of a completely miscible alloy in the temperature range between the liquidus and solidus curves (schematic)

Fig. 2.19. Free enthalpies for the system Ge/Si at different temperatures as a function of the Si-concentration x_{Si} (schematic)

of low quantum energy than a solid. Simply speaking, the degrees of freedom of the transverse sound waves of the solid become free translations in the melt. At a temperature of 1412 °C the free enthalpies of the melt and the solid become equal for the concentration $x_{Si} = 1$. Melt and the solid coexist: we have reached the melting point of pure silicon. For even lower temperatures, e.g. at 1150 °C, one has, according to the common tangent construction, a coexistence of a Si-depleted, liquid phase with a Si-enriched solid phase. Below 940 °C finally, only the solid phase exists for all mixing ratios. If the cooling process is performed using a melt with a mixing ratio of 50 atom% Ge and 50 atom% Si, e.g., crystallites with 80% Si will solidify first at a temperature of about 1270 °C. In the temperature range between the liquidus and solidus curve the equilibrium concentrations of Si in the liquid and the solid state are given by the corresponding values for the concentrations of the liquidus and solidus curves. The ratio of the atom numbers in the two phases obey the lever rule (2.19). With decreasing temperature the solidified fraction of the melt increases, and the Si-concentration in the crystalline phase decreases. It is therefore not possible for continuously miscible alloys to grow a crystal out of a melt that has a homogeneous concentration ratio, unless one confines the crystallization to a small fraction of the melt.

One can utilize, however, the different equilibrium concentrations in the melt and the solid to purify a crystal of undesirable impurities. This is the basis of purification by zone melting: One begins by melting a narrow zone of a crystal rod at one end. In this molten zone the impurity concentration necessarily is as it was in the solid. Then, the molten zone is slowly pulled over the rod. If the liquidus and solidus curves are as in Fig. 2.14 with regard to an impurity (with Si as the base material and Ge as an impurity) then the re-crystallized rod in the cooling zone has a lower concentration of impurities than the (respective) melt. Hence, the impurities are enriched at that end of the rod that is molten last. A large section of the crystal can very effectively be purged of impurities by repeating the process many times.

2.7 Defects in Solids

Mechanical and electrical properties of solids are largely controlled by defects in the periodic structure. This section briefly reviews various known defects. In the previous section we have already learned about a special defect: In a diluted substitutional alloy, minority atoms may assume the sites of the majority atoms. In the context of semiconductors replacing majority atoms by atoms of another kind, preferably by atoms of a higher or lower valence is known as "doping". Doping varies the electrical conductivity of a semiconductor by many orders of magnitude (see Sect. 12.3). Defects consisting only of one or a few atoms, as in the case of doping, are known as *point defects*. Point defects do not necessarily involve foreign atoms. The so-called *Frenkel defect* consists of an atom displaced from its regular site to an *interstitial* site. The atom in the interstitial site and the vacancy in the regular site together are named "Frenkel pair". Since the bond energy of the atom is lower at the interstitial site the formation of a Frenkel pair requires energy. Nevertheless, Frenkel pairs do exist in equilibrium at higher temperatures because the formation of a Frenkel pair increases the entropy. That increase arises from the fact that the atom as well as the vacancy may sit in any possible interstitial or lattice site, respectively, thereby enjoying many distinguishable microscopic realizations. The number of possibilities to distribute n_v vacancies on N regular atom sites is

$N!/[n_v!(N{-}n_v)!]$. Similarly, the number of possibilities to distribute $n_{\rm int}$ interstitial atoms on N' interstitial sites is $N'!/[n_{\rm int}!(N'-n_{\rm int})!]$. For a Frenkel pair the number of interstitial atoms $n_{\rm int}$ necessarily equals the number of vacancies n_v. With $n{=}n_{\rm int}{=}n_v$ one obtains for the entropy S (comp. 2.12)

$$\begin{aligned} S &= k \ln \frac{N!}{n!(N-n)!} + k \ln \frac{N'!}{n!(N'-n)!} \\ &\cong k\,[N \ln N + N' \ln N' - 2n \ln n - (N-n)\ln(N-n) \\ &\quad - (N'-n)\ln(N'-n)]\ . \end{aligned} \tag{2.22}$$

In equilibrium the system is in the state of lowest free energy $F{=}n\Delta E - TS$, in which ΔE is the energy required to create a Frenkel pair. The corresponding equilibrium concentration $\langle n\rangle$ is obtained by differentiating the free energy with respect to n

$$\frac{dF}{dn} = 0 = \Delta E + kT \ln \frac{\langle n\rangle^2}{(N' - \langle n\rangle)(N - \langle n\rangle)}\ . \tag{2.23}$$

Since the concentration of defects is small $(\langle n\rangle \ll N, N')$ $\langle n\rangle$ is approximately

$$\langle n\rangle \cong \sqrt{N N'}\, \mathrm{e}^{-\Delta E/2kT}\ . \tag{2.24}$$

Hence, the concentration rises exponentially with the temperature according to an Arrhenius law. The activation energy in the Arrhenius law is half the energy required for the creation of a Frenkel-pair. The factor of two in the denominator of the exponent arises because vacancies as well as interstitial atoms are distributed independently in the crystal. One therefore has two independent contributions to the entropy. If the atom that is displaced from the regular site diffuses to the surface or into an interface ("Schottky defect"), the full energy of creation for the defect appears in the Arrhenius law. The reason is that for a macroscopic solid the number of available sites on the surface or in an interface are infinitely small compared to the number of sites in the bulk. In that case only the vacancies in the regular crystal sites contribute to the entropy.

Defects of the next higher dimension are line defects. A common intrinsic line defect is the *dislocation*. A simple example is shown in Fig. 2.20, which displays a cross section of a crystal with a dislocation. Around the core of the dislocation atoms are displaced from their regular lattice positions because of the elastic stresses. Most of the energy required to create a dislocation is actually in the elastic strain that decays rather slowly as one moves away from the core. Dislocations are described by the *Burgers vector*. The Burgers vector is constructed by considering the positions of atoms after completing a closed loop of an arbitrary size around the dislocation core for a lattice with and without a dislocation (Fig. 2.20). If the Burgers vector is oriented perpendicular to the dislocation line as in

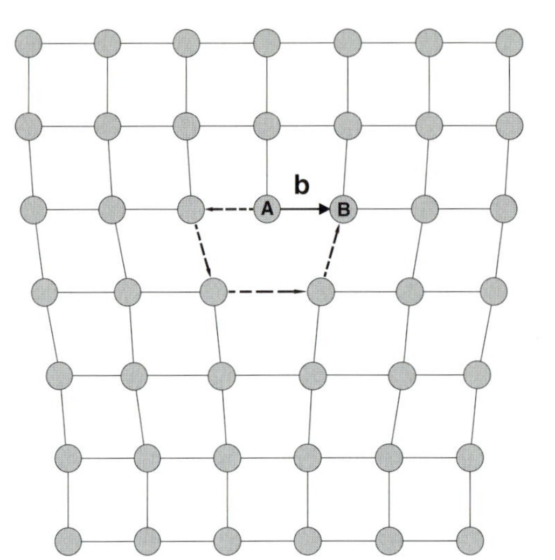

Fig. 2.20. Sectional drawing of a crystal with an edge dislocation (schematic). The dashed line represents a loop around the core of the dislocation. The loop begins with atom A. It would close at atom B if the dislocation were not present. The Burgers vector **b** points from atom *A* to atom *B*. The same Burgers vector is obtained for any arbitrary loop that encloses the dislocation

Fig. 2.20, the dislocation is called an "*edge dislocation*". If the Burgers vector is oriented along the dislocation line, then the dislocation is called a "*screw dislocation*", since by moving along a closed loop around the dislocation one climbs from one lattice plane onto the next. Edge dislocations and screw dislocations represent two limiting cases of a general, intermediate form of a dislocation. Furthermore, the angle between the orientation of the Burgers vector and the dislocation line may vary as one moves along the dislocation line. The modulus of the Burgers vector is equal to the distance of an atom plane for the common screw or edge dislocations. However, dislocations for which the modulus of the Burgers vector is only a fraction of a distance between an atom plane also exist. Such a *partial dislocation* is generated, e.g., if all atoms in a section of an fcc-crystal are displaced along a direction in a densely packed plane so as to produce a stacking fault (Fig. 2.9).

Dislocations play a crucial role in plastic deformation of crystalline material. Consider a shear force acting parallel to an atom plane. It is not feasible to make all the atoms glide simultaneously since the shear force works against the atomic bonds of all atoms in the glide plane at once. A step wise glide is energetically much more favorable. Firstly an edge dislocation is generated at the surface and then the dislocation is shifted through the crystal. Then gliding is effectuated by displacing the atoms row-by-row until the dislocation line has moved through the entire crystal. The required forces are much lower since fewer atoms are affected and bonds need not be broken but must merely be strained and re-oriented. Plastic deformations of a crystalline solid are therefore connected with the generation and wandering of dislocations. In a pure, ideally crystalline material dislocations can move easily, provided the temperature is not too low. For many metals, e.g., room temperature suffices. Such materials have little resistance to plastic deformation. Examples are rods or wires consisting of annealed copper and silver. If the material is polycrystalline, e.g., after cold working, then the wandering of dislocations is hindered by the grain boundaries between the crystallites, and the material resists plastic deformation more effectively.

Problems

2.1 The phase transition from graphite to diamond requires high pressure and high temperature in order to shift the equilibrium in favor of diamond and also to overcome the large activation barrier. Suggest a method of producing diamond (layers) without the use of high pressure.

2.2 Below 910 °C iron exists in the bcc structure (α-Fe). Between 910 °C and 1390 °C it adopts the fcc structure (γ-Fe). Assuming spherical atoms, determine the shape and size of the octahedral interstitial sites in γ-Fe ($a=3.64$ Å) and in α-Fe ($a=2.87$ Å). Sketch the lattices and the interstitial sites. For which phase would you expect the solubility of carbon to be higher? (Hint: The covalent radius of carbon is 0.77 Å.) When molten iron containing a small amount of carbon ($\lesssim 1\%$) is cooled, it separates into a more-or-less ordered phase containing α-Fe with a small concentration of carbon atoms on interstitial sites (ferrite) and a phase containing iron carbide (cementite, Fe_3C). Why does this occur? Why does cementite strengthen the medium against plastic deformation?
(Hint: Fe_3C, like many carbides, is very hard and brittle.)

2.3 Copper and gold form a continuous solid solution with the copper and gold atoms statistically distributed on the sites of an fcc lattice. For what relative concentrations of copper and gold atoms do you expect ordered alloys and what would they look like? Draw the unit cells of these alloys and identify the corresponding Bravais lattices. Can you suggest an experiment which would determine whether the alloy is ordered or not?

2.4 Draw and describe the symmetry elements of all Bravais lattices.

2.5 Draw the primitive unit cell of the fcc lattice and determine the lengths of the primitive lattice vectors $\boldsymbol{a}'$, $\boldsymbol{b}'$, $\boldsymbol{c}'$ (in units of the conventional lattice constant a) and also the angles α', β', γ' between the primitive lattice vectors. (Hint: Express the primitive lattice vectors as a linear combination of the lattice vectors $\boldsymbol{a}$, $\boldsymbol{b}$, $\boldsymbol{c}$ of the face-centered cubic lattice and use elementary vector algebra.) What distinguishes this unit cell from that of the rhombic Bravais lattice?

2.6 Determine the ratio of the lattice constants c and a for a hexagonal close packed crystal structure and compare this with the values of c/a found for the following elements, all of which crystallize in the hcp structure: He (c/a=1.633), Mg (1.623), Ti (1.586), Zn (1.861). What might explain the deviation from the ideal value?

2.7 Supposing the atoms to be rigid spheres, what fraction of space is filled by atoms in the primitive cubic, fcc, hcp, bcc, and diamond lattices?

2.8 Give a two-dimensional matrix representation of a 2-, 3-, 4-, and 6-fold rotation. Which representation is reducible?

2.9 Show that the rhombohedral translation lattice in Fig. 2.3 is equivalent to a hexogonal lattice with two atoms on the main diagonal at a height of $c/3$ and $2c/3$. Hint: Consider the projection of a rhombohedral translation lattice along the main diagonal into the plane perpendicular to this main diagonal. The main diagonal of the rhombohedral lattice is parallel to the c-axis of the hexagonal lattice. How are the a and the c axis of the corresponding hexagonal lattice related to the angle $\alpha=\beta=\gamma$ and the length $a=b=c$ of the rhombohedral lattice?

2.10 Take a piece of copper wire and anneal it by using a torch! Demonstrate that the wire is easily plastically deformed. Then pull the wire hard and suddenly or work it cold using a hammer. How is the plastic behavior now? Explain the observations!

3 Diffraction from Periodic Structures

A direct imaging of atomic structures is nowadays possible using the high-resolution electron microscope, the field ion microscope, or the tunneling microscope. Nonetheless, when one wishes to determine an unknown structure, or make exact measurements of structural parameters, it is necessary to rely on diffraction experiments. The greater information content of such measurements lies in the fact that the diffraction process is optimally sensitive to the periodic nature of the solid's atomic structure. Direct imaging techniques, on the other hand, are ideal for investigating point defects, dislocations, and steps, and are also used to study surfaces and interfaces. In other words, they are particularly useful for studying features that represent a disruption of the periodicity.

For performing diffraction experiments one can make use of X-rays, electrons, neutrons and atoms. These various probes differ widely with respect to their elastic and inelastic interaction with the solid, and hence their areas of application are also different. Atoms whose particle waves have a suitable wavelength for diffraction experiments do not penetrate into the solid and are thus suitable for studying surfaces. The same applies, to a lesser extent, for electrons. Another important quantity which differs significantly for the various probes is the spatial extent of the scattering centers. Neutrons, for example, scatter from the nuclei of the solid's atoms, whereas X-ray photons and electrons are scattered by the much larger ($\sim 10^4$ times) electron shells. Despite this and other differences, which will be treated in more detail in Sect. 3.7, it is possible to describe the essential features of diffraction in terms of a single general theory. Such a theory is not able, of course, to include differences that arise from the polarization or spin polarization of the probes. The theory described in Sect. 3.1 below is quasi classical since the scattering itself is treated classically. The quantum mechanical aspects are treated purely by describing the probe particles as waves. For more detailed treatments including features specific to the various types of radiation, the reader is referred to [3.1–3.3].

3.1 General Theory of Diffraction

In our mathematical description of diffraction we will make the assumption of single scattering: the incoming wave induces at all points $\boldsymbol{r}$ of the target material the emission of spherical waves. A fixed phase relationship is assumed between the primary wave and each of the emitted spherical waves (coherent scattering). Further scattering of the spherical waves, however, is neglected. This is also known as the "kinematic" approximation and corresponds to the first Born approximation of the quantum mechanical scattering theory. The approximation is valid for neutrons and X-rays, and within certain limits also for the scattering of high energy electrons. For highly perfect single crystals, it is possible to observe "nonkinematic" ("dynamical") effects in the scattering of X-rays.

For the derivation of the scattering amplitudes we make use of Fig. 3.1. Here Q is the location of the source of radiation, P is the position of the scattering center, and B the point of observation. As an example of a source we shall take the spherical light waves emitted in conjunction with an electronic transition in an atom. At sufficiently large distances from the source the spherical waves can be approximated as plane waves. The amplitude (or for X-rays more accurately the field strength vector) at position P and time t may thus be written

$$A_P = A_0 \, e^{i \boldsymbol{k}_0 \cdot (\boldsymbol{R}+\boldsymbol{r}) - i \omega_0 t} \; . \tag{3.1}$$

If we follow this wave back to the source Q ($\boldsymbol{R}+\boldsymbol{r}=0$), then its amplitude here as a function of time behaves as $\sim \exp(-i\omega_0 t)$, i.e., it has a well-defined phase at all times. The reasoning, however, can only be applied to a *single* emission process. In real light sources photons are emitted with uncorrelated phases from many atoms (an exception to this is the laser). For other types of incident particle the phases are likewise uncorrelated. Thus, when we use expression (3.1), we must keep in mind that the result for the observed intensity arises by averaging over many individual diffraction events.

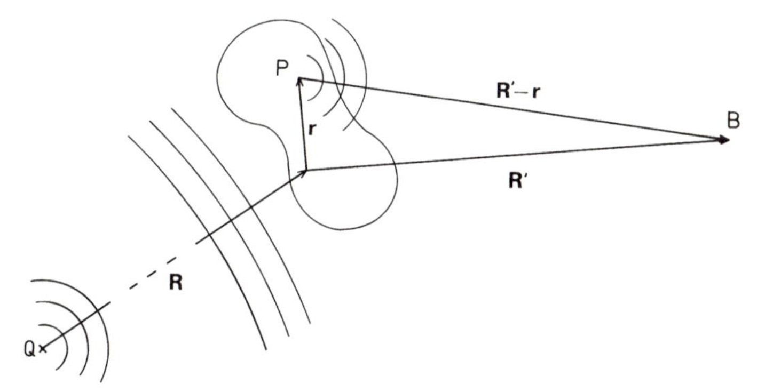

Fig. 3.1. Schematic representation of scattering indicating the parameters used in deriving the scattering kinematics. The source Q is assumed to be sufficiently far away from the target that the spherical waves reaching the target can be well approximated by plane waves. The same condition is assumed for the point of observation relative to the scattering centers

The relative phases of the wave at points P at time t are given by the position-dependent factor in (3.1). We now allow the primary wave to be scattered by the material. Every point P in the scattering material is caused by the primary wave to emit spherical waves, whose amplitude and phase relative to the incident (primary) wave are described by a complex scattering density $\varrho(\boldsymbol{r})$. The time dependence of these spherical waves is determined by the time dependence in (3.1) (forced oscillation). The spherical waves observed at B are therefore described by

$$A_B = A_P(\boldsymbol{r}, t)\, \varrho(\boldsymbol{r}) \frac{\mathrm{e}^{\mathrm{i}k|\boldsymbol{R}'-\boldsymbol{r}|}}{|\boldsymbol{R}'-\boldsymbol{r}|} \,. \tag{3.2}$$

For a fixed position P, the wave vector $\boldsymbol{k}$ is in the direction $\boldsymbol{R}'-\boldsymbol{r}$. Thus we can also write

$$A_B = A_P(\boldsymbol{r}, t)\, \varrho(\boldsymbol{r}) \frac{\mathrm{e}^{\mathrm{i}\boldsymbol{k}\cdot(\boldsymbol{R}'-\boldsymbol{r})}}{|\boldsymbol{R}'-\boldsymbol{r}|} \,. \tag{3.3}$$

At large distances from the scattering center A_B is then given by

$$A_B = A_P(\boldsymbol{r}, t)\, \varrho(\boldsymbol{r}) \frac{1}{R'}\, \mathrm{e}^{\mathrm{i}\boldsymbol{k}\cdot(\boldsymbol{R}'-\boldsymbol{r})} \tag{3.4}$$

where $\boldsymbol{k}$ now has the same direction for *all* positions P in the target material.

Inserting (3.1) into (3.4) we obtain

$$A_B = \frac{A_0}{R'}\, \mathrm{e}^{\mathrm{i}(\boldsymbol{k}_0\cdot\boldsymbol{R}+\boldsymbol{k}\cdot\boldsymbol{R}')}\, \mathrm{e}^{-\mathrm{i}\omega_0 t}\, \varrho(\boldsymbol{r})\, \mathrm{e}^{\mathrm{i}(\boldsymbol{k}_0-\boldsymbol{k})\cdot\boldsymbol{r}} \,. \tag{3.5}$$

The total scattering amplitude is given by integration over the entire scattering region:

$$A_B(t) \propto \mathrm{e}^{-\mathrm{i}\omega_0 t} \int \varrho(\boldsymbol{r})\, \mathrm{e}^{\mathrm{i}(\boldsymbol{k}_0-\boldsymbol{k})\cdot\boldsymbol{r}}\, d\boldsymbol{r} \,. \tag{3.6}$$

For scattering from a rigid lattice, $\varrho(\boldsymbol{r})$ is time independent and the time dependence of A_B only contains the frequency ω_0. In the quantum mechanical picture this corresponds to energy conservation. We thus have elastic scattering. This form of scattering is important for structural analyses. If instead we allow the scattering density $\varrho(\boldsymbol{r})$ to vary with time, then we also obtain scattered waves with $\omega \neq \omega_0$. This inelastic scattering will be dealt with in Sect. 4.4.

In diffraction experiments for structure determination, it is not the amplitude but the intensity of the scattered waves that is measured:

$$I(\boldsymbol{K}) \propto |A_B|^2 \propto \left|\int \varrho(\boldsymbol{r})\, \mathrm{e}^{-\mathrm{i}\boldsymbol{K}\cdot\boldsymbol{r}}\, d\boldsymbol{r}\right|^2 \,. \tag{3.7}$$

In this equation we have introduced the scattering vector $\boldsymbol{K}=\boldsymbol{k}-\boldsymbol{k}_0$.

We see that the intensity is the absolute square of the Fourier transform of the scattering density $\varrho(\boldsymbol{r})$ with respect to the scattering vector $\boldsymbol{K}$. From this we establish an important fact: The smaller the structures to be resolved in the diffraction measurement, the larger is the required value of $\boldsymbol{K}$, and thus also of the $\boldsymbol{k}$-vector of the incident radiation. In studies of the atomic structures of solids the wavelengths should thus roughly correspond to the lattice constants. For such waves it is impossible to measure their amplitude as a function of position and time – only the intensity can be measured. This leads to considerable complications in structural analysis. If it were actually possible to measure the amplitude instead of the intensity, then one could make use of inverse Fourier transformation to obtain the spatial distribution of the scattering density directly from the diffraction pattern. However, since in reality one can only observe the intensities, the information about the phases is lost and the scattering density cannot be calculated directly. To determine a particular structure it is therefore necessary to do the calculation in reverse: One chooses a feasible model structure, calculates the diffraction pattern that it would produce, and then compares this with the experimentally observed diffraction pattern. The structural parameters of the model are varied until optimal agreement with experiment is found.

The analysis of unknown structures is facilitated by invoking the so-called Patterson function, which is the Fourier transform of the intensity. In order to elucidate the meaning of the Patterson function we rewrite (3.7) for the intensity

$$I(\boldsymbol{K}) \propto \int \varrho(\boldsymbol{r})\, \mathrm{e}^{-\mathrm{i}\boldsymbol{K}\cdot\boldsymbol{r}}\, d\boldsymbol{r} \int \varrho(\boldsymbol{r}')\, \mathrm{e}^{\mathrm{i}\boldsymbol{K}\cdot\boldsymbol{r}'}\, d\boldsymbol{r}' \,. \tag{3.8}$$

Since both integrals extend over the entire space the variable $\boldsymbol{r}'$ in the second integral can be replaced by $\boldsymbol{r}+\boldsymbol{r}'$. Hence, one obtains

$$I(\boldsymbol{K}) \propto \int \mathrm{e}^{\mathrm{i}\boldsymbol{K}\cdot\boldsymbol{r}'}\, d\boldsymbol{r}' \int \varrho(\boldsymbol{r})\varrho(\boldsymbol{r}'+\boldsymbol{r})\, d\boldsymbol{r} \,. \tag{3.9}$$

The auto-correlation function of the scattering density

$$P(\boldsymbol{r}') = \int \varrho(\boldsymbol{r})\, \varrho(\boldsymbol{r}'+\boldsymbol{r}) d\boldsymbol{r} \tag{3.10}$$

is the Patterson function. The function has its peaks where $\boldsymbol{r}'$ corresponds to a vector between two atoms of the crystal structure. A peak is particularly strong if the vector connects two atoms with a large scattering cross section. The interatomic distances in an unknown structure are therefore easily obtained by inspection of the Patterson function.

Scattering from disordered systems, i.e. liquids and amorphous solids is most suitably described with the help of the Patterson function $P(\boldsymbol{r}')$. We decompose the scattering density $\varrho(\boldsymbol{r})$ into contributions from individual atoms. For simplicity we assume that the material is made up of a single type of atoms. The scattering density centered at the position $\boldsymbol{r}_i$ is denoted as $\varrho_{\mathrm{at}}(\boldsymbol{r}-\boldsymbol{r}_i)$.

$$\varrho(\boldsymbol{r}) = \sum_i \varrho_{\mathrm{at}}(\boldsymbol{r}-\boldsymbol{r}_i) \,. \tag{3.11}$$

The Patterson-function $P(\boldsymbol{r}')$ can be split into two contributions: one describing the correlation of one atom with itself and the other one describing the correlation of an atom with all other atoms.

$$\begin{aligned} P(\boldsymbol{r}') &= \sum_{i,j} \int \varrho_{\text{at}}(\boldsymbol{r}-\boldsymbol{r}_i)\varrho_{\text{at}}(\boldsymbol{r}-\boldsymbol{r}_j+\boldsymbol{r}')d\boldsymbol{r} \\ &= \sum_{i} \int \varrho_{\text{at}}(\boldsymbol{r}-\boldsymbol{r}_i)\varrho_{\text{at}}(\boldsymbol{r}-\boldsymbol{r}_i+\boldsymbol{r}')d\boldsymbol{r} + \sum_{i} \int \varrho_{\text{at}}(\boldsymbol{r}-\boldsymbol{r}_i) \\ &\quad \times \sum_{j\neq i} \varrho_{\text{at}}(\boldsymbol{r}-\boldsymbol{r}_j+\boldsymbol{r}')d\boldsymbol{r}\,. \end{aligned} \tag{3.12}$$

The second term in (3.12) contains the information on the structure. We assume now that the scattering density is localized at the centers of the atoms. This assumption is particularly well fulfilled for neutron scattering (see also Sect. 3.7 and I.3). In that case, the first integral in (3.12) contributes only at $\boldsymbol{r}' = 0$. The integral can therefore be replaced by $f^2\delta_{0,r'}$ where $\delta_{0,r'}$ denotes the Kronecker symbol and f the "atom factor", which is a measure of the magnitude of the scattering amplitude of an atom. The sum over all (identical) atoms i in the first term can be replaced by a multiplication with the number of atoms N. The second integral in (3.12) vanishes for $\boldsymbol{r}' = 0$ since the probability to have a second atom j at the position of any other atom i is zero. We assume the system to be disordered, but homogeneous on a coarse scale. Because of the disorder, the mean scattering density around each atom is identical and independent of angle. Thus, the mean value of the sum

$$\bar{\varrho}(r) = \left\langle \sum_{j\neq i} \varrho_{\text{at}}(\boldsymbol{r}-\boldsymbol{r}_j) \right\rangle$$

depends only on the distance r from the atom whose environment is being considered.

The Patterson function is then

$$P(r') = Nf^2\delta_{0,r'} + N\int \varrho(r)\bar{\varrho}(r+r')d\boldsymbol{r}\,. \tag{3.13}$$

It is useful to introduce a function $g(r')$ that is a measure of the pair correlation of atoms, independent of their scattering amplitude and density. The limiting value of $g(r')$ for $r' \to \infty$ is one. For large distances away from any particular atom considered the scattering density is $\lim_{r'\to\infty} \bar{\varrho}(r+r') = Nf/V$ with V the volume. We therefore define the function $g(r')$ by

$$\frac{N}{V} f^2 g(r') = \int \varrho(r)\bar{\varrho}(r+r')\,d\boldsymbol{r}\,. \tag{3.14}$$

After inserting (3.14) and (3.13) in (3.9) one obtains for the scattering intensity $I(\boldsymbol{K})$

$$I(\boldsymbol{K}) \propto S(\boldsymbol{K}) = 1 + \frac{N}{V}\int g(r)\,\mathrm{e}^{\mathrm{i}\boldsymbol{K}\cdot\boldsymbol{r}}d\boldsymbol{r}\,. \tag{3.15}$$

$S(\boldsymbol{K})$ is known as the structure factor. The pair correlation function $g(r)$ can be calculated from the Fourier transform of $S(\boldsymbol{K}) - 1$. Here, however,

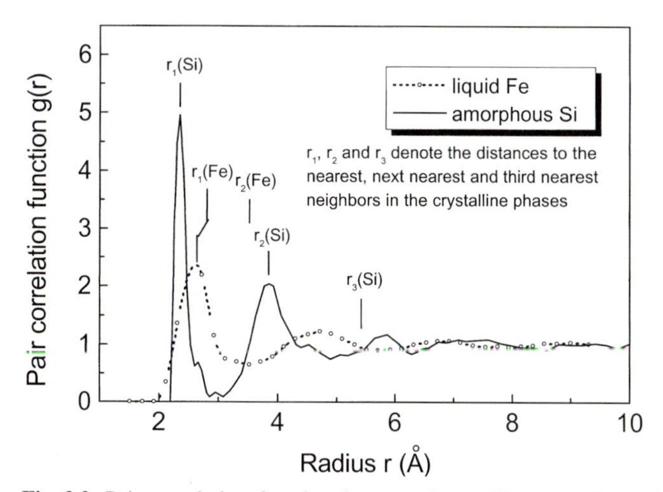

Fig. 3.2. Pair correlation function for amorphous silicon and liquid iron at a temperature of $T = 1833\,\mathrm{K}$ [3.4, 3.5]. The distances of the nearest, next-nearest, and third nearest-neighbors for crystalline silicon are marked as r_1(Si), r_2(Si) and r_3(Si), the corresponding distances for crystalline iron (fcc structure) as r_1(Fe) and r_2(Fe)

one encounters a technical difficulty: The contribution to $g(r)$ that arises from the homogeneous part of the scattering density at large r leads to a forward scattering at $\boldsymbol{K} = 0$. This contribution cannot be separated from the primary beam. One therefore extracts a function $h(r) = g(r) - 1$ from the experimental data that does not include the forward scattering. The function $h(r)$ is therefore the Fourier transform of the structure factor without the forward-scattering contribution (see Problem 3.7). Figure 3.2 shows two examples for the pair correlation function $g(r)$. The dashed and full lines represent $g(r)$ for liquid iron and amorphous silicon, respectively. The first peak corresponds to the distance of nearest-neighbors. The sharpness of the peak indicates that the distance to the nearest-neighbor is well defined even for structurally disordered systems. The mean distances of further neighbors are likewise discernible. Here, we encounter a characteristic difference between the amorphous state and the liquid state of matter. In the amorphous state, the distance to the next nearest-neighbor is nearly the same as in the crystalline state. The lower density of the amorphous state is reflected only in the distance to the third nearest-neighbor. In liquids, the second peak is approximately at twice the distance of the nearest-neighbor peak, hence at a significantly larger distance than, e.g., in a close-packed solid. This larger distance of second nearest-neighbors is an essential feature of the liquid state. Only with a larger distance of second nearest-neighbors can a liquid state be realized, as is easily demonstrated in a model in which the atoms are represented as hard spheres.

3.2 Periodic Structures and the Reciprocal Lattice

For periodic structures, $\varrho(\boldsymbol{r})$ can be expanded in a Fourier series. We first consider a one-dimensional example in which $\varrho(x)$ repeats with period a

$$\varrho(x) = \varrho(x + na) \quad n = 0, \pm 1, \pm 2, \dots \tag{3.16}$$

The corresponding Fourier series then reads

$$\varrho(x) = \sum_n \varrho_n \, \mathrm{e}^{\mathrm{i}(n2\pi/a)x} . \tag{3.17}$$

It is readily seen that a displacement by an arbitrary lattice vector $x_m = ma$ leads to an identical $\varrho(x)$, hence satisfying the required translational invariance. The extension to three dimensions is straightforward and yields

$$\varrho(\boldsymbol{r}) = \sum_{\boldsymbol{G}} \varrho_{\boldsymbol{G}} \, \mathrm{e}^{\mathrm{i}\boldsymbol{G}\cdot\boldsymbol{r}} . \tag{3.18}$$

The vector $\boldsymbol{G}$ must fulfill certain conditions in order to preserve the translational invariance of ϱ with respect to all lattice vectors

$$\boldsymbol{r}_n = n_1 \boldsymbol{a}_1 + n_2 \boldsymbol{a}_2 + n_3 \boldsymbol{a}_3 . \tag{3.19}$$

The conditions are expressed by

$$\boldsymbol{G} \cdot \boldsymbol{r}_n = 2\pi m \tag{3.20}$$

where m is an integer for all values of n_1, n_2, n_3. We now decompose $\boldsymbol{G}$ in terms of three as yet undetermined basis vectors $\boldsymbol{g}_i$,

$$\boldsymbol{G} = h\boldsymbol{g}_1 + k\boldsymbol{g}_2 + l\boldsymbol{g}_3 \tag{3.21}$$

with integer h,k,l. The condition (3.20) now implies for the example of $n_2 = n_3 = 0$

$$(h\boldsymbol{g}_1 + k\boldsymbol{g}_2 + l\boldsymbol{g}_3)\, n_1 \boldsymbol{a}_1 = 2\pi m . \tag{3.22}$$

For an arbitrary choice of n_1 this can only be satisfied if

$$\boldsymbol{g}_1 \cdot \boldsymbol{a}_1 = 2\pi \quad \text{and} \quad \boldsymbol{g}_2 \cdot \boldsymbol{a}_1 = \boldsymbol{g}_3 \cdot \boldsymbol{a}_1 = 0 . \tag{3.23}$$

Expressed in general terms this requirement becomes

$$\boldsymbol{g}_i \cdot \boldsymbol{a}_j = 2\pi \delta_{ij} . \tag{3.24}$$

The basis set $\boldsymbol{g}_1, \boldsymbol{g}_2, \boldsymbol{g}_3$ that we have thus defined spans the so-called reciprocal lattice. For every real lattice there is a corresponding and unambiguously defined reciprocal lattice. Its lattice points are denoted by the numbers h,k,l. The rules for constructing this lattice are given directly by (3.24): the reciprocal lattice vector $\boldsymbol{g}_1$ lies perpendicular to the plane containing $\boldsymbol{a}_2$ and $\boldsymbol{a}_3$ and its length is $2\pi/a_1$ [$\cos \sphericalangle(\boldsymbol{g}_1, \boldsymbol{a}_1)$]. Figure 3.3 shows a planar oblique lattice and its corresponding reciprocal lattice. It should be noted however that, although the reciprocal lattice is drawn here in real space, its dimensions are actually m^{-1}.

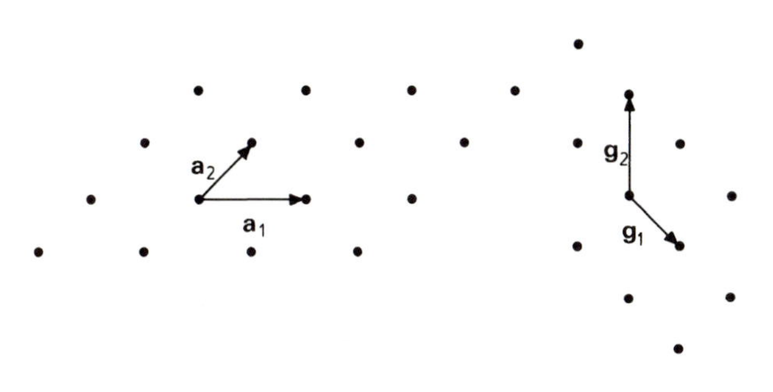

Fig. 3.3. A plane oblique lattice and its corresponding reciprocal lattice. The vectors g_1 and g_2 lie perpendicular to a_2 and a_1 respectively

A useful expression for the basis vectors of the reciprocal lattice is the following

$$g_1 = 2\pi \frac{a_2 \times a_3}{a_1 \cdot (a_2 \times a_3)} \quad \text{and cyclic permutations.} \tag{3.25}$$

It is easy to show that (3.25) satisfies the condition (3.24).

It follows from the one-to-one correspondence of the lattice and its reciprocal lattice that every symmetry property of the lattice is also a symmetry property of the reciprocal lattice. The reciprocal lattice therefore belongs to the same point group as the real-space lattice.

3.3 The Scattering Conditions for Periodic Structures

We now proceed to insert the Fourier expansion of $\varrho(\boldsymbol{r})$ into the equation (3.7) for the scattering intensity. With the notation $\boldsymbol{K} = \boldsymbol{k} - \boldsymbol{k}_0$ we obtain

$$I(\boldsymbol{K}) \propto \frac{|A_0|^2}{R'^2} \left| \sum_{\boldsymbol{G}} \varrho_G \int e^{i(\boldsymbol{G}-\boldsymbol{K})\cdot \boldsymbol{r}}\, d\boldsymbol{r} \right|^2 . \tag{3.26}$$

If the crystal consists of many identical unit cells, the only significant contributions to the integral in (3.26) arise when $\boldsymbol{G} = \boldsymbol{K}$. Expressed in its components, this integral would be, for an infinite volume, a representation of the respective δ-functions. Its value is then equal to the scattering volume V

$$\int e^{i(\boldsymbol{G}-\boldsymbol{K})\cdot \boldsymbol{r}}\, d\boldsymbol{r} = \begin{cases} V & \text{für} \quad \boldsymbol{G} = \boldsymbol{K} \\ \sim 0 & \text{otherwise} \end{cases} . \tag{3.27}$$

Scattering from lattices thus leads to diffracted beams when the difference between the $\boldsymbol{k}$ vectors of the incident and scattered waves is equal to a reciprocal lattice vector $\boldsymbol{G}$. This condition is named the "Laue condition" after Max von Laue. The measured intensity is

$$I(\boldsymbol{K}=\boldsymbol{G}) \propto \frac{|A_0|^2}{R'^2} |\varrho_G|^2 V^2 . \tag{3.28}$$

The apparent proportionality to V^2 needs further comment. An exact analysis of the integral shows in fact that the width of the intensity distribution around a diffraction beam maximum decreases as V^{-1}. Thus, as expected, the total intensity is proportional to the scattering volume.

The vector $\boldsymbol{G}$ is unambiguously defined by its coordinates h, k, l with respect to the basis vectors $\boldsymbol{g}_i$ of the reciprocal lattice. Thus the indices h, k, l can also be used to label the diffraction beams. Negative values of h, k, l are denoted by $\bar{h}, \bar{k}, \bar{l}$

$$I_{hkl} \propto |\varrho_{hkl}|^2 . \tag{3.29}$$

If no absorption of radiation takes place in the target material, $\varrho(\boldsymbol{r})$ is a real function and on account of (3.18) we then have

$$\varrho_{hkl} = \varrho^*_{\bar{h}\bar{k}\bar{l}} . \tag{3.30}$$

This means that the intensities obey

$$I_{hkl} = I_{\bar{h}\bar{k}\bar{l}} \quad \text{(Friedel's rule)} . \tag{3.31}$$

The above rule has an interesting consequence. The X-ray pattern always displays a center of inversion, even when none is present in the structure itself. For structures containing a polar axis, the orientation of this axis cannot be determined from the X-ray diffraction pattern. An exception to this statement is found when one works in a region of strong absorption, i.e., when the above condition of a real scattering density does not hold.

We now devote some more attention to the interpretation of the Laue condition

$$\boldsymbol{K} = \boldsymbol{G} . \tag{3.32}$$

This condition is of fundamental importance for all diffraction phenomena involving periodic structures, regardless of the type of radiation employed. It can be represented pictorially by means of the Ewald construction (Fig. 3.4). One selects an arbitrary reciprocal lattice point as the origin and draws the vector $\boldsymbol{k}_0$ to point towards the origin. Since we are assuming elastic scattering we have $k = k_0 = 2\pi/\lambda$ where λ is the wavelength of the radiation. All points on the sphere of radius $k = k_0$ centered around the starting point of the vector $\boldsymbol{k}_0$ describe the end points of a vector $\boldsymbol{K} = \boldsymbol{k} - \boldsymbol{k}_0$. The condition $\boldsymbol{G} = \boldsymbol{K}$ is satisfied whenever the surface of the sphere coincides with points of the reciprocal lattice. At these points diffraction beams are produced and they are labeled with the indices (hkl) corresponding to the relevant reciprocal lattice point.

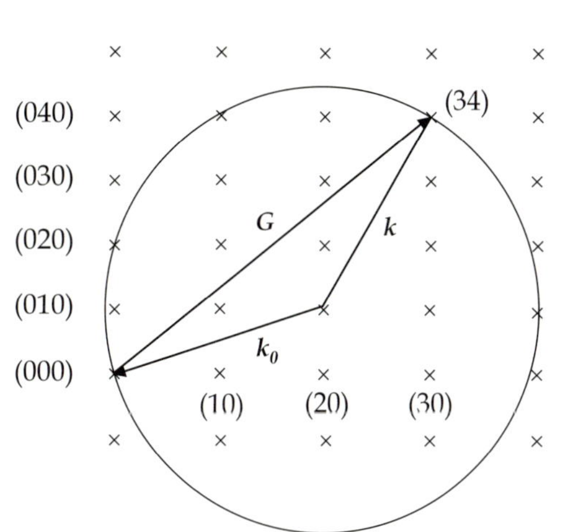

Fig. 3.4. The Ewald sphere of the reciprocal lattice illustrating the Laue condition $\boldsymbol{k} - \boldsymbol{k}_0 = \boldsymbol{G}$. Diffraction beams are produced whenever a reciprocal lattice point coincides with the surface of the sphere. For arbitrary values of the magnitude and direction of $\boldsymbol{k}_0$ this will generally not be the case. In order to observe diffraction one must either use a continuum of incident wavelengths or vary the orientation of the crystal

3.4 The Bragg Interpretation of the Laue Condition

Any three lattice points that do not lie on a straight line can be seen (Fig. 3.5) to define a so-called lattice plane. Such lattice planes may be labeled in a manner that leads to a particularly simple interpretation of the diffraction from the lattice. We assume that the lattice plane intersects the coordinate

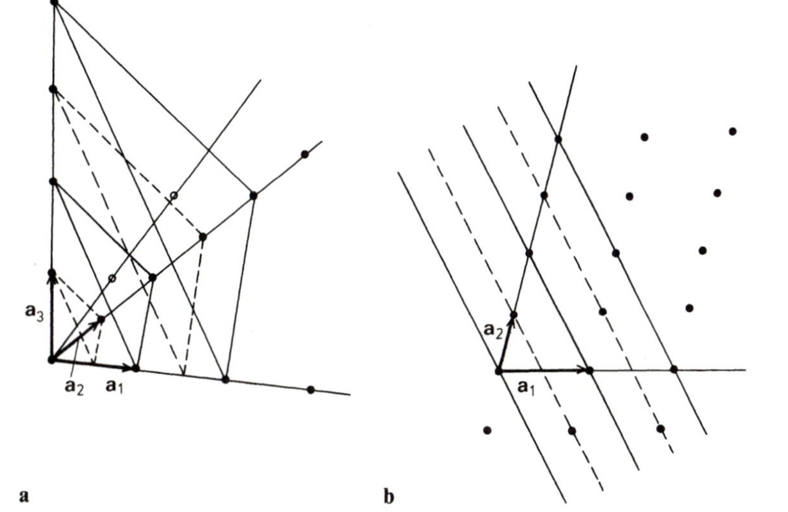

Fig. 3.5 a, b. Sets of crystal lattice planes. The planes illustrated here have the values $m = 1, n = 2, o = 2$. The corresponding Miller indices are derived for the triplet $(1/m, 1/n, 1/o)$ by multiplying this by an integer $p = 2$ to give $(hkl)=(211)$. Between the planes with indices m, n, o there lie additional planes (*dotted lines*). These contain the same density of atoms as can be seen from Fig. 3.4 b, and are thus completely equivalent to the original planes. The perpendicular separation of equivalent lattice plane is exactly a factor of p smaller than the separation of the original planes (*full lines*) constructed from the positions of atoms lying on the coordinate axes

axes at values m, n, o, where each of these numbers denotes an integer multiple of the corresponding basis vector. One then takes the reciprocal values $h' = 1/m, k' = 1/n, l' = 1/o$ and multiplies h', k', l' by an integer p so as to obtain a triplet of coprime integers (h, k, l). The numbers h, k, l are known as the Miller indices of the lattice plane (hkl). Parallel to the planes that intersect each axis at a lattice point (full lines in Fig. 3.5) one can also draw other equivalent lattice planes. The number of these planes is such that every lattice point on each of the three axes lies in one of these lattice planes. This is a consequence of the required translational symmetry (Fig. 3.5 b). The total number of equivalent lattice planes is now exactly p times as many as the number of original planes (full lines). The reciprocal values of the axis intersection of these planes (dotted and full lines in Fig. 3.5) directly supply the required index triplet (hkl) of coprime integers.

We now prove an important relation: The reciprocal lattice vector $\boldsymbol{G}$ with components (hkl) lies perpendicular to the lattice plane with the same indices (hkl). The length of the vector $\boldsymbol{G}_{hkl}$ is equal to 2π times the reciprocal distance between neighboring (hkl) planes.

We begin by proving the first part of this statement. The vectors

$$\frac{\boldsymbol{a}_1}{h'} - \frac{\boldsymbol{a}_2}{k'} \quad \text{and} \quad \frac{\boldsymbol{a}_3}{l'} - \frac{\boldsymbol{a}_2}{k'}$$

span the lattice plane. Their vector product

$$\left(\frac{\boldsymbol{a}_1}{h'} - \frac{\boldsymbol{a}_2}{k'}\right) \times \left(\frac{\boldsymbol{a}_3}{l'} - \frac{\boldsymbol{a}_2}{k'}\right) = -\frac{1}{h'k'}(\boldsymbol{a}_1 \times \boldsymbol{a}_2) - \frac{1}{k'l'}(\boldsymbol{a}_2 \times \boldsymbol{a}_3) - \frac{1}{h'l'}(\boldsymbol{a}_3 \times \boldsymbol{a}_1) \tag{3.33}$$

is normal to the plane (hkl). On multiplying this vector by $-2\pi h'k'l'/[\boldsymbol{a}_1 \cdot (\boldsymbol{a}_2 \times \boldsymbol{a}_3)]$ one obtains

$$2\pi\left(h'\frac{\boldsymbol{a}_2 \times \boldsymbol{a}_3}{\boldsymbol{a}_1 \cdot (\boldsymbol{a}_2 \times \boldsymbol{a}_3)} + k'\frac{\boldsymbol{a}_3 \times \boldsymbol{a}_1}{\boldsymbol{a}_1 \cdot (\boldsymbol{a}_2 \times \boldsymbol{a}_3)} + l'\frac{\boldsymbol{a}_1 \times \boldsymbol{a}_2}{\boldsymbol{a}_1 \cdot (\boldsymbol{a}_2 \times \boldsymbol{a}_3)}\right) . \tag{3.34}$$

This however, apart from the numerical factor p, is equal to $\boldsymbol{G}_{hkl}$ [see (3.21) and (3.25)]. Thus we have demonstrated that $\boldsymbol{G}_{hkl}$ lies perpendicular to the plane (hkl).

We now show that the separation of the planes, d_{hkl}, is equal to $2\pi/G_{hkl}$. The perpendicular distance of the lattice plane (hkl) from the origin of the basis $\boldsymbol{a}_1, \boldsymbol{a}_2, \boldsymbol{a}_3$ is

$$d'_{hkl} = \frac{a_1}{h'} \cos \sphericalangle(\boldsymbol{a}_1, \boldsymbol{G}_{hkl}) \tag{3.35}$$

$$= \frac{a_1}{h'} \frac{\boldsymbol{a}_1 \cdot \boldsymbol{G}_{hkl}}{a_1 G_{hkl}} = \frac{2\pi}{G_{hkl}} \frac{h}{h'} = \frac{2\pi}{G_{hkl}} p . \tag{3.36}$$

The distance to the *nearest* lattice plane is therefore $d_{hkl} = d'_{hkl}/p = 2\pi/G_{hkl}$.

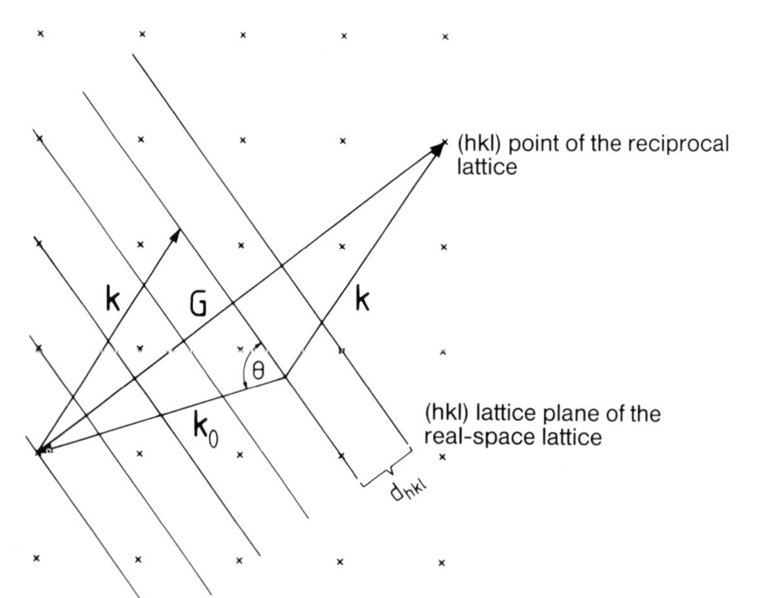

Fig. 3.6. The Bragg interpretation of the scattering condition. Since the vector $\boldsymbol{G}_{hkl}$ lies perpendicular to the lattice planes (hkl) in real space, the scattering appears to be a mirror reflection from these planes. It should be noted that real space and reciprocal space are shown here superposed

With the help of the lattice planes it is possible to obtain an intuitively clear interpretation of the scattering conditions. We take the modulus of the equation $\boldsymbol{G}=\boldsymbol{K}$:

$$G_{hkl} = \frac{2\pi}{d_{hkl}} = 2k_0 \sin\Theta \quad \text{(Fig. 3.6)} \tag{3.37}$$

and thereby obtain the Bragg equation

$$\lambda = 2d_{hkl}\sin\Theta\,. \tag{3.38}$$

This equation implies that the waves behave as if they were reflected from the lattice planes (hkl) (Fig. 3.6). It is from this interpretation that the expression "Bragg reflection" stems. The scattering condition then amounts to the requirement that the path difference between waves reflected from successive lattice planes should be an integer multiple of the wavelength of the radiation, such as is needed to produce constructive interference (Fig. 3.7).

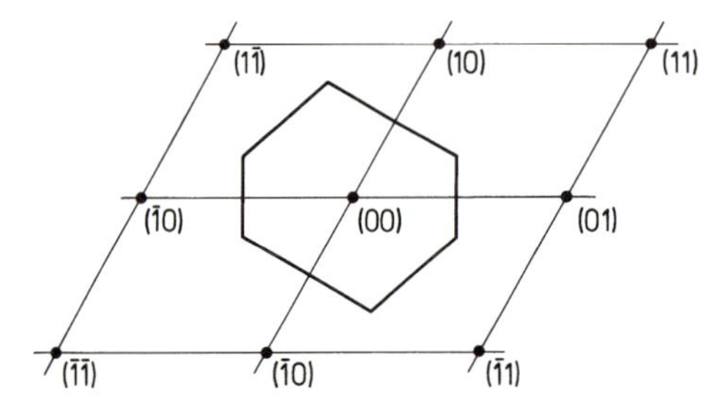

Fig. 3.7. Derivation of the Bragg condition. The path difference of the two reflected waves is $2\,d_{hkl}\sin\Theta$

3.5 Brillouin Zones

The condition for the occurrence of a Bragg reflection was $\boldsymbol{k} - \boldsymbol{k}_0 = \boldsymbol{G}_{hkl}$. The end points of all vector pairs $\boldsymbol{k}, \boldsymbol{k}_0$ that satisfy this condition lie on the perpendicular bisector of $\boldsymbol{G}_{hkl}$ (Fig. 3.4). The smallest polyhedron centered at the origin and enclosed by perpendicular bisectors of reciprocal lattice vectors is called the Brillouin zone (also first Brillouin zone). The construction of the Brillouin zone is best demonstrated for the case of an oblique planar lattice (Fig. 3.8).

The Brillouin zones for a few simple three-dimensional lattices are shown in Fig. 3.9. The symbols denoting points in the Brillouin zone originate from group theory and characterize the symmetry. Like the reciprocal lattice, the Brillouin zone also possesses the same point symmetry as the respective lattice type.

The points on the zone boundary are special because every wave with a $\boldsymbol{k}$-vector extending from the origin to the zone boundary gives rise to a Bragg-reflected wave. In the case of weak scattering and small crystals this wave has only a small intensity. For large single crystals however, the intensities of the primary and Bragg-reflected waves may be equal. These interfere to produce a standing wave field. The position of the nodes and antinodes is determined by the relative phases of the two waves and can be varied by changing the angle of incidence of the primary beam. This effect

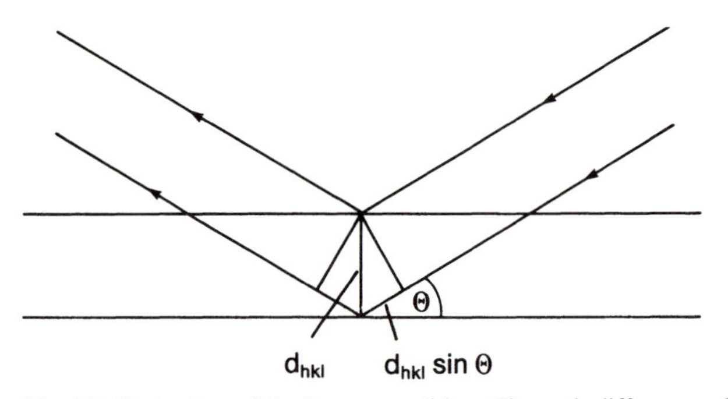

Fig. 3.8. Construction of the first Brillouin zone for a plane oblique lattice. Further zones can be constructed from the perpendicular bisectors of larger reciprocal lattice vectors

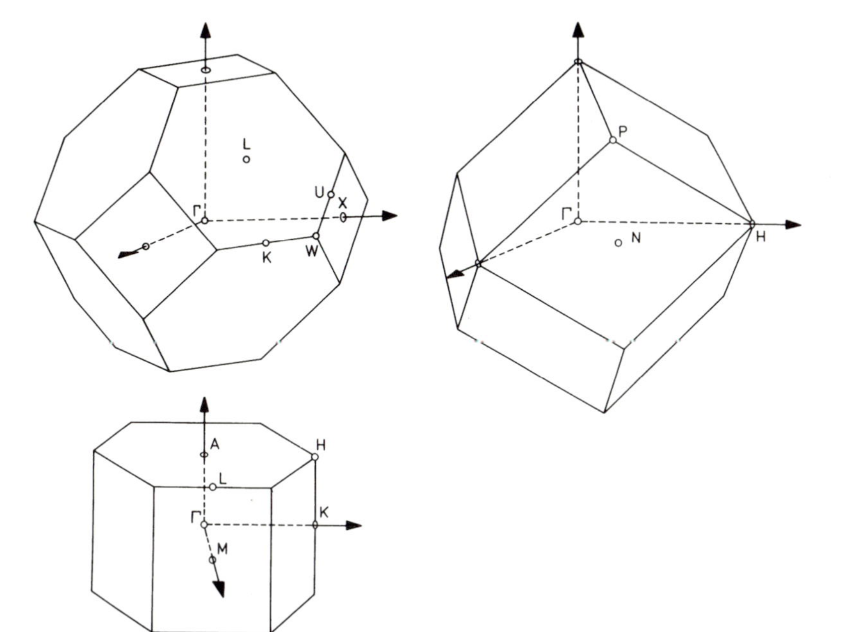

Fig. 3.9. The Brillouin zones of the face-centered cubic, body-centered cubic and hexagonal lattices. Points of high symmetry are denoted by Γ, L, X etc. The surfaces enclosing the Brillouin zones are parts of the planes that perpendicularly bisect the smallest reciprocal lattice vectors. The polyhedra that are produced by these rules of construction can be drawn about every point of the reciprocal lattice. They then fill the entire reciprocal space. The cell produced by the equivalent construction in real space is known as the Wigner-Seitz cell. It can be used to describe the volume that one may assign to each point of the real crystal lattice

can be used, for example, to determine the position of impurity atoms in a lattice via the observation of their X-ray fluorescence. The production of two waves of equal intensity and a fixed phase relation can also be used to construct an X-ray interferometer with which individual lattice defects can be imaged (Panel II).

For the case of electrons in a periodic solid, the production of Bragg-reflected waves and their significance for the electron bands of the solid will be discussed at greater length in Chap. 7.

3.6 The Structure Factor

The scattering condition (3.27) predicts only the positions at which diffraction beams appear. To obtain their intensity from (3.26) we first need to calculate the Fourier coefficients ϱ_{hkl} of the scattering density

$$\varrho_{hkl} = \frac{1}{V_c} \int\limits_{\text{cell}} \varrho\,(\boldsymbol{r}) e^{-i\boldsymbol{G}\cdot\boldsymbol{r}}\, d\boldsymbol{r}\ . \tag{3.39}$$

In this the integral extends over the unit cell. By substituting the Fourier expansion (3.18) of $\varrho(\boldsymbol{r})$ one can convince oneself of the validity of this equation. The scattering of X-rays is due to the electrons of the atoms. Except in the case of light elements, the majority of the solid's electrons (the core electrons) are concentrated in a small region around the atoms. Scattering from the valence electrons, which extend into the region between the atoms, can be neglected in comparison. The integral over the scattering density $\varrho(\boldsymbol{r})$ can therefore be divided into single integrals over the individual atoms: these must then be added together with the appropriate phases. For this it is convenient to divide the position vector $\boldsymbol{r}$ into a vector $\boldsymbol{r}_n$ that gives the position of the origin of the nth unit cell, a vector $\boldsymbol{r}_\alpha$ defining the position of each atom within the unit cell, and a new position vector $\boldsymbol{r}'$ which points away from the center of each atom: $\boldsymbol{r} = \boldsymbol{r}_n + \boldsymbol{r}_\alpha + \boldsymbol{r}'$ (Fig. 3.10). With this notation, the Fourier coefficients of the scattering density can be expressed as

$$\varrho_{hkl} = \frac{1}{V_c} \sum_\alpha e^{-i\boldsymbol{G}\cdot\boldsymbol{r}_\alpha} \int \varrho_\alpha\,(\boldsymbol{r}')\, e^{-i\boldsymbol{G}\cdot\boldsymbol{r}'}\, d\boldsymbol{r}'\ . \tag{3.40}$$

The integral now extends only over the volume of a single atom. It is evident that it describes the interference of the spherical waves emanating from different points within the atom. This integral is known as the atomic scattering factor.

Since the scattering density is essentially spherically symmetric about each atom, the integral can be further evaluated by introducing spherical polar coordinates

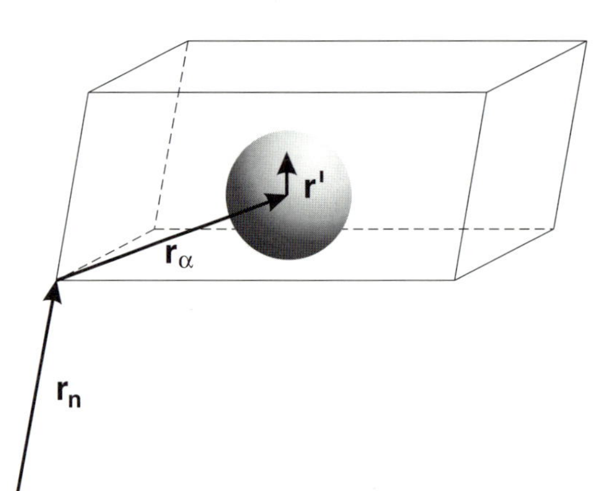

Fig. 3.10. Definition of the vectors r_n, r_α and r'. The vector r_n points to the origin of the nth unit cell, which is described by the triplet $n = n_1, n_2, n_3$, r_α points to the center of an atom within the cell, and r' from the center to a point within the atom

$$f_\alpha = \int \varrho_\alpha(\boldsymbol{r}')\mathrm{e}^{-\mathrm{i}\boldsymbol{G}\cdot\boldsymbol{r}'}d\boldsymbol{r}' = -\int \varrho_\alpha(\boldsymbol{r}')\,\mathrm{e}^{-\mathrm{i}Gr'\cos\vartheta}\, r'^2 dr'\, d(\cos\vartheta)\, d\varphi\,. \tag{3.41}$$

Here ϑ is the polar angle between $\boldsymbol{G}$ and $\boldsymbol{r}'$. On integrating over ϑ and φ one obtains

$$f_\alpha = 4\pi \int \varrho_\alpha(r')\, r'^2 \frac{\sin G r'}{G r'}\, dr'\,. \tag{3.42}$$

If we denote the angle between $\boldsymbol{k}$ and $\boldsymbol{k}_0$ as the scattering angle 2Θ (forward scattering: $\Theta = 0$) then on account of the relation

$$G = 2\,k_0 \sin\Theta\,, \tag{3.43}$$

it follows that

$$f_\alpha = 4\pi \int \varrho_\alpha(r')\, r'^2 \frac{\sin[4\pi r'(\sin\Theta/\lambda)]}{4\pi r'(\sin\Theta/\lambda)}\, dr'\,. \tag{3.44}$$

Thus the atomic scattering factor is a function $f(\sin\Theta/\lambda)$ which has its maximum value for forward scattering. For $\Theta = 0$ we have $f = 4\pi \int \varrho(r')r'^2 dr'$, i.e. it is equal to the integral of the scattering density over the atomic volume. For X-ray scattering this is proportional to Z, the total number of electrons per atom.

The summation over α in (3.40) leads to the so-called structure factor S_{hkl}. This describes the interference between waves scattered from the different atoms within the unit cell,

$$S_{hkl} = \sum_\alpha f_\alpha\, \mathrm{e}^{-\mathrm{i}\boldsymbol{G}_{hkl}\cdot\boldsymbol{r}_\alpha}\,. \tag{3.45}$$

For primitive lattices, i.e., those with only one atom per unit cell, $S = f$. Other special cases arise for centered lattices. To show this we describe the vector $\boldsymbol{r}_\alpha$ in units of the basis vectors of the lattice

$$\boldsymbol{r}_\alpha = u_\alpha \boldsymbol{a}_1 + v_\alpha \boldsymbol{a}_2 + w_\alpha \boldsymbol{a}_3\,. \tag{3.46}$$

Since $\boldsymbol{r}_\alpha$ is within the unit cell, we have $u, v, w < 1$. Using the definition of the reciprocal lattice vectors (3.24) the structure factor may be written

$$S_{hkl} = \sum_\alpha f_\alpha\, \mathrm{e}^{-2\pi\mathrm{i}(h u_\alpha + k v_\alpha + l w_\alpha)}\,. \tag{3.47}$$

As an example we consider the body-centered cubic lattice. The two atoms in the unit cell occupy the positions

$$r_1 = (0, 0, 0) \quad \text{and} \quad r_2 = (1/2, 1/2, 1/2)\,.$$

Both have the same atomic scattering factor f.

For S it follows that

$$S_{hkl} = f\left(1 + \mathrm{e}^{-\mathrm{i}\pi(h+k+l)}\right) = \begin{cases} 0 & \text{for } h+k+l \text{ odd} \\ 2f & \text{for } h+k+l \text{ even} \end{cases}. \tag{3.48}$$

This lattice therefore gives rise to systematic extinctions. For example there is no (100) reflection. The (100) planes form the faces of the unit cell and the reason for the destructive interference is the presence of the additional intermediate lattice planes containing the atoms in the body-centered position (Fig. 2.10). A prerequisite for the complete extinction of the Bragg reflections is that the central atom is identical to the corner atoms, in other words, one must have a true body-centered Bravais lattice. The CsCl structure, for example, does not produce extinctions, except in the case of CsI where the electron numbers of Cs^+ and I^- are identical.

It is easy to show that other centered lattices also lead to systematic extinctions.

Even when the complete extinction produced by centered lattices is not observed, the intensities of the diffraction beams are nonetheless modulated by the presence of additional atoms within the unit cell. This fact enables one to determine the positions and types of atoms in the unit cell. We summarize by stressing an important point: the *shape* and the *dimensions* of the unit cell can be deduced from the position of the Bragg reflections; the *content* of the unit cell, on the other hand, must be determined from the intensities of the reflections.

3.7 Methods of Structure Analysis

Types of Probe Beam

For structure investigations one can employ electrons, neutrons, atoms and X-ray photons. In each case, the wavelength must lie in a region that allows Bragg reflections to be produced. It is this condition that determines the respective energy ranges of the beams (Fig. 3.11). These are

10 eV – 1 keV for electrons,
10 meV – 1 eV for neutrons and light atoms,
1 keV – 100 keV for photons.

The actual uses of these various probes in structure analyses are determined by the cross sections for elastic and inelastic scattering, and also by the availability and intensity of the sources.

For electrons between 10 eV and 1 keV the scattering cross sections are so large that only 10–50 Å of solid material can be penetrated by the beam. Thus electrons are frequently employed to gather information about the atomic structure of surfaces. Diffraction experiments with atoms offer a further method of investigating surfaces (Panel I).

In the case of photons, depending on the nature of the target material and the type of radiation, it is possible to investigate the bulk structure of targets up to several mm in thickness. As a source of radiation one generally employs the characteristic X-ray lines emitted by solids under electron bombardment (X-ray tubes). Such sources also produce a continuous

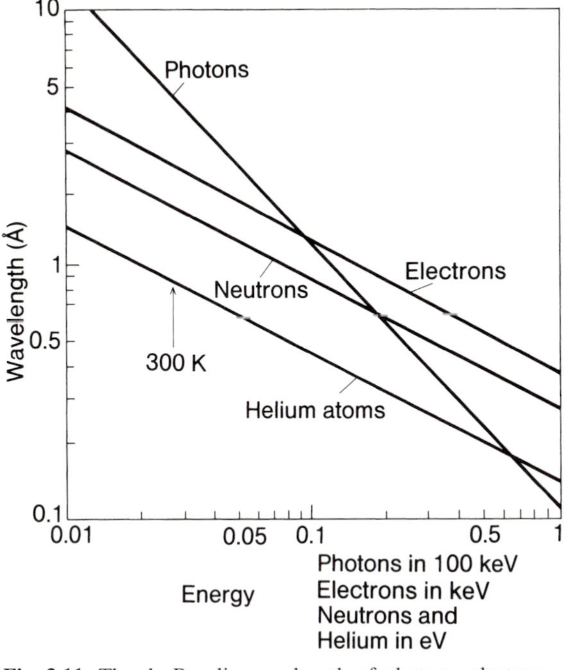

Fig. 3.11. The de Broglie wavelength of photons, electrons, neutrons and helium atoms as a function of the particle energies. The arrow shows the energy for a thermal beam at room temperature (eV scale)

bremsstrahlung spectrum. The spectrum of characteristic lines is caused by the ionization of atoms, which is followed by the emission of light when electrons from higher energy levels fall into the vacant state. Another excellent source of X-rays possessing high intensity and strongly collimated beams with 100% polarization is the electron synchrotron (e.g., “DESY” in Hamburg, “BESSY” in Berlin, ESRF in Grenoble, NSLS in Brookhaven, or ALS in Berkeley; see Panel XI). Because of the ready availability of X-ray sources, the majority of structure investigations to date have been carried out with X-ray beams. However, there are a number of questions that cannot be answered with X-ray studies. We have seen that the atomic structure factors are proportional to the nuclear charge Z. Thus the scattering intensities vary as Z^2. Therefore, when hydrogen, for example, occurs in combination with heavy elements, it is very difficult to detect it with X-rays. Here it is better to use neutron beams. The scattering cross section for neutrons lies within a single order of magnitude for all elements. On the other hand, the neutron cross sections for elements with adjacent atomic numbers, which are difficult to distinguish with X-ray diffraction, are quite different. Hence it is possible, for example, to easily distinguish iron, cobalt and nickel in neutron scattering (Panel I). A difficulty with neutrons, however, is that intense beams can only be obtained from nuclear reactors or, more recently, also from so-called spallation sources. Furthermore, the cross sections are small and the detection of the beams is less straightforward. Thus the necessary experimental effort is far greater than for X-ray or electron scattering. Neutron beams are utilized particularly in circumstances where one can take advantage of their specific properties, e.g., for investigating the structure of organic materials and polymers.

Procedures for Determining Structure

The incidence of a monochromatic plane wave on a crystal does not, in general, lead to any diffracted beams. We can see this immediately from the Ewald construction (Fig. 3.4). Only for particular choices of the wavelength (i.e. the magnitude of $\boldsymbol{k}_0$), or for particular angles of incidence, will a point of the reciprocal lattice fall on the Ewald sphere. The various procedures for determining structures differ in the method by which this condition is obtained. One can, for example, simply turn the crystal (preferably about a principal axis oriented perpendicular to the incident beam). Since the reciprocal lattice is associated in a fixed manner to the real crystal lattice, a rotation of the crystal corresponds to a rotation of the reciprocal lattice through the Ewald sphere (whose position in space is defined by the incident beam and is therefore fixed). One after another the points of the reciprocal lattice pass through the surface of the Ewald sphere. Thus, for particular rotation angles diffracted beams emerge in certain directions and these can be imaged by placing a photographic film around the crystal. This is the so-called rotating crystal procedure. To obtain an unambiguous indexing of the beams, the crystal is in addition translated along the axis of rotation (Weissenberg method). Together, these two procedures can be used to determine unknown crystal structures.

Using the powder method developed by Debye and Scherrer, it is possible to measure the lattice constant to an accuracy of five decimal places. In this technique the beam is directed at a powder composed of tiny single crystals whose random orientation ensures that all possible reflections are produced. In terms of the Ewald construction in reciprocal space (Fig. 3.4) one can determine the allowed reflections by imagining the reciprocal lattice to be rotated about the origin through the Ewald sphere. Since all orientations are present in the powder, a reflection is produced for every lattice point that passes through the sphere in the course of the rotation. In other words, one observes all reflections that lie within a radius of $2k_0$ from the origin of the reciprocal lattice. The powder method can be used, for example, to measure the change of lattice constant with temperature or with varying composition of an alloy.

The simplest method of producing diffraction is to use an incident beam containing a continuous spectrum of wavelengths, for example, the X-ray bremsstrahlung spectrum. In this case one observes all reflections whose lattice points lie between the Ewald spheres corresponding to the minimum and maximum k_0 values of the incident radiation.

This so-called Laue method has the advantage that, for a suitable orientation of the crystal, one can determine the crystal symmetry directly from the diffraction pattern. For instance, if the incident beam is directed along an n-fold symmetry axis, then the diffraction pattern also displays n-fold symmetry. Thus the Laue method is often used to determine the orientation of crystals of known structure and plays an important role in the preparation of crystals that are to be used as targets in other investigations. It cannot be applied, however, for the determination of structure.

Problems

3.1 a) Show that the reciprocal lattice of the reciprocal lattice is the original real-space lattice:
Hint: let $\boldsymbol{G} = m_1\boldsymbol{g}_1 + m_2\boldsymbol{g}_2 + m_3\boldsymbol{g}_3$ be reciprocal lattice vectors of the real-space lattice and the $\boldsymbol{g}_i$ the corresponding basis vectors. Then, by definition, the reciprocal lattice vectors of the reciprocal lattice $\boldsymbol{G}^* = n_1\boldsymbol{g}_1^* + n_2\boldsymbol{g}_2^* + n_3\boldsymbol{g}_3^*$ must satisfy $\boldsymbol{G}^* \cdot \boldsymbol{G} = 2\pi k$, where k is an integer. This is the case for $\boldsymbol{g}_1^* = 2\pi \dfrac{\boldsymbol{g}_2 \times \boldsymbol{g}_3}{\boldsymbol{g}_1 \cdot (\boldsymbol{g}_2 \times \boldsymbol{g}_3)}$ and for cyclic permutations of the indices.
b) Let the function $f(\boldsymbol{r})$ be lattice periodic. Show that the vectors $\boldsymbol{k}$ occurring in the Fourier series $f(\boldsymbol{r}) = \sum_k \hat{f}_k \exp(\mathrm{i}\boldsymbol{k} \cdot \boldsymbol{r})$ are reciprocal lattice vectors $\boldsymbol{G}$.

3.2 a) Calculate the structure factor $S_{hkl} = \sum_\alpha f_\alpha \exp(-\mathrm{i}\,\boldsymbol{G}_{hkl} \cdot \boldsymbol{r}_\alpha)$ for the face-centered cubic structure. For which indices hkl does one find extinction of the diffracted beams?
b) The fcc structure arises as the result of an appropriate superposition of four interpenetrating primitive simple cubic (sc) structures, all of which have the same lattice constant as the fcc structure. Interpret the result of part (a) for the case of the (001) diffraction beam (extinction compared with sc), and for the (111) beam (enhancement compared with sc). To do this, consider the Bragg reflection from the corresponding lattice planes of the sc and the fcc structures, making use of relevant sketches.

3.3 Calculate the structure factor for the diamond structure.

3.4 Show that the reciprocal lattice of a two-dimensional lattice can be represented by rods. Discuss the Ewald construction for diffraction from a two-dimensional lattice and determine the diffracted beam for a particular orientation and magnitude of $\boldsymbol{k}_0$. Why does one observe a diffraction pattern of electrons from a surface for all values and orientations of $\boldsymbol{k}_0$ above a critical value? Calculate the critical energy at which the first diffracted beam appears, when the electrons are incident perpendicular to a (100) surface of a Cu crystal.

3.5 Calculate the diffraction intensities from a rectangular lattice formed by the lattice vector $\boldsymbol{r}_n = 2an\boldsymbol{e}_x + am\boldsymbol{e}_y$ with a basis of atoms at (0, 0) and $(\frac{1}{2}, \varrho)$. Where do you find (glide plane) extinctions? Discuss the result with regard to the dependence on ϱ.

3.6 Consider the matrix element for the absorption of an X-ray photon by an atom $\langle i|x|f\rangle$ where $\langle i|$ is the initial localized state of an electron residing on the atom and $\langle f|$ is the final s-wave $\Psi = \mathrm{e}^{\mathrm{i}kr}/r$ with $\hbar^2 k^2/2m = \hbar\nu - E_\mathrm{I}$, where E_I is the threshold energy for ionization. Now assume that the atom is surrounded by a shell of nearest neighbor atoms that scatter the emitted wave. The final state thus consists of the emitted spherical wave plus spherical waves scattered from the ensemble of nearest neighbor atoms. What do you expect for the X-ray absorption above threshold? Describe a method to determine the distance of nearest neighbors in an amorphous material from the oscillations in the absorption coefficient of X-rays above threshold (**E**xtended **X**-ray **A**bsorption **F**ine **S**tructure: EXAFS). What type of light source is needed? Why does the method require the investigation of the oscillatory structure at photon energies considerably above the threshold? Why is it (experimentally) difficult to determine the local environment of carbon in an amorphous matrix?

3.7 a) Show that the contribution of large r to the correlation function $g(r)$ (3.14) leads to forward scattering! How can one determine the correlation function $g(r)$ from the experimentally observed intensity $I(\boldsymbol{K})$ even when the intensity of the forward scattering $I(\boldsymbol{K} = \boldsymbol{0})$ is not known?

b) Liquids are characterized by vanishing shear forces. By virtue of the Pauli-principle atoms behave nearly as hard spheres. Demonstrate with the help of hard spheres (e.g. tennis balls) that the distance of next nearest neighbors in liquids must be on the average at least $2r\sqrt{3}$ with r the radius of the spheres, or half the distance to the next neighbors. Compare this result to Fig. 3.2! Why do these considerations not apply to water?

3.8 Calculate the scattered intensity from a linear chain of atoms with ordered domains of N atoms. Assume that there is no phase correlation between atoms in different domains.

3.9 Discuss qualitatively the atomic scattering factor (as a function of scattering angle) for electron, X-ray, and neutron diffraction by a crystalline solid.

3.10 Elastic scattering by an infinite periodic crystal lattice yields infinitely sharp Bragg reflection spots according to (3.26). Discuss, on the basis of the Fourier transform representation of the scattered intensity (3.26), diffraction from crystallites of finite size. How can the average size of a crystallite be estimated from the diffraction pattern?

Panel I
Diffraction Experiments with Various Particles

I.1 Electrons

We describe here an experiment in which low energy (10–1000 eV) electrons are diffracted (LEED – Low Energy Electron Diffraction). In solids, low energy electrons are absorbed before they have penetrated more than a few atomic lattice planes. Thus diffraction experiments can only be performed in a reflection mode and they deliver information about the structure of the topmost atomic layers of a crystal. The first electron diffraction experiment was carried out by Davisson and Germer [I.1] in 1927, and served to demonstrate the wave nature of electrons. An experimental arrangement that conveniently enables the diffraction beams to be imaged on a fluorescent screen is shown in Fig. I.1. Because of the surface sensitivity of the method, it is necessary to perform LEED experiments in ultra-high vacuum ($p<10^{-8}$ Pa) and on atomically "clean" surfaces. An example showing the diffraction pattern obtained from a (111) surface of nickel is given

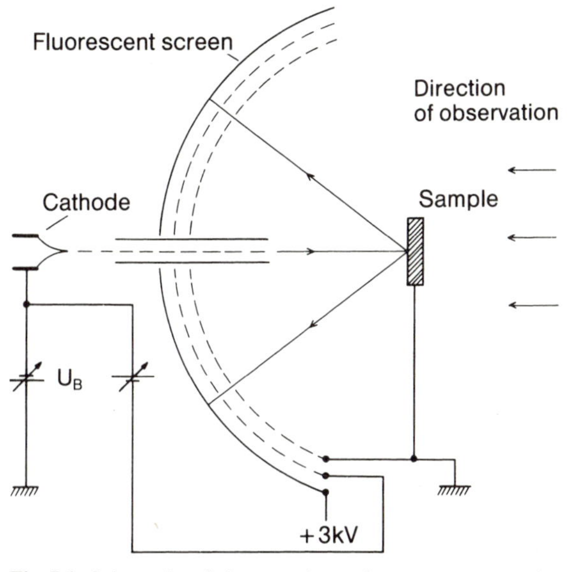

Fig. I.1. Schematic of the experimental arrangement used to observe LEED reflections from the surface of a single crystal

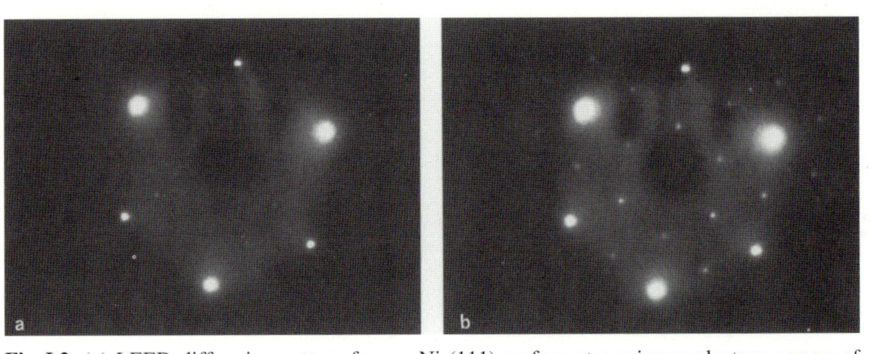

Fig. I.2. (**a**) LEED diffraction pattern from a Ni (111) surface at a primary electron energy of 205 eV, corresponding to a wavelength of 0.86 Å. The position of the spots can be used to determine the lattice constant. Of perhaps greater interest are adsorption experiments since adsorbates often form a variety of overlayer structures. (**b**) The diffraction pattern observed after the adsorption of hydrogen. The extra spots indicate the formation of a so-called (2×2) adsorbate superstructure, i.e., the elementary mesh of the adsorbate structure is, in both directions, twice as large as that of the nickel substrate

in Fig. I.2. The strong absorption of the electrons means that the third Laue condition for the constructive interference between electrons scattered from the atomic planes parallel to the surface is of little importance. As a result it is possible to observe diffraction at all electron energies. It should be noted that the diffraction pattern of Ni(111) displays the true 3-fold symmetry of the body of an fcc crystal since the scattering is not only from the surface layer but includes contributions from deeper layers. Figure I.2b shows the diffraction pattern of the same surface after the adsorption of hydrogen. The additional diffraction spots indicate that the hydrogen – like many other adsorbates on surfaces – creates an overlayer with a new structure. In this case the elementary mesh of the hydrogen overlayer is exactly twice the size of that of the Ni(111) surface. The additional spots therefore lie halfway between those of the nickel substrate.

I.2 Atomic Beams

The diffraction of He and H_2 beams from solid surfaces was first detected in the experiments of Estermann and Stern [I.2]. At the time, this result provided vital confirmation of the validity of quantum mechanics! From a modern viewpoint it must be judged a very lucky chance that Estermann and Stern chose to work with alkali halide crystals (NaCl and LiF). Indeed most other surfaces, especially those of metals would not have led to the observation of the diffraction phenomenon. The reason is that He atoms with an appropriate wavelength (Fig. 3.11) only interact with the outermost surface of the solid where the interaction potential is very smooth and thus they are hardly sensitive to the atomic structure of the crystal. Furthermore,

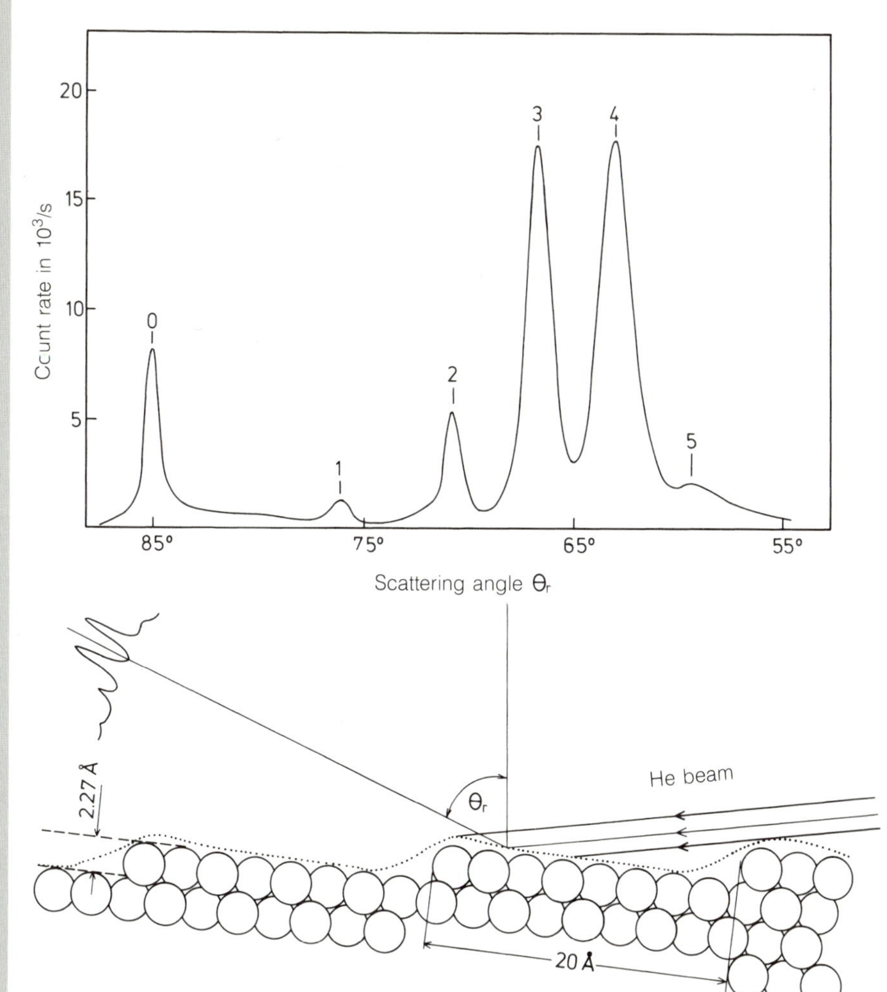

Fig. I.3. Diffraction of a He beam from a stepped platinum surface [I.3]. The Miller indices of this surface are (997). As for an optical echelon, one obtains maximum intensity in the diffraction orders that correspond to specular reflection from the contours of the interaction potential. In this case it should be noted that these potential contours are not exactly parallel to the terraces

diffraction experiments on metals and most other materials could never have been successful prior to the development of ultra-high vacuum technology, which today allows us to prepare extremely clean surfaces.

We show here as an example the diffraction of He atoms from a stepped platinum surface (Fig. I.3) as measured by Comsa et al. [I.3]. Surfaces with regularly spaced monatomic steps can be produced by cutting the crystal at the appropriate angle and annealing in vacuum. The atomic beam used in the diffraction experiments is produced by a supersonic expansion of the gas from a nozzle. The interaction between the atoms in the expanding gas produces a velocity distribution that is significantly sharper than the Maxwell distribution present before the expansion. Here one can make an analogy to vehicles traveling on a crowded freeway that have to adjust their forwards velocity to that of the other vehicles.

In Fig. I.3 the diffracted intensity is shown as a function of scattering angle. The angle of incidence is 85° to the surface normal of the macroscopic surface. The intensity maxima correspond to the diffraction orders of the periodic lattice of terraces (and not the lattice of individual atoms!). As is the case for an optical echelon grating, the direction corresponding to specular (mirror) reflection from the terraces is favored in the intensity distribution. Here, however, there is a slight bump in the otherwise flat potential of the terraces near to the step edge. This leads to a slight shift in the intensity maximum towards smaller angles.

I.3 Neutrons

The first diffraction experiments with neutrons were carried out as long ago as the 1930s. However, only since about 1945, when high neutron fluxes became available with the advent of nuclear reactors, has it been possible to employ neutron beams for structure investigations. The majority of the neutrons produced are so-called thermal neutrons ($T \sim 400$ K). As can be seen from Fig. 3.11, the de Broglie wavelength then falls in a favorable range for atomic structure studies. In order to select neutrons of a certain well-defined wavelength to be used in the Debye-Scherrer or rotating crystals procedures (Sect. 3.7), one makes use of a crystal monochromator (Fig. I.4).

The most important applications of neutron diffraction are the determination of the location of hydrogen atoms in solids and biological systems, the investigation of magnetic structures, and the study of order-disorder phase transitions [I.5]. We will discuss an example from the field of phase transitions. The alloy FeCo composed of an equal number of Fe and Co atoms crystallizes in the bcc structure. At high temperatures and for rapidly quenched samples, the Fe and Co atoms are distributed randomly among the sites of the bcc lattice (Fig. I.5 a). If, however, the crystal is cooled slowly, an ordered phase is produced in which the corner and body-center positions are each occupied by only one of the two elements. This arrangement corresponds to a CsCl lattice. Similar ordering phenomena are also found for other alloy systems and also include other lattice types. They can be detected and studied by means of diffraction experiments.

In Sect. 3.6 we met the systematic extinctions of the face-centered cubic lattice. These extinctions affect all diffraction beams for which the sum $h+k+l$ is odd, i.e., the (100), (111), (210) etc. For this disordered alloy phase (Fig. I.5 a, left) these beams are thus absent. For the ordered phase,

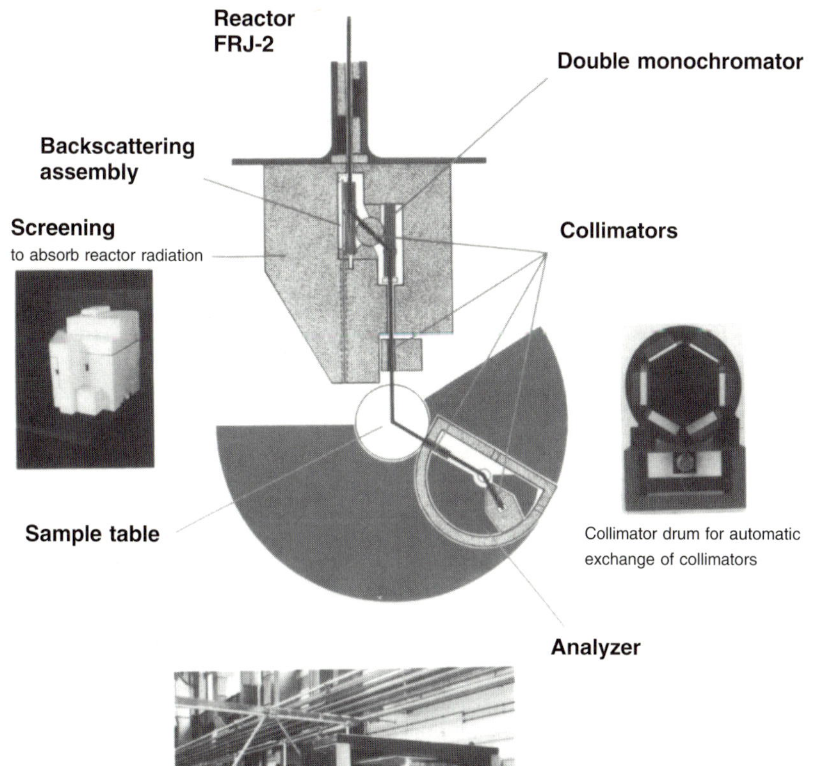

Fig. I.4. The Jülich neutron diffractometer.

The sample table is for positioning and orientation of the sample. The analyzer is driven around the sample table on air cushions in order to vary the scattering angle.

Double monochromator: Neutrons with wavelengths between λ_i and $\lambda_i+\Delta\lambda_i$ are filtered out of the reactor spectrum with the help of Bragg reflection from single crystals.

Collimators determine the direction and divergence of the beam and thereby influence the resolution of the spectrometer.

Analyzer: Neutrons with wavelengths between λ_f and $\lambda_f+\Delta\lambda_f$ are filtered out of the scattered spectrum by Bragg reflection from a single crystal and are registered in the detector. λ_f may be varied by rotating the crystal and the detector. The entire analyzer can be rotated about its own axis by 90° or 180° in order to extend the range of the analyzer angle.

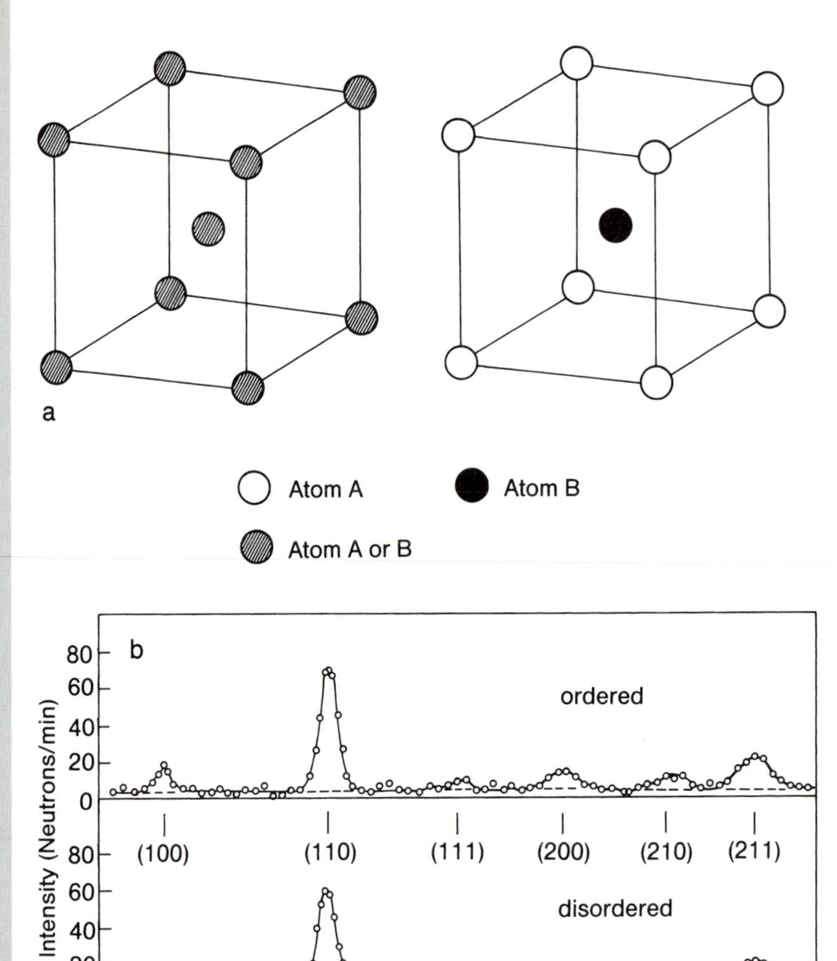

Fig. I.5. (**a**) The disordered and ordered phases of FeCo. (**b**) Neutron diffractogram of the ordered and disordered phases; after [I.6]. Note the low count rates which are typical for neutron diffraction experiments. To obtain good statistics it is necessary to measure for long periods of time

however (Fig. I.5 b, right), the beams are present since the atomic scattering factors of Fe and Co are not equal. In principle, this behavior is to be expected for every type of radiation. For X-rays, however, the atomic scattering factors of Fe and Co only differ slightly since the two elements are neighbors in the periodic table and since the atomic scattering factor for X-rays varies systematically approximately in proportion to the nuclear charge Z. As is readily seen from (3.47), the intensity of the (100) reflection of the CsCl structure (ordered phase) is proportional to $(f_{Fe} - f_{Co})^2$. The degree of order in the alloy is thus generally hard to determine from X-ray diffraction. Quite different is the case of neutron scattering: Here there is a factor 2.5 difference between the atomic scattering factors of Fe and Co. Figure I.5b shows neutron diffraction scans from powders of ordered and disordered FeCo. The forbidden reflections in the bcc structure, (100), (111), (210), are clearly visible for the ordered phase with the CsCl structure.

References

I.1 C.J. Davisson, L.H. Germer: Nature 119, 558 (1927); Phys. Rev. 30, 705 (1927)
I.2 I. Estermann, O. Stern: Z. Phys. 61, 95 (1930)
I.3 G. Comsa, G. Mechtersheimer, B. Poelsema, S. Tomoda: Surface Sci. 89, 123 (1979)
I.4 H.H. Stiller: Private communication
I.5 G.F. Bacon: Neutron Diffraction, 2nd edn. (Oxford Univ. Press, Oxford 1962)
I.6 C.G. Shull, S. Siegel: Phys. Rev. 75, 1008 (1949)

Panel II
X-Ray Interferometry and X-Ray Topography

X-ray beams may be used not only to determine the structural parameters of single crystals, but also to determine deviations from periodic structure and to observe defects in the structure. With X-ray interferometry it is possible, for example, to image even slight strains within a crystal.

As is well known, interference phenomena can be observed in the superposition of waves of equal frequency and with a fixed phase relation to one another. In optics, this is achieved by the division and subsequent recombination of a light beam. It is possible to proceed analogously with X-rays. To do so one makes use of an arrangement such as that shown in Fig. II.1. The blocks labeled S, M and A represent perfect single crystals that are aligned exactly parallel to one another. In the scatterer S a suitable incident angle gives rise to Bragg reflection from the lattice planes. As described in Sects. 3.4 and 7.2, two types of waves are induced in the crystal, one which has intensity nodes at the atomic sites, and a second for which the nodes lie between the atoms (see also Fig. 7.4). For large crystals, only

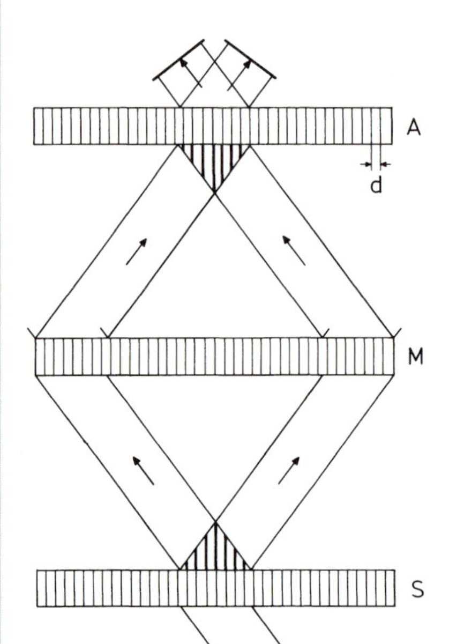

Fig. II.1. An arrangement for measuring X-ray interference [II.1]. The length d (not to scale!) represents the separation of lattice planes

Fig. II.2. An X-ray interferometer made from a single crystal silicon disk

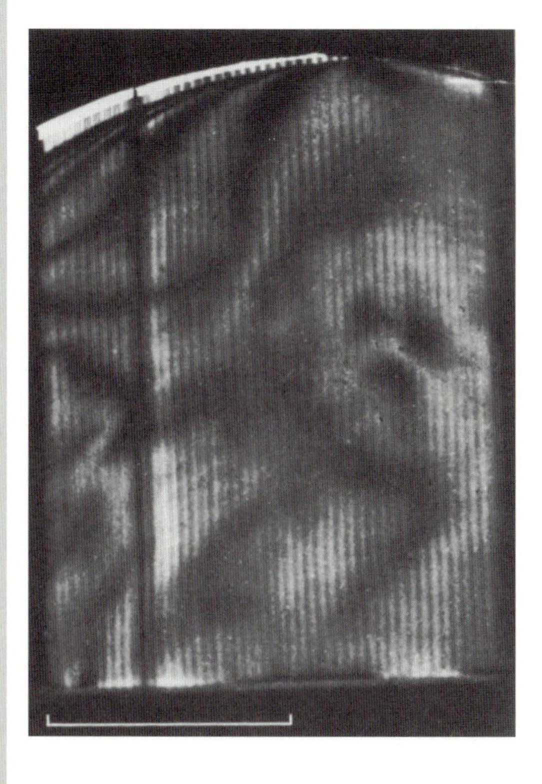

Fig. II.3. Moiré topography of the silicon interferometer. The vertical lines are due to the fact that the pattern is recorded in strips

the first type of wave survives, whereas the second is strongly absorbed (anomalous transmission and absorption). What emerges from the crystal are two beams of equal intensity, one being the transmitted beam and the other the Bragg reflected beam. These beams are recombined by means of a further Bragg reflection at the "mirror" M. The nodes and antinodes of the standing wave have the same positions here as in the scattering crystal. If the lattice planes in the analyser A have the same position relative to the nodes as those in S, then the beams are transmitted by A and we observe brightness. If, however, the analyser A is displaced by half a lattice constant the wave is strongly absorbed and we observe darkness. When the lattices of S and A are not identical this gives rise to a Moiré pattern.

Figures II.2 and 3 illustrate an example taken from Bonse by [II.1]. The interferometer (Fig. II.2) is carved from a single silicon crystal of diameter 8 cm. The corresponding Moiré topograph (Fig. II.3) indicates that the crystal contains extended distortions.

The alternating bright and dark conditions that are obtained when A is gradually displaced relative to S (Fig. II.1) can also be used to provide a direct determination of the lattice constant. One simply needs to divide the total mechanical displacement by the number of bright/dark phases observed and thereby obtains the lattice constant without needing to know the wavelength of the X-rays. This procedure can be used to achieve an exact relation between the X-ray wavelength scale and the definition of the meter

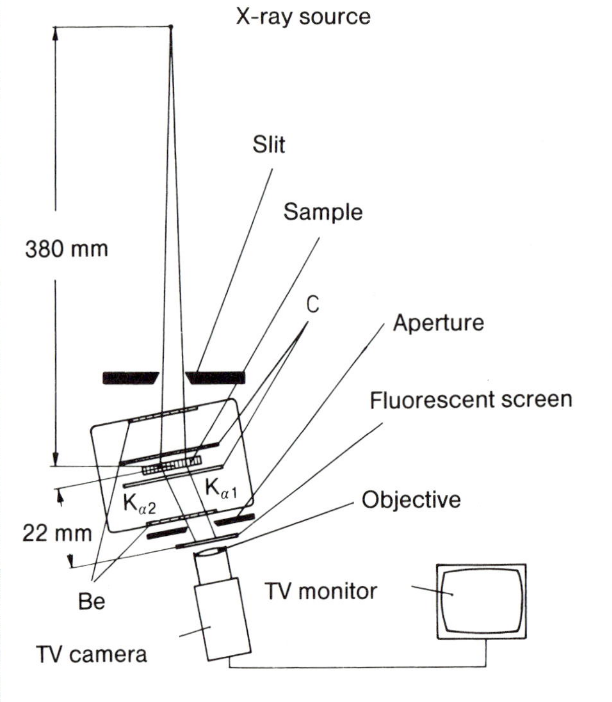

Fig. II.4. X-ray topography for the direct imaging of lattice defects [II.3]. The crystal is situated in an oven with beryllium windows. Beryllium, on account of its small nuclear charge, displays only weak X-ray absorption. The oven can be heated by means of graphite rods. Lattice defects appear as bright patches on the fluorescent screen and their development in time can be followed using a TV monitor

by way of the red emission line of ^{86}Kr. Whereas X-ray interferometry produces a Moiré pattern of lattice strain and defects, X-ray topography further enables one to make direct obervations. We consider here the experimental arrangement of Hartmann [II.2] (Fig. II.4). The X-rays from a Mo-K_{α} source are incident on a crystal and after being Bragg reflected are made visible on a fluorescent screen. For an ideal point source, only the two beams illustrated in Fig. II.4 would satisfy the Bragg condition. For a source of finite extent, however, one obtains bright patches on the screen. By a suitable choice of the size of the source these patches can be made sufficiently large that the areas of illumination of $K_{\alpha 1}$ and $K_{\alpha 2}$ just overlap. For fixed wavelengths of the radiation, this arrangement implies that a single point on the screen corresponds to a single point of the source. If, however, the crystal contains imperfections, then the Bragg condition for a point on the screen is satisfied not for a single point on the source, but for a large area or even for the entire source. Thus crystal imperfections lead

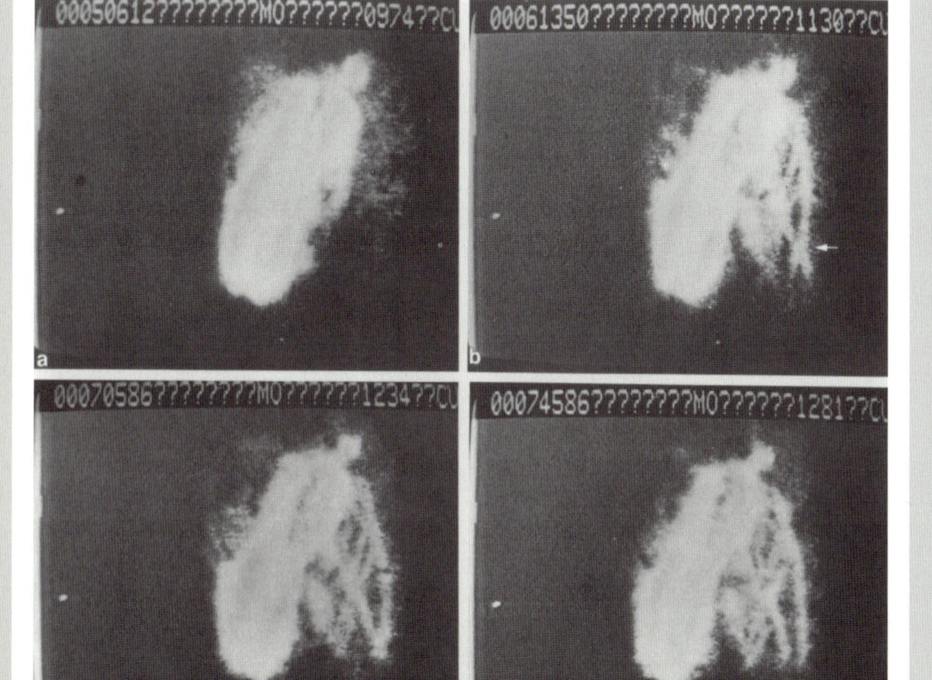

Fig. II.5. (**a**) X-ray topograph of a silicon crystal damaged by pressure from a diamond [II.3]. (**b**) After annealing at 1130 °C the formation of dislocations is observed. Two dislocations (*arrow*) have an intersection. (**c**) The intersection has wandered to the surface. The temperature is now 1234 °C. (**d**) The dislocations have separated and continue to move away from one another

to additional brightness on the screen. Figure II.5 a illustrates this for the case of a silicon crystal upon which an indentation has been made with a diamond. At high temperatures it is possible for lattice defects to be partially repaired. There remain, however, so-called dislocations in the crystal. A schematic representation of a dislocation is shown in Fig. 2.20. A dislocation gives rise to a strain field along a line and this, too, can be imaged by topography. Figure II.5 b shows such dislocation lines produced after annealing of the silicon crystal. At even higher temperatures the dislocations become mobile and move away from one another (Fig. II.5 c).

References

II.1 U. Bonse, W. Graeff, G. Materlik: Rev. Phys. Appl. 11, 83 (1976); U. Bonse: Private communication (1979)

II.2 W. Hartmann: In *X-Ray Optics*, ed. by H.J. Queisser, Topics Appl. Phys. Vol. 22 (Springer, Berlin, Heidelberg 1977) p. 191

II.3 W. Hartmann: Private communication

4 Dynamics of Atoms in Crystals

The physical properties of a solid can be roughly divided into those that are determined by the electrons and those that relate to the movement of the atoms about their equilibrium positions. In the latter category are, for example, the sound velocity and also the thermal properties: specific heat, thermal expansion, and – for semiconductors and insulators – the thermal conductivity. The hardness of a material is also determined, in principle, by the movement of the atoms about their equilibrium positions. Here, however, structural defects generally play a decisive role.

The division of solid properties into atom dynamics and electronic properties is qualitatively easy to justify: the motions of atomic nuclei are, due to their high mass, much slower than the motions of the electrons. If the atoms are displaced from their equilibrium positions, then the electrons adopt a new distribution (with higher total energy). The electron system, however, remains thereby in a ground state. If the initial positions of the nuclei are restored, then the energy expended is recovered in full and there remains no excitation of the electron system. The total energy as a function of the coordinates of all atomic nuclei thus plays the role of a potential for the atomic motion. This approach is, of course, only an approximation. There are also effects for which the interaction between the atom dynamics and the electron system become significant (Chap. 9). The so-called "adiabatic" approximation to be discussed here was introduced by Born and Oppenheimer [4.1].

Since the potential for the motion of the atomic nuclei is given by the total energy and thus, in essence, by the properties of the electron system, one might try initially to describe all the details of the electronic properties and derive from these the potential for the atomic motion. Finally, one could deduce from this all those properties of the solid that are determined by the atomic motion. This approach is indeed possible, but for our purposes in this text book it involves excessive mathematical effort. Fortunately, it is possible to obtain many important predictions about the thermal behavior of solids and their interaction with electromagnetic radiation without needing to know the explicit form of the potential for atomic motion. One simply needs a general formalism which enables equations of motion to be formulated and solved for an arbitrary potential. We will deal with such a formalism in the following sections. The concepts presented here will be a necessary prerequisite for an understanding of Chap. 5 in which we discuss the thermal properties of solids.

4.1 The Potential

First of all we require a suitable indexing system to refer to the individual atoms. Unfortunately, this is rather complicated due to the many degrees of freedom present. As in the past, we number the unit cells by the triplets $\boldsymbol{n} = (n_1, n_2, n_3)$ or $\boldsymbol{m} = (m_1, m_2, m_3)$ and the atoms within each cell by α, β. The ith component of the equilibrium position vector of an atom is then denoted by $r_{\boldsymbol{n}\alpha i}$ and the displacement from the equilibrium position by $u_{\boldsymbol{n}\alpha i}$ (Fig. 4.1). We now expand the total energy of the crystal Φ, which is a function of all nuclear coordinates, in a Taylor series about the equilibrium positions $r_{\boldsymbol{n}\alpha i}$

$$\Phi\left(r_{\boldsymbol{n}\alpha i}+s_{\boldsymbol{n}\alpha i}\right)=\Phi(r_{\boldsymbol{n}\alpha i})+\frac{1}{2}\sum_{\substack{\boldsymbol{n}\alpha i\\ \boldsymbol{m}\beta j}}\frac{\partial^2\Phi}{\partial r_{\boldsymbol{n}\alpha i}\,\partial r_{\boldsymbol{m}\beta j}}\,u_{\boldsymbol{n}\alpha i}u_{\boldsymbol{m}\beta j}\ldots . \tag{4.1}$$

The terms that are linear in $u_{\boldsymbol{n}\alpha i}$ disappear since the expansion is about the equilibrium (minimum energy) position. The summation indices $\boldsymbol{n},\boldsymbol{m}$ run over all unit cells; α,β over the atoms in the cell; and i,j over the three spatial coordinate directions. Higher terms in the expansion will be neglected for the time being. Equation (4.1) then represents an extension of

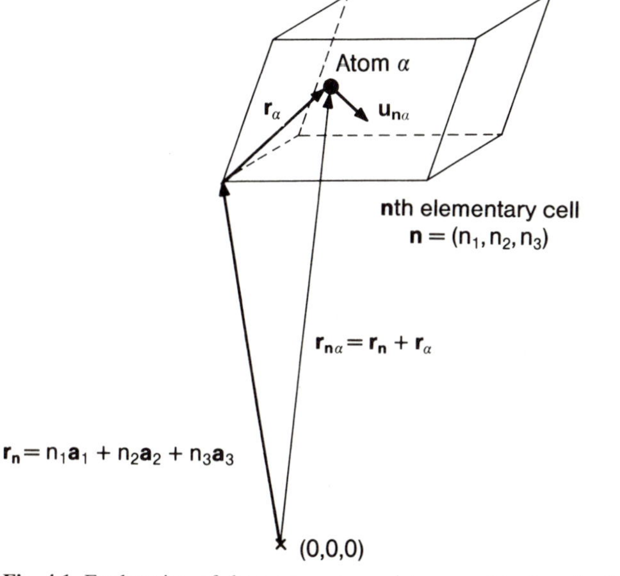

Fig. 4.1. Explanation of the vector nomenclature used to describe lattice vibrations in a three-dimensional periodic crystal: The lattice vector $\boldsymbol{r}_{\boldsymbol{n}}$ extends from an arbitrarily chosen lattice point $(0,0,0)$ to the origin of the $\boldsymbol{n}$th unit cell $\boldsymbol{n} = (n_1, n_2, n_3)$, from which the positions of the atoms α are described by the vector $\boldsymbol{r}_{\alpha}$. The displacement from equilibrium of atom α in cell n is then $\boldsymbol{u}_{\boldsymbol{n}\alpha}$. Thus the time-dependent position of this atom relative to $(0,0,0)$ is $\boldsymbol{r}_{\boldsymbol{n}\alpha}+\boldsymbol{u}_{\boldsymbol{n}\alpha}(t)$ where $\boldsymbol{r}_{\boldsymbol{n}\alpha}=\boldsymbol{r}_{\boldsymbol{n}}+\boldsymbol{r}_{\alpha}$

the harmonic oscillator potential to the case of many particles. The neglect of the higher order terms in (4.1) is therefore known as the "harmonic" approximation. Effects that can only be described by the inclusion of additional terms (e.g., the thermal expansion of solids; Chap. 5) are referred to as anharmonic effects.

The derivatives of the potential

$$\frac{\partial^2 \Phi}{\partial r_{\boldsymbol{n}\alpha i}\,\partial r_{\boldsymbol{m}\beta j}} = \Phi^{\boldsymbol{m}\beta j}_{\boldsymbol{n}\alpha i} \tag{4.2}$$

are called "coupling constants". They have the dimensions of spring constants and serve to generalize the spring constants of the harmonic oscillator to a system with many degrees of freedom. The quantity $-\Phi^{\boldsymbol{m}\beta j}_{\boldsymbol{n}\alpha i} u_{\boldsymbol{m}\beta j}$ is thus the force exerted on atom α in the unit cell $\boldsymbol{n}$ in the i-direction when the atom β in unit cell $\boldsymbol{m}$ is displaced by a distance $u_{\boldsymbol{m}\beta j}$ in the j-direction. For positive values of $\Phi^{\boldsymbol{m}\beta j}_{\boldsymbol{n}\alpha i}$ the force acts in the direction opposite to that of u. We see that this description allows for interactions between all atoms regardless of their separation from one another. In simple models, one often includes only the interaction between nearest neighbor atoms.

The coupling constants must satisfy a number of conditions that arise from the isotropy of space, the translation invariance, and the point group symmetry [2.2]. The translation invariance upon displacement of the lattice by an arbitrary lattice constant implies, for example, that the quantity $\Phi^{\boldsymbol{m}\beta j}_{\boldsymbol{n}\alpha i}$ can only depend on the difference between $\boldsymbol{m}$ and $\boldsymbol{n}$:

$$\Phi^{\boldsymbol{m}\beta j}_{\boldsymbol{n}\alpha i} = \Phi^{(\boldsymbol{m}-\boldsymbol{n})\beta j}_{\boldsymbol{0}\alpha i} \; . \tag{4.3}$$

4.2 The Equation of Motion

For the displacement s of atom α in cell $\boldsymbol{n}$ in direction i, the sum of the coupling forces and the force of inertia must be equal to zero (Newton's law):

$$M_\alpha \ddot{u}_{\boldsymbol{n}\alpha i} + \sum_{\boldsymbol{m}\beta j} \Phi^{\boldsymbol{m}\beta j}_{\boldsymbol{n}\alpha i} u_{\boldsymbol{m}\beta j} = 0 \; . \tag{4.4}$$

For N unit cells each with r atoms, this gives rise to $3rN$ differential equations which describe the motion of the atoms. Fortunately, for periodic structures, it is possible to use a suitable ansatz in order to achieve a significant decoupling. This involves writing the displacements $u_{\boldsymbol{n}\alpha i}$ in terms of a plane wave with respect to the cell coordinates:

$$u_{\boldsymbol{n}\alpha i} = \frac{1}{\sqrt{M_\alpha}} u_{\alpha i}(\boldsymbol{q}) \mathrm{e}^{\mathrm{i}(\boldsymbol{q}\cdot\boldsymbol{r}_n - \omega t)} \; . \tag{4.5}$$

In contrast to a normal plane wave, this wave is only defined at the lattice points $\boldsymbol{r}_n$. On substitution of this form into (4.4) one obtains an equation for the amplitude $u_{\alpha i}$:

$$-\omega^2 u_{\alpha i}(\boldsymbol{q}) + \sum_{\beta j} \underbrace{\sum_{\boldsymbol{m}} \frac{1}{\sqrt{M_\alpha M_\beta}} \Phi^{\boldsymbol{m}\beta j}_{\boldsymbol{n}\alpha i}\, \mathrm{e}^{\mathrm{i}\boldsymbol{q}\cdot(\boldsymbol{r}_m - \boldsymbol{r}_n)}}_{D^{\beta j}_{\alpha i}(\boldsymbol{q})} u_{\beta j}(\boldsymbol{q}) = 0 \,. \tag{4.6}$$

Due to the translational invariance, the terms of the sum depend, as in (4.3), only on the difference $\boldsymbol{m}$–$\boldsymbol{n}$. After performing the summation over $\boldsymbol{m}$, one obtains a quantity $D^{\beta j}_{\alpha i}(\boldsymbol{q})$ that is independent of $\boldsymbol{n}$. It couples the amplitudes to one another in a manner that does not depend on $\boldsymbol{n}$. This justifies the fact that the amplitudes in the ansatz (4.5) were written without the index $\boldsymbol{n}$. The quantities $D^{\beta j}_{\alpha i}(\boldsymbol{q})$ form the so-called dynamical matrix. The system of equations

$$-\omega^2 u_{\alpha i}(\boldsymbol{q}) + \sum_{\beta j} D^{\beta j}_{\alpha i}(\boldsymbol{q})\, u_{\beta j}(\boldsymbol{q}) = 0 \tag{4.7}$$

is a linear homogeneous system of order $3r$. In the case of a primitive unit cell we have $r=1$ and for every wave vector $\boldsymbol{q}$ we have a system of merely three equations to solve. This is a convincing demonstration of the simplifications brought about by the translational symmetry.

A system of linear homogeneous equations only has solutions (eigensolutions) if the determinant

$$\mathrm{Det}\,\{D^{\beta j}_{\alpha i}(\boldsymbol{q}) - \omega^2 \mathbf{1}\} = 0 \tag{4.8}$$

vanishes. This equation has exactly $3r$ different solutions, $\omega(\boldsymbol{q})$, for each $\boldsymbol{q}$. The dependence $\omega(\boldsymbol{q})$ is known as the *dispersion relation.* The $3r$ different solutions are called the *branches* of the dispersion relation. It is possible to make a number of general statements about these branches. However, rather than deriving these mathematically from (4.8) for the general case, we will discuss the special case of a diatomic linear chain. We can then make use of the results to present an overview of the dispersion branches of a three-dimensional crystal.

4.3 The Diatomic Linear Chain

The formalism developed above can be illustrated most readily in terms of the diatomic linear chain model. Although this model has little in common with a real solid, it is frequently discussed because of its simple mathematics. We consider a linear chain in which all nearest neighbors are connected by identical springs with force constant f. The unit cell contains two atoms of masses M_1 and M_2 (Fig. 4.2).

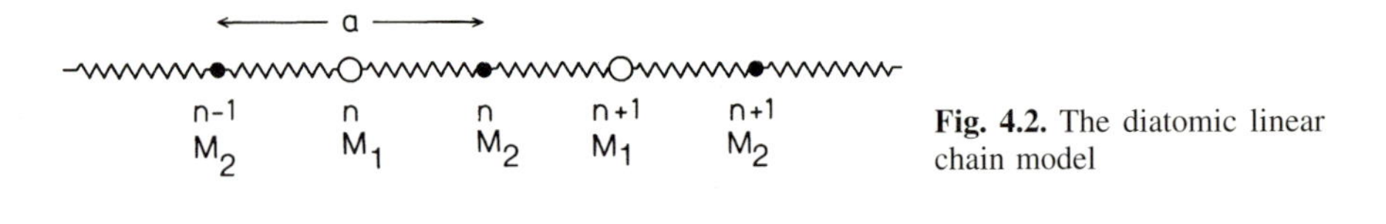

Fig. 4.2. The diatomic linear chain model

The indices α, β in (4.4) thus take on the possible values 1 and 2; the index i has only one value, since the system is one dimensional, and can therefore be omitted. Since it is assumed that only nearest neighbors interact, the index m in the sum in (4.4) can take only the values $n+1, n, n-1$. One thus obtains the following equations

$$\begin{aligned} M_1 \ddot{u}_{n1} + \Phi_{n1}^{n-1,2} u_{n-1,2} + \Phi_{n1}^{n1} u_{n1} + \Phi_{n1}^{n2} u_{n2} &= 0 \, , \\ M_2 \ddot{u}_{n2} + \Phi_{n2}^{n1} u_{n1} + \Phi_{n2}^{n2} u_{n2} + \Phi_{n2}^{n+1,1} u_{n+1,1} &= 0 \, . \end{aligned} \tag{4.9}$$

The values of these remaining coupling constants are

$$\begin{aligned} &\Phi_{n1}^{n-1,2} = \Phi_{n1}^{n2} = \Phi_{n2}^{n1} = \Phi_{n2}^{n+1,1} = -f \quad \text{and} \\ &\Phi_{n1}^{n1} = \Phi_{n2}^{n2} = +2f \, . \end{aligned} \tag{4.10}$$

Thus we obtain

$$\begin{aligned} M_1 \ddot{u}_{n1} + f(2u_{n1} - u_{n2} - u_{n-1,2}) &= 0 \, , \\ M_2 \ddot{u}_{n2} + f(2u_{n2} - u_{n1} - u_{n+1,1}) &= 0 \, . \end{aligned} \tag{4.11}$$

The plane-wave ansatz (4.5) then reads

$$u_{n\alpha} = \frac{1}{\sqrt{M_\alpha}} u_\alpha(q) \, \mathrm{e}^{\mathrm{i}(qan - \omega t)} \, . \tag{4.12}$$

We insert (4.12) into (4.11) to give

$$\begin{aligned} &\left(\frac{2f}{M_1} - \omega^2 \right) u_1 - f \frac{1}{\sqrt{M_1 M_2}} (1 + \mathrm{e}^{-\mathrm{i}qa}) u_2 = 0 \, , \\ &-f \frac{1}{\sqrt{M_1 M_2}} (1 + \mathrm{e}^{\mathrm{i}qa}) u_1 + \left(\frac{2f}{M_2} - \omega^2 \right) u_2 = 0 \, . \end{aligned} \tag{4.13}$$

The dynamical matrix $D_{\alpha i}^{\beta j}(\boldsymbol{q})$ is therefore

$$\begin{pmatrix} \dfrac{2f}{M_1} & \dfrac{-f}{\sqrt{M_1 M_2}} (1 + \mathrm{e}^{-\mathrm{i}qa}) \\ \dfrac{-f}{\sqrt{M_1 M_2}} (1 + \mathrm{e}^{\mathrm{i}qa}) & \dfrac{2f}{M_2} \end{pmatrix} . \tag{4.14}$$

Setting the determinant of the system (4.13) equal to zero leads to the dispersion relation

$$\omega^2 = f\left(\frac{1}{M_1}+\frac{1}{M_2}\right) \pm f\left[\left(\frac{1}{M_1}+\frac{1}{M_2}\right)^2 - \frac{4}{M_1 M_2}\sin^2\frac{qa}{2}\right]^{1/2} . \tag{4.15}$$

This dispersion relation is clearly periodic in q with a period given by

$$\frac{qa}{2} = \pi ,$$
$$q = \frac{2\pi}{a} . \tag{4.16}$$

The periodic repeat distance in q is thus exactly one reciprocal lattice vector. It can in fact be shown that this property holds for all lattices. To do so we simply need to go back to the definition of the dynamical matrix. We see that on account of (3.15)

$$D^{\beta j}_{\alpha i}(\boldsymbol{q}) = D^{\beta j}_{\alpha i}(\boldsymbol{q}+\boldsymbol{G}) \quad \text{with} \quad \boldsymbol{G}\cdot\boldsymbol{r}_n = 2\pi m . \tag{4.17}$$

Thus the eigensolution (4.7 or 4.8) must satisfy the condition

$$\omega(\boldsymbol{q}) = \omega(\boldsymbol{q}+\boldsymbol{G}) . \tag{4.18}$$

Furthermore we have

$$\omega(-\boldsymbol{q}) = \omega(\boldsymbol{q}) , \tag{4.19}$$

since $u(-\boldsymbol{q})$ represents a wave identical to $u(\boldsymbol{q})$ but propagating in the opposite direction. However, the forward- and backward-propagating waves are related to one another by time reversal. Since the equations of motion are invariant with respect to time reversal, it follows that the eigenfrequencies for $+\boldsymbol{q}$ and $-\boldsymbol{q}$ must be equal. It is also possible to derive the inversion symmetry of ω in $\boldsymbol{q}$-space (4.19) from the corresponding symmetry of the dynamical matrix (4.6). If in (4.6) one replaces $\boldsymbol{q}$ by $-\boldsymbol{q}$, then this corresponds in the definition of the dynamical matrix merely to a renaming of the indices $\boldsymbol{m}$ and $\boldsymbol{n}$. However, the dynamical matrix is not dependent on these indices. Taking these facts together we see that it suffices to represent $\omega(\boldsymbol{q})$ in the region $0 \le q \le G/2$. The point $q=G/2$ lies precisely on the edge of the Brillouin zone introduced in Sect. 3.5. Thus the function $\omega(\boldsymbol{q})$ can be fully specified by giving its values in one octant of the Brillouin zone.

For the example of the diatomic linear chain, Fig. 4.3 shows the two branches of the dispersion relation for a mass ratio of $M_1/M_2=5$. The branch that goes to zero at small q is known as the *acoustic* branch. For this branch, at small $q(q \ll \pi/a)$ the angular frequency ω is proportional to the wave vector q. Here the acoustic branch describes the dispersion-free propagation of sound waves.

The branch for which $\omega(q) \neq 0$ at $q=0$ is called the *optical* branch. Its frequencies at $q=0$ and $q=\pi/a$ have a simple interpretation. For $q=0$ the displacements of the atoms in every unit cell are identical. The sublattices of light and heavy atoms are vibrating against one another. In this case the problem can be reduced to a system of two masses with force constant $2f$

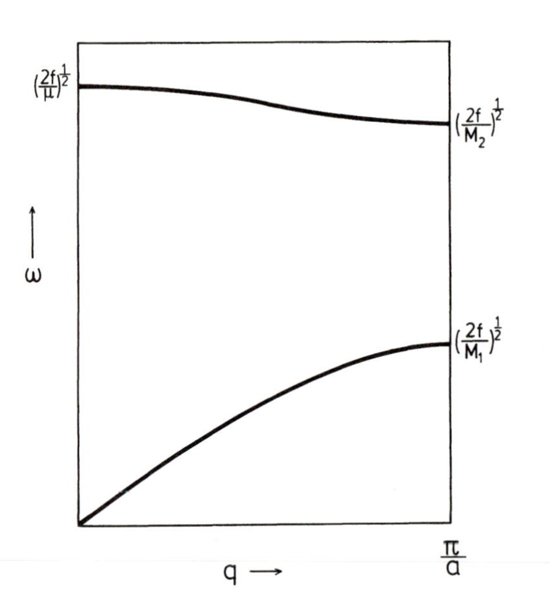

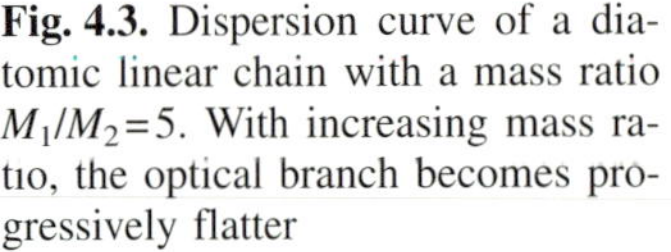

Fig. 4.3. Dispersion curve of a diatomic linear chain with a mass ratio $M_1/M_2=5$. With increasing mass ratio, the optical branch becomes progressively flatter

and the reduced mass $1/\mu = 1/M_1 + 1/M_2$. At $q=\pi/a$, one or other of the two sublattices is at rest and thus the two frequencies at this wave vector are $(2f/M_2)^{1/2}$ and $(2f/M_1)^{1/2}$.

The diatomic linear chain model is a popular tool for describing vibrations in ionic crystals, in which neighboring sites are occupied by ions of opposite charge. If an optical mode is present at $q \simeq 0$ then the positive ions within the unit cell move in the opposite direction to the negative ones, i.e. an oscillating dipole moment is created. This can couple to an external oscillating electric field (e.g., infrared radiation) which means that such vibrations are "infrared active", i.e. they cause infrared light to be absorbed (Chap. 11).

For the diatomic linear chain we only allowed displacements of the atoms along the direction of the chain. Such waves are called longitudinal waves. For a three-dimensional crystal there are two additional transverse waves. However, the clear separation of the vibrational modes into longitudinal and transverse is only possible in certain symmetry directions of the crystal. For an arbitrary direction the waves have a mixed character. Every crystal has three acoustic branches. For small q values (long wavelengths) these correspond to the sound waves of elasticity theory (4.5). For every additional atom in the unit cell one obtains a further three optical branches. Here the enumeration of the atoms must be referred to the smallest possible unit cell. For the fcc structure, in which many metals crystallize, the conventional unit cell contains four atoms, but the smallest possible cell contains only one atom (Fig. 2.8). Such crystals therefore possess only acoustic branches. The same is true for the bcc structure.

Although a crystal may be said to have three acoustic branches, this does not necessarily mean that these must everywhere have different frequencies. In cubic structures for example, the two transverse branches are degenerate along the [001] and [111] directions. This also holds for the diamond structure (Fig. 4.4). In the latter case, the smallest possible unit cell contains two atoms and thus, along with the acoustic branches, there are also optical branches.

It should be noted that "optical" is the term used to describe all branches that have a non-zero frequency at $q=0$. It does not necessarily imply optical activity, as can be demonstrated for the case of the diamond

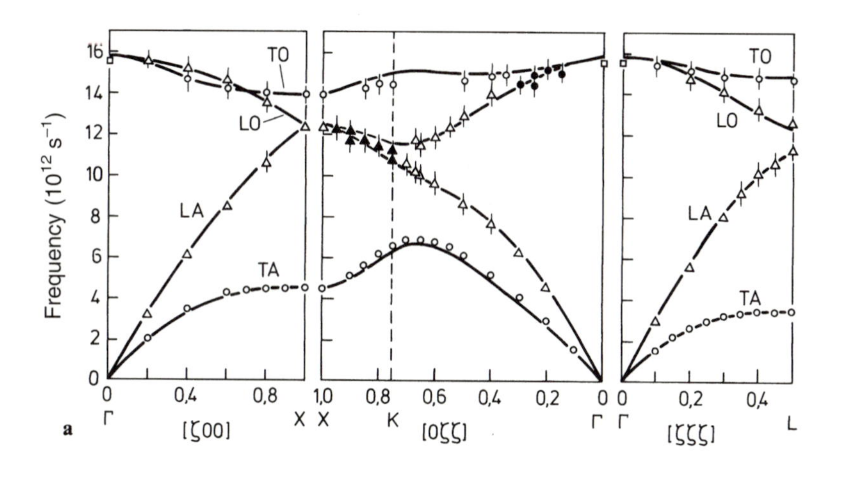

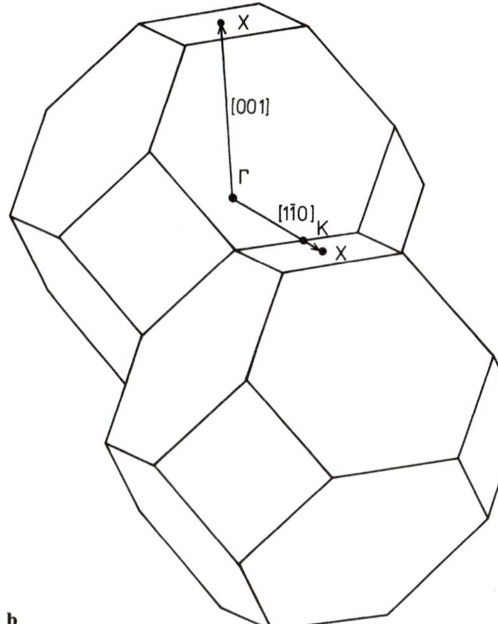

structure. At $q=0$ in the optical branch the two fcc substructures of the diamond structure vibrate against one another. However, since these two substructures are occupied by identical atoms, this vibration does not give rise to a dipole moment and hence it cannot interact with light. The optical modes are 3-fold degenerate at $q=0$. In Sect. 2.4 we saw that such 3-fold degeneracy is only possible for the point groups of cubic structures.

In contrast to the diamond structure, the zinc-blende structure consists of two substructures that are occupied by different atoms. "Optical" vibrations of this structure do lead to an oscillating dipole moment, which in turn gives rise to the absorption of electromagnetic radiation in this frequency range. In Chap. 11 we will see that this also lifts the degeneracy between the longitudinal- and transverse-optical waves.

4.4 Scattering from Time-Varying Structures – Phonon Spectroscopy

The solutions of the equations of motion for the atoms have the form of plane waves. In analogy to the wave-particle dualism of quantum mechanics, one might ask whether these waves can also be interpreted as particles. Any such particle aspect would manifest itself, for example, in the interaction with other particles, i.e., electrons, neutrons, atoms and photons. Thus we shall now extend the scattering theory developed in Chap. 3 to the case of structures that are time varying. The problem will again be treated in a quasi-classical framework. Later on, when we consider electron-phonon scattering (Chap. 9), we shall also meet the quantum-mechanical formalism.

We return to the scattering amplitude A_B derived in (3.6):

$$A_B \propto \mathrm{e}^{-\mathrm{i}\omega_0 t} \int \varrho(\boldsymbol{r}(t))\mathrm{e}^{-\mathrm{i}\boldsymbol{K}\cdot\boldsymbol{r}(t)} d\boldsymbol{r} \,. \tag{4.20}$$

To simplify the mathematics we consider a primitive structure and assume the atoms to be point-like scattering centers situated at the time-dependent positions $\boldsymbol{r}_n(t)$. Thus we write $\varrho(\boldsymbol{r},t) \propto \sum_n \delta(\boldsymbol{r}-\boldsymbol{r}_n(t))$

Fig. 4.4. (**a**) Phonon dispersion curves of Si. The circles and triangles are measured points and the solid lines are the result of a model calculation by Dolling and Cowley [4.3]. Instead of the wave vector $\boldsymbol{q}$, one often uses the reduced wave vector $\zeta=\boldsymbol{q}/(2\pi/a)$ as the coordinate for such plots. The relative lengths of the abscissae correspond to the actual separation of the points in the Brillouin zone. The branches of the dispersion curves carry the notation TA (transverse acoustic), LA (longitudinal acoustic), TO (transverse optical) and LO (longitudinal optical). Along the [100] and [111] directions the transverse branches are degenerate. Concerning the degeneracy of LO and TO at Γ, see also Sect. 11.4. (**b**) A sketch of two neighboring Brillouin zones showing that by moving along [110] from Γ to K one can arrive at X by continuing along the adjoining Brillouin zone boundary. Thus the point X can be described either by the wave vector $\boldsymbol{q}=2\pi/a\,[001]$ or by $\boldsymbol{q}=2\pi/a\,[1\bar{1}0]$. By studying the fcc lattice (Figs. 2.8, 12), one can convince oneself that these two $\boldsymbol{q}$-vectors describe the same atomic motion

$$A_B \propto e^{-i\omega_0 t} \sum_n e^{-i\boldsymbol{K}\cdot\boldsymbol{r}_n(t)} . \tag{4.21}$$

We separate each of the time-dependent vectors $\boldsymbol{r}_n$(t) into a lattice vector $\boldsymbol{r}_n$ and a displacement from the lattice site $\boldsymbol{u}_n$(t)

$$\boldsymbol{r}_n(t) = \boldsymbol{r}_n + \boldsymbol{u}_n(t) . \tag{4.22}$$

With this one obtains

$$A \propto \sum_n e^{-i\boldsymbol{K}\cdot\boldsymbol{r}_n} e^{-i\boldsymbol{K}\cdot\boldsymbol{u}_n(t)} e^{-i\omega_0 t} . \tag{4.23}$$

For small displacements $\boldsymbol{u}_n(t)$ we can make the expansion

$$A \propto \sum_n e^{-i\boldsymbol{K}\cdot\boldsymbol{r}_n} [1 - i\boldsymbol{K}\cdot\boldsymbol{u}_n(t) \ldots] e^{-i\omega_0 t} . \tag{4.24}$$

With the most general form of the expansion in terms of plane waves

$$\boldsymbol{u}_n(t) = \boldsymbol{u}\frac{1}{\sqrt{M}} e^{\pm i[\boldsymbol{q}\cdot\boldsymbol{r}_n - \omega(\boldsymbol{q})t]} \tag{4.25}$$

we obtain, besides the familiar elastic scattering, the terms

$$A_{\text{inel}} \propto \sum_n e^{-i(\boldsymbol{K}\mp\boldsymbol{q})\cdot\boldsymbol{r}_n} i\boldsymbol{K}\cdot\boldsymbol{u}\frac{1}{\sqrt{M}} e^{-i[\omega_0 \pm \omega(\boldsymbol{q})]t} . \tag{4.26}$$

Thus there are scattered waves for which the frequency ω differs from that of the primary wave by exactly the frequency of the crystal vibration. These scattered waves must obey a further condition relating to the wave vector since the sum over $\boldsymbol{n}$ only yields contributions when $\boldsymbol{K}\mp\boldsymbol{q}$ is equal to a reciprocal lattice vector $\boldsymbol{G}$

$$\begin{aligned} &\omega = \omega_0 \pm \omega(\boldsymbol{q}) , \\ &\boldsymbol{k} - \boldsymbol{k}_0 \mp \boldsymbol{q} = \boldsymbol{G} . \end{aligned} \tag{4.27}$$

On multiplying both equations by $\hbar$:

$$\begin{aligned} &\hbar\omega - \hbar\omega_0 \mp \hbar\omega(\boldsymbol{q}) = 0 , \\ &\hbar\boldsymbol{k} - \hbar\boldsymbol{k}_0 \mp \hbar\boldsymbol{q} - \hbar\boldsymbol{G} = \boldsymbol{0} , \end{aligned} \tag{4.28}$$

one sees that the first of these classical equations can be interpreted quantum mechanically as the conservation of energy. The plus sign corresponds to the excitation of a crystal vibration by the scattered particle; the minus sign applies to processes in which a crystal vibration loses energy to the scattered particle. The latter possibility can of course only occur if the crystal vibration has sufficient initial energy (amplitude of excitation); see (5.8). The second of the conditions in (4.28) can be interpreted as the conservation of quasimomentum, if it is assumed that $\hbar\boldsymbol{q}$ is the quasimomentum of the wave-like crystal vibration. Thus, in the sense of the conservation equations (4.28), one can regard these waves as particles. The term commonly used to describe such "particles" is *phonons*. The quasimomentum of phonons, however, is unlike a normal momentum because it is only defined to within an arbitrary reciprocal lattice vector. Furthermore, it has nothing to do with the momentum of the individual atoms. For these we have $\sum_i m_i v_i \equiv 0$ at all times because of the nature of the solutions of the equations of motion (Problem 4.2). Hence the term "quasimomentum" for the quantity $\hbar \boldsymbol{q}$.

It should be emphasized that the derivation of the conservation equations (4.28) proceeded via a purely classical treatment of the motion of the crystal atoms. Thus the particle model of phonons that emerges here is not a quantum mechanical result. It can, however, be quantum mechanically justified if one begins with the general quantization rules for equations of motion.

The momentum and energy conservation laws that govern the inelastic interaction of light and particle waves with phonons can be used to advantage for the experimental determination of phonon dispersion curves. We first discuss the interaction with light.

The inelastic scattering of light in and around the visible region is known either as Raman scattering, or, when the interaction is with acoustic waves, as Brillouin scattering. The scattering is produced by the polarization of the atoms in the radiation field and the subsequent emission of dipole radiation. In the frequency range of visible light the maximum wave vector transfer is

$$2k_0 = \frac{4\pi}{\lambda} \sim 2 \cdot 10^{-3} \mathrm{\mathring{A}}^{-1} ,$$

i.e., approximately 1/1000 of a reciprocal lattice vector. Thus Raman scattering can only be used to study lattice vibrations near to the center of the Brillouin zone (i.e., around $q=0$) (Panel III).

This would not be the case for the inelastic scattering of X-rays. With X-rays, as we saw in the discussion of diffraction, one can readily have wave-vector transfers of the order of reciprocal lattice vectors. The photon energy of such X-rays is around 10^4 eV. The natural width of characteristic X-ray lines is about 1 eV, whereas phonon energies lie in the range 1–100 meV. To perform phonon spectroscopy one would therefore have to monochromate the X-rays to within about 1 meV; for this one could use a crystal monochromator. The energy selection is achieved, as with an optical grating (Panel XII), by making use of the wavelength dependence of the diffraction. If one differentiates the Bragg equation

$$\begin{aligned} \lambda &= 2d \sin\theta , \\ \Delta\lambda &= 2\Delta d \sin\theta + 2d\Delta\theta \cos\theta , \end{aligned} \tag{4.29}$$

one obtains an expression for the monochromaticity of the beam as a function of the angular aperture $\Delta\theta$ and the deviation Δd of the lattice constant from its mean value (Panel II)

$$\frac{\Delta\lambda}{\lambda} = -\frac{\Delta E}{E} = \frac{\Delta d}{d} + \Delta\theta \operatorname{ctg}\theta \,. \tag{4.30}$$

For X-rays one would require $\Delta\lambda/\lambda \sim 10^{-7}$. The corresponding angular aperture would be so small that only the use of X-rays from synchrotron sources (Panel XI) would provide sufficient intensity. Furthermore, it is very difficult to find crystals that are sufficiently perfect and stress-free to fulfill the condition $\Delta d/d \sim 10^{-7}$ (Panel II). Using synchrotron beams, however, it is possible to obtain an energy resolution of just a few meV.

For neutrons and atoms, however, the prerequisites for phonon spectroscopy are far more realistic. The primary energies needed to give the required wave-vector transfer now lie in the range 0.1–1 eV ($\Delta\lambda/\lambda \sim 10^{-2}$–$10^{-3}$). Such neutrons are readily available in the form of the thermal neutrons produced by reactors. The principle of a complete neutron spectrometer, with monochromator, probe and analyzer, is shown in Fig. 1.4. Inelastic neutron scattering has been used to determine the phonon dispersion curves of most materials. As an example, Fig. 4.4 shows the dispersion curves for Si.

4.5 Elastic Properties of Crystals

In the limit of long wave length ($q \to 0$) the frequency of phonons in the "acoustic branches" is proportional to the wave vector q. For these acoustical phonons of long wave length, the sound waves, the displacement vectors in neighboring unit cells are nearly equal. The state of deformation of the crystal can be therefore described in the framework of a continuum theory. We consider the transition from atom dynamics to the continuum theory of elasticity for a particular example, namely a simple cubic crystal with nearest-neighbor bonds in the form of springs (Fig. 4.5). The model is an unrealistic representation of the elastic properties of a real solid since it has no stability with regard to shearing. However, it suffices for a general consideration of longitudinal waves along a cubic axis. The coordinate

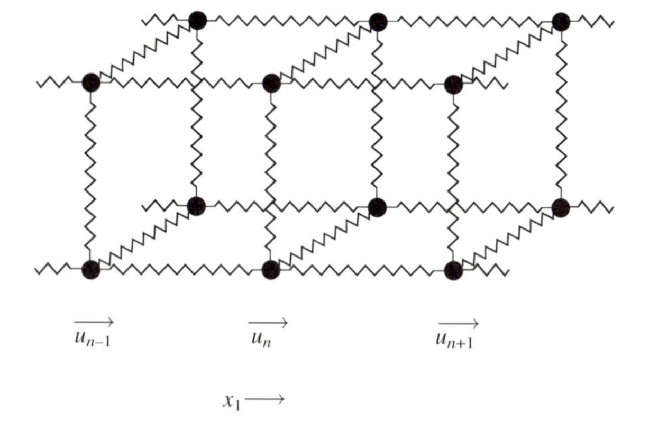

Fig. 4.5. Generalization of the linear chain spring model into three dimensions to represent a cubic primitive crystal. While the model with springs to nearest-neighbors is lacking stability against shear it suffices for a simple ansatz for longitudinal sound waves along a cubic axis

along the cubic axis and the components of the displacement vectors in the nth elementary cell in the direction of the axis are denoted as x_1 and u_{n1}, respectively. For a longitudinal wave along a cubic axis all displacement vectors perpendicular to this axis are equal. Furthermore, only the springs connecting atoms along the axis of propagation are strained. The equation of motion therefore becomes one-dimensional as for the linear chain (4.11).

$$M\ddot{u}_{n1} = f(u_{(n+1),1} - u_{n1}) - f(u_{n1} - u_{(n-1),1}) \,. \tag{4.31}$$

If the displacements vary little from cell to cell the differences can be replaced by the differential quotients.

$$\begin{aligned}(u_{(n+1),1} - u_{n1}) - (u_{n1} - u_{(n-1),1}) &= a\left.\frac{\partial u_1}{\partial x_1}\right|_{x=(n+1/2)a} - a\left.\frac{\partial u_1}{\partial x_1}\right|_{x=(n-1/2)a} \\ &= a^2 \left.\frac{\partial^2 u_1}{\partial x_1^2}\right|_{x=na} \,.\end{aligned} \tag{4.32}$$

The derivative

$$\varepsilon_{11} = \frac{\partial u_1}{\partial x_1} \,, \tag{4.33}$$

which is called the *strain,* is already a continuum quantity. The strain describes the change in the displacements of the atoms per length, hence a stretch and a compression of the material for $\varepsilon_{11} > 0$ and $\varepsilon_{11} < 0$, respectively. From (4.32) we see that the restoring elastic force in the x_1-direction is proportional to the derivative of the local strain ε_{11} with respect to the coordinate x_1. If, furthermore, the masses of atoms are replaced by the mass density $\varrho = M/a^3$, then the continuum equation of motion for a longitudinal sound wave is obtained.

$$\varrho\ddot{u}_1 = c_{11}\frac{\partial^2 u_1}{\partial x_1^2} \quad \text{with} \quad c_{11} = \frac{f}{a} \,. \tag{4.34}$$

The quantity c_{11} is an elastic modulus that describes the force per unit area in the x_1-direction in response to a deformation along the same axis. The sound velocity of the longitudinal wave is

$$c_L = \sqrt{\frac{c_{11}}{\varrho}} \,. \tag{4.35}$$

The continuum equation of motion (4.34) for the longitudinal sound wave along a cubic axis is valid for all cubic crystals. Only the relation between the elastic modulus and the interatomic potential is specific for the structure and the interatomic force field. Thus, the relation is different for fcc, bcc, or the diamond structure.

Up to now we were interested in the forces along the cubic axis that occurred in response to a deformation along the same axis. Now general deformations and forces are considered. We begin with a generalization of (4.33), the definition of a deformation

$$\varepsilon_{ij} = \frac{\partial u_i}{\partial x_j} \; . \tag{4.36}$$

The quantity ε_{ij} is a second rank-tensor. As a matter of convenience the components of the tensor are expressed in terms of particular cartesians that are chosen to agree as much as possible with the crystallographic axes. The diagonal elements of the tensor ε_{ii} describe infinitesimal distortions associated with a change in volume (Fig. 4.6 a). The magnitude of the (infinitesimal) change in the volume is given by the trace of the deformation tensor

$$\frac{\Delta V}{V} = \sum_i \varepsilon_{ii} \; . \tag{4.37}$$

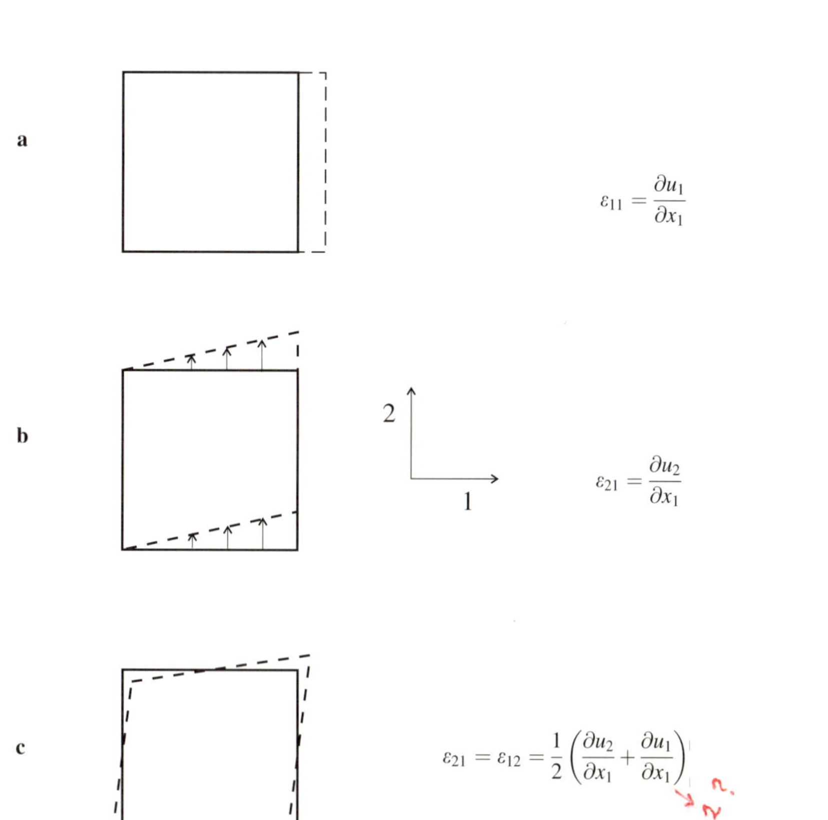

Fig. 4.6. Illustrations to elucidate the terminology in the theory of elasticity. (**a**) Strain along the x_1-axis; (**b**) shear along the x_2-axis, without separation of the rotational part of ε_{21}; (**c**) the same shear after splitting off the rotational component by symmetrizing the strain tensor

The nondiagonal elements ε_{ij} describe the deformation of a volume element in the i-direction as one moves along the j-direction and hence correspond to a shear distortion (Fig. 4.6b). It is useful to split ε_{ij} into a symmetric and an antisymmetric part with respect to an exchange of i and j.

$$\varepsilon_{ij} = \frac{\partial u_i}{\partial x_j} \equiv \frac{1}{2}\left(\frac{\partial u_i}{\partial x_j} + \frac{\partial u_j}{\partial x_i}\right) + \frac{1}{2}\left(\frac{\partial u_i}{\partial x_j} - \frac{\partial u_j}{\partial x_i}\right) . \tag{4.38}$$

The antisymmetric part $((\partial u_i/\partial x_j) - (\partial u_j/\partial x_i))$ describes a rotation whereas the deformation of the material is given by the symmetric tensor (Fig. 4.6c)

$$\varepsilon_{ij} = \frac{1}{2}\left(\frac{\partial u_i}{\partial x_j} + \frac{\partial u_j}{\partial x_i}\right) . \tag{4.39}$$

A solid resists deformations, hence deformations generate forces. For a homogeneous material the forces in response to a strain or shear are proportional to the area upon which the deformation is acting. One therefore relates all forces to the areas upon which they act. For a definition of these area-related forces, the "stresses", one considers a section through the crystal perpendicular to the x_l-axis and removes, in thought, the material on the right hand side of the intersection. The forces per area in the direction k that are necessary to keep the crystal in balance without distortions are the components of the stress tensor τ_{kl} (Fig. 4.7). The stress tensor is symmetric just as the strain tensor: the antisymmetric part of the stress tensor represents a torque, and in equilibrium all torques must vanish inside a solid.

To first order the relation between stress and strain is linear (Hooks law). In its most general form Hooks law reads

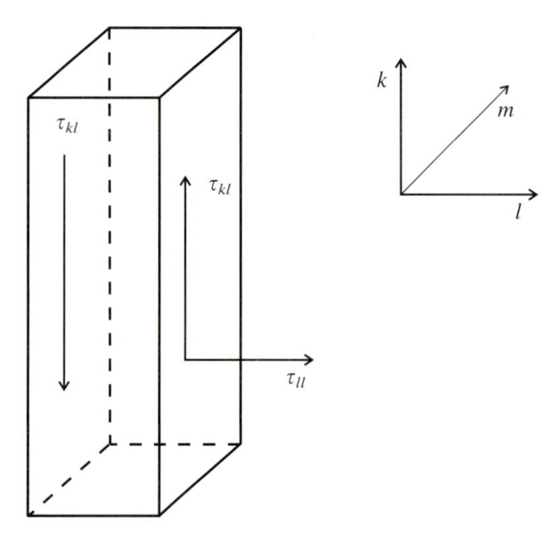

Fig. 4.7. Definition of the shear stress τ_{kl} and the normal stress τ_{ll}

$$\tau_{kl} = \sum_{ij} c_{klij} \varepsilon_{ij} \tag{4.40}$$

with the components of the elastic tensor (modules) c_{klij}. Because of the symmetry of the stress and strain tensors τ_{kl} and ε_{ij} one has the relations

$$c_{klij} = c_{lkij} = c_{klji} \; . \tag{4.41}$$

The number of independent components of the elastic tensor is further reduced by the requirement that the elastic energy be a unique function of the state of strain [4.2]. The energy density u is

$$u = \sum_{kl} \int \tau_{kl} d\varepsilon_{kl} = \frac{1}{2} \sum_{ijkl} c_{klij} \varepsilon_{ij} \varepsilon_{kl} \; . \tag{4.42}$$

This equation yields the same result independent of the chosen indices for the axes if

$$c_{klij} = c_{ijkl} \; . \tag{4.43}$$

The relations (4.41) and (4.43) permit a short-hand notation introduced by Voigt. In this notation a number between 1 and 6 is attributed to each pair of indices i and j. The assignment follows the scheme

$$\begin{aligned} &11 \to 1 \quad 23 \to 4 \\ &22 \to 2 \quad 13 \to 5 \\ &33 \to 3 \quad 12 \to 6 \; . \end{aligned} \tag{4.44}$$

Components of the stress and strain tensor can also be denoted using Voigt's notation. In order to ensure that all non diagonal elements of the strain and stress tensor in the energy density (4.42) are properly accounted for (i.e. ε_{ij} and ε_{ji}) a complete transition to Voigt's notation requires the introduction of redefined elastic modules. For our purpose here it is easier to use Voigt's notation only as an abbreviation for the indices in the elastic modules and stay with the standard tensor notation and summation otherwise. In the short-hand notation the elastic tensor becomes a 6×6 symmetric tensor with 21 independent components, at most. The number of independent components is further reduced by the crystal symmetry. For crystals with cubic symmetry the elastic tensor has only three independent components

$$\begin{pmatrix} c_{11} & c_{12} & c_{12} & 0 & 0 & 0 \\ c_{12} & c_{11} & c_{12} & 0 & 0 & 0 \\ c_{12} & c_{12} & c_{11} & 0 & 0 & 0 \\ 0 & 0 & 0 & c_{44} & 0 & 0 \\ 0 & 0 & 0 & 0 & c_{44} & 0 \\ 0 & 0 & 0 & 0 & 0 & c_{44} \end{pmatrix} . \tag{4.45}$$

It is easy to see that the elastic tensor must have this form, even without a formal proof. For example, the cubic axes are equivalent. Therefore, the diagonal components for normal and shear distortions must be equal ($c_{11}=c_{22}=c_{33}$ and $c_{44}=c_{55}=c_{66}$). A shear strain along one cubic axis cannot give rise to forces that would cause a shear along another cubic axis ($c_{45}=0$, etc.). Furthermore, a shear cannot cause a normal stress ($c_{14}=0$, etc.), and finally the forces perpendicular to a strain along one cubic axis must be isotropic ($c_{12}=c_{13}$, etc.).

For a hexagonal crystal the elastic tensor has the components

$$\begin{pmatrix} c_{11} & c_{12} & c_{13} & 0 & 0 & 0 \\ c_{12} & c_{11} & c_{13} & 0 & 0 & 0 \\ c_{13} & c_{13} & c_{33} & 0 & 0 & 0 \\ 0 & 0 & 0 & c_{44} & 0 & 0 \\ 0 & 0 & 0 & 0 & c_{44} & 0 \\ 0 & 0 & 0 & 0 & 0 & c_{66} \end{pmatrix} . \tag{4.46}$$

A hexagonal crystal is elastically isotropic in its basal plane. The tensor component that describes the stress-strain relation for a shear distortion in the basal plane, c_{66}, is therefore related to the tensor components c_{11} and c_{12} by the "isotropy condition" (see also 4.65)

$$2c_{66} = c_{11} - c_{12} . \tag{4.47}$$

With the help of the elastic tensor we can now generalize the wave equation (4.34). As noted before, we keep the standard double indices for the stress and strain tensors. The force that acts upon an infinitesimal cubicle of the volume $dV=dx_1dx_2dx_3$ in a direction k can be expressed in terms of the forces acting upon the faces of the cubicle

$$\begin{aligned} dF_k &= (\tau_{k1}(x_1+dx_1) - \tau_{k1}(x_1))dx_2dx_3 \\ &\quad + (\tau_{k2}(x_2+dx_2) - \tau_{k2}(x_2))dx_1dx_3 \\ &\quad + (\tau_{k3}(x_3+dx_3) - \tau_{k3}(x_3))dx_1dx_2 \\ &= dV \sum_l \frac{\partial \tau_{kl}}{\partial x_l} \\ &= dV \sum_{ijl} c_{klij} \frac{1}{2}\left(\frac{\partial^2 u_i}{\partial x_l \partial x_j} + \frac{\partial^2 u_j}{\partial x_l \partial x_i}\right) \\ &\equiv dV \sum_{ijl} c_{klij} \frac{\partial^2 u_i}{\partial x_l \partial x_j} . \end{aligned} \tag{4.48}$$

The sum of the elastic forces and the force of inertia $\varrho dV\ddot{u}_k$ must be zero. Hence, the generalized equation for the propagation of elastic waves in a crystalline solid of arbitrary symmetry is:

$$\varrho \ddot{u}_k = \sum_{ijl} c_{klij} \frac{\partial^2 u_i}{\partial x_l \partial x_j} . \tag{4.49}$$

In general (4.49) represents three coupled equations with three independent solutions. The equations decouple for particular high-symmetry orientations where entries in the elastic tensor vanish. A simple example are the sound waves along the axis of a cubic crystal. The solutions of (4.49) are a longitudinal and two degenerate transverse sound waves. If one chooses the x_1-axis as the direction of propagation and considers the motion along the x_2-axis the only non-vanishing derivative is

$$\frac{\partial^2 u_2}{\partial x_1 \partial x_1} \neq 0 , \tag{4.50}$$

and one obtains the equation of motion for the transverse sound wave:

$$\varrho \ddot{u}_2 = c_{2121} \frac{\partial^2 u_2}{\partial x_1^2} = c_{44} \frac{\partial^2 u_2}{\partial x_1^2} , \tag{4.51}$$

with the sound velocity

$$c_\mathrm{T} = \sqrt{\frac{c_{44}}{\varrho}} . \tag{4.52}$$

In many cases it is useful to work with the inverse of the elastic tensor c_{ijkl}. The inverse tensor s_{ijkl} is defined by the equation

$$\varepsilon_{ij} = \sum_{kl} s_{ijkl} \tau_{kl} . \tag{4.53}$$

The relations between the tensor components of $\boldsymbol{c}$ and $\boldsymbol{s}$ can be calculated by a formal tensor inversion, but also by describing certain states of strain. Consider, e.g., an isotropic deformation of a cubic crystal

$$\varepsilon_{ij} = \varepsilon \delta_{ij} . \tag{4.54}$$

Because of the cubic symmetry the stresses must likewise be isotropic

$$\tau_{ij} = \tau \delta_{ij} . \tag{4.55}$$

In that case, the elastic equations (4.43 and 4.53) reduce to

$$\begin{aligned} \tau &= (c_{11} + 2c_{12})\varepsilon , \\ \varepsilon &= (s_{11} + 2s_{12})\tau , \end{aligned} \tag{4.56}$$

and one obtains the relation

$$(c_{11} + 2c_{12})(s_{11} + 2s_{12}) = 1 . \tag{4.57}$$

We learn from (4.57), that a measurement of the volume change of a cubic crystal under hydrostatic pressure yields the combination of the elastic constants $c_{11}+2c_{12}$ and $s_{11}+2s_{12}$, respectively.

A second relation between the elastic modules $\boldsymbol{c}$ and the elastic constants $\boldsymbol{s}$

$$(c_{11} - c_{12})(s_{11} - s_{12}) = 1 \tag{4.58}$$

is obtained by an analogous consideration of the strain state $\varepsilon_{11}=-\varepsilon_{22}$, $\varepsilon_{33}=0$, with $\varepsilon_{ij}=0$ for $i \neq j$. By combining (4.57) and (4.58) one obtains

$$\begin{aligned} s_{11} &= \frac{1}{3}\left(\frac{1}{c_{11}+2c_{12}} + \frac{2}{c_{11}-c_{12}}\right) , \\ s_{12} &= \frac{1}{3}\left(\frac{1}{c_{11}+2c_{12}} - \frac{1}{c_{11}-c_{12}}\right) . \end{aligned} \tag{4.59}$$

Because of the diagonal form of the elastic tensor for shear stresses and strains (4.45) one has furthermore

$$c_{44} = 1/s_{44} . \tag{4.60}$$

The relations (4.58) and (4.60) hold also for hexagonal and the most important tetragonal crystals. Equation (4.57) is replaced by the set of equations

$$\begin{aligned} &(c_{11} + c_{12}) = s_{33}s^{-1} , \\ &c_{13} = -s_{13}s^{-1} , \\ &c_{33} = (s_{11} + s_{12})s^{-1} , \\ &\text{with } s = s_{33}(s_{11} + s_{12}) - 2s_{13}^2 . \end{aligned} \tag{4.61}$$

Furthermore, one has $c_{66}=s_{66}^{-1}$.

Despite its high symmetry, a cubic crystal is not elastically isotropic. The stress arising from a deformation along a cubic axis differs from the stress arising from a deformation along the diagonal. In order to be elastically isotropic the elastic constants of cubic crystals must fulfill a particular condition. In order to derive this isotropy condition we consider a cubic crystal that is strained along an arbitrarily oriented x-axis by the amount ε_{xx} and compressed along the perpendicular y-axis by $\varepsilon_{yy} = -\varepsilon_{xx}$. If the cubic axes x_1 and x_2 are parallel to x- and y-axes, one obtains for the stress τ_{xx}

$$\tau_{xx} = (c_{11} - c_{12})\varepsilon_{xx} . \tag{4.62}$$

If the cubic axes are rotated with respect to the x- and y-axis by 45° then the deformation corresponds to a shear deformation in the cubic axes (Fig. 4.8). By writing the components of the strain tensor in the cubic axes and in the x- and y-axis in terms of the displacement vector s_2 (Fig. 4.8), it is easy to see that $\varepsilon_{23} = \varepsilon_{xx}$. We thus obtain for the stress

$$\tau_{23} = c_{44}\varepsilon_{23} = c_{44}\varepsilon_{xx} . \tag{4.63}$$

We now express τ_{23} by τ_{xx}. On a cube with an edge length a the stress τ_{xx} exerts the force

$$f_x = \tau_{xx}a^2/\sqrt{2} . \tag{4.64}$$

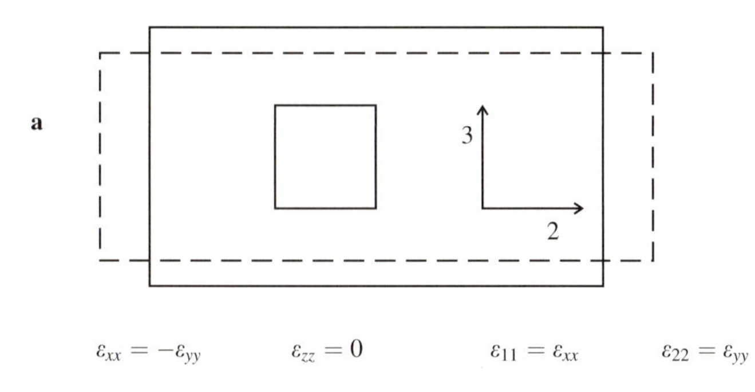

$\varepsilon_{xx} = -\varepsilon_{yy}$ $\varepsilon_{zz} = 0$ $\varepsilon_{11} = \varepsilon_{xx}$ $\varepsilon_{22} = \varepsilon_{yy}$

$\tau_{xx} = \tau_{22} = (c_{11} - c_{12})\varepsilon_{xx}$

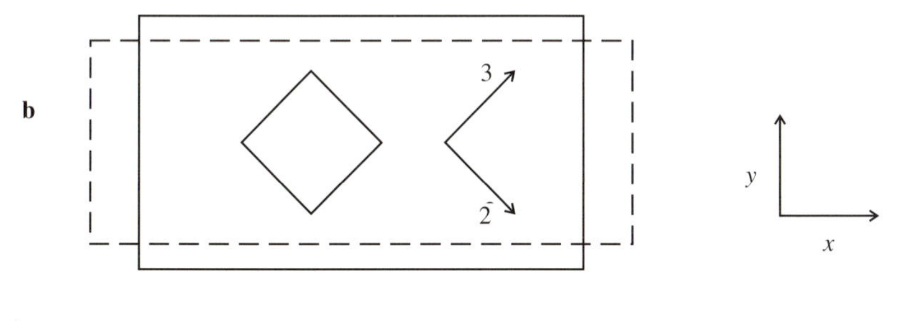

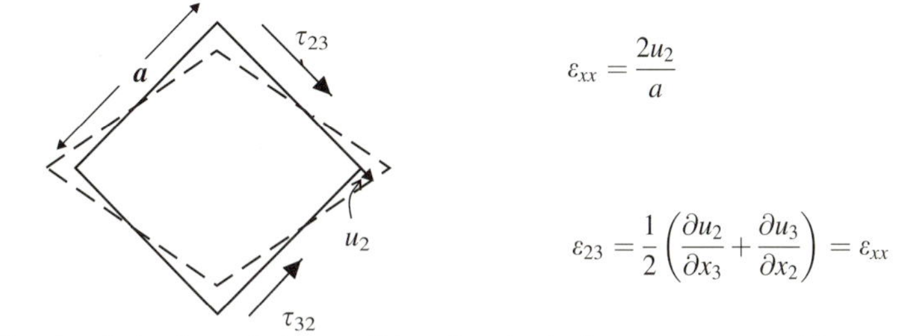

$\varepsilon_{xx} = \frac{2u_2}{a}$

$\varepsilon_{23} = \frac{1}{2}\left(\frac{\partial u_2}{\partial x_3} + \frac{\partial u_3}{\partial x_2}\right) = \varepsilon_{xx}$

Fig. 4.8. For the derivation of the condition for elastic isotropy one considers a volume-conserving deformation of a material under the assumption that (**a**) the deformation is along the cubic axes and (**b**) that the strain directions form an angle of 45° with the cubic axes. The deformation (**b**) corresponds to a shear in the cubic axes. For an isotropic material the resulting stresses must be identical in both cases. This condition yields the isotropy relation between the elastic constants (4.66)

Projected into the direction x_2 one has the force

$$f_2 = \tau_{23}a^2 = f_x/\sqrt{2} = \tau_{xx}a^2/2 \; . \tag{4.65}$$

With (4.63) and (4.65) and by comparison to (4.62) one obtains the condition for elastic isotropic behavior

Table. 4.1. Elastic constants for several cubic crystals at 20 °C (after [4.6]). The moduli c_{ij} are in 10^{10} N/m^2 and the constants (compliances) s_{ij} in 10^{-12} m^2/N. The numbers refer to the cubic axes. For a transformation into arbitrarily rotated axes see [4.4, 4.5]. The reliability and accuracy of the data differs for the various materials because of the different quality of available crystals. The temperature dependence of the elastic constant is particularly large for potassium and sodium. These materials become significantly stiffer at low temperatures. The condition for elastic isotropy, $2c_{44}/(c_{11}-c_{12})=1$, is fulfilled only in exceptional cases. Many metals are surprisingly anisotropic (see last column)

Material	s_{11}	s_{44}	s_{12}	c_{11}	c_{44}	c_{12}	$\frac{2c_{44}}{c_{11}-c_{12}}$
K	1225	530	–560	0.37	0.19	0.31	6.3
Na	590	240	–270	0.74	0.42	0.62	7.0
Ta	6.86	12.1	–2.58	26.7	8.25	16.1	1.56
Cr	3.05	9.9	–0.495	35.0	10.0	6.78	0.71
Mo	2.8	9.1	–0.78	45.5	11.0	17.6	0.79
W	2.53	6.55	–0.726	50.1	15.1	20.5	1.0
Fe	7.7	8.9	–2.8	23.7	11.6	14.1	2.4
Ir	2.28	3.91	–0.67	58.0	25.6	24.2	1.5
Ni	7.7	9.0	–3.0	24.4	11.2	15.4	2.5
Pd	13.6	13.9	–5.95	22.7	7.17	17.6	2.8
Pt	73.4	131	30.8	3.46	0.764	2.5	1.44
Cu	15.0	13.3	–6.3	16.8	7.54	12.1	3.2
Ag	22.9	21.7	–9.8	12.4	4.6	9.35	3.0
Au	23.3	23.8	–10.7	18.6	4.2	16.3	3.7
Al	15.7	35.9	–5.8	11.2	2.8	6.6	1.2
C	1.48	1.74	–0.517	107.6	57.6	12.5	1.21
Si	7.68	12.56	2.14	16.57	7.96	6.39	1.56
Ge	9.75	14.9	–2.66	12.9	6.71	4.83	1.66
GaAs	12.6	18.6	–4.23	11.9	5.4	6.0	1.83
LiF	11.35	15.9	–3.1	11.1	6.3	4.2	1.82
NaCl	22.9	79.4	–4.65	4.87	1.26	1.24	0.69

$$2c_{44} = c_{11} - c_{12} \,. \tag{4.66}$$

Because of this condition an elastic isotropic solid has only two independent constants that are denoted as

$$\lambda = c_{12} \quad \mu = c_{44} \,. \tag{4.67}$$

Hooks law (4.40) then becomes

$$\tau_{ik} = \lambda\delta_{ik}\sum_i \varepsilon_{ii} + 2\mu\varepsilon_{ik} \,. \tag{4.68}$$

In mechanical engineering Young's modulus Y (also denoted by E) and the Poisson number ν are commonly used. Young's modulus describes the change in length of a rod in response to a pull and ν describes the relative contraction perpendicular to the direction of pull. The constants Y, ν, λ μ, and the elastic constants s_{ij} in the cubic system are related by:

$$\nu = \frac{\lambda}{2(\lambda+\mu)} = -\frac{s_{12}}{s_{11}} \quad Y = \frac{\mu(2\mu+3\lambda)}{\mu+\lambda} = \frac{1}{s_{11}} \ . \tag{4.69}$$

Note that Y is merely the inverse of s_{11} and not a component of the inverse tensor of modules $\boldsymbol{c}$, although it has the dimension of a modulus!

Problems

4.1 Localized vibrations in a crystal can be represented by a superposition of phonon modes with different wave vectors. Show that the center of gravity of such a wave packet moves with the group velocity $v_g = d\omega/dq$.

4.2 Write down the dynamic equation for a one-dimensional linear chain of N (large number) atoms (atomic distance a, restoring force f, atomic mass m) and solve this with an ansatz $u_n(t) = u(q)\exp[\mathrm{i}(qna - \omega t)]$.
a) Compare the obtained dispersion $\omega(q)$ with that of the diatomic linear chain (4.15).
b) Show that the total momentum $\sum_{n=1}^{N} m\dot{u}_n(t)$ of a phonon vanishes.
c) Show that for long wavelengths ($q \ll a^{-1}$) the dynamic equation for the chain transforms into a wave equation for elastic waves when the displacements $u_n(t) = u(x = na, t)$, $u_{n+1}(t)$ and $u_{n-1}(t)$ are evaluated in a Taylor series.
d) Compare the resulting wave velocity with that of sound waves in a long rod and determine the effective modulus of elasticity. (Hint: for the long rod the velocity of sound is $c = \sqrt{E/\varrho}$ where E is the modulus of elasticity and ϱ the density.)

4.3 Calculate the eigenfrequency of a mass defect $M \neq m$ in a linear chain at the position $n = 0$ by invoking the ansatz $u_n = u_0 \exp(-\kappa|n| - \mathrm{i}\omega t)$ for the displacements. For which range of M do localized vibrations exist?

4.4 Calculate the dispersion relation for longitudinal and transverse phonons along the [100] direction of a fcc crystal whose atoms are joined to their nearest neighbors by springs. Using symmetry arguments, first identify any possible degeneracy. In which other high-symmetry direction does a similar consideration apply?

Then try to describe the phonons such that the equation of motion for a linear chain becomes applicable. For this it is important to be aware of the effect of the position of the phase planes on the displacements of the atoms.

Draw the displacements of the atoms for various phonons at the edge of the Brillouin zone.

4.5 Calculate the sound velocity of longitudinal and transverse acoustic modes along the [100] direction of a fcc crystal using the model of problem

4.4. According to crystal elasticity theory the sound velocities are $c_{\mathrm{long}} = (c_{11}/\varrho)^{1/2}$ and $c_{\mathrm{trans}} = (c_{44}/\varrho)^{1/2}$, where ϱ is the density. Calculate the elastic constant c_{11} and the shear elastic constant c_{44} in terms of the nearest neighbor force constant. Make the force constant such that the maximum vibrational frequency corresponds to 8.85 THz (representative of nickel) and calculate the numerical values of the sound velocities. The experimental values are 5300 m/s and 3800 m/s for the longitudinal and transverse waves, respectively.

4.6 Calculate the frequency of the surface phonon with odd parity (with respect to the mirror plane spanned by the wave vector and the surface normal) at the zone boundary of a (100) surface of an fcc crystal in the [110] direction using the nearest neighbor central force model. What makes this calculation so easy? Can you find another strictly first-layer mode on the same surface?

4.7 Derive the wave vector conservation for inelastic phonon scattering from a 2D periodic layer of atoms. Do the same problem for inelastic scattering from a surface when the incoming wave is damped inside the solid according to $\exp(-z/\lambda)$, with λ the effective mean free path. Assume a primitive lattice for simplicity. Develop an Ewald construction for the wave vector conservation law when the lattice has 2D periodicity.

4.8 Carry the expansion (4.24) one step further and calculate the time average of the scattered amplitude. Rewrite the expansion

$$1 - \tfrac{1}{2}\langle(\boldsymbol{K} \cdot \boldsymbol{u}_n)^2\rangle_t \approx \mathrm{e}^{-\frac{1}{2}\langle(\boldsymbol{K}\cdot\boldsymbol{u}_n)^2\rangle_t} .$$

By equating the time average with the ensemble average, calculate the scattered intensity for a primitive lattice.

In the prefactor $\exp(-\langle(\boldsymbol{K} \cdot \boldsymbol{u}_n)^2\rangle_t$ the quantity $W = \frac{1}{2}\langle(\boldsymbol{K} \cdot \boldsymbol{u}_n)^2\rangle$ is known as the Debye-Waller factor. For a harmonic lattice the result is correct even for arbitrarily large values of $\langle(\boldsymbol{K} \cdot \boldsymbol{u}_n)^2\rangle$. The proof, however, is not straightforward. Calculate the temperature dependence of the Debye-Waller factor assuming that all atoms vibrate as independent oscillators of a frequency $\hbar\omega$ by using (5.2 and 5.15). Carry out the same calculation based on the Debye model.

4.9 Calculate the elastic constants c_{11}, c_{12} and c_{44} of a face-centered cubic crystal under the assumption of spring forces to the nearest-neighbors. Show that the model crystal is not elastically isotropic! Show that $c_{12} = c_{44}$! This is the Cauchy-relation for cubic crystals. Cauchy relations among the elastic constants hold if the interatomic force field involves only central forces (forces acting between pairs of atoms along the bond direction). In reality, non-central forces such as angle bending valence forces and many-body forces cause deviations from the Cauchy relation. For which materials is the Cauchy-relation approximately fulfilled (see Table 4.1)?

4.10 Calculate the elastic energy per area in a thin epitaxial, pseudomorphic Cu film that is deposited on the (100) surface of Ni! Hint: the strain in the film ε_m is the misfit between the lattice constants, $\varepsilon_m = (a_{\mathrm{Ni}} - a_{\mathrm{Cu}})/a_{\mathrm{Ni}}$. The elastic energy density is $u = t\sum_{ij}\int_0^{\varepsilon_m} \tau_{ij} d\varepsilon_{ij}$, with t the film thickness. Why is there a limit to the thickness of epitaxial, pseudomorphic film growth and what happens, once a critical film thickness is reached?

Panel III
Raman Spectroscopy

Since the development of the laser, Raman Spectroscopy [III.1] has become an important method for investigating elementary excitations in solids, for example, phonons and plasmons. In this type of spectroscopy one studies the inelastic scattering of light by the elementary excitations of interest. This inelastic scattering was already mentioned in Sect. 4.5 in connection with scattering from phonons. As for all scattering from time-varying structures (e.g. vibrations of atoms in a crystal), energy must be conserved and, to within a reciprocal lattice vector, wave vector too, i.e. we have

$$\hbar\omega_0 - \hbar\omega \pm \hbar\omega(\boldsymbol{q}) = 0 , \qquad \text{(III.1)}$$

$$\hbar\boldsymbol{k}_0 - \hbar\boldsymbol{k} \pm \hbar\boldsymbol{q} + \hbar\boldsymbol{G} = \boldsymbol{0} , \qquad \text{(III.2)}$$

where $\omega_0, \boldsymbol{k}_0$ and $\omega, \boldsymbol{k}$ characterize the incident and scattered light waves respectively; $\omega(\boldsymbol{q})$ and $\boldsymbol{q}$ are the angular frequency and the wave vector of the elementary excitation, e.g. phonon. For light in the visible region of the spectrum, $|\boldsymbol{k}_0|$ and $|\boldsymbol{k}|$ are of the order of 1/1000 of a reciprocal lattice vector (Sect. 4.5), which means that only excitations in the center of the Brillouin zone ($|\boldsymbol{q}| \approx 0$) can take part in Raman scattering.

The interaction of visible light with the solid occurs via the polarizability of the valence electrons. The electric field $\mathscr{E}_0$ of the incident light wave induces, via the susceptibility tensor $\underset{\sim}{\boldsymbol{\chi}}$ a polarization $\boldsymbol{P}$, i.e.

$$\boldsymbol{P} = \varepsilon_0 \underset{\sim}{\boldsymbol{\chi}} \mathscr{E}_0 \quad \text{or} \quad P_i = \varepsilon_0 \sum_j \chi_{ij} \mathscr{E}_{j0} . \qquad \text{(III.3)}$$

The periodic modulation of $\boldsymbol{P}$ leads, in turn, to the emission of a wave – the scattered wave. In a classical approximation, the scattered wave can be regarded as dipole radiation from the oscillating dipole $\boldsymbol{P}$. From the laws of electrodynamics one obtains the energy flux density in direction $\hat{\boldsymbol{s}}$, i.e. the Poynting vector $\boldsymbol{S}$, at distance r from the dipole as

$$\boldsymbol{S}(t) = \frac{\omega^4 P^2 \sin^2\vartheta}{16\pi^2 \varepsilon_0 r^2 c^3} \hat{\boldsymbol{s}} . \qquad \text{(III.4)}$$

Here ϑ is the angle between the direction of observation $\hat{\boldsymbol{s}}$ and the direction of the vibration of $\boldsymbol{P}$. The electronic susceptibility $\underset{\sim}{\boldsymbol{\chi}}$ in (III.3) is now a function of the nuclear coordinates and thus of the displacements associated with the vibration $[\omega(\boldsymbol{q}), \boldsymbol{q}]$. Similarly, $\underset{\sim}{\boldsymbol{\chi}}$ can also be a function of some other collective excitations $X[\omega(\boldsymbol{q}), \boldsymbol{q}]$, for example, the density variations associated with a longitudinal electron plasma wave (Sect. 11.9), or the travelling-wave-like variations of magnetization in an otherwise perfectly ordered ferromagnet (magnons). These "displacements" $X[\omega(\boldsymbol{q}),\boldsymbol{q}]$ can be regarded as perturbations and in a formal expansion in X. It suffices to retain the first two terms:

$$\underset{\sim}{\chi} = \underset{\sim}{\chi}^0 + (\partial\underset{\sim}{\chi}/\partial X)X \,. \tag{III.5}$$

As we only need to consider excitations with $q \simeq 0$, we can simplify matters by writing $X = X_0 \cos[\omega(\boldsymbol{q})t]$ and, if the electric field $\mathscr{E}_0$ of the incident wave is described by $\mathscr{E}_0 = \hat{\mathscr{E}}_0 \cos\omega_0 \cdot t$, we obtain from (III.3) the polarization appearing in (III.4) as

$$\begin{aligned}\boldsymbol{P} &= \varepsilon_0 \underset{\sim}{\chi}^0 \hat{\mathscr{E}}_0 \cos\omega_0 t + \varepsilon_0 \frac{\partial\underset{\sim}{\chi}}{\partial X} X_0\, \hat{\mathscr{E}}_0 \cos[\omega(\boldsymbol{q})t] \cos\omega_0 t \\ &= \varepsilon_0 \underset{\sim}{\chi}^0 \hat{\mathscr{E}}_0 \cos\omega_0 t + \frac{1}{2}\varepsilon_0 \frac{\partial\underset{\sim}{\chi}^0}{\partial X} X_0 \hat{\mathscr{E}}_0 \{\cos[\omega_0 + \omega(\boldsymbol{q})]t + \cos[\omega_0 - \omega(\boldsymbol{q})]t\} \,.\end{aligned} \tag{III.7}$$

The scattered radiation expressed by (III.4) therefore contains, along with the elastic contribution of frequency ω_0 (the Rayleigh scattering), further terms known as Raman side bands with the frequencies $\omega_0 \pm \omega(\boldsymbol{q})$ (Fig. III.1). The plus and minus signs correspond to scattered light quanta that have, respectively, absorbed the energy of, and lost energy to, the relevant elementary excitation $[\omega(q),\boldsymbol{q}]$. The lines with frequency smaller than ω_0 are called the Stokes lines; those with higher frequency are the anti-Stokes lines. For the latter lines to be present it is necessary that the elementary excitation, e.g. phonon, is already excited in the solid. Thus at low temperatures the intensity of the anti-Stokes lines is much reduced because the relevant elementary excitation is largely in its ground state. The intensity of the inelastically scattered radiation is typically a factor of 10^6 weaker than that of the primary radiation.

A prerequisite for the observation of a Raman line is that the susceptibility $\underset{\sim}{\chi}$ (III.5) has a non-vanishing derivative with respect to the coordinate X of the elementary excitation. On account of the crystal symmetry and the resulting symmetry properties of the elementary excitation that determine the vanishing or nonvanishing of the quantities $(\partial\chi_{ij}/\partial X)$, the observability of the corresponding Raman lines depends on the geometry of the experiment. This is illustrated for the example of two Raman spectra measured from a Bi_2Se_3 single crystal (Fig. III.2). Bi_2Se_3 possesses a trigonal c-axis along which the crystal is built up of layers of Bi and Se. This crystal symmetry means, among other things, that the normal susceptibility tensor has the following form when referred to the principal axes:

$$\underset{\sim}{\chi}^0 = \begin{bmatrix} \chi^0_{xx} & 0 & 0 \\ 0 & \chi^0_{xx} & 0 \\ 0 & 0 & \chi^0_{zz} \end{bmatrix} \,. \tag{III.7}$$

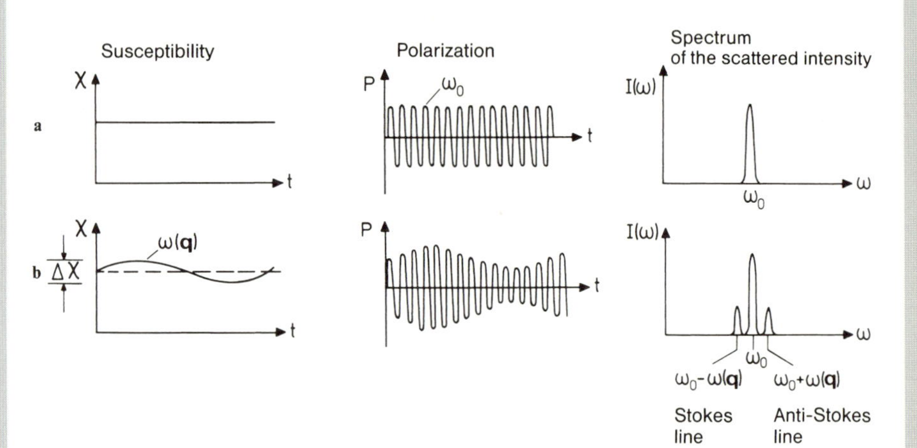

Fig. III.1. Schematic representation of the mechanics of elastic (**a**) and inelastic (**b**) light scattering (Raman scattering): (**a**) if the electronic susceptibility is assumed to be constant in time, the polarization P oscillates with the frequency ω_0 of the incident light and, in turn, radiates only at this frequency (elastic process); (**b**) if the susceptibility itself oscillates with the frequency $\omega(\boldsymbol{q})$ of an elementary excitation (e.g. phonon), then the oscillation of the polarization induced by the primary radiation (frequency ω_0) is modulated with frequency $\omega(\boldsymbol{q})$. This modulated oscillation of the polarization leads to contributions in the scattered light from the so-called Raman side bands of frequencies $\omega_0 \pm \omega(\boldsymbol{q})$

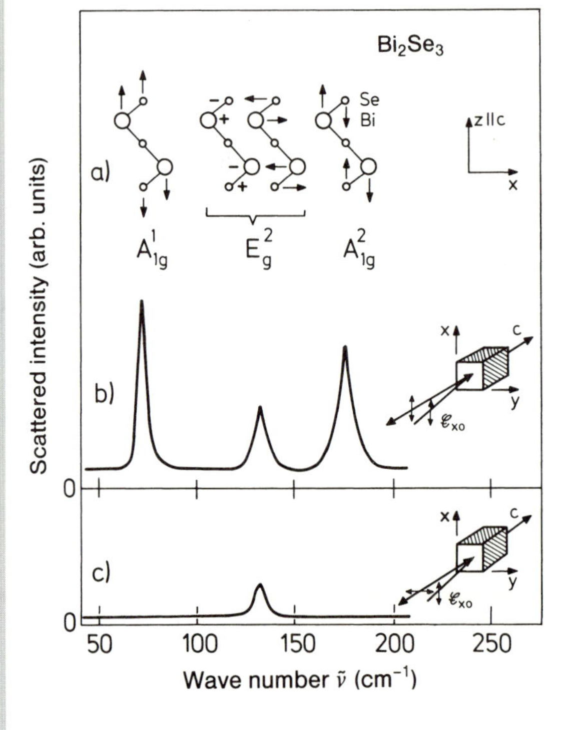

Fig. III.2 a–c. Raman spectra of phonons of types A_{1g} and E_g^2 measured from a single crystal of Bi_2Se_3. The c-axis of the crystal lies parallel to the z-axis of the coordinate system. (**a**) The displacement patterns of the A_{1g} and E_g^2 phonons for one of the three basis atom configurations in the nonprimitive unit cell. Arrows and +/– signs give a snapshot view of the atomic displacements. (**b**) Raman spectrum taken in the geometry $z(xx)\bar{z}$, i.e., the primary wave is incident in the z-direction and is polarized in the x-direction; the Raman scattered light is detected in the $-z$ (or $\bar{z}$) direction, and analysed for its x-polarized component. (**c**) Raman spectrum taken in the geometry $z(xy)\bar{z}$ [III.2]

In the measurements of Fig. III.2 the beam was incident along the c-axis (z-axis of the coordinate system), and the scattering was analyzed in the backscattering direction, i.e., also along the z-direction. If a polarization of the x-direction is present for both the incident and scattered radiation, then one will observe the phonons denoted by A^1_{1g}, A^2_{1g} and E^2_g (Fig. III.2 b). If, however, one measures scattered light with a polarization in the y-direction, then only E^2_g will appear in the Raman spectrum (Fig. III.2 c). This can be understood by considering the nature of the atomic displacements associated with the two types of phonon (Fig. III.2 a): if a phonon of type A_{1g} is excited, the symmetry of the crystal remains unchanged; thus the change in the susceptibility χ_{ij} induced by the phonon displacement, i.e., $(\partial\chi_{ij}/\partial X)$ leads to a tensor that thas the same form as χ^0_{ij} (III.7). Such a tensor implies that the polarization induced by the incident electric field $\mathscr{E}_0 = (\mathscr{E}_{x0}, \mathscr{E}_{y0}, 0)$, has the same direction as $\mathscr{E}_0$. In other words, for phonons of the type A_{1g} one has $(\partial\chi_{xy}/\partial X) = 0$.

According to Fig. III.2 a, a general phonon of the type E_g possesses displacements in both the x- and y-directions. The trigonal crystal symmetry is therefore broken by this phonon. The phonon-induced modification of the susceptibility in the x-direction is coupled to a modification in the y-direction. An incident electric field $\mathscr{E}_{x0}$ thus induces polarization changes in both the x- and y-directions. The scattered light that results contains polarization components in both these directions, i.e., $(\partial\chi_{xx}/\partial X) \neq 0$, $(\partial\chi_{xy}/\partial X) \neq 0$.

For crystals with centers of inversion (e.g. the NaCl and CsCl structures) there is a general exclusion principle which states that infrared-active transverse optical (TO) phonons (Sects. 4.3, 11.3, 11.4) are not Raman active and vice versa.

As a further example of an experimental Raman spectrum, Fig. III.3 shows the spectrum measured for an n-doped GaAs crystal with a free electron density of $n = 10^6\,\mathrm{cm}^{-3}$ (Sect. 10.6). Besides the strong lines between wavenumbers 250 and $300\,\mathrm{cm}^{-1}$ (wavenumber $\tilde{\nu} = \lambda^{-1}$) attributable to the excitation of TO and LO phonons, one also observes a structure at $40\,\mathrm{cm}^{-1}$ very close to the elastic peak ($\tilde{\nu} = 0$). This structure is essentially the result of excitation of collective vibrations of the "free" electron gas, so-called plasmons (Sect. 11.9). A weak coupling between the plasmons and the LO phonons leads to a small frequency shift in both these peaks.

Also of interest is the dependence of the Raman spectra on the primary energy $\hbar\omega_0$. If the incident photon energy $\hbar\omega_0$ is exactly equal to the energy of an electronic transition, i.e. if it corresponds to a resonance in χ or in the dielectric constant $\varepsilon(\omega)$, then one observes an enormous enhancement of the Raman scattering cross section, or so-called *resonant Raman scattering*. By varying the primary energy in order to find such resonances in the Raman cross section, it is also possible to study electronic transitions.

From (III.4) it follows that, for frequencies below the electronic resonance, the intensity varies as ω^4 or λ^{-4} as a function of the frequency

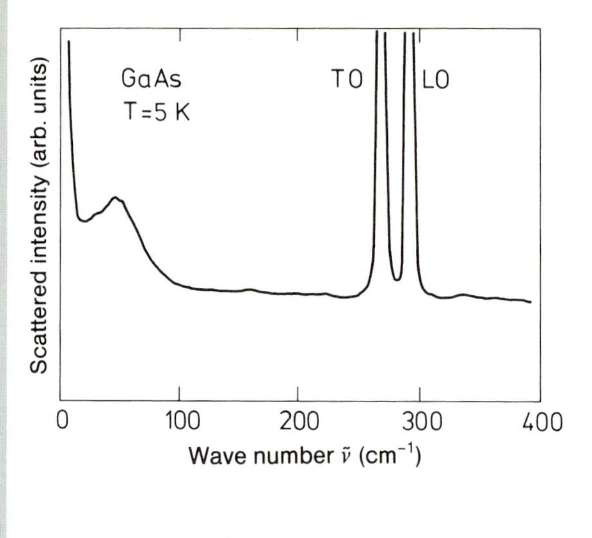

Fig. III.3. Raman spectrum of an n-doped GaAs crystal at a sample temperature of 5 K. The concentration of free electrons is $n \simeq 10^{16}\,\mathrm{cm}^{-3}$. TO and LO denote transverse and longitudinal optical phonons. The band at 40 cm^{-1} stems essentially from plasmon excitations [III.3]

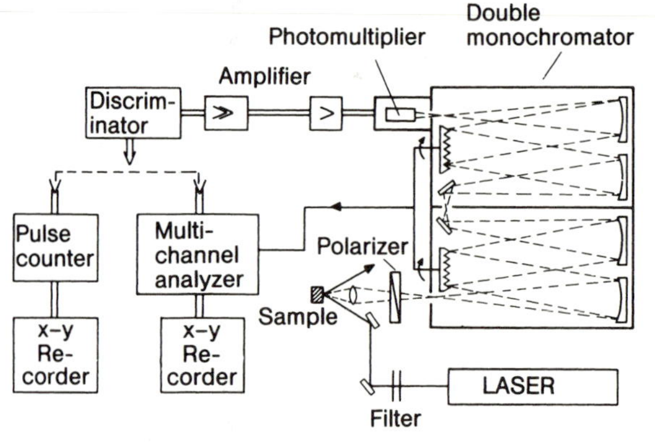

Fig. III.4. Schematic of experimental arrangement used to observe the Raman effect. To minimize the background due to internal scattering in the measuring device, a double monochromator is used. Furthermore, because of the weak signal a pulse counting technique is applied. The path of the Raman-scattered light is shown by the dashed lines

or wavelength of the incident light; it is thus desirable to use as short a wavelength as possible. Today, high-power lasers (neodymium, krypton, argon-ion, etc.) are used for this purpose. For resonance Raman spectroscopy in particular one can employ tunable dye lasers. Emission powers of up to several watts in the violet and near-UV spectral range are applied. To detect the scattered radiation in the visible and near-UV one uses highly sensitive photomultipliers. High demands are made of the spectrometer used to analyse the scattered radiation: whereas the primary photon energy is in the range 2–4 eV, i.e. has frequency ν of the order of 10^{15} Hz, one needs to measure frequency differences between this and the Raman side bands that lie anywhere from a few Hertz to 10^{14} Hz ($\widehat{=} 3000\,\mathrm{cm}^{-1}$). For scattering from sound waves in particular, a resolution of $\omega_0/\Delta\omega = 10^8$ is desirable. This can be achieved with Fabry-Pérot interferometers. In this case the method is commonly known as *Brillouin scattering.* Because of the low intensity of the Raman lines, it is important that there is no background intensity in the region close to primary line produced by primary light that has been scattered within the instrument, i.e. high contrast is an important prerequisite. Modern experiments often employ double or triple spectrometers (Fig. III.4). The gratings used are produced holographically in order to avoid spurious diffraction peaks (ghosts) in the spectral background intensity. Figure III.4 shows a modern experimental setup for Raman spectroscopy.

References

III.1 D.A. Long: *Raman Spectroscopy* (McGraw-Hill, New York 1977)
W. Hayes, R. Loudon: *Scattering of Light by Crystals* (Wiley, New York 1987)
III.2 W. Richter, H. Köhler, C.R. Becker: Phys. Status Solidi B 84, 619 (1977)
III.3 A. Mooradian: In: *Light Scattering Spectra of Solids*, ed. by G.B. Wright (Springer, Berlin Heidelberg 1969), p. 285

5 Thermal Properties

In Sect. 4.2 we saw how the $3rN$ equations of motion of a periodic solid can be largely decoupled by means of the plane-wave ansatz and the assumption of harmonic forces. With (4.7) we arrived at a system of equations that, for a given wave vector $\boldsymbol{q}$, couples the wave amplitudes of the atoms within a unit cell. It can be shown mathematically that within the harmonic approximation the equations of motion, even for a nonperiodic solid, can be completely decoupled by means of a linear coordinate transformation to so-called normal coordinates. We thereby obtain a total of $3rN$ independent forms of motion of the crystal, each with a harmonic time dependence and a specific frequency which, in the case of a periodic solid, is given by the dispersion relation $\omega(\boldsymbol{q})$. Any one of these "normal modes" can gain or lose energy independently of the others. The amount of energy that can be exchanged is quantized, however, as for a single harmonic oscillator:

$$E_n = \left(n + \frac{1}{2}\right)\hbar\omega \qquad n = 0, 1, 2, \ldots \, . \tag{5.1}$$

Classically, the quantum number n corresponds to the amplitude of the vibration according to

$$M\omega^2 \langle s^2 \rangle_t = \left(n + \frac{1}{2}\right)\hbar\omega \, . \tag{5.2}$$

where $\langle s^2 \rangle_t$ denotes the time average. If, for example, one wishes to calculate in the harmonic approximation the thermal energy per unit volume of a solid, one needs to know firstly the eigenfrequency spectrum of the solid, and secondly the energy of a harmonic oscillator in equilibrium with a heat bath. We begin by considering how to obtain, in principle at least, the frequency spectrum of the solid.

5.1 The Density of States

The $3rN$ equations of motion (4.4) have exactly $3rN$, in general distinct, solutions. In contrast, the plane-wave ansatz (4.5) for the periodic solid would appear to give a continuous manifold of solutions. This contradiction stems from the assumptions, on the one hand, of complete translational symmetry, i.e., an infinitely extended lattice, and, on the other hand, of a finite number N of unit cells. We can overcome this problem by considering a finite crystal of volume V containing N unit cells, which, however, we imagine to be part of an infinitely extended periodic continuation of the same. In this way we obtain a finite crystal while simultaneously preserving the full translational symmetry that is a prerequisite for the plane-wave solutions. If we were to consider only the finite crystal, this would lead to complications due to the additional localized solutions that are produced by its surfaces. For very small crystals, where the number of surface atoms is comparable with the number of bulk atoms, it is indeed necessary to consider such localized solutions when calculating the thermal properties.

The requirement that all properties of the lattice should be repeated in each direction after every $N^{1/3}$ unit cells means that the displacements of the atoms $\boldsymbol{s}_n$ must also repeat. According to (4.5) this leads to the condition

$$e^{iN^{1/3}\boldsymbol{q}\cdot(\boldsymbol{a}_1+\boldsymbol{a}_2+\boldsymbol{a}_3)} = 1 . \tag{5.3a}$$

If the wave vector $\boldsymbol{q}$ is separated into components in terms of the basis vectors of the reciprocal lattice $\boldsymbol{g}_i$ (3.21), the individual components q_i must satisfy the equation

$$q_i = \frac{n_i}{N^{1/3}} \quad \text{with}$$

$$\begin{cases} n_i = 0, 1, 2, \ldots N^{1/3} - 1 \\ n_i = 0, \pm 1, \pm 2, \ldots \quad \text{with the condition} \quad \boldsymbol{G}\cdot\boldsymbol{q} \leq \frac{1}{2}G^2 \end{cases} . \tag{5.3b}$$

The series of integers n_i can either be chosen so that $\boldsymbol{q}$ takes values within the unit cell of the reciprocal lattice, or such that it always lies within the first Brillouin zone introduced in Sect. 3.5, which indeed has the same volume as the unit cell. In the latter case the maximum values of n_i are determined by the condition $\boldsymbol{G}\cdot\boldsymbol{q} \leq \frac{1}{2}G^2$ (Fig. 3.8). This procedure of introducing a finite lattice while retaining the full translational symmetry thus leads to discrete $\boldsymbol{q}$-values. The total number of $\boldsymbol{q}$-values is equal to the number of unit cells N. The density of allowed $\boldsymbol{q}$-values in reciprocal space is N divided by the volume of the unit cell of the reciprocal lattice $\boldsymbol{g}_1\cdot(\boldsymbol{g}_2\times\boldsymbol{g}_3)$. On applying (3.25) one obtains the density of states in reciprocal space to be $V/(2\pi)^3$. In a cubic lattice the separation of allowed $\boldsymbol{q}$-values is thus simply $2\pi/L$ $(=g/N^{1/3})$ where L is the repeat distance in real space. This result can also be deduced directly from (5.3) (Fig. 5.1).

For the large N, the states in $\boldsymbol{q}$-space are densely packed and form a homogeneous *quasicontinuous* distribution. The number of states in a frequency interval $d\omega$ is then given by the volume of $\boldsymbol{q}$-space between the surfaces $\omega(\boldsymbol{q})=\text{const}$ and $\omega(\boldsymbol{q})+d\omega(\boldsymbol{q})=\text{const}$, multiplied by the $\boldsymbol{q}$-space density of states

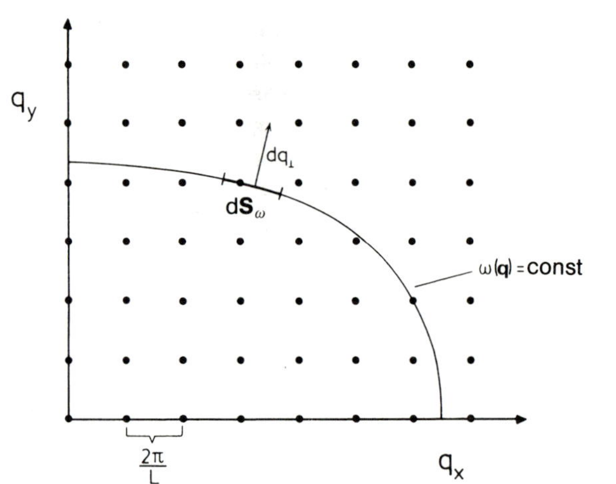

Fig. 5.1. Allowed values of $\boldsymbol{q}$ in reciprocal space for a square lattice. L is the repeat distance in real space

$$Z(\omega)d\omega = \frac{V}{(2\pi)^3} \int_{\omega}^{\omega+d\omega} d\boldsymbol{q} \;. \tag{5.4}$$

The function $Z(\omega)$ is also called the density of states. The density of states is a concept of central importance in solid-state physics, also for electronic properties (Sect. 6.1). We separate the wave vector volume element $d\boldsymbol{q}$ into a length perpendicular to the surface $\omega(\boldsymbol{q})$=const and an element of surface area

$$d\boldsymbol{q} = df_{\omega}\, dq_{\perp} \;.$$

With $d\omega = |\mathrm{grad}_{\boldsymbol{q}}\, \omega| dq_{\perp}$ one obtains

$$Z(\omega)\, d\omega = \frac{V}{(2\pi)^3} d\omega \int_{\omega=\mathrm{const}} \frac{df_{\omega}}{|\mathrm{grad}_{\boldsymbol{q}}\, \omega|} \;. \tag{5.5}$$

The density of states is high in regions where the dispersion curve is flat. For frequencies at which the dispersion relation has a horizontal tangent, the derivative of the density of states with respect to frequency has a singularity (van Hove singularity; Fig. 5.2). For the case of a linear chain, even the density of states itself is singular. We note that the concept of density of states does not presuppose a periodic structure. Amorphous solids also have a density of states, which is usually not very different from that of the corresponding periodic solid. However, in the case of amorphous solids there are no van Hove singularities.

As an example we shall calculate the density of states for an *elastic isotropic medium* with sound velocity c_{L} for longitudinal waves and c_{T} for the two (degenerate) transverse branches. For each branch the surface

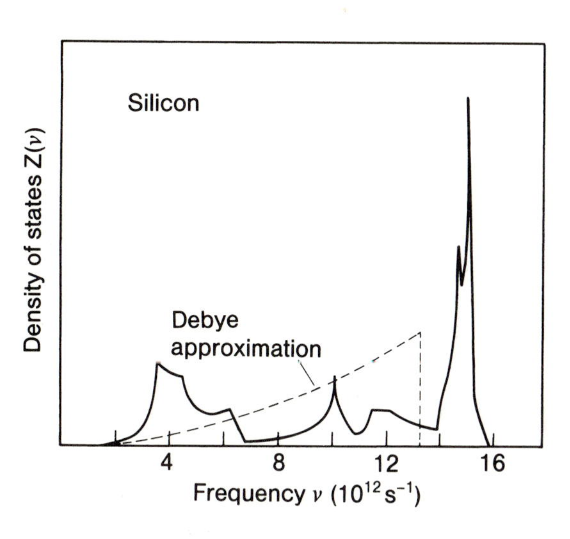

Fig. 5.2. Phonon density of states of Si [5.1] (Fig. 4.4). The dashed line is the density of states that one would obtain for an elastic isotropic continuum (Debye approximation with Θ=640 K; Sect. 5.3)

$\omega(\boldsymbol{q})$=const is a sphere. Thus $|\mathrm{grad}_{\boldsymbol{q}}\omega|$ is equal to the sound velocity c_i for each branch i and is independent of $\boldsymbol{q}$. The surface integral in (5.5) is therefore simply the surface area of the sphere $4\pi q^2$. For each branch we then have the result

$$Z_i(\omega)\,d\omega = \frac{V}{2\pi^2}\frac{q^2}{c_i}d\omega = \frac{V}{2\pi^2}\frac{\omega^2}{c_i^3}d\omega \tag{5.6}$$

and for the total density of states

$$Z(\omega)\,d\omega = \frac{V}{2\pi^2}\left(\frac{1}{c_\mathrm{L}^3}+\frac{2}{c_\mathrm{T}^3}\right)\omega^2\,d\omega\ . \tag{5.7}$$

Thus the density of states for an elastic isotropic medium, and likewise for a crystal at small frequencies and wave vectors, increases quadratically with frequency. With increasing frequency it would become ever larger. However, using the example of the linear chain (Fig. 4.3), we can see that for every solid there is a maximum possible frequency. This is also true for nonperiodic structures.

5.2 The Thermal Energy of a Harmonic Oscillator

We now consider an oscillator in equilibrium with a heat bath at temperature T. The oscillator cannot be assumed to be in a fixed and known quantum state n with energy $E_n=(n+\frac{1}{2})\hbar\omega$. Instead one can only state the probability P_n that the oscillator is found in state n. The appropriate probability is given by the *Boltzmann distribution* (known also as the *canonical distribution*)

$$P_n \propto \mathrm{e}^{-E_n/k T} \quad (k\text{: Boltzmann constant})\ . \tag{5.8}$$

The constant of proportionality is given by the condition that the oscillator must be in one of the possible states

$$\sum_{n=0}^{\infty} P_n = 1 \, ,$$

$$\sum_{n=0}^{\infty} \mathrm{e}^{-E_n/k T} = \mathrm{e}^{-\hbar\omega/2k T} \sum_{n=0}^{\infty} (\mathrm{e}^{-\hbar\omega/k T})^n$$

$$= \mathrm{e}^{-\hbar\omega/2k T} (1 - \mathrm{e}^{-\hbar\omega/k T})^{-1} \, . \tag{5.9}$$

Therefore we have

$$P_n = \mathrm{e}^{-n\hbar\omega/k T} (1 - \mathrm{e}^{-\hbar\omega/k T}) \, . \tag{5.10}$$

The average energy $\varepsilon(\omega, T)$ is thus given by

$$\varepsilon(\omega, T) = \sum_{n=0}^{\infty} E_n P_n = (1 - \mathrm{e}^{-\hbar\omega/k T})\hbar\omega \sum_{n=0}^{\infty} \left(n + \frac{1}{2}\right) (\mathrm{e}^{-\hbar\omega/k T})^n \, . \tag{5.11}$$

By differentiating the summation formula for the geometric series

$$\sum_{n=0}^{\infty} x^n = \frac{1}{1-x} \tag{5.12}$$

to give

$$\sum_{n=0}^{\infty} n x^n = \frac{x}{(1-x)^2} \, , \tag{5.13}$$

it can be shown that the mean energy is

$$\varepsilon(\omega, T) = \hbar\omega \left(\frac{1}{2} + \frac{1}{\mathrm{e}^{\hbar\omega/k T} - 1} \right) \, . \tag{5.14}$$

This expression has a form similar to the energy levels (5.1) of a single oscillator. Thus one can write

$$\langle n \rangle_T = \frac{1}{\mathrm{e}^{\hbar\omega/k T} - 1} \tag{5.15}$$

for the expected value of the quantum number n of an oscillator in thermal equilibrium at temperature T.

As was discussed in Sect. 4.3, it is possible to regard the wave-like motion of the atoms as noninteracting particles (phonons) whose state is determined by the wave vector $\boldsymbol{q}$ and the branch j. The number n then corresponds to the number of particles in a state $\boldsymbol{q}$,j, and $\langle n \rangle_T$ is the expected value of this number. The statistics of such noninteracting particles for situations where there is no limit on the number of particles in a given state is called *Bose statistics*. The wave quanta therefore behave as Bose particles (bosons).

It should be noted that the two different statistical distributions P_n in (5.8) and $\langle n \rangle_T$ in (5.15), i.e., the Boltzmann and the Bose distributions, result from two different ways of examining the problem: the Boltzmann distribution gives us the probability that a single particle occupies a certain state; Bose statistics, on the other hand, tells us the average number of noninteracting particles to be found in a certain state that can be occupied by any number of particles.

5.3 The Specific Heat Capacity

We now know the thermal energy $\varepsilon(\omega, T)$ of an oscillator with frequency ω. This also gives us the energy content of a normal vibration of the solid of frequency ω. The total energy of the solid in thermal equilibrium, i.e., the internal energy $U(T)$, is obtained by summing over all the eigenfrequencies. Using the density of states Z introduced in Sect. 5.1, the internal energy may be written

$$U(T) = \frac{1}{V}\int_0^\infty Z(\omega)\,\varepsilon(\omega, T)\,d\omega\,. \tag{5.16}$$

The derivative of the internal energy with respect to temperature is the specific heat capacity. We should immediately remark that, in the harmonic approximation, the specific heat at constant volume and that at constant pressure are identical, and thus it is not necessary to consider any additional parameters in the derivatives.

The thermal energy of a crystal and its corresponding specific heat capacity can thus be calculated from the density of states $Z(\omega)$ using (5.16). In turn, the density of states can be deduced, in principle, from the coupling matrices. In order to understand the qualitative behavior of the specific heat as a function of temperature, it is sufficient to consider a simple model for the density of states. For this we will take the previously calculated density of states of the elastic isotropic medium. The dispersion relation is then simply $\omega = cq$ and the typical dispersion due to the discrete lattice is neglected. Using (5.7) and (5.16) this model leads to

$$c_v(T) = \frac{1}{2\pi^2}\left(\frac{1}{c_L^3} + \frac{2}{c_T^3}\right)\int_0^{\omega_D} \omega^2 \frac{d}{dT}\varepsilon(\omega, T)\,d\omega\,. \tag{5.17}$$

The Debye cutoff frequency ω_D is determined by the requirement that the total number of states is equal to $3rN$, i.e.,

$$3\,rN = \frac{V}{2\,\pi^2}\left(\frac{1}{c_{\mathrm{L}}^3} + \frac{2}{c_{\mathrm{T}}^3}\right)\int\limits_0^{\omega_{\mathrm{D}}} \omega^2\,d\omega\,. \tag{5.18}$$

The imposition of a common cutoff frequency for all three acoustic modes represents a certain inconsistency of the model. Nonetheless, it actually leads to a better agreement with the experimental values of $c_v(T)$ than is obtained by the introduction of separate cutoff frequencies for the longitudinal and two transverse branches.

From (5.14, 5.17) we have

$$c_v = \frac{9\,rN}{V}\frac{1}{\omega_{\mathrm{D}}^3}\frac{d}{dT}\int\limits_0^{\omega_{\mathrm{D}}} \frac{\hbar\,\omega^3\,d\omega}{e^{\hbar\omega/k T} - 1}\,. \tag{5.19}$$

Introducing the Debye temperature Θ according to the definition

$$\hbar\,\omega_{\mathrm{D}} = k\,\Theta \tag{5.20}$$

we obtain, with the integration variable $y = \hbar\,\omega/k\,T$,

$$c_v = \frac{3\,rNk}{V}3\left(\frac{T}{\Theta}\right)^3\int\limits_0^{\Theta/T} \frac{y^4\,e^y\,dy}{(e^y - 1)^2}\,. \tag{5.21}$$

The form of $c_v(T)$ is depicted in Fig. 5.3. As is readily seen from (5.19), for $k\,T > \hbar\,\omega_{\mathrm{D}}$, the specific heat is given by

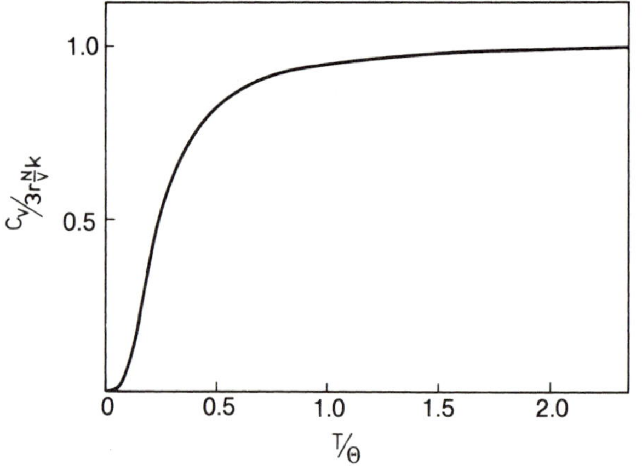

Fig. 5.3. The specific heat capacity per unit volume according to the Debye model. The specific heat is normalized to the Boltzmann constant k, the density of unit cells N/V and the number of atoms in the unit cell r. In this model different materials are only distinguished by their values of Debye temperature Θ

Table 5.1. Debye temperatures of selected materials in K [5.2]

Cs	38	Hg	72	Se	90	K	91	Ar	93
Pb	105	In	108	Te	153	Au	165	KCl	235
Pt	240	Nb	275	ZnS	315	NaCl	321	Cu	343
Li	344	Ge	370	W	400	C	420	Ir	420
LiCl	422	Al	428	Mo	450	Ni	450	Fe	467
Cr	630	Si	640	LiF	732	Be	1440	C	2230

$$c_v = \frac{1}{V} 3\, rN k$$

and is thus temperature independent. In relation to the density it is also identical for all solids since the characteristic temperature Θ is no longer involved. This is only true however within the framework of the harmonic approximation. Experimentally, one observes an additional slight increase in the specific heat roughly proportional to T. For low temperatures the integration limit Θ/T in (5.21) can be replaced by $+\infty$ and one obtains the result

$$c_v(T) = \frac{1}{V} 3\, rN k \frac{4\pi^4}{5} \left(\frac{T}{\Theta}\right)^3 \qquad T \ll \Theta\,. \tag{5.22}$$

Since at sufficiently low temperatures only elastic waves are excited, for which the density of states in real solids actually varies as $\propto \omega^2$, the T^3-law is valid for the vibrational contribution to the specific heat for all solids. The temperature range for which the T^3-law holds can however lie below 1 K.

Within the Debye approximation, the specific heat of a solid is completely determined at all temperatures by the characteristic temperature Θ. Thus to compare various materials with one another a knowledge of their Debye temperatures is useful (Table 5.1). Since in reality the specific heat deviates from that of the Debye model, it is not quite clear how best to define Θ. It is usual to determine Θ according to (5.22) using the experimentally measured value of c_v at low temperatures. This Θ-value, however, can differ markedly from the value obtained for higher temperatures from (5.20).

5.4 Effects Due to Anharmonicity

Until now we have only considered the atomic motion within the harmonic approximation. Higher terms in the expansion of the potential (4.1) have been neglected. However, many important properties of the solid are not described by this approximation. Some examples are the thermal expansion, the temperature dependence of the elastic constants, and the (weak) increase in the specific heat above Θ. A perfectly "harmonic" solid would also have an infinitely large thermal conductivity. This arises from the fact that a wave packet of elastic waves, once created, would have an infinite lifetime. The associated heat transport would thus proceed unhindered.

Unfortunately, the description of anharmonic effects is not simple. An exact treatment as in the harmonic case is not possible since one no longer has the nice decoupling of the equations of motion with the plane-wave ansatz. Thus in the anharmonic case one considers the solutions for the harmonic potential, the phonons, as a first approximation to the true solution. The phonons, however, are now no longer the exact eigensolutions to the equations of motion. Even if one could describe the state of motion of the crystal at a particular time by a plane wave (a phonon), this description would, in contrast to the harmonic case, become progressively less accurate with time. Instead, one would have to describe the time development by introducing a spectrum of other phonons. This feature is also known as "phonon decay".

A phonon can decay into two or more other phonons. An exact quantum mechanical treatment of this problem using perturbation theory shows that the decay of one phonon into two phonons, and also the corresponding inverse process, derive from the inclusion of the third term in the expansion of the potential. Processes that involve four phonons stem from the next highest term, and so on. Since the magnitude of higher terms generally decreases monotonically, the probability for such multiphonon processes also becomes very small. This is important for example in the inelastic interaction of phonons with light or particle waves (Sect. 4.4): the largest inelastic cross section is that for the excitation of a single phonon. The first anharmonic term of the expansion allows the simultaneous excitation of two phonons. Absorption involving the excitation of three phonons is very weak in comparison. It is only because of this that is it possible to make measurements of dispersion curves such as those shown in Fig. 4.4; these measurements rely on the predominance of single phonon excitation and absorption.

Another interesting question in this context is whether stationary solutions are also possible for nonlinear force laws. In certain special cases it is indeed possible to find such stationary solutions, known as solitons. Solitons are important particularly for the electrodynamics of nonlinear media [5.3].

In the following two sections we discuss the two most important anharmonic effects, the thermal expansion and the thermal conductivity due to phonons, and present models to describe these.

5.5 Thermal Expansion

All substances change their volume or dimensions with temperature. Although these changes are relatively small for solids, they are nonetheless of great technical importance, particularly in situations where one wishes to permanently join materials with differing expansion coefficients. In order to arrive at a definition that is independent of the length l of the sample, one defines the linear expansion coefficient, α, by

$$\alpha = \frac{1}{l}\frac{dl}{dT} \,. \tag{5.23}$$

For isotropic substances and cubic crystals, α is equal to one-third of the volume expansion coefficient

$$\alpha_V = 3\,\alpha = \frac{1}{V}\frac{dV}{dT} \,. \tag{5.24}$$

Typical values for linear expansion coefficients of solids are of the order of $10^{-5}\,\mathrm{K}^{-1}$. The expansion coefficient can clearly only be measured if the sample is maintained in a stress-free state. Thermodynamically, this means that the derivative of the free energy with respect to the volume, i.e., the pressure p, must be equal to zero for all temperatures:

$$-\left(\frac{\partial F}{\partial V}\right)_T = p = 0 \,. \tag{5.25}$$

This equation can be used to calculate the thermal expansion coefficient: Provided one can express the free energy as a function of the volume, then the condition of zero stress for every temperature yields a relation between volume and temperature and thus the thermal expansion. We will use this approach and begin by considering the free energy of a single oscillator. The generalization to a lattice is then straightforward.

The free energy of a system can be expressed in terms of the partition function Z

$$F = -k T \ln Z \quad \text{with} \quad Z = \sum_i \mathrm{e}^{-E_i/k T} \,. \tag{5.26}$$

The index i runs over all the quantum mechanically distinct states of the particular system. For a harmonic oscillator we have

$$Z = \sum_n \mathrm{e}^{-\hbar\omega(n+1/2)/k T} = \frac{\mathrm{e}^{-(\hbar\omega/k T)/2}}{1 - \mathrm{e}^{-\hbar\omega/k T}} \,. \tag{5.27}$$

The vibrational contribution to the free energy is therefore

$$F_s = \frac{1}{2}\hbar\omega + k T \ln\left(1 - \mathrm{e}^{-\hbar\omega/k T}\right) . \tag{5.28}$$

The total free energy also includes the value Φ of the potential energy in the equilibrium position

$$F = \Phi + \frac{1}{2}\hbar\omega + k T \ln\left(1 - \mathrm{e}^{-\hbar\omega/k T}\right) . \tag{5.29}$$

For a *harmonic* oscillator it is easy to convince oneself that the frequency ω is unaffected by a displacement s from the equilibrium position. Correspondingly, one finds that application of the equilibrium condition (5.25) yields no thermal expansion.

We now proceed to the case of the anharmonic oscillator in that we allow the frequency to change with a displacement from the equilibrium position. We assume that the energy levels are still given by $E_n = (n + \frac{1}{2}\hbar\omega)$. This procedure is known as the quasi-harmonic approximation. For a *single* oscillator it is easy to express the frequency change in terms of the third coefficient of the potential expansion (4.1). The actual calculation need not be performed here (Problem 5.6). For the simple calculation of the derivative (5.25) we consider the free energy expanded about the equilibrium position. The position of the potential minimum will be denoted by a_0. In the anharmonic case, the time-averaged position of the oscillator is no longer equal to a_0, and will be denoted a. Then, with force constant f, we obtain for the expansion

$$\Phi = \Phi_0(a_0) + \frac{1}{2} f (a - a_0)^2 ,$$

$$F_s = F_s(a_0) + \left.\frac{\partial F_s}{\partial a}\right|_{a=a_0} (a - a_0) . \tag{5.30}$$

The equilibrium condition (5.25), together with (5.29), then yields

$$f(a - a_0) + \frac{1}{\omega}\frac{\partial \omega}{\partial a}\varepsilon(\omega, T) = 0 . \tag{5.31}$$

With this equation we already have the relation between the average displacement and the temperature. The displacement is proportional to the thermal energy $\varepsilon(\omega, T)$ of the oscillator. Thus, for the linear expansion coefficient, we obtain

$$\alpha(T) = \frac{1}{a_0}\frac{da}{dT} = -\frac{1}{a_0^2 f}\frac{\partial \ln \omega}{\partial \ln a}\frac{\partial}{\partial T}\varepsilon(\omega, T) . \tag{5.32}$$

To generalize this to solids we simply need to replace $\alpha = a_0^{-1}(da/dT)$ by $\alpha_v = V^{-1}(dV/dT)$ and to sum over all phonon wave vectors $\boldsymbol{q}$ and all branches j. In place of $a_0^2 f$ one has $V\kappa$, where $\kappa = V(\partial p/\partial V)$ is the bulk modulus of compressibility

$$\frac{1}{V}\frac{dV(T)}{dT} = \alpha_V = \frac{1}{V\kappa}\sum_{q,j} -\frac{\partial \ln \omega(\boldsymbol{q},j)}{\partial \ln V}\frac{\partial}{\partial T}\varepsilon[\omega(\boldsymbol{q},j), T] . \tag{5.33}$$

This is the thermal equation of state of a lattice. One can immediately recognize that in the low- and high-temperature limits, the expansion coefficient shows the same behavior as the specific heat capacity, i.e., it is proportional to T^3 at low temperatures, and is constant (within this approxima-

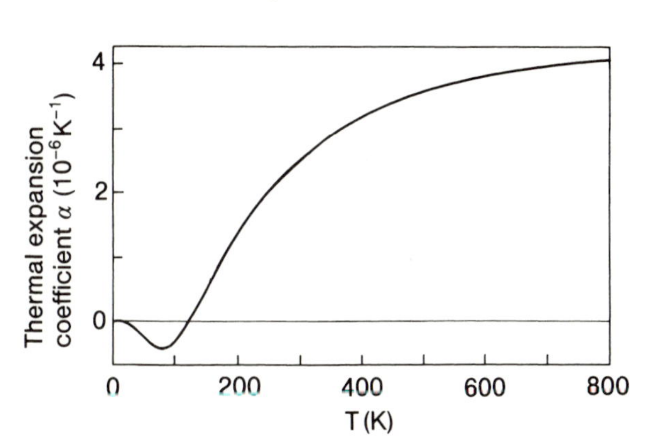

Fig. 5.4. Linear expansion coefficient of silicon as a function of temperature [5.4]

tion) at high temperatures. For many lattice types, even the "Grüneisen number"

$$\gamma = -\frac{\partial \ln \omega\,(\boldsymbol{q},j)}{\partial \ln V} \tag{5.34}$$

shows only weak dependence on the frequency $\omega(\boldsymbol{q},j)$. The Grüneisen number can then be assigned an average value and taken out of the sum in (5.33). The expansion coefficient thereby becomes approximately proportional to the specific heat at all temperatures. Typical values of this average Grüneisen parameter $\langle\gamma\rangle$ are around 2, and are relatively independent of the material. On account of the bulk modulus appearing in the denominator of (5.33), one can claim, as a rule of thumb, that soft materials with their small bulk moduli have a high thermal expansion coefficient.

The proportionality between α_V and the specific heat does not hold, however, for all crystal classes. For structures with tetrahedral coordination, the expansion coefficient changes sign at low temperatures. The expansion coefficient of silicon shown in Fig. 5.4 serves as an example.

We have implicitly assumed in our derivation of the thermal equation of state that we are dealing with a cubic structure. Hexagonal structures have different expansion coefficients parallel and perpendicular to the c-axis. These coefficients can even have different signs as is the case for tellurium: with increasing temperature a tellurium crystal expands perpendicular to the c-axis, but shrinks – albeit only slightly – in the direction parallel to the c-axis. Crystals with triclinic, monoclinic and rhombic lattices have three different expansion coefficients.

5.6 Heat Conduction by Phonons

In solids, heat is transported by phonons and by free electrons. For metals, it is the electronic contribution that dominates the thermal conductivity. However, this does not mean that insulators are necessarily poor conductors of heat. At low temperatures the thermal conductivity of crystalline Al_2O_3 and SiO_2 is higher than that of copper. This juxtaposition of properties – electrical insulation together with good thermal conductivity – makes these materials useful for experiments in low-temperature physics.

In contrast to the thermal properties discussed so far, thermal conduction is a nonequilibrium phenomenon. A thermal current only arises in a temperature gradient and the thermal current density $\boldsymbol{Q}$ is proportional to the temperature gradient

$$\boldsymbol{Q} = -\lambda \operatorname{grad} T \, . \tag{5.35}$$

where λ is the thermal conductivity.

The fact that we are dealing with deviations from thermal equilibrium and with spatially varying temperatures complicates the description somewhat: the thermal quantities $\varepsilon(\omega, T)$ and mean phonon number $\langle n \rangle$ (Sect. 5.2) have, until now, been defined only for systems at a single temperature. We must therefore assume that the spatial variation of T is small, such that in a sufficiently large region (i.e., one containing many atoms) the temperature can be considered homogeneous and the phonon number $\langle n \rangle$ can be defined. Neighboring regions will then have a slightly different temperature. In this way the phonon number now becomes a function of position. To calculate the thermal conductivity we must first express the thermal current density $\boldsymbol{Q}$ in terms of the properties of the phonons. As illustrated in Fig. 5.5, the thermal current passing through the area A in the x-direction in a time τ is equal to the energy density times the volume of the cylinder of length $v_x\tau$. Here v is the energy transport velocity of the phonons. This is not equal to the phase velocity ω/q of the phonon waves

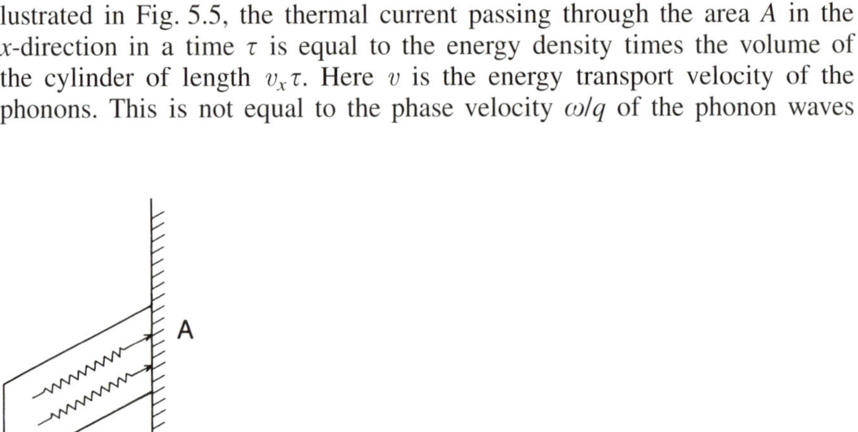

Fig. 5.5. Schematic representation of the thermal current through a cross-sectional area A. In the time interval τ all phonons travelling in the x direction within the cylinder of length $v_x\tau$ pass through the surface A

but, as is shown in electrodynamics text books for light and in quantum mechanical texts for electrons, it is given by the velocity of a wave packet $\partial\omega/\partial q$ (Sect. 9.1)

$$Q_x = \frac{1}{V}\sum_{\boldsymbol{q},j} \hbar\omega \langle n \rangle v_x , \quad v_x = \frac{\partial \omega}{\partial q_x} . \tag{5.36}$$

Here and in the following we shall drop the indices $\boldsymbol{q}$ and j in ω, $\langle n \rangle$ and v_x for the sake of brevity. In thermal equilibrium the thermal current density $\boldsymbol{Q}$ is of course zero. This can also be seen from the expression (5.36) for $\boldsymbol{Q}$ since, in equilibrium, the phonon occupation numbers $\langle n \rangle$ are equal for positive and negative q-values. And because of the symmetry of the dispersion curve, we have $v_x(\boldsymbol{q})=-v_x(-\boldsymbol{q})$. Thus the summation gives a vanishing thermal current. A thermal current can therefore only arise when the phonon number $\langle n \rangle$ deviates from the equilibrium value $\langle n \rangle^0$. This gives us a further expression for the thermal current in terms of the deviation in phonon occupation numbers from their equilibrium values:

$$Q_x = \frac{1}{V}\sum_{\boldsymbol{q},j} \hbar\omega \left(\langle n \rangle - \langle n \rangle^0\right) v_x . \tag{5.37}$$

A time variation of $\langle n \rangle$ in a particular region can arise in two ways: More or fewer phonons may diffuse into than out of the region from neighboring regions, or phonons may decay within the region into other phonons:

$$\frac{d\langle n \rangle}{dt} = \left.\frac{\partial \langle n \rangle}{\partial t}\right|_{\text{diff.}} + \left.\frac{\partial \langle n \rangle}{\partial t}\right|_{\text{decay}} . \tag{5.38}$$

This is a special form of the so-called Boltzmann equation, which is also applicable to problems concerning electron transport (Sect. 9.4). We shall consider the particular case of steady-state thermal currents in which the temperature is constant in time and thus also the phonon number. The total time derivative $d\langle n \rangle/dt$ is therefore zero.

For the time variation due to phonon decay, one can introduce a relaxation time τ such that

$$\left.\frac{\partial \langle n \rangle}{dt}\right|_{\text{decay}} = -\frac{\langle n \rangle - \langle n \rangle^0}{\tau} . \tag{5.39}$$

According to this expression, the more the phonon number deviates from its equilibrium value, the greater its time variation. The diffusion term is related to the temperature gradient. In a time interval Δt all the phonons that were originally within the region $x{-}v_x\Delta t$, will arrive in the region of interest around x. We thus have

$$\left.\frac{\partial \langle n \rangle}{\partial t}\right|_{\text{diff.}} = \lim_{\Delta t \to 0} \frac{1}{\Delta t} \left[\langle n\,(x - v_x\,\Delta t) \rangle - \langle n\,(x) \rangle \right]$$

$$= -v_x \frac{\partial \langle n \rangle}{\partial x} = -v_x \frac{\partial \langle n \rangle^0}{\partial T} \frac{\partial T}{\partial x} \,. \tag{5.40}$$

Because we have supposed steady-state conditions and local thermal equilibrium, having introduced the temperature gradient, we have replaced $\langle n \rangle$ by $\langle n \rangle^0$. If we now substitute (5.38–5.40) into (5.37), we obtain

$$Q_x = -\frac{1}{V} \sum_{\boldsymbol{q},j} \hbar\,\omega\,(\boldsymbol{q},j)\,\tau\,(\boldsymbol{q},j)\,v_x^2\,(\boldsymbol{q},j) \frac{\partial \langle n\,(\boldsymbol{q},j) \rangle^0}{\partial T} \frac{\partial T}{\partial x} \,. \tag{5.41}$$

For cubic or isotropic systems we have, in addition,

$$\langle v_x^2 \rangle = \frac{1}{3}\, v^2 \,. \tag{5.42}$$

Comparing this with the phenomenological equation (5.35), we obtain for the thermal conductivity

$$\lambda = \frac{1}{3\,V} \sum_{\boldsymbol{q},j} v\,(\boldsymbol{q},j)\,\Lambda\,(\boldsymbol{q},j) \frac{\partial}{\partial T}\, \varepsilon\,[\omega\,(\boldsymbol{q},j),\, T] \,. \tag{5.43}$$

Here $\Lambda = v\tau$ is the mean free path of a phonon. An analogous relation holds for the thermal conductivity of a gas and of the electron gas (Sect. 9.7). As expected, the specific heat capacity of the individual phonons plays an important role in heat transport. A further significant quantity is the group velocity: phonons close to the zone boundary and optical phonons contribute little to the thermal current. However, the temperature dependence of λ is also determined by the mean free path. Here, according to the temperature range of interest, one has to consider a variety of processes. These will be discussed in more detail in the following.

We must firstly take a closer look at phonon decay. For the decay due to anharmonic interactions that was described in Sect. 5.4, one has conservation of quasimomentum and energy:

$$\boldsymbol{q}_1 = \boldsymbol{q}_2 + \boldsymbol{q}_3 + \boldsymbol{G} \,, \quad \hbar\,\omega_1 = \hbar\,\omega_2 + \hbar\,\omega_3 \,. \tag{5.44}$$

At low temperatures, where only sound waves are thermally excited, the momentum and energy conservation can be satisfied with $\boldsymbol{G} = 0$. Such processes are illustrated in Fig. 5.6a. One sees that the projections of $\boldsymbol{q}_1$ and of $\boldsymbol{q}_2 + \boldsymbol{q}_3$ onto an arbitrary direction are in this case equal. Since for elastic waves the magnitude of the group velocity is independent of $\boldsymbol{q}$, the thermal current is not disturbed by the decay process. Therefore, at low temperatures (in practice those below ~ 10 K) the anharmonic interaction does not influence the mean free path in (5.43). In this case only processes for which $\boldsymbol{q}$-conservation does not hold contribute to the thermal resistivity. These processes include the scattering of phonons by crystal defects, or – for a highly perfect single crys-

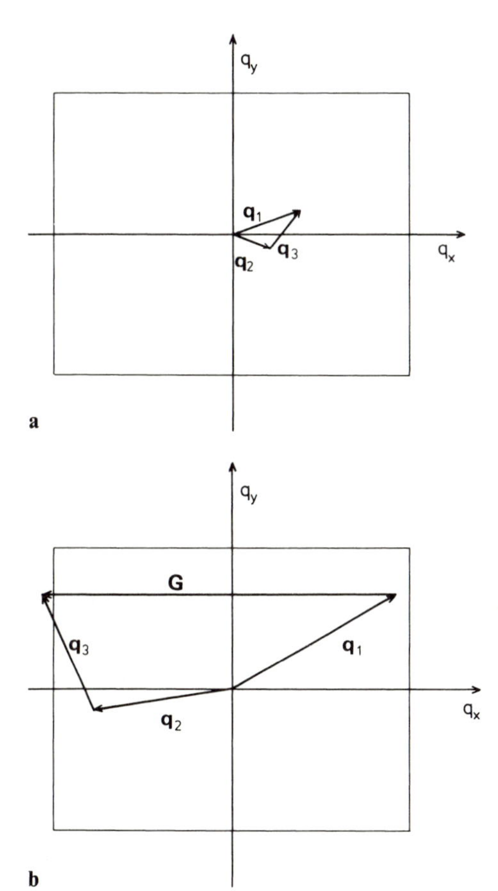

Fig. 5.6. A normal decay process (**a**) and an umklapp process (**b**) in q-space. In case (**b**) the vector $\boldsymbol{q}_1$ is split with the help of the vector $\boldsymbol{G}$ into two vectors $\boldsymbol{q}_2$ and $\boldsymbol{q}_3$, for which the group velocity is in the negative q_x-direction. This causes a reversal of the direction of energy flow

tal – their scattering at the surface of the crystal. We then have the seemingly improbable, but nonetheless observed, phenomenon of the thermal conductivity that depends on the external dimensions of the crystal and the condition of its surface. The temperature dependence of λ is determined here by the specific heat and is thus proportional to T^3.

At higher temperatures, momentum and energy conservation may also involve a reciprocal lattice vector. Such processes can reverse the direction of energy transport (Fig. 5.6b). They are therefore also known as "Umklapp" processes (from the German term for "folding over"). The condition for their occurrence is that phonons with sufficiently large $\boldsymbol{q}$-vectors are excited. The decaying phonon must have a wave vector q_1 of roughly half the diameter of the Brillouin zone and therefore possesses an energy of $\sim k\,\Theta/2$. The probability for this is proportional to $\exp(-\Theta/bT)$, with $b=2$. The mean free path Λ thus obeys

$$\Lambda \propto e^{\Theta/bT} . \tag{5.45}$$

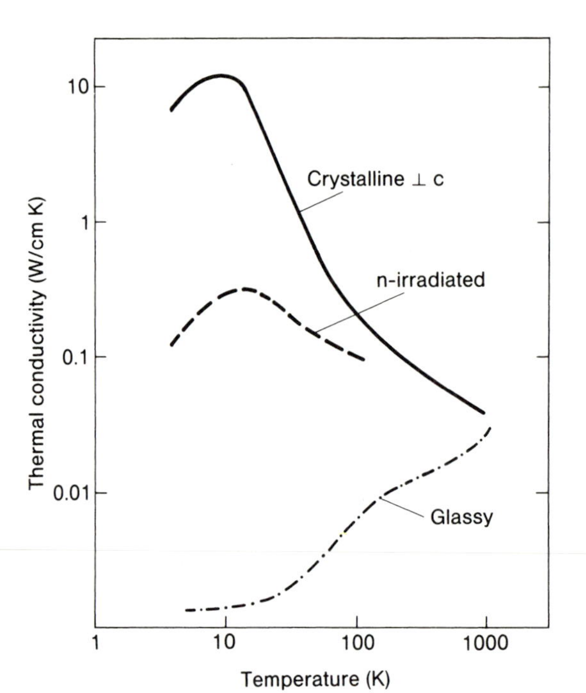

Fig. 5.7. Thermal conductivity of simple-crystal SiO_2 (quartz) perpendicular to the crystal *c*-axis. Also shown are the corresponding curves for the same crystal with defects induced by neutron bombardment, and for quartz glass [5.5, 5.6]

This strong exponential dependence on temperature determines the behavior of λ in the region of intermediate temperatures.

At high temperatures Λ only drops slowly with temperature ($\propto T^{-1}$). The full characteristic behavior of the thermal conductivity of a (non-conducting) single crystal is shown in Fig. 5.7 for the example of SiO_2 (quartz). For comparison the figure also displays the totally different behavior observed for the same material in the amorphous state (quartz glass). Here the scattering from defects is dominant even at the Debye temperature, and λ drops rapidly with decreasing temperature without showing the intermediate maximum typical for single crystals. Radiation damage and other defects also significantly reduce the thermal conductivity of single crystals.

Problems

5.1 Calculate the density of states and the specific heat at high and low temperatures for a one-dimensional and a two-dimensional elastic continuum. Are there physical realizations of such systems?

5.2 Calculate the thermal energy and specific heat for
a) a system of two harmonic oscillators,

b) a system with two energy levels.
Explain the difference in the two results. Are there any physical realizations of case (b)?

5.3 Assume a tetragonal lattice with a base of two atoms at (0, 0, 0) and (0, 0, $\frac{1}{2}$) carrying equal charges of opposite sign.
a) Calculate the static polarization of the lattice.
b) How large a surface charge is needed in order to compensate the static polarization?
c) Calculate the piezoelectric constant $\partial P_3/\partial\tau_3$ where τ_3 is the stress along the polar c-axis and assuming central forces to nearest neighbors.
d) For which direction of the ZnS and the wurtzite structures do you expect a longitudinal piezoeffect?

5.4 Calculate and plot the phase and group velocity of phonons for a diatomic linear chain with a mass ratio of 1:5. Estimate the contribution of the optical mode to the thermal conductivity.

5.5 Show that the equation of motion for an anharmonic oscillator

$$M\ddot{u} + fu - \frac{1}{2}gu^2 = 0$$

is solved by an approximate solution involving multiples of the harmonic frequency $\omega_0^2 = f/M$:

$$u = \sum_{n=1}^{\infty} a_n e^{in\omega_0 t} .$$

Discuss the result in relation to phonon decay. What is the analogy to electrical circuits and to signal transmission in nonlinear media?

5.6 Calculate the thermal expansion of an anharmonic oscillator following the procedure of Sect. 5.5. The frequency shift for a displacement s_{stat} can be found by evoking the ansatz $u(t) = u_{stat} + u_1 \sin \omega t$.

Panel IV
Experiments at Low Temperatures

In the history of solid state physics, advances in the production and measurement of low temperatures have often been associated with the discovery of new physical phenomena. For example, in 1911, shortly after the first successful production of liquid ^{4}He in 1908, Kamerlingh Onnes discovered superconductivity [IV.1]. Indeed, it is a feature of the many-particle systems that we know as "solids", that they possess elementary excitations with very small energies. However, the quantum character of the excitation spectrum only becomes particularly noticeable when kT is small compared to the quantum energies. In the endeavor to produce ever lower temperatures, modern research has reached the micro-Kelvin region (12 μK [IV.2]). To obtain such temperatures requires the simultaneous application of many sophisticated techniques. For example, the heat flow to the cold sample must not exceed 10^{-9} W. It is even necessary to avoid incident electromagnetic radiation in the radio frequency range and mechanical vibrations. Besides the application of liquid ^{4}He (T=4.2–1.2 K) and ^{3}He (T=3.2–0.3 K) for preliminary cooling, the chief method for the production of the lowest temperatures is the so-called "adiabatic demagnetization" of nuclear spin systems.

In this cooling process one begins with a set of nuclear spins in the milli-Kelvin range, which are split in energy by the presence of a magnetic field. The removal of heat causes the spins to adopt lower energy states. The magnetic field is then gradually reduced, which causes a corresponding reduction of the energy level splitting of the spin system. Eventually, at the appropriate temperature, a few of the spins are able to enter higher energy states of the nuclear spin system. The energy required in this process is supplied as heat from the electron and (at not too low temperatures) phonon systems of the solid.

As in all frontier areas of physics, not only the production, but also the measurement of the lowest temperatures, presents a problem. Even the equilibration of nuclear spin and electron temperatures can take hours.

In this experimental section we will introduce the reader to two experimental arrangements that allow one to measure the specific heat capacity and thermal conductivity of solids down to about 0.3 K. Compared to the work in the μK region, these are simple experiments. Nonetheless, they serve to illustrate the essential elements of low-temperature techniques.

Figure IV.1 depicts a so-called Nernst calorimeter [IV.3] used for the measurement of specific heat capacity. The calorimeter consists of an evac-

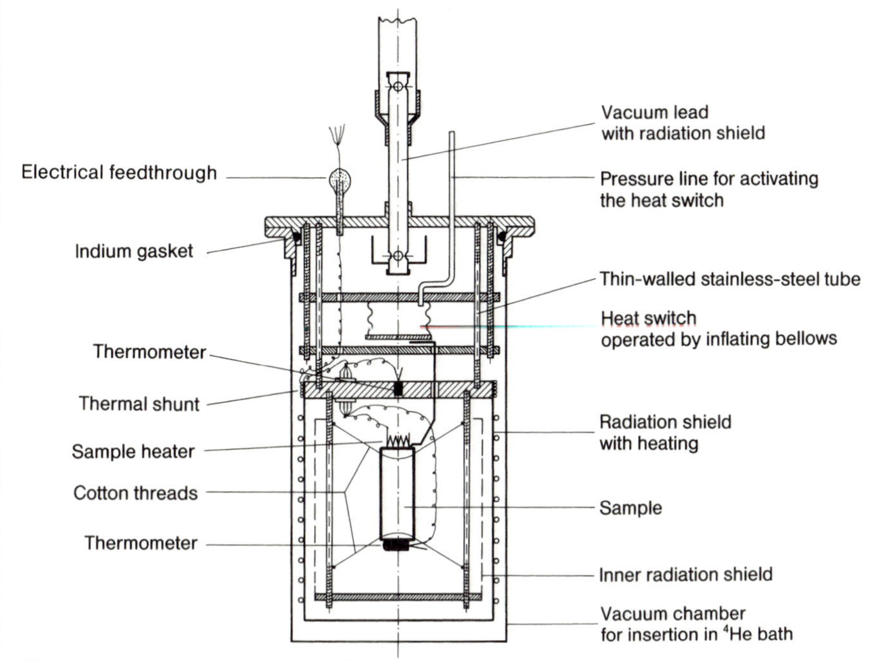

Fig. IV.1. An adiabatic Nernst calorimeter [IV.3]

uated vessel to prevent heat conduction by gas, which is submersed in the helium bath of a conventional cryostat. The helium bath in turn is surrounded by a mantle a liquid nitrogen temperature which serves to reduce thermal radiation. The principle of the specific heat measurement is to record the temperature rise of the sample upon supplying a known amount of energy, usually in the form of electrically produced heat. The main problem is the undesired extraneous heat reaching the sample. This stems from three sources: thermal conduction by the background gas in the calorimeter, thermal radiation, and conductivity of the leads. Heat conduction by the background gas can be largely avoided by evacuation, preferably at high temperatures. The influence of radiation is kept to a minimum by surrounding the sample with a radiation shield whose temperature is maintained close to that of the sample (the so-called "adiabatic" calorimeter). The sample itself is held by cotton or nylon threads which provide good thermal isolation. The conduction of heat through the leads cannot be totally avoided, but can be minimized by careful choice of materials and by ensuring good thermal contact between the leads and the outer radiation shield. To establish the desired sample temperature, particularly for cooling, one can use a heat switch. In the calorimeter of Gmelin shown in Fig. IV.1, this heat switch is a pneumatically switched heat bridge, with which the sample can be coupled to the temperature of the helium bath. The temperature of the sample is measured by the resistance of carbon resistors, or, even more re-

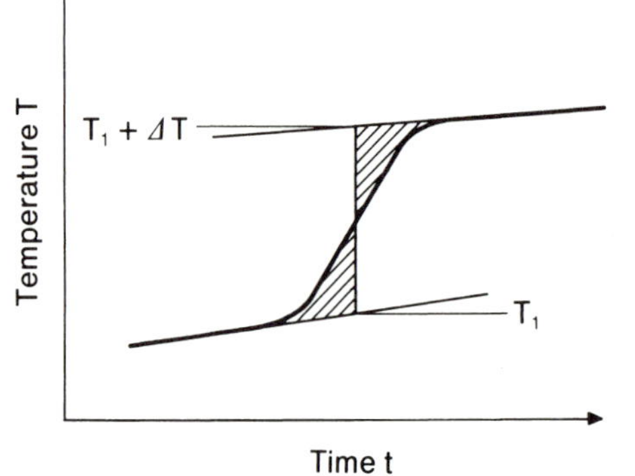

Fig. IV.2. Typical temperature variation of a sample in an experiment to measure specific heat from the temperature rise produced by supplying a known amount of heat. To determine the true temperature rise, the curves before and after the measurement must be extrapolated as shown in the figure

producibly, by the resistance of a doped germanium crystal which falls exponentially with increasing temperature (Chap. 12). Such resistance thermometers must be calibrated against the thermodynamical fixed points of ^{3}He and ^{4}He or, better still, against a vapor pressure thermometer with these gases. As a small example of the sophistication involved in the experiment of Fig. IV.1, one should note how the special form of the vacuum lead prevents the 300 K radiation from escaping from the lead into the calorimeter.

Despite careful screening, it is not possible to avoid a slight temperature drift of the sample (Fig. IV.2). After supplying a quantity of heat ΔQ one must therefore determine the true temperature rise by extrapolation (Fig. IV.2). The specific heat capacity can then be calculated from

$$c_\mathrm{p} = \frac{1}{m}\frac{\Delta Q}{\Delta T} \; . \qquad \text{(IV.1)}$$

With a calorimeter of the type shown in Fig. IV.2 it is also possible, in principle, to measure the thermal conductivity of a sample. A somewhat different set-up, from the laboratory of Pohl [IV.4], developed especially for thermal conductivity measurements, is shown in Fig. IV.3. The entire apparatus can again be submerged in a bath of ^{4}He (4.2 K at atmospheric pressure), which in turn is shielded by a radiation shield at liquid nitrogen temperature. The apparatus of Fig. IV.3 possesses in addition a tank for ^{3}He. By pumping down this tank one can exploit the latent heat of evaporation of ^{3}He to obtain temperatures of about 0.3 K. This apparatus for measuring thermal conductivity contains two heating elements. One serves to set the temperature of the sample and the other sends a stationary thermal current through the crystalline sample from its upper end. The temperature difference is registered by the two carbon resistors. The thermal conductivity can then be calculated from

$$\lambda = \frac{L}{F}\frac{\dot{Q}}{\Delta T} \qquad \text{(IV.2)}$$

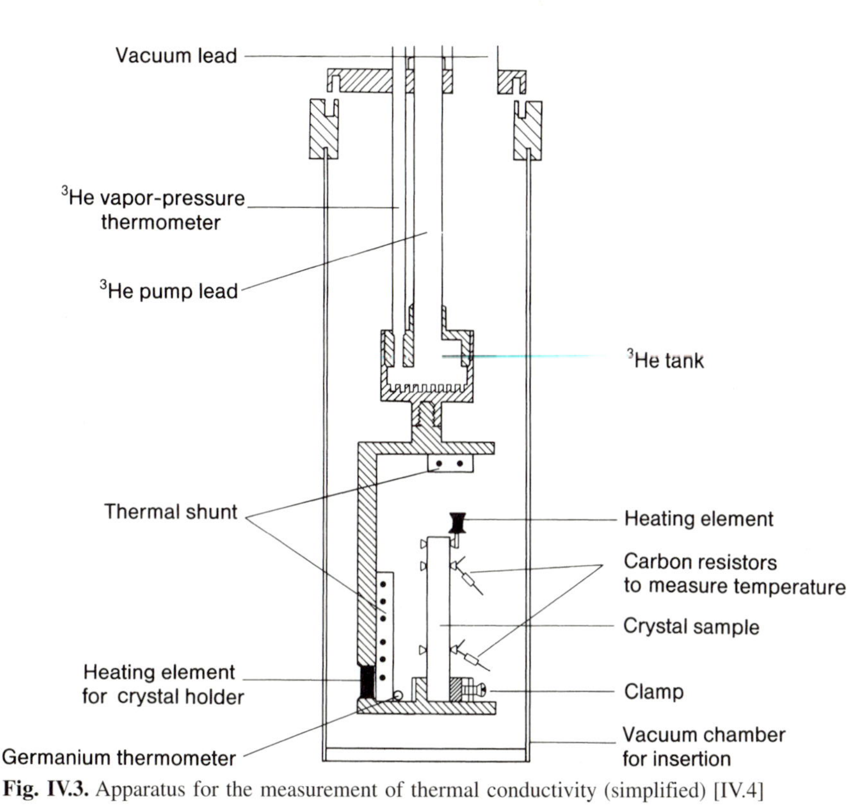

Fig. IV.3. Apparatus for the measurement of thermal conductivity (simplified) [IV.4]

where L is the distance between the carbon resistors, A is the cross-sectional area of the sample, and $\dot{Q}$ is the power of the heating element.

References

IV.1 W. Buckel: *Supraleitung*, 2nd. edn. (Physik Verlag, Weinheim 1977)

IV.2 K. Gloos, P. Smeibidel, C. Kennedy, A. Singsaas, P. Sekowski, R.M. Muelle, F. Pobell: J. Low Temp. Phys. 73, 101 (1988)

IV.3 E. Gmelin: Thermochimica Acta 29, 1 (1979)

IV.4 W.D. Seward, V. Narayanamurti: Phys. Rev. 148, 463 (1966)

6 “Free” Electrons in Solids

To a good approximation, the properties of solids can be divided into vibrational dynamics and electronic properties. This so-called *adiabatic approximation* (Chap. 4) is based on the fact that for the dynamics of the heavy nuclei, or of the nuclei together with their strongly bound core electrons (this combination is known as the “atomic core”), the energy can be expressed as a function of the nuclear or core coordinates in terms of a time-independent potential: the electron system, because of its very much smaller mass, follows the motion of the nuclei or cores almost instantaneously. From the viewpoint of the electron system this also means that for the electron dynamics one can regard the nuclear or core motion as extremely slow and, in the limiting case, as nonexistent. Within the adiabatic approximation one can then determine the excitation states of the electron system in the static potential of the positively charged, periodically arranged nuclei or atomic cores. In doing so, one neglects any interactions between the moving atomic cores and the remaining electrons of the crystal. In order to treat electronic transport phenomena (Sects. 9.3–9.5) in crystals, one has to reintroduce these so-called electron-lattice interactions in the form of a perturbation.

Even the adiabatic approximation of stationary nuclei or cores does not enable a quantitative treatment of the excitation states of electrons; one would still have to solve the Schrödinger equation for about 10^{23} electrons (which also interact with one another) in a periodic, static core potential. The problem must therefore be further simplified:

One considers just a simple electron in an effective periodic and time-independent potential. This potential is the one produced by the stationary nuclei in their equilibrium positions and by all the other electrons. These electrons shield the nuclear charge to a large extent and one obtains a potential which, in a section through an atomic row of the crystal, appears qualitatively as shown in Fig. 6.1 (full line). In this so-called *one electron approximation* one neglects all electron-electron interactions that cannot be represented as a local potential for the single electron under consideration, for example interactions arising from the exchange of two electrons. However, such correlations between electrons are important, for example, for understanding magnetism and superconductivity. We will thus be returning later to the subject of electron correlations. For the time being, however, we shall confine ourselves to the assumption of a local periodic potential and will solve the Schrödinger equation for a single electron in this potential. For this electron we

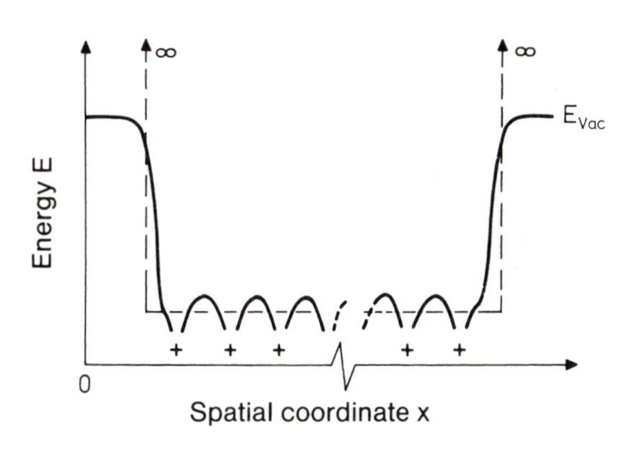

Fig. 6.1. Qualitative form of the potential for an electron in a periodic lattice of positive cores (+). The vacuum level E_{vac} is the level to which the electron must be promoted in order for it to leave the crystal and escape to infinity. The simplest approximation to describe this system is that of the square well potential (– – –) with infinitely high walls at the surfaces of the crystal

shall find a series of one electron quantum states that will be successively filled with the available electrons. In this procedure the Pauli principle demands that each state contain only a single electron.

6.1 The Free-Electron Gas in an Infinite Square-Well Potential

An even simpler model, first considered by *Sommerfeld* and *Bethe* in 1933 [6.1], also ignores the periodic potential within the crystal. Despite its simplicity, this model yielded a much improved understanding of many of the electronic properties of solids, in particular those of metals. In this model a metal crystal (a cube of side L) is described by a three-dimensional potential box with an infinite barrier at the surfaces (Fig. 6.1); in other words, the electrons are unable to leave the crystal, which is clearly a gross oversimplification given that work function values lie in the region of 5 eV (Sect. 6.6). The time-independent Schrödinger equation for the electron in the one-electron approximation in the infinite square well is

$$-\frac{\hbar^2}{2m}\Delta\psi(\boldsymbol{r}) + V(\boldsymbol{r})\,\psi(\boldsymbol{r}) = E'\,\psi(\boldsymbol{r}) \,, \tag{6.1}$$

where the potential $V(\boldsymbol{r})$ is given by

$$V(x,y,z) = \begin{cases} V_0 = \text{const} & \text{for} \quad 0 \leqq x,y,z \leqq L \\ \infty \ \text{otherwise} \,. \end{cases} \tag{6.2}$$

With $E = E' - V_0$ this yields

$$-\frac{\hbar^2}{2m}\Delta\psi(\boldsymbol{r}) = E\,\psi(\boldsymbol{r}) \,. \tag{6.3}$$

Since the electrons, due to the infinite barrier at the surfaces ($x, y, z=0$ and L), cannot leave the crystal, we have so-called *fixed boundary conditions* (cf. the periodic boundary conditions adopted in Sect. 5.1). These read

$$\begin{aligned} \psi = 0 \quad \text{für} \quad & x = 0 \text{ and } L\,; && \text{for} \quad y, z \text{ between } 0 \text{ and } L\,; \\ & y = 0 \text{ and } L\,; && \text{for} \quad x, z \text{ between } 0 \text{ and } L\,; \\ & z = 0 \text{ and } L\,; && \text{for} \quad x, y \text{ between } 0 \text{ and } L\,. \end{aligned} \tag{6.4}$$

The electron is certain to be found somewhere within the potential box and thus the normalization condition for $\psi(\boldsymbol{r})$ is written

$$\int_{\text{box}} d\boldsymbol{r}\psi^*(\boldsymbol{r})\psi(\boldsymbol{r}) = 1\,. \tag{6.5}$$

The Schrödinger equation (6.3) together with the boundary conditions (6.4) yield the solution

$$\psi(\boldsymbol{r}) = \left(\frac{2}{L}\right)^{3/2} \sin k_x x \sin k_y y \sin k_z z\,. \tag{6.6}$$

The possible energy states are found by substituting (6.6) into (6.3) as

$$E = \frac{\hbar^2 k^2}{2m} = \frac{\hbar^2}{2m}\left(k_x^2 + k_y^2 + k_z^2\right)\,. \tag{6.7}$$

The energies are, as expected, those of a free electron (de Broglie relation), where, however, the condition $\psi=0$ at $x, y, z=L$ (6.4) leads to the following constraints on the wave vector k_x, k_y, k_z:

$$\begin{aligned} k_x &= \frac{\pi}{L} n_x\,, \\ k_y &= \frac{\pi}{L} n_y\,, \\ k_z &= \frac{\pi}{L} n_z \quad \text{with} \quad n_x, n_y, n_z = 1, 2, 3, \,\ldots\,. \end{aligned} \tag{6.8}$$

Solutions with n_x, n_y or $n_z=0$ cannot be normalized over the volume of the box and must therefore be excluded. Negative wave vectors give no new linearly independent solutions in (6.6). The possible states of an electron in a three-dimensional infinite square well (standing waves, Fig. 6.2) can be listed according to their quantum numbers (n_x, n_y, n_z) or (k_x, k_y, k_z). A representation of the allowed values in three-dimensional wave-vector space yields constant energy surfaces, $E=\hbar^2 k^2/2m=\text{const}$, that are spherical.

For the fixed boundary conditions described here, the possible $\boldsymbol{k}$-values are confined to the positive octant of $\boldsymbol{k}$-space. In comparison with the case of periodic boundary conditions (Sect. 5.1), however, the states are twice as dense in every axis direction. Thus every state corresponds to a volume $V_k=(\pi/L)^3$. For macroscopic dimensions L one can again consider the states to be quasi-continuous, so that for many purposes one can replace sums over $\boldsymbol{k}$-space by integrals.

As in the case of phonons, we can calculate a density of states. We simply take the volume of a thin shell of the octant bounded by the energy

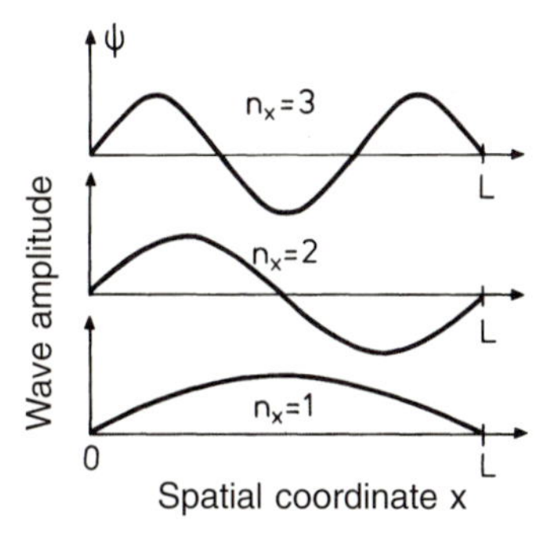

Fig. 6.2. Spatial form of the first three wavefunctions of a free electron in a square well potential of length L in the x-direction. The wavelengths corresponding to the quantum numbers $n_x=1, 2, 3, \ldots$ are $\lambda=2L, L, 2L/3, \ldots$

surfaces $E(\boldsymbol{k})$ and $E(\boldsymbol{k})+dE$ and divide this by the volume V_k associated with a single $\boldsymbol{k}$-point:

$$dZ' = \frac{1}{8} 4\pi k^2 dk/(\pi/L)^3 \ . \tag{6.9}$$

Since $dE = (\hbar^2 k/m)dk$, we have for the number of states per unit volume of the crystal:

$$dZ = \frac{(2m)^{3/2}}{4\pi^2\hbar^3} E^{1/2} dE \ . \tag{6.10}$$

In Schrödinger's wave mechanics as used up to now, no provision has been made for the intrinsic angular momentum, i.e. the spin, of the electron. As can be seen from the construction of the periodic table (Sect. 1.1), one must attribute a spin to the electron, such that it has two possible orientations in an external magnetic field. In the absence of an external field, the energy levels of these two orientations are degenerate. This means that every $\boldsymbol{k}$-space point in Fig. 6.3 describes two possible electron states when one takes the electron spin into account. Thus, for the density of states $D(E)=dZ/dE$ of the free electron gas in the infinite potential well, we finally obtain

$$D(E) = \frac{(2m)^{3/2}}{2\pi^2\hbar^3} E^{1/2} \ . \tag{6.11}$$

$D(E)$ is usually expressed in units of $\mathrm{cm^{-3}\,eV^{-1}}$. The same density of states (Fig. 6.4), and thus the same expressions for the macroscopic properties of the crystal, are obtained if one uses periodic boundary conditions:

$$\psi(x+L,\ y+L,\ z+L) = \psi(x,y,z) \ . \tag{6.12}$$

These conditions yield propagating electron waves as the solutions of (6.3):

$$\psi(\boldsymbol{r}) = \left(\frac{1}{L}\right)^{3/2} e^{i\boldsymbol{k}\cdot\boldsymbol{r}} \ .$$

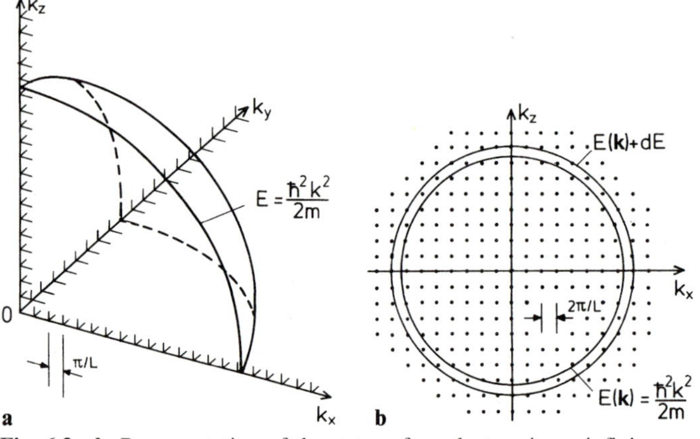

Fig. 6.3a,b. Representation of the states of an electron in an infinite square well by means of a lattice of allowed wave vector values in $\boldsymbol{k}$-space. Because of the two possible spin orientations, each point corresponds to two states. (**a**) For fixed boundary conditions the states all lie in one octant and have a linear separation of π/L. (**b**) For periodic boundary conditions the allowed states span the whole of $\boldsymbol{k}$-space, but with a linear separation that is now $2\pi/L$. The figure shows a cross-section perpendicular to k_y (cf. Fig. 5.1). For both (**a**) and (**b**) spherical surfaces of constant energy $E(\boldsymbol{k})$ are also shown

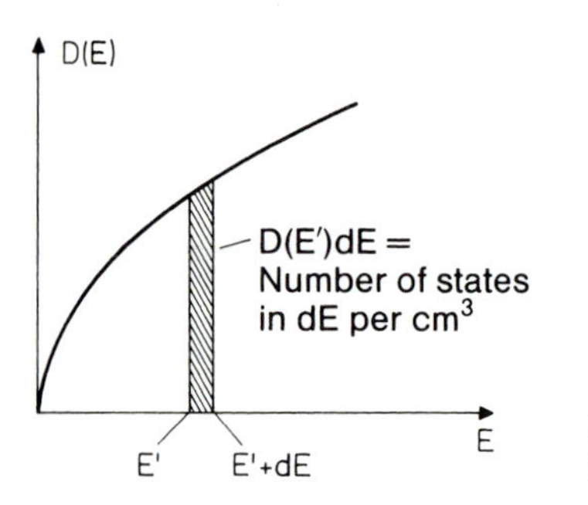

Fig. 6.4. Density of one-particle states $D(E)$ for a free electron gas in three dimensions

In this case, positive and negative $\boldsymbol{k}$-values represent linearly independent solutions and, furthermore, the solution with $\boldsymbol{k}=0$ can be normalized. Thus the possible states now extend throughout $\boldsymbol{k}$-space and possess the k-values

$$\begin{aligned} k_x &= 0, \ \pm 2\pi/L, \ \pm 4\pi/L, \ \dots, \ \pm 2\pi n_x/L, \dots \\ k_y &= 0, \ \pm 2\pi/L, \dots, \ \pm 2\pi n_y/L, \dots \\ k_z &= 0, \ \pm 2\pi/L, \dots, \ \pm 2\pi n_z/L, \dots . \end{aligned} \tag{6.13}$$

The separation of neighboring points is now $2\pi/L$, and the volume associated with each point (=two electron states because of spin) is

$$(2\pi/L)^3 = 8V_k \ .$$

However, instead of an octant in $\boldsymbol{k}$-space we must now consider the full solid angle of 4π when calculating the density of states. This leads to the same expression, (6.11), for $D(E)$ as was obtained for the case of fixed boundary conditions.

If the model is modified to allow for a finite potential barrier at the crystal surface (finite work function), the resulting expressions are also modified: the electron waves now decay exponentially outside the crystal, i.e., there is a nonvanishing probability of finding electrons in the vacuum just outside the crystal surface. It is also possible for certain localized surface states to occur. Here, however, we are interested in the bulk properties of relatively large crystals, and for these one may neglect such surface effects.

6.2 The Fermi Gas at $T=0$ K

The states that an electron can occupy within the one-electron approximation for the square well potential are distributed along the energy axis according to the density of states $D(E)$. The occupation of these states by the available electrons of the crystal must be such that their total energy corresponds to the mean thermal energy of the system. In other words there has to be a temperature-dependent occupation probability $f(T,E)$ that governs the distribution of the available electrons among the possible states. The electron density per unit volume can therefore be expressed as

$$n = \int_0^\infty D(E) f(T,E) dE \; . \tag{6.14}$$

For a gas of classical particles this distribution function $f(T,E)$ would be the familiar Boltzmann exponential, which would require that at temperatures $T \to 0$ K all electrons should occupy the lowest available states.

However, for all fermions, i.e., particles with half-integral spin such as electrons, the Pauli principle applies. Within the one-particle approximation for noninteracting particles this can be formulated as follows: in an atomic system no two fermions may possess identical sets of quantum numbers. This exclusion principle therefore demands that in the lowest energy state, i.e., for $T \to 0$ K, the available electrons occupy successive energy levels starting with the lowest and ending at some upper limit. This limiting energy, which, at $T \to 0$ K, separates occupied from unoccupied states, is known as the Fermi energy E_F^0 for zero temperature. In the free-electron-gas model with a square-well potential, this energy corresponds to the spherical surface $E_F^0(\boldsymbol{k}_F) = \hbar^2 k_F^2 / 2m$ in $\boldsymbol{k}$-space with the Fermi wave vector k_F as its radius.

The occupation probability for electrons in the potential well at $T=0$ K is a step function with $f=1$ for $E<E_F^0$ and $f=0$ for $E>E_F^0$ (Fig. 6.5). The spherical form of the Fermi surface $E_F^0(\boldsymbol{k})$ at $T \to 0$ K leads immediately to a simple relationship between the electronic density n and the Fermi radius k_F or Fermi energy E_F^0:

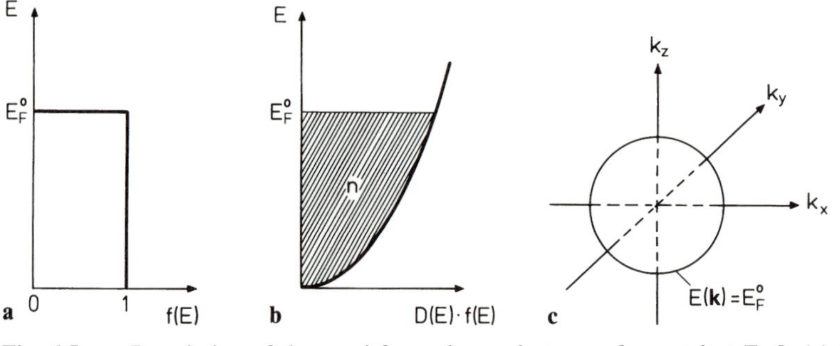

Fig. 6.5 a–c. Description of the quasi-free valence electrons of a metal at $T=0$. (**a**) $f(E)$ is a step function. (**b**) The concentration n of valence electrons is given by the area under the density of states curve up to the Fermi energy E_F^0. (**c**) In $\boldsymbol{k}$-space the Fermi sphere $E(\boldsymbol{k})=E_F^0$ separates occupied from unoccupied states

$$nL^3 = \frac{L^3 k_F^3}{3\pi^2} , \tag{6.15}$$

$$E_F^0 = \frac{\hbar^2}{2m}(3\pi^2 n)^{2/3} . \tag{6.16}$$

The magnitude of the Fermi energy can thus be estimated by using the number of valence electrons per atom to determine the electron concentration n. A few values of E_F^0 are listed in Table 6.1. From these we see that at normal temperatures the Fermi energy is always very large compared to kT. To make this more obvious one can define a Fermi temperature $T_F = E_F^0/k$; this temperature lies about two orders of magnitude above the melting point of the metals.

Table 6.1. Fermi energy E_F^0, radius of the Fermi sphere in k-space k_F, Fermi velocity $v_F = \hbar k_F/m$, and Fermi temperature $T_F = E_F^0/k$ for a few typical metals. n is the concentration of conduction electrons deduced from the structural data of the elements [6.2]. It should be noted that the electron configuration of Cu, Ag and Au is $3d^{10}4s^1$ and so every atom contributes one free electron (Fig. 7.12). The characteristic radius r_s is also often used in this context. It is defined via the volume of a hypothetical sphere containing one electron, $4\pi r_s^3/3 = a_0^{-3} n^{-1}$, where a_0 is the Bohr radius so that r_s is dimensionless. Values for r_s lie between 2 and 6 for typical metals

Metal	n (10^{22} cm^{-3})	r_s (–)	k_F (10^8 cm^{-1})	v_F (10^8 cm/s)	E_F^0 (eV)	T_F (10^4 K)
Li	4.62	3.27	1.11	1.29	4.70	5.45
Na	2.53	3.99	0.91	1.05	3.14	3.64
Cs	0.86	5.71	0.63	0.74	1.53	1.78
Al	18.07	2.07	1.75	2.03	11.65	13.52
Cu	8.47	2.67	1.36	1.57	7.03	8.16
Ag	5.86	3.02	1.20	1.39	5.50	6.38
Au	5.9	3.01	1.20	1.39	5.52	6.41

An interesting consequence of the Pauli exclusion principle is that the Fermi gas, in contrast to a classical gas, has a nonvanishing internal energy at $T=0\,\mathrm{K}$. It is well known that the internal energy density U of a system is the average value over all states; thus at $T=0\,\mathrm{K}$ we obtain

$$U = \int_0^{E_\mathrm{F}^0} D(E) E \, dE = \frac{3}{5} n E_\mathrm{F}^0 \, . \tag{6.17}$$

As has already been established, this value lies many orders of magnitude above the internal energy of a classical gas at $T=300\,\mathrm{K}$. In order to treat the conduction electrons in a metal it is therefore sufficient for many purposes to use a zero-temperature description (Fig. 6.5).

6.3 Fermi Statistics

We now move on to consider the Fermi gas at finite temperatures. We shall need to derive the distribution function or occupation probability $f(E,T)$ for non-zero temperatures. This is a thermodynamical problem, since we are inquiring about the distribution that arises when various quantum mechanical states are in equilibrium with one another. To derive the distribution $f(E,T)$ we must therefore apply some basic concepts of thermodynamics.

We consider an atomic system with single-particle energy levels E_j. We assume that the energy levels E_j lie very close to one another as in a solid. We can then consider new energy levels E_i each of which consists of many E_j. The degeneracy of these new levels is denoted by g_i and their occupation number by n_i where both g_i and n_i are large numbers. On account of the Pauli principle we must have $n_i \leq g_i$. From thermodynamics we know the conditions that the system must fulfill in order for all energy levels to be in equilibrium: the free energy F of the total system must be stationary with respect to a variation in the relative occupation numbers of the levels. In other words we must have

$$\delta F = \sum_i \frac{\partial F}{\partial n_i} \delta n_i = 0 \tag{6.18}$$

with the subsidiary condition of conservation of particle number

$$\sum_i \delta n_i = 0 \, . \tag{6.19}$$

For the specific case of exchange of electrons between two arbitrary levels k and l the equilibrium conditions read

$$\frac{\partial F}{\partial n_k}\delta n_k + \frac{\partial F}{\partial n_l}\delta n_l = 0 \ , \tag{6.20}$$

$$\delta n_k + \delta n_l = 0 \ . \tag{6.21}$$

From this it follows immediately that the derivatives of the free energy with respect to the occupation numbers must be equal

$$\frac{\partial F}{\partial n_k} = \frac{\partial F}{\partial n_l} \ . \tag{6.22}$$

Since the two levels were selected at random, at equilibrium all $\partial F/\partial n_i$ must be equal and we denote this quantity by a new constant, μ, defined as the "chemical potential" of the electrons.

We will now calculate the free energy of the system of electrons. From thermodynamics we have the relation

$$F = U - TS \tag{6.23}$$

with the internal energy U

$$U = \sum_i n_i E_i \tag{6.24}$$

and the entropy S. The entropy is given by

$$S = k \ln P \ , \tag{6.25}$$

where P represents the number of possible ways of distributing the electrons among the states. The number of ways of accommodating one electron in the level E_i is g_i. For a second electron, also in level E_i, the number of possibilities is g_i–1, and so on. There would therefore be

$$g_i(g_i-1)(g_i-2)\ldots(g_i-n_i+1) = \frac{g_i!}{(g_i-n_i)!} \tag{6.26}$$

possible ways of accommodating n_i electrons at definite positions within the energy level E_i. However, arrangements which differ only in the exchange of electrons within the energy level are not distinguishable. Since there are $n_i!$ such possibilities, the total number of distinguishable ways of accommodating n_i electrons in the level E_i is given by

$$\frac{g_i!}{n_i!(g_i-n_i)!} \ . \tag{6.27}$$

The number of ways P of realizing the total system is then the product over all possibilities for occupying each level:

$$P = \prod_i \frac{g_i!}{n_i!(g_i-n_i)!} \ . \tag{6.28}$$

Thus the entropy can be expressed as

$$S = k \sum_i [\ln g_i! - \ln n_i! - \ln(g_i - n_i)!] \ , \tag{6.29}$$

where the factorials can be replaced by using Stirling's approximate formula

$$\ln n! \approx n \ln n - n \quad \text{(for large } n) \ . \tag{6.30}$$

It now becomes a straightforward matter to calculate the chemical potential, i.e., the derivative of the free energy F with respect to the occupation number of an arbitrary level i:

$$\mu = \frac{\partial F}{\partial n_i} = E_i + kT \ln \frac{n_i}{g_i - n_i} \ . \tag{6.31}$$

We can rearrange this expression to obtain the occupation number n_i:

$$n_i = g_i (e^{(E_i - \mu)/kT} + 1)^{-1} \ . \tag{6.32}$$

The probability that a quantum mechanical state (degenerate states are also considered distinct here) is occupied is given by the distribution function $f(E, T)$, which, from (6.32), can be seen to be (Fig. 6.6)

$$f(E, T) = \frac{1}{e^{(E-\mu)/kT} + 1} \ . \tag{6.33}$$

This function is also known as the Fermi distribution. It is the distribution function for particles, only one of which may occupy each quantum state. For electrons and all particles with half-integral spin, i.e., fermions, this distribution function guarantees that the Pauli principle is obeyed. However,

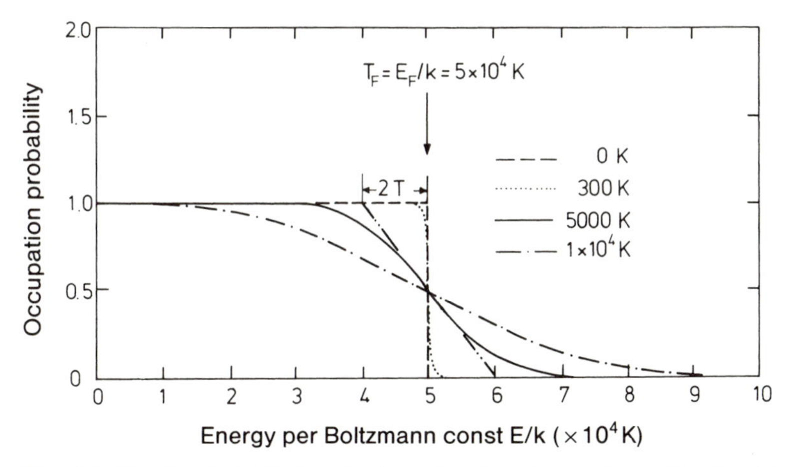

Fig. 6.6. The Fermi distribution function at various temperatures. The Fermi temperature $T_F = E_F^0/k$ has been taken as 5×10^4 K. The tangent at the point of inflection (–··–) intersects the energy axis at $2kT$ above E_F^0 at all temperatures

it would be wrong to claim that the Fermi distribution is only valid for particles with spin $\frac{1}{2}$; it is equally valid for atoms or molecules that are distributed among predetermined fixed positions, whenever only one atom or molecule can occupy such a position. Corresponding situations arise in the thermodynamics of defects (Sect. 2.7), the solubility of gases in solids, and adsorption processes.

The significance of the chemical potential μ in the Fermi distribution is most readily seen in the limiting case of $T=0$ K. At zero temperature the Fermi function becomes identical with the step function introduced previously. It has a value of 1 for $E<\mu$ and of 0 for $E>\mu$. Thus at $T=0$ K the chemical potential of the electrons is equal to the Fermi energy:

$$\mu(T = 0\,\mathrm{K}) = E_\mathrm{F}^0 . \tag{6.34}$$

Because of this equality, one often speaks of the "Fermi level" in place of the chemical potential and uses the symbol E_F. This Fermi level, however, is then a temperature-dependent quantity!

At higher temperatures the sharp edge of the Fermi distribution becomes more rounded; states below E_F have a finite probability of being unoccupied, while those slightly above E_F may be occupied (Fig. 6.6). The size of the region, over which the Fermi function deviates significantly from the step function, is of the order of $2kT$ to each side, as shown by the tangent to $f(T,E)$ at E_F drawn in Fig. 6.6. One sees that as the temperature is raised, only a small fraction of the electrons is able to gain energy. This has important consequences, e.g., for the specific heat capacity of the electron gas (Sect. 6.4).

If one is interested in the occupation probability for energies or temperatures in the range $|E-E_\mathrm{F}| \gg 2kT$, it is possible to use approximations to the Fermi function (6.33).

The condition $E-E_\mathrm{F} \gg 2kT$ is often fulfilled, for example for the conduction electrons in semiconductors (Sect. 12.2). In this regime, with energies E far above the Fermi edge, the Fermi function $f(E,T)$ can be approximated by the classical Boltzmann distribution, $f(E,T) \sim \exp(E_\mathrm{F} - E)/kT$; see (12.5).

6.4 The Specific Heat Capacity of Electrons in Metals

The application of the square-well potential model to the conduction electrons allows a very simple description of the specific heat capacity c_v of these metal electrons. In fact this is an age-old problem, and one which seemed insoluble prior to the development of quantum mechanics. For a typical conduction electron density of $n=10^{22}\,\mathrm{cm}^{-3}$ one would have expected, in addition to the lattice specific heat, an electronic contribution according to the equidistribution law of $c=3nk/2$, at least at elevated temperatures. Experiments on metals, however, showed no deviation from the

Dulong-Petit value. The reason is simple: electrons, in contrast to a classical gas, can only gain energy if they can move into free states in their energetic neighborhood. The number of such electrons, expressed as a fraction of the total density n, is only of the order of 1/100, as demonstrated by the following simple estimate:

The “smeared out” region of the Fermi function has a width on the order of $4kT$, i.e., according to Fig. 6.7, the Pauli principle dictates that only a fraction of about $4kT/E_F$ of all “free” electrons (density n) can absorb thermal energy. The energy per electron is around kT and so the total energy of these thermally excited electrons is on the order of

$$U \sim 4(kT)^2 n/E_F \ . \tag{6.35}$$

With $T_F = E_F/k$ as the Fermi temperature, one obtains the following order of magnitude estimate for the specific heat of the electrons

$$c_v = \partial U/\partial T \sim 8knT/T_F \ . \tag{6.36}$$

As seen in Table 6.1, the Fermi temperatures are typically on the order of 10^5 K and this, on account of the factor T/T_F in (6.36), explains the vanishingly small contribution of the conduction electrons to the specific heat capacity.

The exact calculation to determine the specific heat capacity of the gas of free electrons is as follows:

Upon heating the Fermi electron gas from 0 K to a temperature T, the internal energy per unit volume is increased by an amount U given by

$$U(T) = \int_0^\infty dE \cdot ED(E) f(E,T) - \int_0^{E_F^0} dE \cdot ED(E) \ . \tag{6.37}$$

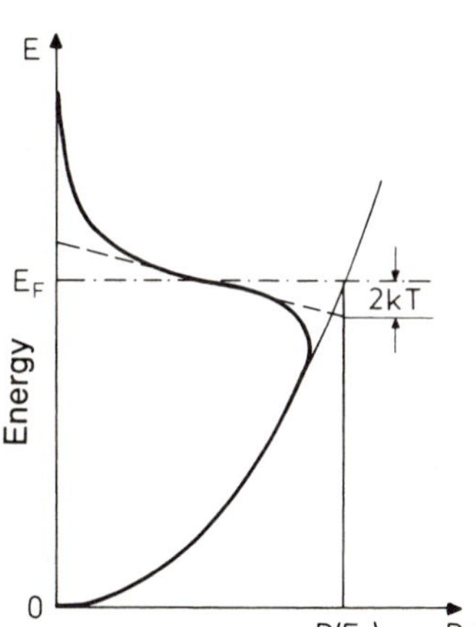

Fig. 6.7. Explanation of the specific heat capacity of quasi-free metal electrons. The effect of raising the temperature from 0 K to T is to allow electrons from $\leq 2kT$ below the Fermi energy to be promoted to $\leq 2kT$ above E_F. The tangent (– – –) intersects the energy axis at $E_F + 2kT$

We also have the relation

$$E_{\mathrm{F}} \cdot n = E_{\mathrm{F}} \int_0^\infty dE\, D(E) f(E,T) \,, \tag{6.38}$$

where n is the total concentration of free electrons. Differentiating (6.37) and (6.38) yields

$$c_{\mathrm{v}} = \partial U/\partial T = \int_0^\infty E\, D(E)(\partial f/\partial T) dE \,, \tag{6.39}$$

$$0 = E_{\mathrm{F}}(\partial n/\partial T) = \int_0^\infty E_{\mathrm{F}} D(E)(\partial f/\partial T) dE \,. \tag{6.40}$$

Subtracting (6.40) from (6.39), the specific heat capacity c_{v} of the electrons is obtained as

$$c_{\mathrm{v}} = \partial U/\partial T = \int_0^\infty dE (E - E_{\mathrm{F}}) D(E)(\partial f/\partial T) \,. \tag{6.41}$$

The derivative $\partial f/\partial T$ (Fig. 6.7) has significant values only in the "smeared out" region of $\pm 2\not{k} T$ about E_{F}. The density of states $D(E)$ does not vary a great deal in this region and may be approximated by $D(E_{\mathrm{F}})$, i.e.,

$$c_{\mathrm{v}} \approx D(E_{\mathrm{F}}) \int_0^\infty dE (E - E_{\mathrm{F}})(\partial f/\partial T) \,, \tag{6.42}$$

where the temperature derivative of the Fermi function (6.33) is given by

$$\frac{\partial f}{\partial T} = \frac{E - E_{\mathrm{F}}}{\not{k} T^2} \frac{\exp[(E - E_{\mathrm{F}})/\not{k} T]}{\{\exp[(E - E_{\mathrm{F}})/\not{k} T] + 1\}^2} \,. \tag{6.43}$$

With the abbreviation $x = (E - E_{\mathrm{F}})/\not{k} T$ we then have

$$c_{\mathrm{v}} \approx \not{k}^2\, T D(E_{\mathrm{F}}) \int_{-E_{\mathrm{F}}/\not{k} T}^\infty dx\, x^2 \exp x (\exp x + 1)^{-2} \,. \tag{6.44}$$

Since the factor of $\exp x$ in the integrand is negligible for $x \leq -E_{\mathrm{F}}/\not{k} T$, the lower integration limit can be extended to minus infinity. The resulting integral

$$\int_{-\infty}^\infty dx\, x^2 \exp x (\exp x + 1)^{-2} = \pi^2/3 \tag{6.45}$$

is a standard integral whose value can be found in tables.

One thus has the following general result for the specific heat capacity of the “free” electrons in metals:

$$c_V \approx \frac{\pi^2}{3} D(E_F) k^2 T \ . \tag{6.46}$$

In deriving (6.46) no use has been made of the explicit form of the density of states $D(E)$. Equation (6.46) is therefore also valid for cases where the density of states deviates from that of the free electron gas, as indeed is expected in the majority of cases. For metals, measurements of the electronic heat capacity are therefore used as a method of determining the density of states $D(E_F)$ at the Fermi level.

In the free electron gas model $D(E_F)$ can be very simply expressed in terms of the electron concentration. In the case of metals, the validity of the relation $T \ll T_F$ implies

$$n = \int_0^{E_F} D(E) dE \ , \tag{6.47}$$

and the density of states for this case can be written

$$D(E) = D(E_F)(E/E_F)^{1/2} \ . \tag{6.48}$$

It follows that

$$n = \frac{2}{3} D(E_F) E_F \ , \tag{6.49}$$

and from (6.46) we have

$$c_V \approx \frac{\pi^2}{2} n k \frac{kT}{E_F} = \frac{\pi^2}{2} n k \frac{T}{T_F} \ . \tag{6.50}$$

Thus the only difference between this exact calculation and the previous rough estimate (6.36) is the appearance of the factor $\pi^2/2$ in place of the factor 8.

The predicted linear dependence of the electronic specific heat capacity on temperature is well confirmed experimentally. For low temperatures, where the phonon contribution to c_V displays the Debye T^3 dependence, one expects to observe

$$c_V = \gamma T + \beta T^3 \ , \quad \gamma, \beta = \text{const.} \tag{6.51}$$

The experimental results in Fig. 6.8 show the linear dependence expected from (6.51) for a plot of c_V/T against T^2.

For the transition metals in particular, the experimentally determined values of γ bear little resemblance to those calculated from the electron gas model, as is seen in Table 6.2.

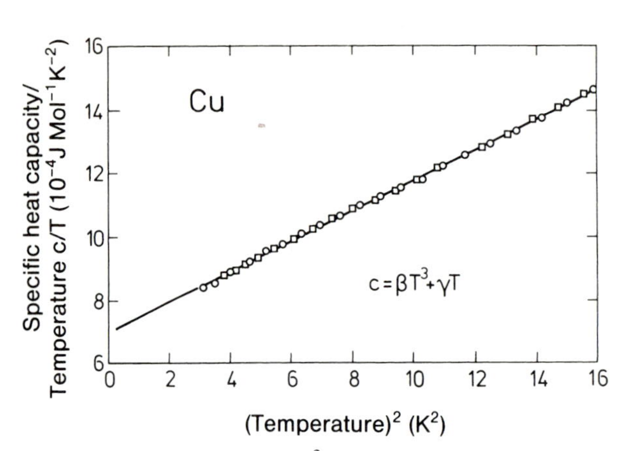

Fig. 6.8. Plot of c_V/T against T^2 for copper. The experimental points (○ and □) stem from two separate measurements [6.3]

Table 6.2. Comparison of experimentally determined values of the coefficient γ of electronic specific heat with values calculated using the free-electron-gas model. At low temperatures one has $c_V = \gamma T + \beta T^3$ for the combined electronic ($\propto T$) and ($\propto T^3$) contributions to the specific heat

Metal	γ_{exp} (10^{-3} J/Mol K^2)	$\gamma_{exp}/\gamma_{theo}$
Li	1.7	2.3
Na	1.7	1.5
K	2.0	1.1
Cu	0.69	1.37
Ag	0.66	1.02
Al	1.35	1.6
Fe	4.98	10.0
Co	4.98	10.3
Ni	7.02	15.3

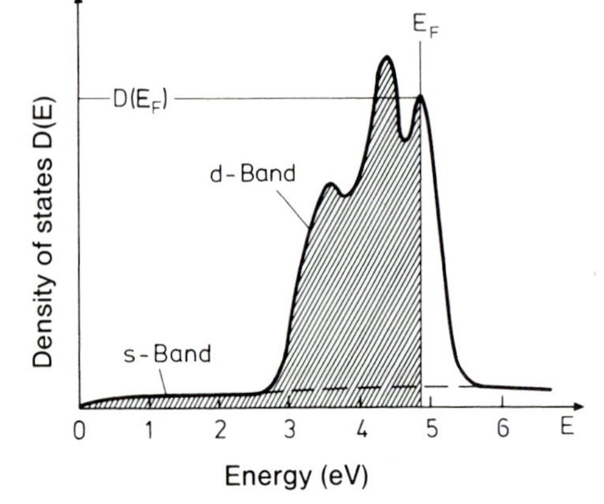

Fig. 6.9. Qualitative behavior of the density of states $D(E)$ for the conduction band of a transition metal. The strong contribution of the d-electrons in the vicinity of the Fermi level lies on top of that of the s-band (*partially dashed*)

The large deviations for Fe, Co, and Ni are attributed to the partially filled d-shells of these transition metals, whose d-bands thus lie at the Fermi energy. Because of the strong localization of the d-electrons at the atoms, the overlap of their wavefunctions is small. This means that the corresponding energy band is relatively narrow and therefore yields a large contribution to the density of states as shown in Fig. 6.9.

6.5 Electrostatic Screening in a Fermi Gas – The Mott Transition

If an electric charge is introduced into a metal, e.g. by the inclusion of a charged defect, then in the vicinity of this charge there is a perturbation in the otherwise homogeneous electron concentration which compensates or screens the electric field of the charge.

This problem can be treated approximately using the model of a quasi-free electron gas in a potential well:

A local perturbation potential δU (assumed to obey $|e\delta U| \ll E_F$) produces a local raising of the density of states parabola $D(E)$ by an amount $e\delta U$ (Fig. 6.10). If one imagines the perturbation potential to be switched on, it is clear that some electrons must immediately leave this region in order for the Fermi level to remain constant throughout the crystal. This homogeneity is necessary since the Fermi level is a thermodynamic function of state (equal to the electrochemical potential). For not too large δU the change in the electron concentration is given in terms of the density of states at the Fermi level (in analogy to the specific heat capacity) by

$$\delta n(\boldsymbol{r}) = D(E_F)|e|\delta U(\boldsymbol{r}) \ . \tag{6.52}$$

Except in the immediate vicinity of the perturbation charge, one can assume that $\delta U(\boldsymbol{r})$ is caused essentially by the induced space charge. Thus $\delta n(\boldsymbol{r})$ is related to δU via the Poisson equation:

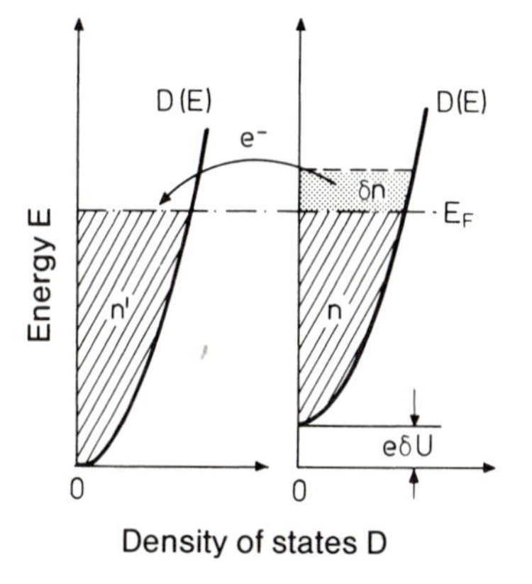

Fig. 6.10. Effect of a local perturbation potential δU on the Fermi gas of "free" electrons. Immediately after the perturbation is switched on, δn electrons must move away such that the Fermi level E_F is homogeneous throughout the crystal in thermal equilibrium

$$\nabla^2(\delta U) = \frac{-\delta\varrho}{\varepsilon_0} = \frac{e}{\varepsilon_0}\delta n = \frac{e^2}{\varepsilon_0}D(E_F)\delta U \;, \tag{6.53}$$

where ε_0 is the dielectric constant.

With $\lambda^2 = e^2 D(E_F)/\varepsilon_0$, this differential equation for the screening potential δU has a nontrivial solution in spherical coordinates

$$\left(\nabla^2 = \frac{1}{r^2}\frac{\partial}{\partial r}r^2\frac{\partial}{\partial r}\right):$$

$$\delta U(r) = \alpha\, e^{-\lambda r}/r \;. \tag{6.54}$$

Spherical coordinates are the obvious choice when dealing with a point-like defect. For a point charge e one would have $\alpha = e/(4\pi\varepsilon_0)$ since for $\lambda \to 0$ the screening effect would disappear and one must obtain the Coulomb potential of this point charge (Fig. 6.11). The quantity $r_{TF} = 1/\lambda$ is known as the Thomas-Fermi screening length:

$$r_{TF} = [e^2 D(E_F)/\varepsilon_0]^{-1/2} \;. \tag{6.55}$$

For the special case of the free electron gas model, (6.49) and (6.16) give

$$D(E_F) = \frac{3}{2}n/E_F \quad \text{and} \quad E_F = \frac{\hbar^2}{2m}(3\,\pi^2 n)^{2/3} \;,$$

i.e.

$$D(E_F) = \frac{1}{2\pi^2}\frac{2m}{\hbar^2}(3\,\pi^2 n)^{1/3} \;. \tag{6.56}$$

For the Thomas-Fermi screening length in the square-well model it follows that

$$\frac{1}{r_{TF}^2} = \lambda^2 = \frac{me^2}{\pi^2\hbar^2\varepsilon_0}(3\,\pi^2 n)^{1/3} = \frac{4}{\pi}\,(3\,\pi^2)^{1/3}\,\frac{n^{1/3}}{a_0} \;, \tag{6.57}$$

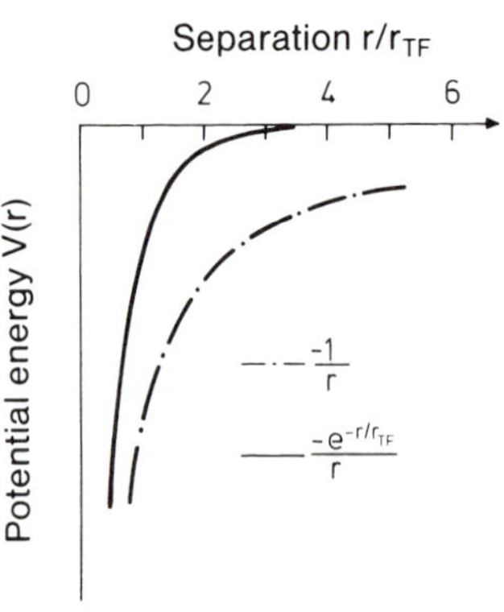

Fig. 6.11. Screened (——) and unscreened (–·–) Coulomb potential of a positive unit charge in a Fermi gas of free electrons. The distance r is given as a multiple of the Thomas-Fermi screening length r_{TF}

$$\frac{1}{r_{\mathrm{TF}}} \simeq 2\frac{n^{1/6}}{a_0^{1/2}} \quad \text{or} \quad r_{\mathrm{TF}} \simeq 0.5\left(\frac{n}{a_0^3}\right)^{-1/6}, \tag{6.58}$$

where $a_0 = 4\pi\hbar^2\varepsilon_0/(m e^2)$ is the Bohr radius. Copper, for example, with an electron concentration of $n=8.5\times10^{22}\,\mathrm{cm}^{-3}$, has a screening length of $r_{\mathrm{TF}}=0.55$ Å.

The screening process described here is responsible for the fact that the highest energy valence electrons of a metal are not localized. These electrons cannot be held in the field of the ion core potential. As the electron density decreases, the screening length r_{TF} becomes even larger.

Using arguments related to screening, it is possible to understand the sharp transition between metallic and insulating or semiconducting properties known as the *Mott transition* [6.4].

Above a certain critical electron density n_c the screening length r_{TF} becomes so small that electrons can no longer remain in a bound state; this produces metallic behavior. Below this critical electron concentration the potential well of the screened field extends far enough for a bound state to be possible. The electron is then localized in a covalent or ionic bond. Such localized states correspond, by definition, to insulating properties where the highest occupied states form localized bonds. To make a simple estimate of when a bound state becomes possible in a screened potential, we assume that the screening length must be significantly larger than the Bohr radius a_0, i.e., that the potential well of a positive center extends sufficiently far to bind an electron:

$$r_{\mathrm{TF}}^2 \simeq \frac{1}{4}\frac{a_0}{n^{1/3}} \gg a_0^2\,, \tag{6.59}$$

i.e.

$$n^{-1/3} \gg 4a_0\,. \tag{6.60}$$

This estimate, originally proposed by Mott, predicts that a solid will lose its metallic character when the average electron separation $n^{-1/3}$ becomes significantly larger than four Bohr radii. One then expects an abrupt transition to insulating properties.

It is today believed that sharp jumps observed in the conductivity of transition metal oxides, glasses, and amorphous semiconductors, may be explicable in terms of the above mechanism.

6.6 Thermionic Emission of Electrons from Metals

If a metal is made sufficiently hot it is found to emit electrons. This phenomenon is exploited in all electron tubes.

In the circuit shown in Fig. 6.12a one observes a saturation current j_s in the current-voltage characteristic (Fig. 6.12b) that is dependent on the cathode temperature T.

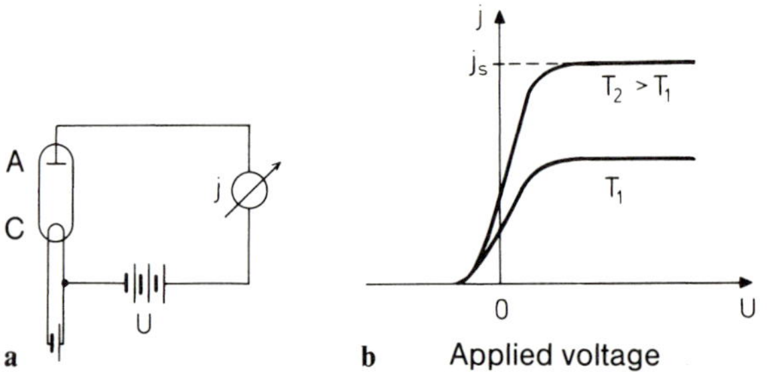

Fig. 6.12. (**a**) Schematic drawing of a diode circuit for observing thermionic emission of electrons from the heated cathode C (A = anode). (**b**) Qualitative behavior of the current-voltage curve at two different temperatures T_1 and $T_2 > T_1$. As a consequence of their thermal energy, electrons can even overcome a countervoltage (A negative with respect to C)

The existence of this effect demonstrates that the assumption of an infinite square well to describe metal electrons is too simple. The potential well clearly has only a finite barrier height. The energy difference $E_{vac} - E_F = \Phi$ is known as the work function. This is the energy barrier that an electron must overcome in order to reach the energy level of the vacuum (far away from the metal) from the "Fermi sea". If the electron also possesses a sufficient momentum perpendicular to the surface, it can leave the metal and will contribute to the saturation current j_S.

We will calculate the temperature-dependent saturation current for the free-electron-gas model. If the drift velocity $\boldsymbol{v}$ of the charge carriers is homogeneous, the current density is given by $\boldsymbol{j} = ne\boldsymbol{v}$, where n is the concentration of charge carriers. (Strictly speaking a minus sign is also required, but this can be omitted for our present purposes.) We can generalize this expression to the case where the electron velocity is a function of the wave vector $\boldsymbol{k}$:

$$j_x = \frac{e}{V} \sum_{\boldsymbol{k}} v_x(\boldsymbol{k}) = \frac{e}{(2\pi)^3} \int\limits_{\substack{E > E_F + \Phi \\ v_x(\boldsymbol{k}) > 0}} v_x(\boldsymbol{k}) \, d\boldsymbol{k} \; . \tag{6.61}$$

This form includes the fact that the density of states in $\boldsymbol{k}$-space is $V/(2\pi)^3$. Both the summation and the integral extend only over occupied states, as dictated by Fermi statistics. This condition can be included explicitly by multiplying by the occupation probability given in (6.33). Thus

$$j_x = \frac{2e\hbar}{(2\pi)^3 m} \int\limits_{-\infty}^{\infty} dk_y dk_z \int\limits_{k_{x\,\mathrm{min}}}^{\infty} dk_x k_x f(E(\boldsymbol{k}), T) \; . \tag{6.62}$$

Here we have written $mv_x = \hbar k_x$, and have taken into account that for a free-electron gas all states are doubly degenerate. Since the work function Φ is large compared to kT, we can approximate the Fermi statistics by Boltzmann statistics:

$$j_x = \frac{e\hbar}{4\pi^3 m} \int_{-\infty}^{\infty} dk_y e^{-\hbar^2 k_y^2/2m k T} \int_{-\infty}^{\infty} dk_z e^{-\hbar^2 k_z^2/2m k T}$$
$$\times \int_{k_{x\,\min}}^{\infty} dk_x k_x e^{-\hbar^2 k_x^2/2m k T} e^{E_F/k T} . \quad (6.63)$$

The integrals are thus factorized and can be readily evaluated. In the third integral we must also take into account that the kinetic energy in the $+x$-direction must be greater than $E_F + \Phi$:

$$\int_{k_{x\,\min}}^{\infty} dk_x k_x e^{-\hbar^2 k_x^2/2m k T} e^{E_F/k T} = \int_{(E_F+\Phi)2m/\hbar^2}^{\infty} \frac{1}{2} dk_x^2 e^{-\hbar^2 k_x^2/2m k T} e^{E_F/k T}$$
$$= \frac{m k T}{\hbar^2} e^{-\Phi/k T} . \quad (6.64)$$

One thus arrives at the so-called Richardson-Dushman formula for the saturation current density:

$$j_s = \frac{4\pi m e}{h^3} (k T)^2 e^{-\Phi/k T} . \quad (6.65)$$

The universal factor $4\pi m e k^2/h^3$ has the value 120 A/(K^2 cm^2). In this derivation we have made the simplifying assumption that electrons arriving at the surface with an energy $\hbar^2 k_x^2/2m \geqslant E_F + \Phi$ have a 100% probability of escaping from the solid. Even in the free-electron-gas model this assumption is not correct. The well-known quantum mechanical treatment of the reflection and transmission of electrons at a potential step tells us that electrons whose energy exactly equals the energy of the potential step have zero probability of transmission. The effect of the potential step can be included by introducing a factor $\sqrt{\pi k T/(E_F + \Phi)}$, which significantly reduces the saturation current density. The Richardson-Dushman formula can also be applied to the ballistic transport of charge carriers in semiconductor multilayer structures (cf. Sect. 12.7).

In the particular case of thermionic emission, it is also necessary to consider the dependence of the work function on the external field $\mathscr{E}$.

The appropriate correction is the replacement of the constant Φ in the exponent by the field-dependent quantity

$$\Phi' = \Phi - \sqrt{\frac{e^3 \mathscr{E}}{4\pi\varepsilon_0}} = \Phi - \Delta\Phi . \quad (6.66)$$

The correction term $\Delta\Phi$ is derived simply by assuming that essential contributions to the work function stem from the Coulomb force due to the im-

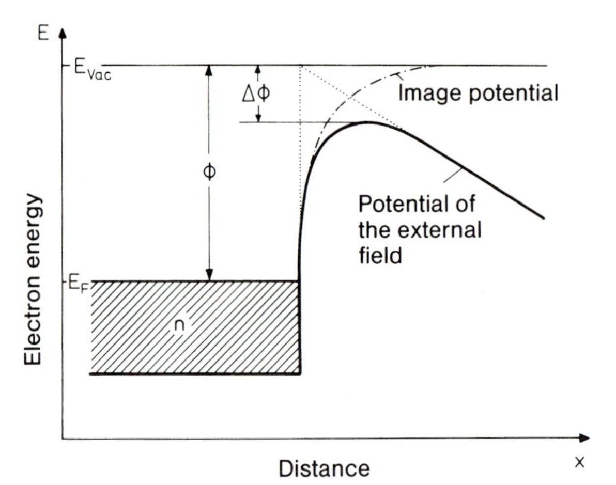

Fig. 6.13. Schematic representation of the thermionic emission of free electrons (density n) from a metal. An electron in the potential well must overcome the work function barrier $\Phi = E_{vac} - E_F$ in order to reach the energy level E_{vac} of the vacuum and escape from the crystal. An important part of the work function is assumed to be the Coulomb potential between the escaping electron and its positive image charge in the metal (image potential –·–). If an external electric field is applied, then Φ is reduced by an amount $\Delta\Phi$. Reductions of the work function of ~ 1 eV as shown here can only be achieved with extremely strong external fields of 10^7–10^8 V/cm

age charge of an electron lying outside the surface and the effect of the external field in reducing the potential barrier. This is illustrated by the superposition of the external applied potential $\mathscr{E}x$ and the Coulomb image potential as shown in Fig. 6.13.

The Richardson-Dushman formula can be used in this extended form to determine the work functions of metals. In order to do this, one must first determine the saturation emission current j_{s0} for $\mathscr{E} = 0$ by extrapolating the measured current j_s at finite fields $\mathscr{E}$. A semi-log plot of j_{s0}/T^2 against $1/T$ yields the work function.

Table 6.3. Work functions of elements in eV (polycrystalline samples, after Michaelson [6.5])

Li	Be											B	C		
2.9	4.98											4.45	5.0		
Na	Mg											Al	Si	P	S
2.75	3.66											4.28	4.85	–	–
K	Ca	Sc	Ti	V	Cr	Mn	Fe	Co	Ni	Cu	Zn	Ga	Ge	As	Se
2.30	2.87	3.5	4.33	4.3	4.5	4.1	4.5	5.0	5.15	4.65	4.33	4.2	5.0	3.75	5.9
Rb	Sr	Y	Zr	Nb	Mo	Tc	Ru	Rh	Pd	Ag	Cd	In	Sn	Sb	Te
2.16	2.59	3.1	4.05	4.3	4.6	–	4.71	4.98	5.12	4.26	4.22	4.12	4.42	4.55	4.95
Cs	Ba	La	Hf	Ta	W	Re	Os	Ir	Pt	Au	Hg	Tl	Pb	Bi	Po
2.14	2.7	3.5	3.9	4.25	4.55	4.96	4.83	5.27	5.65	5.1	4.49	3.84	4.25	4.22	–

Work functions of the elements (polycrystalline) are listed in Table 6.3.

As a concluding remark we note that the work function depends strongly on the crystallographic orientation of the surface and on the degree of contamination.

Problems

6.1 a) Calculate the density of states for a two-dimensional gas of free electrons in a so-called quantum well. The boundary conditions for the electronic wavefunction are: $\psi(x,y,z)=0$ for $|x|>a$, where a is of atomic dimensions.

b) Calculate the density of states for a one-dimensional gas of free electrons in a so-called quantum wire with the boundary conditions: $\psi(x,y,z)=0$ for $|x|>a$ and $|y|>b$, where a and b are of atomic dimensions.

c) Can such electron gases be realized physically?

6.2 Calculate to lowest order the temperature dependence of the chemical potential for a gas of free electrons whose electron concentration remains constant.

Hint: Write down an expression for the electron concentration at finite temperature. You will encounter an integral

$$F(x) = \int_0^\infty \frac{\sqrt{y}\,dy}{1+e^{y-x}} \,.$$

For $x \geq 1.5$

$$F(x) \cong 2/3x^{3/2}\left(1+\frac{\pi^2}{8x^2}\right)$$

is a good approximation.

6.3 The bulk modulus κ is given by the second derivative of the total energy E_{tot} with respect to the volume: $\kappa = V^{-1}\partial^2 E_{tot}/\partial V^2$. Estimate the bulk moduli of alkali metals by assuming that the total energy is equivalent to the kinetic energy of the Fermi gas. What has been neglected in this estimate?

6.4 At what temperature T_0 does the specific heat of the free electrons become larger than the specific heat of the lattice. Express this temperature in terms of the Debye temperature Θ and the electron concentration. Calculate T_0 for copper.

6.5 The benzene molecule C_6H_6 exhibits characteristic optical absorption bands near to the wavelengths 180, 200 and 256 nm. These are explained in terms of $\pi \to \pi^*$ transitions of the carbon ring skeleton. The occupied π

orbitals in the LCAO picture have, respectively: one and the same sign around the C-ring (1st π orbital); one node line through two opposite C–C bonds (2nd π orbital); and one node line through two opposite C atoms (3rd π orbital). From the equal C–C bond length around the ring (~ 1.39 Å) it is inferred that the 6π electrons are not localized in a particular bond but delocalized around the ring. As a simple model for the electronic properties of the π-system a closed, ring-like, one-dimensional electron gas consisting of the 6π electrons is assumed (periodic boundary conditions). Calculate the possible energy eigenvalues E_n for this π system.

a) Successively fill the states with six electrons and determine the optical absorption as the transition between the highest occupied and the lowest unoccupied state. Compare the result with the experimental absorption bands.
b) Sketch the wavefunctions of the free electrons in the π-orbitals and match these to the LCAO orbitals.
c) Discuss reasons why the calculated optical absorption deviates from experimental values.

6.6 A free electron plasma fills the half-space ($z<0$) up to the surface $z=0$, where it is bounded by vacuum. Show that the surface wave

$$\phi = \phi_0 \exp(-k|z|) \exp[\mathrm{i}(kx - \omega t)]$$

is a solution of the Laplace equation $\Delta\phi=0$ and that the Maxwell boundary conditions for the $\mathscr{E}$ and D fields can be fulfilled at the surface $z=0$, thus giving the condition $\varepsilon(\omega_{\mathrm{sp}})=-1$ for the existence of the surface plasma wave with frequency ω_{sp}. Derive the dielectric function $\varepsilon(\omega)$ for the undamped free electron plasma and find a relation between the frequencies ω_{p} and ω_{sp} of the bulk and surface plasmons.

6.7 Normal (main sequence) stars are stabilized by a balance between gravitational pressure and radiation pressure arising from nuclear fusion processes occurring in the interior. After burn out and exhaustion of the nuclear fuel a new equilibrium state, that of a white dwarf, can be reached following some loss of mass and a gravitational contraction of the residual star. This contraction is accompanied by the production of a large amount of heat, which causes disintegration of atoms. A plasma of free electrons and positive H and He nuclei is formed whose internal (Fermi) pressure counteracts further gravitational contraction.

a) Calculate in the non-relativistic limit (electron velocity $\ll c$) the average energy of an electron and the average pressure in the electron plasma as a function of electron density n.
b) Making use of the equilibrium between Fermi pressure and gravitational pressure, calculate for the non-relativistic case the mass spectrum $M(\varrho)$ of possible white dwarfs (ϱ is the density of the star).

Hint: It is easy to see that the gravitational energy is proportional to GM^2/R where G is the gravitation constant. However, since the distribution of density in a star is very inhomogeneous, a better approximation for the gravitational pressure at the center of the star is $p_{GR}(R=0) \simeq 2^{-1/3} G M^{2/3} \varrho^{4/3}$.

c) Investigate whether white dwarfs for which the mass-density relationship of part (b) holds are stable. Plot the total energy for a fixed mass and particle-number against the radius. At which masses is the white dwarf stable?
d) With decreasing radius, the Fermi velocity of the electrons increases. For the relativistic case ($v \cong c$) calculate the average energy per electron and the pressure of the electron plasma as a function of the electron density.

Hint: When the kinetic energy becomes significantly larger than the rest-mass energy, then $E \cong pc = \hbar k c$.

e) For the relativistic case calculate, in analogy to part (b), the equilibrium between Fermi pressure and gravitational pressure and determine the mass M_{crit} for which this is possible. How does M_{crit} compare to the mass of the sun $M_s = 2\times 10^{30}$ kg?
f) Describe explicitly what happens to stars that have masses $M > M_{crit}$, $M = M_{crit}$, and $M < M_{crit}$. For each of the three cases plot the total energy at fixed mass and particle number as a function of the radius.
g) Calculate the electron density n_c at which the non-relativistic approximation must be replaced by the relativistic. This can be achieved with good accuracy by equating the relativistic and non-relativistic electron momenta at the Fermi edge:

$$p_F^{rel}(n_c) \cong p_F^{non\text{-}rel}(n_c) \ .$$

Compare the mean separation of electrons at this density with their compton wavelength $\lambda_c = \hbar/mc$. What star mass (expressed as a multiple of the sun's mass, M_s) corresponds to this density?

Remark: Realistic calculations give an upper limit for the mass of white dwarfs of $M_{crit} \approx 1.4 M_s$. This upper limit is named the Chandrasekhar limit after the Indian-American astrophysicist Subrahmanyon Chandrasekhar (Nobel prize 1984).

6.8 When the gravitational collapse of a burnt out star leads to even higher interior temperatures, inverse β decay processes ($e + p \to n$) cause the generation of neutrons (n). In its final state such a star consists only of neutrons (mass m_n). These form a Fermi gas similar to that of the electrons in a white dwarf. Apply the stability criterion derived in Problem 6.7 (Chandrasekhar limit) to the neutron Fermi gas of a neutron star and estimate the critical mass M_{crit}, the critical density ϱ_{crit}, and the critical radius R_{crit} of a neutron star. Compare these with the values for the sun.

7 The Electronic Bandstructure of Solids

Despite the success of the free-electron-gas model in describing electrons in crystals (Chap. 6), it must be recognized that the assumptions of the one-electron approximation and of a square-well potential, are oversimplifications. Thus one cannot expect this model to explain, for example, the fundamentally important optical and electronic properties of semiconductors. If one imagines, as discussed briefly in Chap. 1, that a solid is created by allowing initially free atoms to gradually approach one another, then the discrete nature of the energy levels of the isolated atoms should clearly be reflected in the properties of the solid. Indeed, discrete energy levels must be present in order to explain, e.g., the sharp resonance-like structures observed in the optical spectra of solids. An explanation of such features is beyond the scope of the free-electron-gas model. Furthermore, this model is unable to shed any light on the nature of semiconductors and insulators. To make further progress one has to take into account that the electronic states in solids form so-called bands. One can think of these as deriving from the states of the free atom.

In our present approximation, all deviations from perfect periodicity, be they static perturbations of the lattice or dynamic vibrations of the atoms, will be neglected. The assumption of an infinitely extended potential also means the neglect of all surface effects. To arrive at a finite crystal, i.e., one with a finite number of degrees of freedom, that is compatible with the infinite periodicity, one again makes use of the periodic boundary conditions introduced in Sect. 5.1.

7.1 General Symmetry Properties

We are now faced with the task of solving the time-independent Schrödinger equation for a single electron under the assumption that the potential $V(\boldsymbol{r})$ is periodic:

$$\mathscr{H}\psi(\boldsymbol{r}) = \left[-\frac{\hbar^2}{2m}\nabla^2 + V(\boldsymbol{r})\right]\psi(\boldsymbol{r}) = E\psi(\boldsymbol{r}) , \tag{7.1}$$

where

$$V(\boldsymbol{r}) = V(\boldsymbol{r}+\boldsymbol{r}_n); \quad \boldsymbol{r}_n = n_1\boldsymbol{a}_1 + n_2\boldsymbol{a}_2 + n_3\boldsymbol{a}_3 . \tag{7.2}$$

As in Sect. 3.2, $\boldsymbol{r}_n$ represents an arbitrary translation vector of the three-dimensional periodic lattice, i.e., $\boldsymbol{r}_n$ consists of multiples (n_1,n_2,n_3) of the three basis vectors $\boldsymbol{a}_1,\boldsymbol{a}_2,\boldsymbol{a}_3$ of the real-space lattice.

Since the potential $V(\boldsymbol{r})$ has the same periodicity as the lattice, it can be expanded in the following Fourier series:

$$V(\boldsymbol{r}) = \sum_{\boldsymbol{G}} V_{\boldsymbol{G}}\, \mathrm{e}^{\mathrm{i}\boldsymbol{G}\cdot\boldsymbol{r}} , \tag{7.3}$$

where the vector $\boldsymbol{G}$ must be a reciprocal lattice vector

$$\boldsymbol{G} = h\boldsymbol{g}_1 + k\boldsymbol{g}_2 + l\boldsymbol{g}_3 , \quad h,k,l \text{ integers.} \tag{7.4}$$

(in the one-dimensional case $\boldsymbol{G} \rightarrow G{=}h2\pi/a$). The most general plane-wave expansion of the required wavefunction $\psi(\boldsymbol{r})$ is

$$\psi(\boldsymbol{r}) = \sum_{\boldsymbol{k}} C_{\boldsymbol{k}}\, \mathrm{e}^{\mathrm{i}\boldsymbol{k}\cdot\boldsymbol{r}} . \tag{7.5}$$

Here $\boldsymbol{k}$ is a point in reciprocal space that is compatible with the periodic boundary conditions (Sects. 5.1 and 6.1). Substituting the expansions (7.3) and (7.5) into the Schrödinger equation (7.1) we obtain:

$$\sum_{\boldsymbol{k}} \frac{\hbar^2k^2}{2m} C_{\boldsymbol{k}} \mathrm{e}^{\mathrm{i}\boldsymbol{k}\cdot\boldsymbol{r}} + \sum_{\boldsymbol{k}'\boldsymbol{G}} C_{\boldsymbol{k}'} V_{\boldsymbol{G}}\, \mathrm{e}^{\mathrm{i}(\boldsymbol{k}'+\boldsymbol{G})\cdot\boldsymbol{r}} = E \sum_{\boldsymbol{k}} C_{\boldsymbol{k}}\, \mathrm{e}^{\mathrm{i}\boldsymbol{k}\cdot\boldsymbol{r}} . \tag{7.6}$$

After renaming the summation indices this becomes

$$\sum_{\boldsymbol{k}} \mathrm{e}^{\mathrm{i}\boldsymbol{k}\cdot\boldsymbol{r}} \left[\left(\frac{\hbar^2k^2}{2m} - E \right) C_{\boldsymbol{k}} + \sum_{\boldsymbol{G}} V_{\boldsymbol{G}} C_{\boldsymbol{k}-\boldsymbol{G}} \right] = 0 . \tag{7.7}$$

Since this condition is valid for every position vector $\boldsymbol{r}$, the expression in brackets, which is independent of $\boldsymbol{r}$, must vanish for every $\boldsymbol{k}$, i.e.,

$$\left(\frac{\hbar^2k^2}{2m} - E \right) C_{\boldsymbol{k}} + \sum_{\boldsymbol{G}} V_{\boldsymbol{G}} C_{\boldsymbol{k}-\boldsymbol{G}} = 0 . \tag{7.8}$$

This set of algebraic equations, which is simply a representation of the Schrödinger equation (7.1) in reciprocal space, couples only those expansion coefficients $C_{\boldsymbol{k}}$ of $\psi(\boldsymbol{r})$ (7.5), whose $\boldsymbol{k}$-values differ from one another by a reciprocal lattice vector $\boldsymbol{G}$. Thus $C_{\boldsymbol{k}}$ is coupled to $C_{\boldsymbol{k}-\boldsymbol{G}}$, $C_{\boldsymbol{k}-\boldsymbol{G}'}$, $C_{\boldsymbol{k}-\boldsymbol{G}''}$, ...

The original problem thus separates into N problems (N=number of unit cells), each corresponding to a $\boldsymbol{k}$-vector in the unit cell of the reciprocal lattice. Each of the N systems of equations yields a solution that can be represented as a superposition of plane waves whose wave vectors $\boldsymbol{k}$ differ only by reciprocal lattice vectors $\boldsymbol{G}$. The eigenvalues E of the Schrödinger equation (7.1) can thus be indexed according to $E_{\boldsymbol{k}}{=}E(\boldsymbol{k})$, and the wavefunction belonging to $E_{\boldsymbol{k}}$ is

$$\psi_k(\boldsymbol{r}) = \sum_{G} C_{k-G}\, \mathrm{e}^{\mathrm{i}(\boldsymbol{k}-\boldsymbol{G})\cdot \boldsymbol{r}} \tag{7.9}$$

or

$$\psi_k(\boldsymbol{r}) = \sum_{G} C_{k-G}\, \mathrm{e}^{-\mathrm{i}\boldsymbol{G}\cdot\boldsymbol{r}}\, \mathrm{e}^{\mathrm{i}\boldsymbol{k}\cdot\boldsymbol{r}} = u_k(\boldsymbol{r})\, \mathrm{e}^{\mathrm{i}\boldsymbol{k}\cdot\boldsymbol{r}} . \tag{7.10 a}$$

The function $u_k(\boldsymbol{r})$ introduced here is a Fourier series over reciprocal lattice points $\boldsymbol{G}$ and thus has the periodicity of the lattice. The wave vector $\boldsymbol{k}$, which, for periodic boundary conditions, can take the values (Sect. 6.1)

$$\begin{aligned} k_x &= 0,\ \pm 2\pi/L,\ \pm 4\pi/L, \ldots,\ \pm 2\pi n_x/L \\ k_y &= 0,\ \pm 2\pi/L,\ \pm 4\pi/L, \ldots,\ \pm 2\pi n_y/L \\ k_z &= 0,\ \pm 2\pi/L,\ \pm 4\pi/L, \ldots,\ \pm 2\pi n_z/L \end{aligned} \tag{7.10 b}$$

(L=macroscopic dimension of the crystal), yields the correct quantum numbers k_x, k_y, k_z or n_x, n_y, n_z, according to which the energy eigenvalues and quantum states may be indexed. In other words, we have shown that the solution of the one-electron Schrödinger equation for a periodic potential can be written as a modulated plane wave

$$\psi_k(\boldsymbol{r}) = u_k(\boldsymbol{r})\, \mathrm{e}^{\mathrm{i}\boldsymbol{k}\cdot\boldsymbol{r}} \tag{7.10 c}$$

with a modulation function

$$u_k(\boldsymbol{r}) = u_k(\boldsymbol{r} + \boldsymbol{r}_n) \tag{7.10 d}$$

that has the periodicity of the lattice. This result is known as *Bloch's theorem*, and the wavefunctions given in (7.10 a–d) are called the *Bloch waves* or Bloch states of an electron (Fig. 7.1).

The strict periodicity of the lattice potential has further consequences that follow directly from the properties of the Bloch states. From the general representation of a Bloch wave (7.10 a), and by renaming the reciprocal lattice vectors $\boldsymbol{G}''=\boldsymbol{G}'-\boldsymbol{G}$, it follows that

$$\begin{aligned} \psi_{k+G}(\boldsymbol{r}) &= \sum_{G'} C_{k+G-G'}\, \mathrm{e}^{-\mathrm{i}\boldsymbol{G}'\cdot\boldsymbol{r}}\, \mathrm{e}^{\mathrm{i}(\boldsymbol{k}+\boldsymbol{G})\cdot\boldsymbol{r}} \\ &= \left(\sum_{G''} C_{k-G''}\, \mathrm{e}^{-\mathrm{i}\boldsymbol{G}''\cdot\boldsymbol{r}} \right) \mathrm{e}^{\mathrm{i}\boldsymbol{k}\cdot\boldsymbol{r}} = \psi_k(\boldsymbol{r}) , \end{aligned} \tag{7.11 a}$$

i.e.

$$\psi_{k+G}(\boldsymbol{r}) = \psi_k(\boldsymbol{r}) . \tag{7.11 b}$$

Thus Bloch waves whose wave vectors differ by a reciprocal lattice vector are identical. The Schrödinger equation (7.1):

$$\mathscr{H}\, \psi_k = E(\boldsymbol{k})\, \psi_k \tag{7.12}$$

and that for the same problem displaced by $\boldsymbol{G}$:

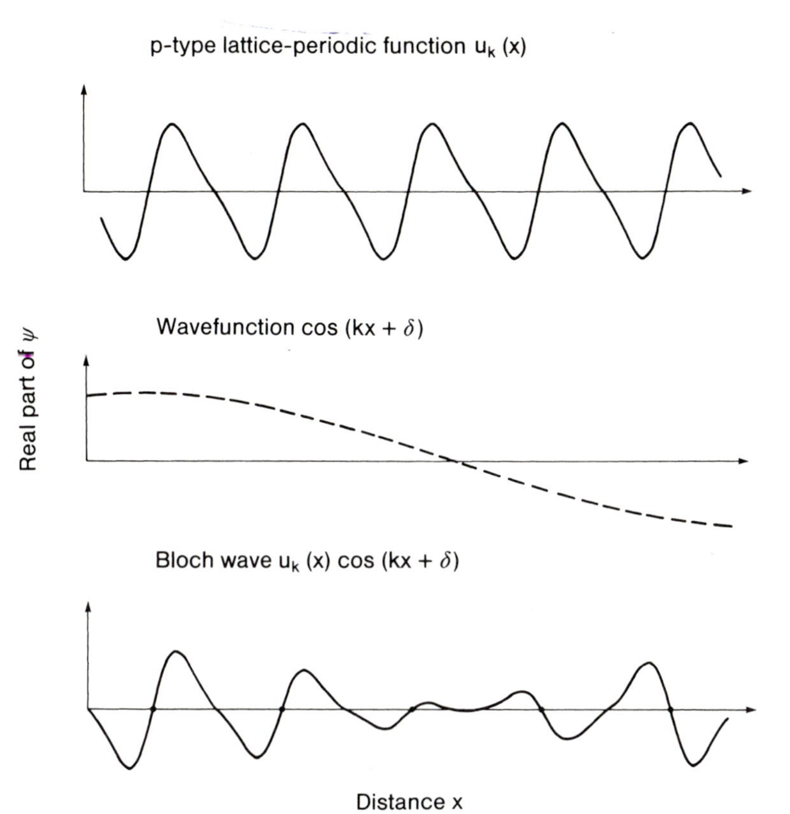

Fig. 7.1. Example of the construction of a Bloch wave $\psi_k(\boldsymbol{r}) = u_k(\boldsymbol{r})\mathrm{e}^{\mathrm{i}\boldsymbol{k}\cdot\boldsymbol{r}}$ from a lattice-periodic function $u_k(\boldsymbol{r})$ with p-type bonding character and a plane wave

$$\mathscr{H}\,\psi_{k+G} = E\,(\boldsymbol{k}+\boldsymbol{G})\psi_{k+G} \tag{7.13}$$

together with (7.11 b) then yield

$$\mathscr{H}\,\psi_k = E\,(\boldsymbol{k}+\boldsymbol{G})\psi_k\ . \tag{7.14}$$

Comparing (7.12) with (7.14) we see that

$$E\,(\boldsymbol{k}) = E\,(\boldsymbol{k}+\boldsymbol{G})\ . \tag{7.15}$$

Thus the energy eigenvalues $E(\boldsymbol{k})$ are a periodic function of the quantum numbers $\boldsymbol{k}$, i.e., of the wave vectors of the Bloch waves.

Similar to the case of phonons, whose $\omega(\boldsymbol{q})$ can be described by dispersion surfaces in reciprocal $\boldsymbol{q}$-space, the one-electron states of a periodic potential can be represented by energy surfaces $E{=}E(\boldsymbol{k})$ that are a periodic function of the wave vector (quantum number) in reciprocal $\boldsymbol{k}$-space. Taken together, these energy surfaces form the *electronic bandstructure* of the crystal. Since both $\psi_k(\boldsymbol{r})$ and $E(\boldsymbol{k})$ are periodic in reciprocal space, one only needs to know these functions for $\boldsymbol{k}$-values in the first Brillouin zone

(Sect. 3.5). A periodic continuation of the functions then provides the values throughout the whole of $\boldsymbol{k}$-space.

7.2 The Nearly Free-Electron Approximation

To understand the general concept of electronic bands it is particularly instructive to consider the limiting case of a vanishingly small periodic potential. We therefore imagine that the periodic potential starts at zero and is gradually "switched on". What happens then to the energy states of the free electrons which, in the square-well potential, were described by the energy parabola $E=\hbar^2 k^2/2m$? In the extreme case where the potential is still zero, i.e., where all Fourier coefficients $V_{\boldsymbol{G}}$ (7.3) vanish, one must nonetheless consider the symmetry requirements of the periodicity, since the requirements will be a decisive factor even for the smallest nonvanishing potential. This general demand of periodicity immediately implies, from (7.15), that the possible electron states are not restricted to a single parabola in $\boldsymbol{k}$-space, but can be found equally well on parabolas shifted by any $\boldsymbol{G}$-vector:

$$E(\boldsymbol{k}) = E(\boldsymbol{k}+\boldsymbol{G}) = \frac{\hbar^2}{2m}\left|\boldsymbol{k}+\boldsymbol{G}\right|^2 . \tag{7.16}$$

For the one-dimensional case ($\boldsymbol{G} \rightarrow G=h2\pi/a$) this is depicted in Fig. 7.2.

Since the behavior of $E(\boldsymbol{k})$ is periodic in $\boldsymbol{k}$-space, it is sufficient to represent this in the first Brillouin zone only. To achieve this one simply displaces the part of the parabola of interest by the appropriate multiple of $G=2\pi/a$. This procedure is called "*Reduction to the first Brillouin zone*".

In three dimensions, the $E(\boldsymbol{k})$ bands are already more complicated, even in the case of a vanishing potential, since in (7.16) one now has $\boldsymbol{G}$ contributions in all three coordinate directions. Figure 7.3 shows the $E(\boldsymbol{k})$ curves along k_x in the first Brillouin zone for a simple cubic lattice with vanishing potential.

The effect of a finite but very small potential can now be discussed with reference to Figs. 7.2 and 7.3.

In the one-dimensional problem of Fig. 7.2 there is a degeneracy of the energy values at the edges of the first Brillouin zone, i.e., at $+G/2=\pi/a$ and

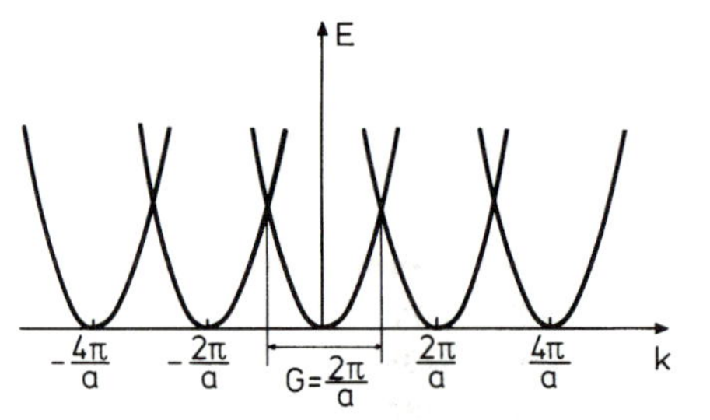

Fig. 7.2. The parabolic energy curves of a free electron in one dimension, periodically continued in reciprocal space. The periodicity in real space is a. This $\boldsymbol{E}(\boldsymbol{k})$ dependence corresponds to a periodic lattice with a vanishing potential ("empty" lattice)

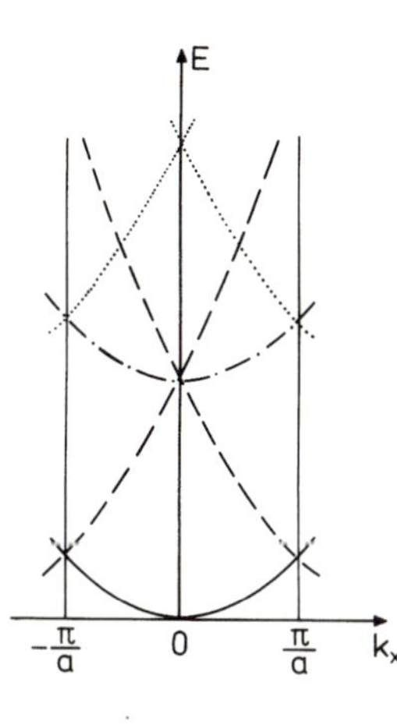

Fig. 7.3. Bandstructure for a free electron gas in a primitive cubic lattice (lattice constant a), represented on a section along k_x in the first Brillouin zone. The periodic potential is assumed to be vanishing ("empty" lattice). The various branches stem from parabolas whose origin in reciprocal space is given by the Miller indices hkl. (——) 000; (– – –) 100, $\bar{1}00$; (–·–) 010, $0\bar{1}0$, 001, $00\bar{1}$; (···) 110, 101, $1\bar{1}0$, $10\bar{1}$, $\bar{1}10$, $\bar{1}01$, $\bar{1}\bar{1}0$, $\bar{1}0\bar{1}$

$-G/2=-\pi/a$, where two parabolas intersect. The description of the state of an electron with these k-values is necessarily a superposition of at least two corresponding plane waves. For a vanishing potential (zeroth-order approximation) these waves are

$$e^{iGx/2} \quad \text{and} \quad e^{i[(G/2)-G]x} = e^{-iGx/2} \,. \tag{7.17}$$

Equation (7.8) implies that waves with G-values larger than $2\pi/a$ must also be taken into account. However, on dividing (7.8) by $[(\hbar^2k^2/2m)-E]$, it follows that C_k is particularly large when E_k and E_{k-G} are both approximately equal to $\hbar^2k^2/2m$, and that the coefficient C_{k-G} then has approximately the same absolute magnitude as C_k. This is precisely the case for the two plane waves at the zone boundaries (7.17), and thus, to a first approximation, one can neglect contributions from other reciprocal lattice vectors. The appropriate expressions for a perturbation calculation of the influence of a small potential are therefore of the form

$$\psi_+ \sim (e^{iGx/2} + e^{-iGx/2}) \sim \cos \pi \frac{x}{a} \,, \tag{7.18 a}$$

$$\psi_- \sim (e^{iGx/2} - e^{-iGx/2}) \sim \sin \pi \frac{x}{a} \,. \tag{7.18 b}$$

These are standing waves possessing zeros at fixed positions in space. As seen in the discussion of diffraction from periodic structures (Chap. 3), these standing waves can be represented as a superposition of an incoming wave and a counter-propagating "Bragg-reflected" wave. The probability densities, corresponding to ψ_+ and ψ_-,

$$\varrho_+ = \psi_+^* \psi_+ \sim \cos^2 \pi \frac{x}{a} \,, \tag{7.19 a}$$

$$\varrho_- = \psi_-^* \psi_- \sim \sin^2 \pi \frac{x}{a} \,, \tag{7.19 b}$$

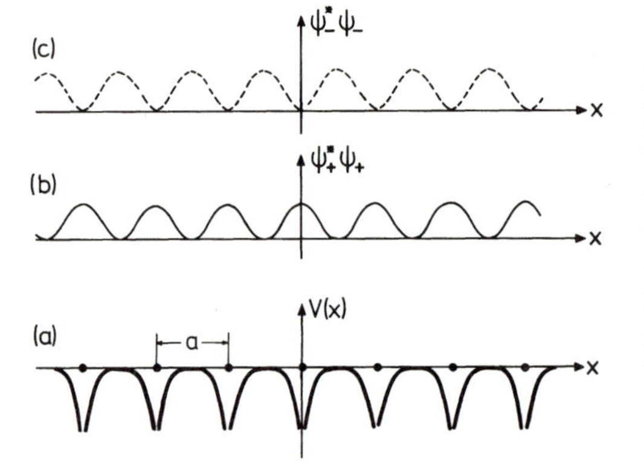

Fig. 7.4. (**a**) Qualitative form of the potential energy $V(x)$ of an electron in a one-dimensional lattice. The positions of the ion cores are indicated by the points with separation a (lattice constant). (**b**) Probability density $\varrho_+ = \psi_+^* \psi_+$ for the standing wave produced by Bragg reflection at $k = \pm\pi/a$ (upper edge of band ① in Fig. 7.5). (**c**) Probability density $\varrho_- = \psi_-^* \psi_-$ for the standing wave at the lower edge of band ② (Fig. 7.5) at $k = \pm\pi/a$

are depicted in Fig. 7.4 together with a qualitative sketch of the potential. For an electron in the state ψ_+, the charge density is maximum at the position of the positive cores and minimum in between; for ψ_- the charge density is maximum between the cores. In comparison with the travelling plane wave e^{ikx}, which is a good approximation to the solution further away from the zone boundary, ψ_+ thus has a lower total energy (particularly potential energy), and ψ_- a higher energy than that of a free electron on the energy parabola (zero potential case). This increase and decrease in the energy of the states at the zone boundary represents a deviation from the free-electron energy parabola (Fig. 7.5).

Having gained insight into the problem from this qualitative discussion, it is now easy to carry out a formal calculation of the magnitude of the so-called band splitting or *energy gap* shown in Fig. 7.5.

Starting from the general representation of the Schrödinger equation in $\boldsymbol{k}$-space (7.8), translation by a reciprocal lattice vector yields

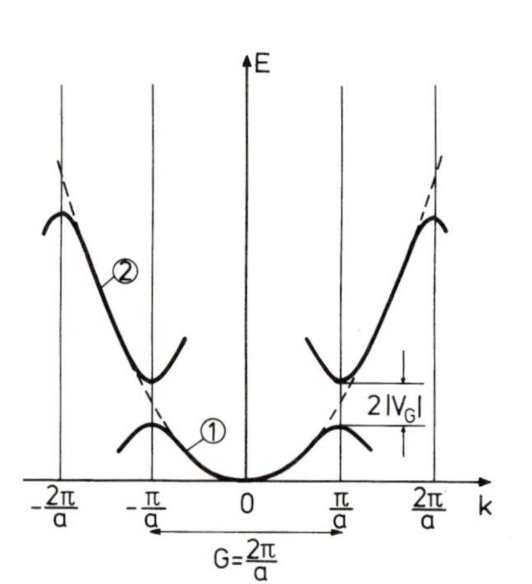

Fig. 7.5. Splitting of the energy parabola of the free electron (– – –) at the edges of the first Brillouin zone $\boldsymbol{k} = \pm\pi/a$ in the one-dimensional case). To a first approximation the gap is given by twice the corresponding Fourier coefficient V_G of the potential. Periodic continuation over the whole of $\boldsymbol{k}$-space gives rise to continuous bands ① and ②, shown here only in the vicinity of the original energy parabola

$$\left(E - \frac{\hbar^2}{2m}|\boldsymbol{k} - \boldsymbol{G}|^2\right) C_{\boldsymbol{k}-\boldsymbol{G}} = \sum_{\boldsymbol{G}'} V_{\boldsymbol{G}'} C_{\boldsymbol{k}-\boldsymbol{G}-\boldsymbol{G}'}$$
$$= \sum_{\boldsymbol{G}'} V_{\boldsymbol{G}'-\boldsymbol{G}} C_{\boldsymbol{k}-\boldsymbol{G}'} \,, \quad \text{i.e.} \tag{7.20 a}$$

$$C_{\boldsymbol{k}-\boldsymbol{G}} = \frac{\sum_{\boldsymbol{G}'} V_{\boldsymbol{G}'-\boldsymbol{G}} C_{\boldsymbol{k}-\boldsymbol{G}'}}{E - \frac{\hbar^2}{2m}|\boldsymbol{k} - \boldsymbol{G}|^2} \,. \tag{7.20 b}$$

For small perturbations, a first approximation to the calculation of $C_{\boldsymbol{k}-\boldsymbol{G}}$ can be made by setting the true eigenvalue E that we are seeking equal to the energy of the free electron ($\hbar^2 k^2/2m$). Furthermore, in this first approximation, only the largest coefficients $C_{\boldsymbol{k}-\boldsymbol{G}}$ are of interest; in other words, we expect the greatest deviation from free-electron behavior when the denominator in (7.20 b) vanishes, i.e., for

$$\boldsymbol{k}^2 \simeq |\boldsymbol{k} - \boldsymbol{G}|^2 \,. \tag{7.21}$$

This is identical to the Bragg condition (3.32). The strongest perturbations to the energy surface of the free electron (spheres in $\boldsymbol{k}$-space), produced by the periodic potential, occur when the Bragg condition is satisfied, i.e., for the $\boldsymbol{k}$-vectors at the edge of the first Brillouin zone. It follows from (7.20 b), however, that besides $C_{\boldsymbol{k}-\boldsymbol{G}}$, the coefficient $C_{\boldsymbol{k}}$ is equally important. Thus, in the system of (7.20 a), for this approximation we only need to consider two relations ($V_0 = 0$):

$$\begin{aligned} \left(E - \frac{\hbar^2}{2m}k^2\right) C_{\boldsymbol{k}} - V_{\boldsymbol{G}} C_{\boldsymbol{k}-\boldsymbol{G}} &= 0 \\ \left(E - \frac{\hbar^2}{2m}|\boldsymbol{k} - \boldsymbol{G}|^2\right) C_{\boldsymbol{k}-\boldsymbol{G}} - V_{-\boldsymbol{G}} C_{\boldsymbol{k}} &= 0 \,. \end{aligned} \tag{7.22}$$

We thus obtain the secular equation for the energy value

$$\begin{vmatrix} \left(\frac{\hbar^2}{2m}k^2 - E\right) & V_{\boldsymbol{G}} \\ V_{-\boldsymbol{G}} & \left(\frac{\hbar^2}{2m}\right)|\boldsymbol{k} - \boldsymbol{G}|^2 - E \end{vmatrix} = 0 \,. \tag{7.23}$$

With $E^0_{\boldsymbol{k}-\boldsymbol{G}} = (\hbar^2/2m)|\boldsymbol{k} - \boldsymbol{G}|^2$ as the energy of the free electrons, the two solutions to this secular equation may be written

$$E^{\pm} = \tfrac{1}{2}(E^0_{\boldsymbol{k}-\boldsymbol{G}} + E^0_{\boldsymbol{k}}) \pm \left[\tfrac{1}{4}(E^0_{\boldsymbol{k}-\boldsymbol{G}} - E^0_{\boldsymbol{k}})^2 + |V_{\boldsymbol{G}}|^2\right]^{1/2} \,. \tag{7.24}$$

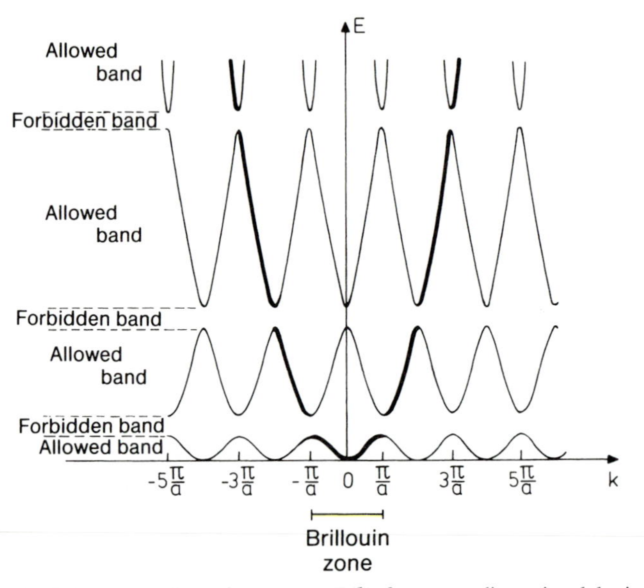

Fig. 7.6. Energy dispersion curves $E(\boldsymbol{k})$ for a one-dimensional lattice (lattice constant a) in the extended zone scheme. As can be seen, the quasi-free-electron approximation gives rise to forbidden and allowed energy regions due to the opening of band gaps, as shown in Fig. 7.5 (cf. the vanishing potential case of Fig. 7.2). The parts of the bands corresponding to the free-electron parabola are indicated by the thick lines

Therefore, at the zone boundary itself, where the contributions of the two waves with $C_{\boldsymbol{k}}$ and $C_{\boldsymbol{k}-\boldsymbol{G}}$ are equal – see (7.21) – and where $E^0_{\boldsymbol{k}-\boldsymbol{G}}=E^0_{\boldsymbol{k}}$, the energy gap has a value

$$\Delta E = E_+ - E_- = 2|V_{\boldsymbol{G}}| \ , \qquad (7.25)$$

i.e., twice the $\boldsymbol{G}$th Fourier component of the potential.

Near to the zone boundary, the form of the two energy surfaces that are separated by this gap is described by (7.24) (in which one again sets $E^0_{\boldsymbol{k}}=\hbar^2 k^2/2m$). Figure 7.5 illustrates this for the one-dimensional case near to the zero boundary at $k=G/2$.

The correspondence between the energy parabola of the free electrons and the periodic bandstructure, with its energy gaps due to the realistic potential, is depicted in Figs. 7.5 and 7.6, in both cases for the one-dimensional problem.

7.3 The Tight-Binding Approximation

The electrons that occupy the energetically low-lying core levels of a free atom are strongly localized in space. They naturally retain this strong localization when the atom participates in the formation of a crystal. It thus appears that the description of a solid's electronic structure in terms of quasi-free electrons must be inadequate. Since these core electrons largely retain the properties that they had in the free atom, an obvious approach is to describe the crystal electrons in terms of a linear superposition of atomic eigenfunctions. This procedure, also known as the LCAO method (Linear Combination of Atomic Orbitals), was already discussed qualitatively in Chap. 1 in relation to chemical bonding, in order to explain the existence of electronic bands in solids.

In formulating the problem, one assumes that the solutions to the Schrödinger equation for the free atoms that form the crystal

$$\mathcal{H}_A(\boldsymbol{r}-\boldsymbol{r}_n)\varphi_i(\boldsymbol{r}-\boldsymbol{r}_n) = E_i\varphi_i(\boldsymbol{r}-\boldsymbol{r}_n) \tag{7.26}$$

are known. $\mathcal{H}_A(\boldsymbol{r}-\boldsymbol{r}_n)$ is the Hamiltonian for a free atom at the lattice position $\boldsymbol{r}_n=n_1\boldsymbol{a}_1+n_2\boldsymbol{a}_2+n_3\boldsymbol{a}_3$ and $\varphi_i(\boldsymbol{r}-\boldsymbol{r}_n)$ is the wavefunction for an electron in the atomic energy level E_i. One imagines the entire crystal to be built up of single atoms, i.e., the Hamiltonian for an electron (one-electron approximation!) in the total potential of all the atoms can be written:

$$\mathcal{H}=\mathcal{H}_A+v=-\frac{\hbar^2}{2m}\Delta+V_A(\boldsymbol{r}-\boldsymbol{r}_n)+v(\boldsymbol{r}-\boldsymbol{r}_n)\,. \tag{7.27}$$

The influence of atoms in the neighborhood of $\boldsymbol{r}_n$, where the electron of interest is relatively strongly localized, is described by a perturbation $v(\boldsymbol{r}-\boldsymbol{r}_n)$ to the potential V_A of the free atom. This perturbation can thus be expressed as

$$v(\boldsymbol{r}-\boldsymbol{r}_n)=\sum_{m\neq n}V_A(\boldsymbol{r}-\boldsymbol{r}_m) \tag{7.28}$$

i.e., as a sum over the potentials of all atoms apart from that at $\boldsymbol{r}_n$, at the position $\boldsymbol{r}$ of the electron (Fig. 7.7).

We now seek solutions of the Schrödinger equation

$$\mathcal{H}\psi_{\boldsymbol{k}}(\boldsymbol{r})=E(\boldsymbol{k})\psi_{\boldsymbol{k}}(r)\,, \tag{7.29}$$

where $\mathcal{H}$ is the Hamiltonian of (7.27) for the crystal electron and $\psi_{\boldsymbol{k}}(\boldsymbol{r})$ are Bloch waves with the general properties discussed in Sect. 7.1.

Multiplying (7.29) by $\psi_{\boldsymbol{k}}^*$ and integrating over the range in which $\psi_{\boldsymbol{k}}$ is defined, one readily obtains

$$E(\boldsymbol{k})=\frac{\langle\psi_{\boldsymbol{k}}|\mathcal{H}|\psi_{\boldsymbol{k}}\rangle}{\langle\psi_{\boldsymbol{k}}|\psi_{\boldsymbol{k}}\rangle}\,, \tag{7.30}$$

where $\langle\psi_{\boldsymbol{k}}|\psi_{\boldsymbol{k}}\rangle=\int d\boldsymbol{r}\,\psi_{\boldsymbol{k}}^*\psi_{\boldsymbol{k}}$ and $\langle\psi_{\boldsymbol{k}}|\mathcal{H}|\psi_{\boldsymbol{k}}\rangle=\int d\boldsymbol{r}\,\psi_{\boldsymbol{k}}^*\mathcal{H}\psi_{\boldsymbol{k}}$ (cf. Sect. 1.2 and Problem 1.8). If, instead of the true wavefunction, one inserts a trial wavefunction $\Phi_{\boldsymbol{k}}$ into (7.30), then one obtains an energy E' that is always larger than $E(\boldsymbol{k})$. The better $\Phi_{\boldsymbol{k}}$ approximates the true wavefunction, the closer $E'(\boldsymbol{k})$ lies to $E(\boldsymbol{k})$. This circumstance provides the basis for the Ritz procedure.

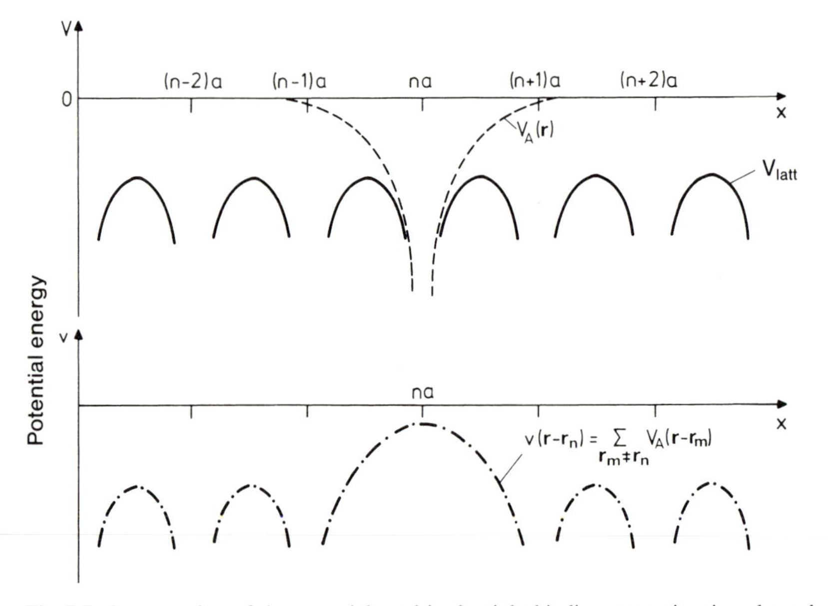

Fig. 7.7. Cross section of the potential used in the tight-binding approximation along the x-direction. The lattice potential V_{latt} (——) is obtained by summing the potentials V_A of the free atoms (– – –). The perturbation potential $v(\boldsymbol{r}-\boldsymbol{r}_n)$ used in the approximate calculation is given by the dash-dotted line in the lower part of the figure

In the present case, when we want to calculate the crystal electron energy states $E(\boldsymbol{k})$ that derive from the energy level E_i of the free atom, we assume that a good approximation to ψ_k is provided by a linear combination of atomic eigenfunctions $\varphi_i(\boldsymbol{r}-\boldsymbol{r}_n)$, i.e.,

$$\psi_k \approx \Phi_k = \sum_n a_n \varphi_i(\boldsymbol{r}-\boldsymbol{r}_n) = \sum_n \mathrm{e}^{\mathrm{i}\boldsymbol{k}\cdot\boldsymbol{r}_n}\varphi_i(\boldsymbol{r}-\boldsymbol{r}_n) \, . \tag{7.31}$$

The expansion coefficients are determined as $a_n = \exp(\mathrm{i}\boldsymbol{k}\cdot\boldsymbol{r}_n)$ by the requirement that Φ_k should be a Bloch wave. It can easily be shown that Φ_k in (7.31) possesses all the required properties of Bloch waves (Sect. 7.1), e.g.,

$$\Phi_{k+G} = \sum_n \mathrm{e}^{\mathrm{i}\boldsymbol{k}\cdot\boldsymbol{r}_n}\mathrm{e}^{\mathrm{i}\boldsymbol{G}\cdot\boldsymbol{r}_n}\varphi_i(\boldsymbol{r}-\boldsymbol{r}_n) = \Phi_k \, . \tag{7.32}$$

An approximate calculation of $E(\boldsymbol{k})$ can now be made by inserting in (7.30) the trial function of (7.31); the denominator in (7.30) becomes:

$$\langle \Phi_k | \Phi_k \rangle = \sum_{n,m} \mathrm{e}^{\mathrm{i}\boldsymbol{k}\cdot(\boldsymbol{r}_n-\boldsymbol{r}_m)} \int \varphi_i^*(\boldsymbol{r}-\boldsymbol{r}_m)\varphi_i(\boldsymbol{r}-\boldsymbol{r}_n)d\boldsymbol{r} \, . \tag{7.33}$$

For a sufficiently localized electron, $\varphi_k(\boldsymbol{r}-\boldsymbol{r}_m)$ only has significant values in the vicinity of $\boldsymbol{r}_m$. Thus, to a first approximation, we only retain terms in (7.33) with $\boldsymbol{n}=\boldsymbol{m}$, and obtain

$$\langle \Phi_k | \Phi_k \rangle \simeq \sum_n \int \varphi_i^*(\boldsymbol{r} - \boldsymbol{r}_n)\varphi_i(\boldsymbol{r} - \boldsymbol{r}_n)d\boldsymbol{r} = N \ , \tag{7.34}$$

where N is the number of atoms in the crystal.

Making use of the fact we know the solutions to (7.26) for the free atom, we write

$$E(\boldsymbol{k}) \approx \frac{1}{N}\sum_{n,m} e^{i\boldsymbol{k}\cdot(\boldsymbol{r}_n - \boldsymbol{r}_m)} \int \varphi_i^*(\boldsymbol{r} - \boldsymbol{r}_m)[E_i + v(\boldsymbol{r} - \boldsymbol{r}_n)]\varphi_i(\boldsymbol{r} - \boldsymbol{r}_n)d\boldsymbol{r} \tag{7.35}$$

where E_i is the energy eigenvalue of the isolated atom. In the term containing E_i we have again neglected the overlap between nearest neighbors (i.e., only terms with $\boldsymbol{n}=\boldsymbol{m}$ are considered). For the term containing the perturbation $v(\boldsymbol{r} - \boldsymbol{r}_n)$, overlap is included only up to nearest neighbors. In the simple case when the relevant atomic state φ_i possesses spherical symmetry, i.e., s-character, the result can be readily represented with the help of the following two quantities:

$$A = -\int \varphi_i^*(\boldsymbol{r} - \boldsymbol{r}_n)v(\boldsymbol{r} - \boldsymbol{r}_n)\varphi_i(\boldsymbol{r} - \boldsymbol{r}_n)d\boldsymbol{r} \ , \tag{7.36a}$$

$$B = -\int \varphi_i^*(\boldsymbol{r} - \boldsymbol{r}_m)v(\boldsymbol{r} - \boldsymbol{r}_n)\varphi_i(\boldsymbol{r} - \boldsymbol{r}_n)d\boldsymbol{r} \tag{7.36b}$$

and reads

$$E(\boldsymbol{k}) \approx E_i - A - B\sum_m e^{i\boldsymbol{k}\cdot(\boldsymbol{r}_n - \boldsymbol{r}_m)} \ . \tag{7.37}$$

The sum over $\boldsymbol{m}$ includes only values for which $\boldsymbol{r}_m$ denotes a nearest neighbor of $\boldsymbol{r}_n$.

In the present case, A is positive since v is negative. Equation (7.37), applied to the case of a primitive cubic lattice with

$$\boldsymbol{r}_n - \boldsymbol{r}_m = (\pm a, 0, 0)\ ; \ \ (0, \pm a, 0)\ ; \ \ (0, 0, \pm a) \ ,$$

gives, for an atomic s-state,

$$E(\boldsymbol{k}) \approx E_i - A - 2B(\cos k_x a + \cos k_y a + \cos k_z a) \ . \tag{7.38}$$

When the atoms are brought together to form a crystal (with primitive cubic lattice), the atomic energy level E_i therefore becomes an electronic band whose center of gravity is reduced by an amount A with respect to E_i and whose width is proportional to B. This situation is illustrated in Fig. 7.8.

The important consequences can be summarized as follows:

i) Since the cosine terms vary between +1 and –1, the width of the energy band is $12B_i$. For small $\boldsymbol{k}$-values, the cosine terms can be expanded such that near to the Γ-point (center of the first Brillouin zone at $\boldsymbol{k}=0$) one obtains

$$E(\boldsymbol{k}) = E_i - A - 6B + B a^2 k^2 \ , \tag{7.39}$$

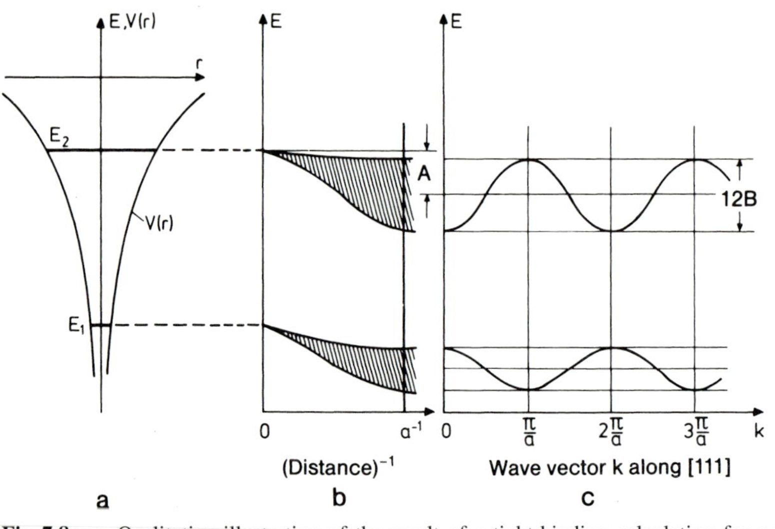

Fig. 7.8a–c. Qualitative illustration of the result of a tight-binding calculation for a primitive cubic lattice with lattice constant a. (**a**) Position of the energy levels E_1 and E_2 in the potential $V(\boldsymbol{r})$ of the free atom. (**b**) Reduction and broadening of the levels E_1 and E_2 as a function of the reciprocal atomic separation r^{-1}. At the equilibrium separation a the mean energy decrease is A and the width of the band is $12B$. (**c**) Dependence of the one-electron energy E on the wave vector $k(1,1,1)$ in the direction of the main diagonal [111]

where $k^2=k_x^2+k_y^2+k_z^2$. This k^2-dependence corresponds to that which results from the quasi-free-electron approximation (Sect. 7.2).

ii) From (7.36b) it follows that the energy width of the band becomes greater as the overlap between the corresponding wavefunctions of neighboring atoms increases. Lower lying bands that stem from more strongly localized states are thus narrower than bands originating from less strongly bound atomic states whose wavefunctions are more extended.

iii) In the framework of the present one-electron approximation, the occupation of the one-electron band states is obtained by placing two of the available electrons of every atom into each band, beginning with the lowest energy band, until all electrons have been accommodated. The Pauli principle allows the double occupation because of the two possible spin states of the electron.

If a crystal with a primitive cubic lattice contains N atoms, and thus N primitive unit cells, then an atomic energy level E_i of the free atom will split, due to the interaction with the other (N–1) atoms, into N states. These then form the corresponding quasi-continuous band. This band can thus be occupied by $2N$ electrons. We obtain the same result by considering this problem in terms of the quasi-free-electron model: In $\boldsymbol{k}$-space each electron state corresponds to a volume $(2\pi)^3/V$ (where V is the macroscopic crystal volume). The volume of the first Brillouin zone, however, is $(2\pi)^3/V_c$

(where V_c is the volume of the unit cell). Thus the part of the band within the first Brillouin zone contains $V/V_c = N$ states, which, when the two spin states are considered, yields $2N$ states available for occupation by electrons.

The existence of a bandstructure arising from the discrete energy levels of isolated atoms upon their joining together to form a crystal was illustrated qualitatively in Fig. 1.1. For sodium, for example, the atomic $3s$- and $3p$-levels give rise, in the crystal (equilibrium atomic separation r_0), to bands that overlap. Since the occupied levels of atomic sodium are $1s^2, 2s^2, 2p^6, 3s^1$, the atomic $3s$-level only contributes one electron per unit cell to the $3s$-band of the crystal, which could, however, accommodate two electrons per unit cell. Thus, even without the $3s$-$3p$ overlap (analogous to the $2s$–$2p$ overlap in Fig. 1.1), the $3s$ band of Na would only be half full. In Sect. 8.2 we will see that this partial occupation of a band is the source of the metallic conductivity of Na. Qualitative arguments concerning the conductivity have already been presented in Sect. 1.4.

It is well known that isolated carbon atoms have the electronic configuration $1s^2, 2s^2, 2p^2$. However, in the diamond crystal, as a result of the forma-

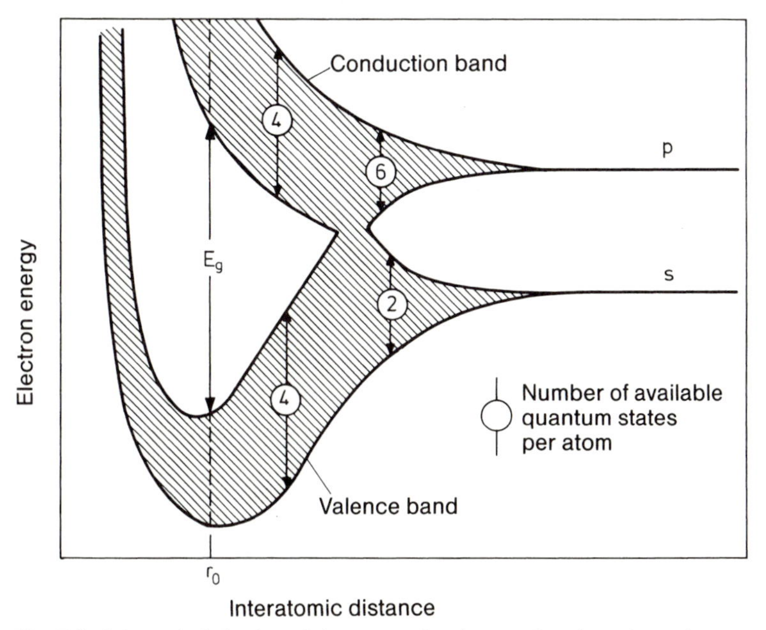

Fig. 7.9. Schematic behavior of the energy bands as a function of atomic separation for the tetrahedrally bound semiconductors diamond (C), Si, and Ge. At the equilibrium separation r_0 there is a forbidden energy gap of width E_g between the occupied and unoccupied bands that result from the sp^3 hybrid orbitals. For diamond, the sp^3 hybrid stems from the $2s$ and $2p^3$ atomic states, for Si from the $3s$ and $3p^3$, and for Ge from the $4s$ and $4p^3$. One sees from this figure that the existence of a forbidden energy region is not tied to the periodicity of the lattice. Thus amorphous materials can also display a band gap. (After [7.1])

tion of the sp^3 hybrid (a mixture of $2s$ and $2p$ wavefunctions with tetrahedral bonding; Chap. 1), there is a modification of the s- and p-levels which manifests itself in a further splitting of the sp^3 hybrid band into two bands, each of which (including spin) can accommodate four electrons (Fig. 7.9).

The four electrons of the atomic $2s$- and $2p$-states thus fill the lower part of the sp^3 band, leaving the upper part unoccupied. Between the two sp^3 subbands there is a forbidden energy gap of width E_g. This is the origin of the insulating property of diamond, as will be shown in Sects. 9.2 and 12.1. The semiconductors Si and Ge are similar cases (Chap. 12).

The form of the bandstructure shown in Fig. 7.9 cannot be derived using the simple approach outlined here. More complex methods are necessary for this calculation and these require the use of modern computing facilities. For further information about such calculations the reader is referred to theoretical reviews and more advanced text books.

7.4 Examples of Bandstructures

In the preceding sections, the origin of the electronic bandstructure, i.e., the existence of allowed and forbidden energy regions for a crystal electron, was attributed to the presence of Bragg reflections, which cause certain regions of the continuous spectrum of free-electron states to become forbidden. The alternative, but equally important, approach starts from the discrete energy levels of the free atom and explains the evolution of bands as a quasi-continuous splitting of the atomic levels due to the interaction with the other atoms of the crystal. In this picture, each band corresponds to an energy level of the free atom and may thus be classified as an s-,p-, or d-band etc. Having given a qualitative picture of a typical metal and a typical insulator in the examples of Figs. 1.1 and 7.9, in this section we will examine a few further examples of realistic bandstructures. Figure 7.10 shows how the highest occupied bands of the ionic crystal KCl can be imagined to evolve from the energy levels of isolated K^+ and Cl^- ions as they approach one another until they reach the equilibrium separation in the crystal. Even at this equilibrium separation, which is known from X-ray diffraction data, the occupied bands are extremely narrow. This indicates that there is relatively little overlap between the charge clouds of the individual ions. If theoretical results such as those shown in Fig. 7.10 are in good agreement with experimental data, then they allow important conclusions to be drawn about the form of the chemical bonding.

The entire information about the one-electron states in the periodic potential is of course contained in a representation of the complete $E(\boldsymbol{k})$ surfaces in wave-vector space. In order to simply portray the often complicated surfaces, one considers cross sections through the energy surfaces along directions of high symmetry in the first Brillouin zone. This is illustrated in Fig. 7.11 a for the example of an Al crystal. The definitions of the

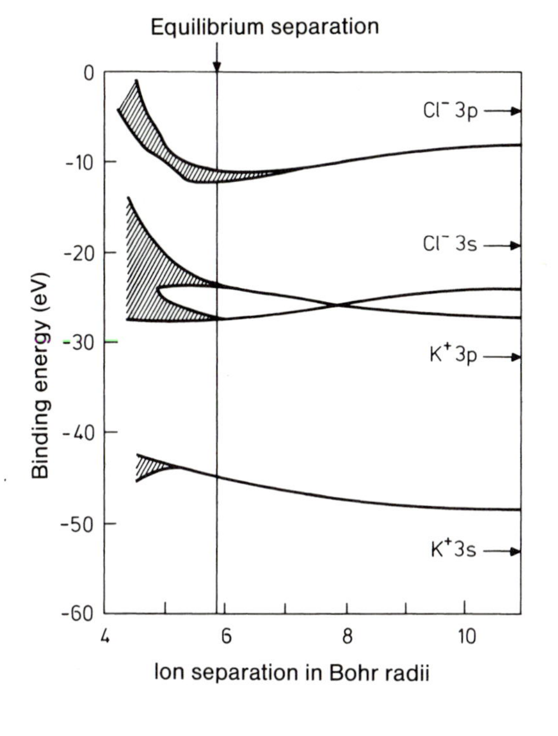

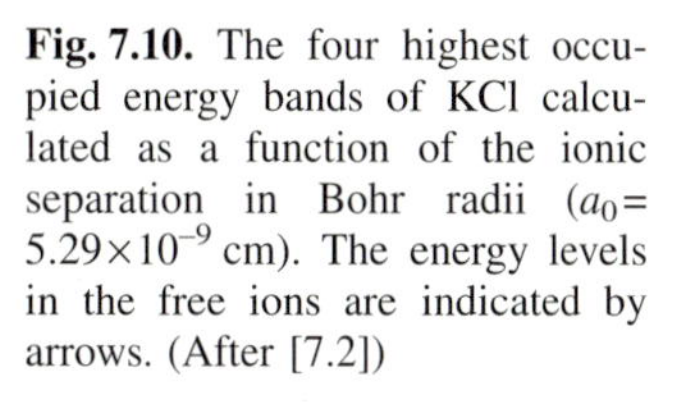

Fig. 7.10. The four highest occupied energy bands of KCl calculated as a function of the ionic separation in Bohr radii ($a_0 = 5.29 \times 10^{-9}$ cm). The energy levels in the free ions are indicated by arrows. (After [7.2])

symmetry directions and symmetry points in the first Brillouin zone of a face-centered cubic lattice are indicated in Figs. 3.8 and 7.11 b.

A striking feature of the Al bandstructure is that it can be described very well by the parabolic dependence of a free-electron gas (dotted lines). The energy gaps at the Brillouin zone edges are relatively small and the

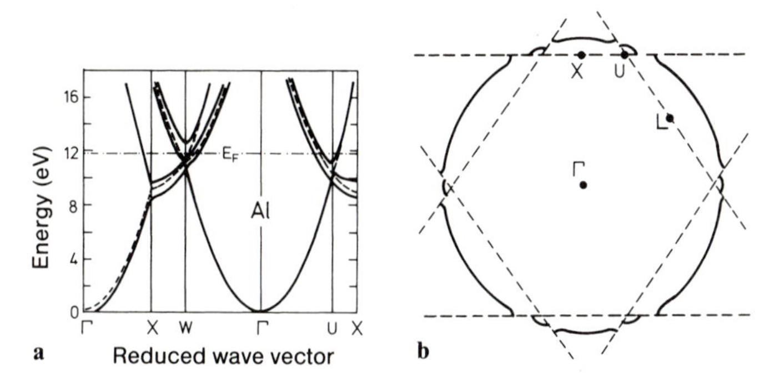

Fig. 7.11. (**a**) Theoretical bandstructure $E(\boldsymbol{k})$ for Al along directions of high symmetry (Γ is the center of the Brillouin zone). The dotted lines are the energy bands that one would obtain if the s- and p-electrons in Al were completely free ("empty" lattice). After [7.3]. (**b**) Cross section through the Brillouin zone of Al. The zone edges are indicated by the dashed lines. The Fermi "sphere" of Al (——) extends beyond the edges of the first Brillouin zone

complexity of the bandstructure stems largely from the fact that the energy parabolas are plotted in the reduced-zone scheme, i.e., "folded" back into the first Brillouin zone. This type of bandstructure is characteristic for simple metals. The similarity to the free electron gas is particularly pronounced for the alkali metals Li, Na and K.

The filling of the bands with the available electrons continues up to the Fermi energy E_F (indicated in Fig. 7.11). It can be seen that the corresponding constant energy surface, the so-called Fermi surface $E(\boldsymbol{k})=E_F$, intersects several bands. Thus, even for Al, the Fermi surface is not a simple continuous surface: whereas the Fermi surfaces of the alkali metals are almost spherical and are contained wholly within the first Brillouin zone, the "Fermi sphere" of Al extends just beyond the edges of the first Brillouin zone. The Bragg reflections occurring at these edges cause a slight deviation from the spherical form in these regions. This is shown qualitatively in Fig. 7.11 b in a cross section through three-dimensional $\boldsymbol{k}$-space.

In comparison to the simple metals, the band structures of the transition metals are considerably more complicated, due to the significant influence

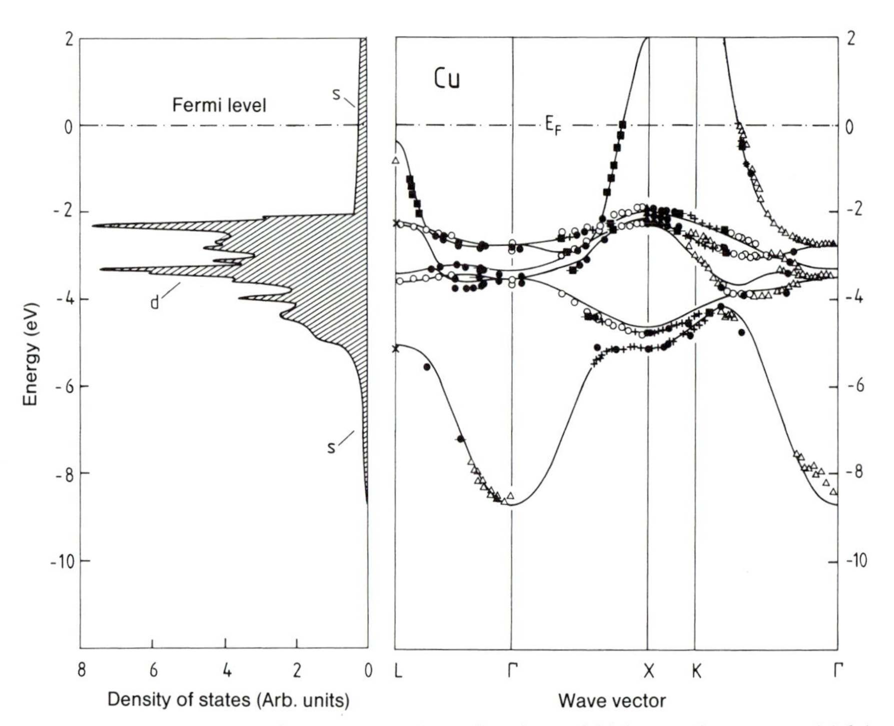

Fig. 7.12. Bandstructure $E(\boldsymbol{k})$ for copper along directions of high crystal symmetry (*right*). The experimental data were measured by various authors and were presented collectively by Courths and Hüfner [7.4]. The full lines showing the calculated energy bands and the density of states (*left*) are from [7.5]. The experimental data agree very well, not only among themselves, but also with the calculation

of the d-bands. Together with the bands that originate from s-levels and resemble the parabolic form of the free-electron gas, there are also very flat $E(\boldsymbol{k})$ bands, whose small energy width (low dispersion) can be attributed to the strong localization of the d-electrons. This is readily seen for the example of copper, whose bandstructure is illustrated in Fig. 7.12. For transition metals such as Pt, W, etc., where the Fermi level intersects the complex d-bands, the Fermi surfaces possess particularly complicated forms.

Other interesting phenomena, such as semiconducting properties (Chap. 12), occur when the bandstructure possesses an absolute gap, i.e., a so-called forbidden region: in this particular energy range and for all $\boldsymbol{k}$-directions in reciprocal space, there are no available electron states. A typical bandstructure of this type is that of germanium (Fig. 7.13). Like diamond and silicon, germanium crystallizes in the diamond structure, whereby the tetrahedral bonding of the individual atoms is a consequence of the formation of sp^3 hybrid orbitals. As was mentioned at the end of Sect. 7.3, the formation of sp^3 hybrids leads to the existence of sp^3 subbands. The lower

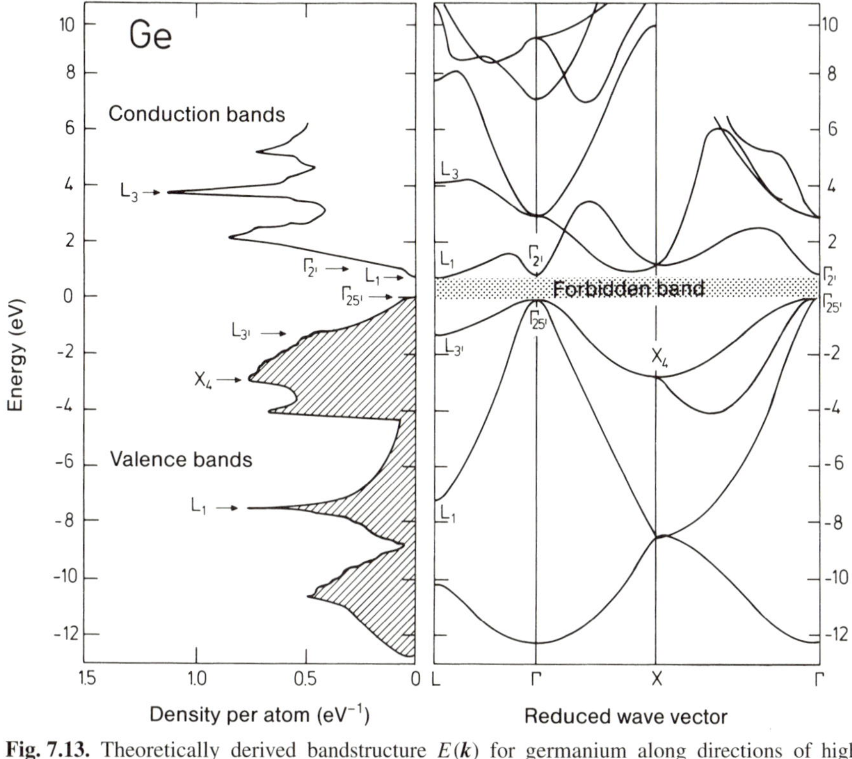

Fig. 7.13. Theoretically derived bandstructure $E(\boldsymbol{k})$ for germanium along directions of high symmetry (*right*), and the corresponding electronic density of states (*left*). A number of critical points, denoted according to their position in the Brillouin zone (Γ, X, L), can be seen to be associated with regions of the bandstructure where $E(\boldsymbol{k})$ has a horizontal tangent. The shaded region of the density of states corresponds to the states occupied by electrons [7.6]

of these (below the forbidden gap) are fully occupied whereas the higher-lying sp^3 subbands above the gap are unoccupied. The Fermi energy must therefore lie within the forbidden gap, a fact that will be important when we come to discuss the semiconducting properties of this crystal in Chap. 12.

7.5 The Density of States

In analogy to the thermal properties of the phonon system (Chap. 5), one finds also in the case of electronic states, that a knowledge of the density of states is sufficient to describe e.g. the energy content of the electron system. For certain electronic excitation mechanisms (e.g. non-angle-resolved photoemission spectroscopy; Panel V), in which the experiment effectively integrates over all $\boldsymbol{k}$-directions, one can often interpret the spectra simply in terms of the number of electron states per energy interval dE.

Once the energy surfaces $E(\boldsymbol{k})$ of the bandstructure are known, then in analogy to (5.4), the density of states is obtained by integrating over an energy shell $\{E(\boldsymbol{k}), E(\boldsymbol{k})+dE\}$ in $\boldsymbol{k}$-space:

$$dZ = \frac{V}{(2\pi)^3} \int_E^{E+dE} d\boldsymbol{k} \; , \tag{7.40}$$

where $V/(2\pi)^3$ is the density of states in $\boldsymbol{k}$-space. If the volume element $d\boldsymbol{k}$ is separated into an area element df_E on the energy surface and a component $dk_\perp$ normal to this surface (Fig. 5.1), i.e., $d\boldsymbol{k} = df_E \, dk_\perp$, then with $dE = |\mathrm{grad}_{\boldsymbol{k}} E| \, dk_\perp$ one has

$$D(E)dE = \frac{1}{(2\pi)^3} \left(\int_{E(\boldsymbol{k})=\mathrm{const}} \frac{df_\mathrm{E}}{|\mathrm{grad}_{\boldsymbol{k}} E(\boldsymbol{k})|} \right) dE \; . \tag{7.41}$$

This density of states $D(E)$ is given here in relation to the real volume V of the crystal in order to obtain a crystal-specific quantity. It should be remembered that, due to spin degeneracy, each state can accommodate two electrons.

The main structure in the function $D(E)$ is again yielded by those points in $\boldsymbol{k}$-space for which $|\mathrm{grad}_{\boldsymbol{k}} E|$ vanishes, i.e., where the energy surfaces are flat. These points are known as van Hove singularities or critical points. In three dimensions $D(E)$ does not become singular near to these critical points because an expansion of $E(\boldsymbol{k})$ about the extremum ($E \sim \boldsymbol{k}^2$) implies that $|\mathrm{grad}_{\boldsymbol{k}} E|^{-1}$ has a k^{-1} singularity. Thus the integration of the $E(\boldsymbol{k})$ surface (7.41) yields a linear k dependence. In three dimensions the density of states near to a critical point therefore has the form shown in Fig. 7.14. For one-dimensional bandstructures, which, to a good approximation, can be used to de-

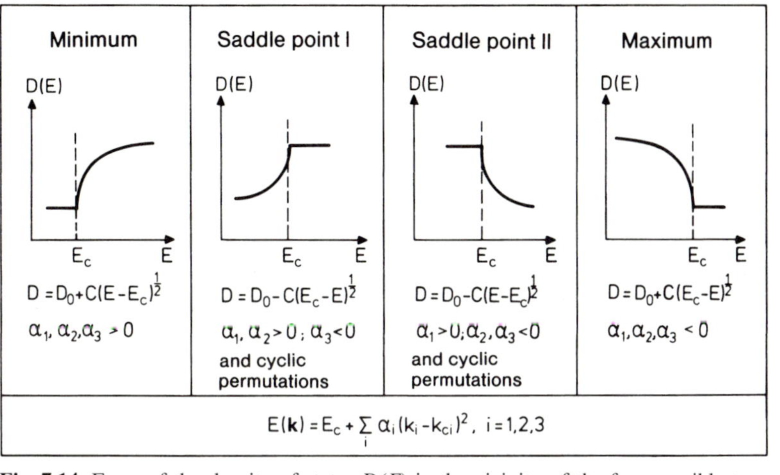

Fig. 7.14. Form of the density of states $D(E)$ in the vicinity of the four possible types of critical point in three dimensions. The energy of the critical points is denoted by E_c and the corresponding $\boldsymbol{k}$-space position by k_{ci} $(i=1,2,3)$. In the parabolic approximation, the energy band has the form $E(\boldsymbol{k})=E_c+\sum_i \alpha_i \cdot (k_i - k_{ci})^2$ in the vicinity of a critical point, where $\alpha_i = \text{const}$. The quantities D_0 and C in the figure are also constants

scribe one-dimensional organic semiconductors, the density of states diverges at the critical points, although its integral remains finite. (See discussion of density of states of lattice vibrations for a linear chain; Sect. 4.3.)

Theoretical densities of states may be obtained by integrating in $\boldsymbol{k}$-space over the first Brillouin zone for a calculated bandstructure, and may then be compared with experimental data, for example from photoemission spectroscopy (Panel V). As such, they provide an important point of connection between calculated bandstructures and the experimental data. In the integration over $\boldsymbol{k}$-space the main contributions to the density of states are derived from the critical points. Since the critical points usually occur along lines of, or at points of high symmetry in $\boldsymbol{k}$-space, this lends further justification for the preferred representation of bandstructures along lines of high symmetry, e.g., $\Gamma K, \Gamma X, \Gamma L$, etc. In the intermediate regions one can expect only minor contributions, and it is sometimes possible to make use of simple mathematical interpolation to obtain the bandstructure there.

The relationship between a calculated bandstructure and the corresponding density of states is nicely demonstrated by the case of the semiconductor germanium as shown in Fig. 7.13. Important contributions, i.e., maxima in the density of states, are clearly correlated with flat portions of the $E(\boldsymbol{k})$ curves along directions of high symmetry. Also evident is the absolute band gap between the fully occupied valence band states and the (at low temperatures) unoccupied conduction band states. This forbidden region has a width of about 0.7 eV for germanium.

As an example for a transition metal, Fig. 7.12 shows the calculated density of states for copper. The density of states is obtained by integrating over the bandstructure $E(\boldsymbol{k})$, which is also shown in the figure. The sharp structures seen between –2 and –6 eV below the Fermi level can be readily attributed to critical points of the relatively flat d-bands. In the $E(\boldsymbol{k})$ plot (Fig. 7.12) one can also recognize the parabola-like shape of the s-band with its minimum at the Γ-point. This s-band is responsible for the structureless contribution to the density of states beginning at about –9.5 eV. Below –6 eV one cannot fail to notice the distinct similarity to the "free-electron-gas" parabolic density of states. At the Fermi level it is again the s-electrons that produce the density of states. This explains why the model of a free-electron gas in a box (Chap. 6) yields relatively good results for copper.

However, as shown in Sect. 6.4, this is far from true for Fe, Ni, Co and the other transition metals. For these metals the Fermi level intersects the high density of states of the d-bands, which are therefore only partially filled. For the ferromagnets Fe, Ni and Co there is an additional complication, which will be dealt with in detail in Sect. 8.3. In these metals there is a ferromagnetic phase at $T < T_C$ (Curie Temperature) in which the atomic spins are aligned. One then has two distinct densities of states – one for electrons whose spin is parallel to the spontaneous magnetization $\boldsymbol{M}$ and one for the electrons with antiparallel spin orientation. Figure 8.6 shows these two densities of states for Ni. To derive such densities of states requires a calculated bandstructure $E(\boldsymbol{k})$ in which the electron–electron interaction is explicitly included.

7.6 Density of States in Non-Crystalline Solids

A crystalline solid is translationally invariant with respect to a lattice vector and the electronic states can be classified according to the components of the wave-vector k_x, k_y, and k_z. Only then are the eigen-values of the electronic states described by a band structure $E(\boldsymbol{k})$. However, the concept of a density of states per energy and volume and of allowed and forbidden bands does not require translational invariance. Thus, a non-crystalline solid possesses a defined density of states, just as the crystalline solid, provided that the non-crystalline solid is sufficiently homogeneous in composition and structure on a mesoscopic length scale. Many materials exist in a crystalline as well as non-crystalline phases, as under-cooled melts (glasses). Examples are SiO_2 and Al_2O_3. As is well known, these two materials are optically transparent in either phase. Both phases must therefore have an energy gap between occupied and unoccupied states larger than 3 eV. Thus it seems that the magnitude of the energy gap does not depend on the existence of crystalline order. The amorphous phases of SiO_2 and Al_2O_3, and likewise the amorphous phases of Si and Ge differ from the crystalline phases only by the missing long-range order. The nearest and next nearest neighbor configuration is rather similar. In Si and Ge the local order is determined by the sp^3-bonding to the nearest-neighbors. Since the electronic structure is dominated by the local bonding, the magnitude of the energy gap is nearly equal for the crystalline and the amorphous phases. Likewise the densities of states for other electron energies are rather similar. Only the sharp features in the band structure result from the critical points (Fig. 7.13) are absent in the amorphous phases.

Amorphous phases of Si and Ge can be fabricated by growth at (relatively) low temperatures. At these temperatures long-range order, which requires diffusion of material, is not established and the state of lowest free enthalpy which is the crystalline phase, is not reached. In order to saturate nevertheless as many bonds as possible the sp^3-tetraeders of the local bonding configuration must be slightly distorted. Hence, instead of defined nearest-neighbor bond angles and bond distances the amorphous phase possesses a (narrow Gaussian) distribution of bond angles and distances. As shown in Fig. 7.9 the magnitude of the energy gap depends on the distance between the nearest-neighbors. The fuzzy distribution of bond distances in amorphous solids therefore leads to a fuzzy band gap: the density of states acquires exponential tails that extend into the forbidden zone (Fig. 7.15). The magnitude of the density of states at any energy reflects the probability for the realization of a particular bond distance and bond angle. States that reach far into the forbidden zone (forbidden in crystals) are due to structural configurations that are realized only rarely. The mean distance between such configurations is therefore large and consequently the wave functions of these states do not overlap. Electronic states deeper in the forbidden

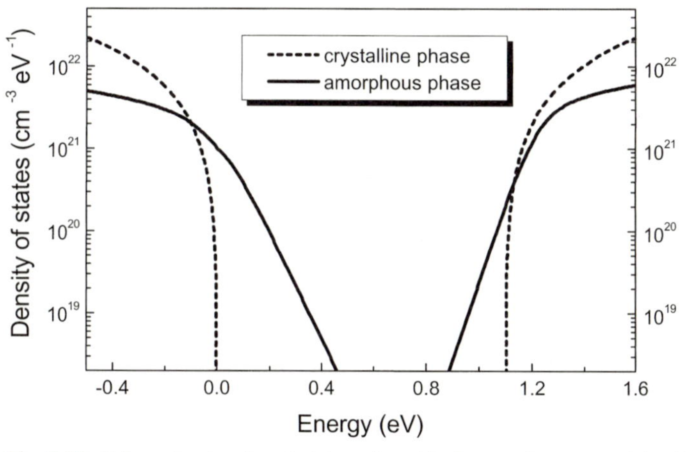

Fig. 7.15. Schematic density of states of an ideal amorphous material with saturated tetrahedral bonds to the nearest-neighbors (*full line*). The numbers for the density of states correspond to amorphous silicon. Compared to crystalline silicon (*dashed line*) the density of states possesses exponential tails into the band gap. Non-saturated bonds in the amorphous network lead to additional states in the forbidden zone. For practical applications of amorphous silicon (e.g. in solar cells) one attempts to reduce the number of unsaturated bonds by adding hydrogen

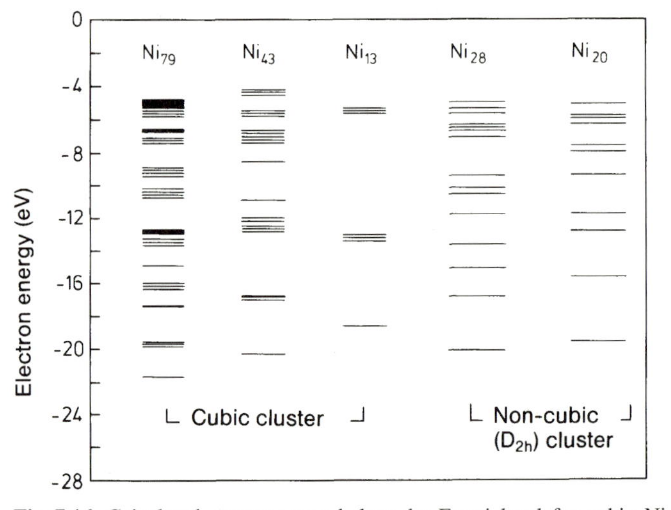

Fig. 7.16. Calculated energy terms below the Fermi level for cubic Ni-clusters consisting of 13, 43, and 79 atoms and non-cubic clusters of the symmetry D_{2h} with 20 and 28 atoms [7.7]. The energy scale refers to the vacuum level. States displayed as thick lines consist of several, nearly degenerate states. The work function of the solid (5.2 eV [6.5]) as well as the high density of *d*-states at the Fermi level is quite well represented by the cluster with 79 atoms.

zone are therefore localized in space. Electrons in these states are not free to move about, unlike the electrons in Bloch states (see also Sect. 9.8).

The calculation of electronic states in "amorphous" solids is significantly more difficult than for crystalline solids. This is in part due to the fact that the characterization of a solid as being amorphous is not a well-defined one. While the atom positions in a crystal are unique, an infinite number of different realizations of an amorphous solid exist. A calculation of the electronic density of states of an amorphous solid requires the input of a particular distribution of bond angles and distances. With that input, one may compose a structure of an amorphous solid and treat this agglomeration of atoms as a large molecule. Such molecules consisting of many atoms or atomic units of the same type are named "clusters". The electronic states of clusters are discrete because of the finite number of atoms that make up the cluster. With increasing number of atoms in the cluster a continuous density of states can be defined. Technically, a density of states is frequently calculated by broadening the individual states artificially (Fig. 7.16). Theoretical methods to extend clusters into all three dimensions in a non-periodic way have also been developed.

Initially, clusters were merely a construct of the theorists to calculate the electronic properties of amorphous solids. Since the experimental realization of clusters with a defined size and composition [7.8], research on clusters has become a field of its own, situated between Moleculer Physics, Solid State Physics, and Material Science.

Problems

7.1 Solve the Schrödinger equation for the potential

$$V(x, y, z) = \begin{cases} \infty, & x < 0 \\ 0, & x \geq 0 \end{cases}$$

and calculate the charge density

$$\varrho(x, y, z) = (-2\,\mathrm{e}) \sum_{k_x, k_y, k_z} |\Psi_{\boldsymbol{k}}|^2 \; ,$$

where the maximum $\boldsymbol{k}$ value is determined by the Fermi energy E_F. Sketch $\varrho(x)$ and discuss the result (Friedel oscillations). Consider the cases of a typical metallic electron density of $10^{22}\,\mathrm{cm}^{-3}$ and of a weakly doped semiconductor with a density of $10^{16}\,\mathrm{cm}^{-3}$.

7.2 A two-dimensional electron gas is described in reciprocal space by a two-dimensional lattice whose Brillouin zone is a primitive square.

a) Show that the kinetic energy of a free electron in a corner of the first Brillouin zone is a factor of two larger than that of an electron with k vector in the middle of the Brillouin-zone edge.
b) How large is the corresponding factor for a primitive lattice in three dimensions?
c) Show that band overlap can occur even in a two-dimensional lattice. Do this by drawing the following dispersion relations for the case of free electrons that are perturbed by a weak periodic potential:

$$E(k_y) \quad \text{for} \quad k_x = 0 \; ,$$

$$E(k_y) \quad \text{for} \quad k_x = \frac{\pi}{a} \; ,$$

$$E(k) \quad \text{for} \quad k_x = k_y \; .$$

7.3 As a simple model of a crystal consider a one-dimensional chain of $2N$ atoms at distances a_0 from each other. Let $\phi_i(x{-}n\,a_0)$ be the correct wavefunction for an electron with energy E_i at the atom located at $n\,a_0$ (eigenvalue for a single-isolated atom). Under what conditions can the one-electron wavefunction $\Psi(x)$ for an electron delocalized over the whole chain of atoms be approximated by a linear combination of atomic orbitals (LCAO)

$$\Psi(x) \cong \sum_{n=-N}^{N} c_n \phi_i(x - n\,a_0) \; ?$$

Choose the coefficients c_n such that for $N \to \infty$ the wave-function $\Psi(x)$ is a Bloch wave.

7.4 a) Consider points in the vicinity of the minimum of an electronic band at sufficiently small k values that $E(\boldsymbol{k})$ can be written in parabolic approximation as

$$E(\boldsymbol{k}) = E_c + \frac{\hbar^2}{2}\left(\frac{k_x^2}{m_x} + \frac{k_y^2}{m_y} + \frac{k_z^2}{m_z}\right),$$

with m_x, m_y and m_z as positive constants. Show that the density of states $D(E)$ is proportional to $(E-E_c)^{1/2}$ around the critical point $E_c(\boldsymbol{k}=\boldsymbol{0})$.

b) Consider the density of states in the neighborhood of a saddle point, where

$$E(\boldsymbol{k}) = E_c + \frac{\hbar^2}{2}\left(\frac{k_x^2}{m_x} + \frac{k_y^2}{m_y} - \frac{k_z^2}{m_z}\right)$$

with positive m_x, m_y, m_z. Show that the density of states can be written near E_c as

$$D(E) \propto \begin{cases} \text{const for } E > E_c \\ D_0 - C(E_c - E)^{1/2} \text{ for } E < E_c \,. \end{cases}$$

Sketch $D(E)$ in the vicinity of (a) a minimum and (b) a saddle point of $E(\boldsymbol{k})$.

7.5 On the basis of the electronic bandstructure and corresponding density of states, explain why copper, in contrast to many other metals, appears colored, i.e., exhibits pronounced spectral structure in its optical constants in the visible spectral range.

7.6 Explain why and how, for diamond and silicon, the energy of the forbidden band E_g changes with increasing temperature.

7.7 Consider an angle resolved UV photoemission spectroscopy (ARUPS) experiment, where UV photons of energy 40.8 eV are incident on the (100) surface of a cubic transition metal with a work function of 4.5 eV. Photoemitted electrons from d-states at 2.2 eV below the Fermi level are detected at an angle of 45° to the surface normal and in the [100] azimuth.

a) Calculate the wave vector $\boldsymbol{k}$ of the emitted electrons.

b) What problem arises in deriving the wave vector $\boldsymbol{k}_i$ of the electronic state from which the electron is released. Consider the components $k_{\parallel}$ and $k_{\perp}$ (parallel and normal to surface) separately.

Panel V
Photoemission Spectroscopy

Photoelectron spectroscopy has become established as one of the most important experimental methods for deriving information about band structures and densities of states [V.1]. The target crystal is irradiated with photons of a relatively high energy $\hbar\omega$. This causes electrons to be excited from occupied (band) states into empty states of the quasi-continuum above the vacuum level E_{vac}. Provided these electrons have sufficient excess kinetic energy E_{kin} they can overcome the work function ϕ and escape from the crystal.

On account of the relation

$$\hbar\omega = \phi + E_{kin} + E_b \tag{V.1}$$

a measurement of the spectrum $N(E_{kin})$ of the photoexcited electrons corresponds directly to the distribution of occupied electronic states (binding energy E_b) in the solid (Fig. V.1 b). In the measured spectrum, the "image" of the occupied state density is superimposed on a background of so-called "true secondary electrons". These are electrons that have lost energy via any of a number of inelastic scattering processes during their escape from the solid. Due to the relatively strong interaction of the electrons with the solid, only those electrons excited within regions close to the surface can escape. (The escape depth is ~ 5 Å for electrons with $50\,\mathrm{eV} < E_{kin} < 100\,\mathrm{eV}$.) The method is thus surface sensitive and thereby provides a powerful tool for studying the electronic properties of solid surfaces. The surface sensitivity also means that for investigations of bulk band structures it is essential that the surface is clean, i.e., the crystals must be prepared under ultra-high vacuum conditions (pressure $\leq 10^{-8}$ Pa).

Commonly used photon sources are the gas discharge lamps with the following spectral lines: He (21.2 eV and 40.8 eV) and Ne (16.8 eV and 26.9 eV). The method is then called UPS (UV photoemission spectroscopy). Higher photon energy can be obtained with X-ray tubes (Al: 1486 eV and Mg: 1253 eV), in which case the method is termed XPS (X-ray photoemission spectroscopy) or ESCA (electron spectroscopy for chemical analysis). To a large extent, however, these photon sources are now being superceded by synchrotron radiation sources (Panel XI), which, with the aid of UV monochromators, allow a continuous variation of the photon energy. In this way one may also fix the kinetic energy of the analyzed electrons and record the spectrum by varying $\hbar\omega$. As energy selectors for the measurement of $N(E_{kin})$, a variety of different electrostatic electron analyzers are available. Figure V.1 a shows a so-called 127° analyzer. By apply-

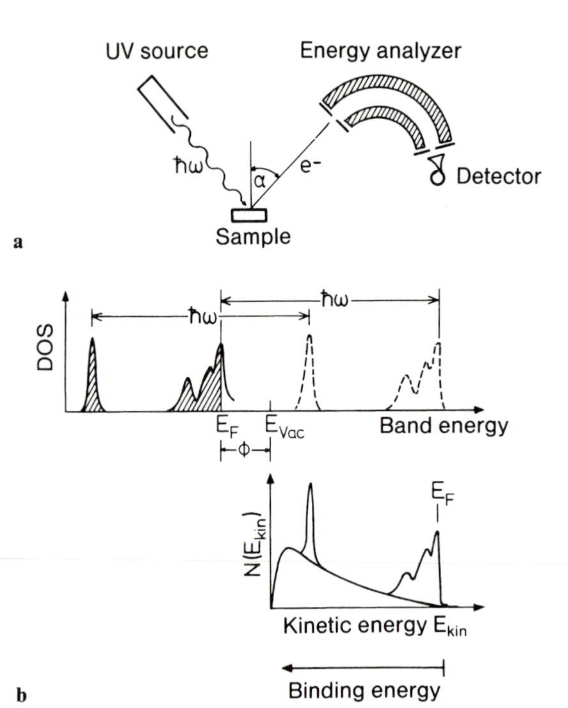

Fig. V.1. (**a**) Schematic arrangement for the measurement of photoemission spectra. The target, energy analyzer, and detector are in a UHV chamber to which the discharge lamp serving as a UV source is connected with no window but with differential pumping: (**b**) Schematic illustration of the measurement of photoemission for a transition metal whose Fermi level E_F lies in the upper region of the *d*-bands (occupied states are shaded). The work function is $\phi = E_{vac} - E_F$. The electrons that are excited into the quasi-continuous empty states can escape from the crystal and are detected in the vacuum with their excess kinetic energy E_{kin}

ing a variable voltage across the cylindrical sector plates, one can adjust the energy of the electrons passing through the exit slit. The $\boldsymbol{k}$ vector of the photoemitted electrons in vacuum $\boldsymbol{k}_v$ is determined by the kinetic energy and by the azimuthal and polar exit angles. For ordered surfaces, the translational symmetry leads to conservation of the component of the $\boldsymbol{k}$ vector parallel to the surface when an electron passes through the surface. Thus the parallel component of the photoemitted electron also determines that of the initial state. This is not so for the perpendicular component, which is not conserved because the electron has to overcome a potential barrier when leaving the solid. Therefore the initial state cannot be deduced directly from an angle-resolved photoemission spectrum.

A method for determining the perpendicular component of the $\boldsymbol{k}$ vector of the initial state within the solid is demonstrated in Fig. V.2. This shows three photoemission spectra for copper *d*-electrons. Spectrum (a) was measured from a Cu(111) surface, the direction perpendicular to the surface, i.e. in the [111] direction. This corresponds to the ΓL direction in the Brillouin zone (see also Fig. 7.12). The magnitude of the $\boldsymbol{k}$-vector still has to be determined. This is achieved in a second step by recording the spectra from a (110) surface at various polar angles in a plane that contains the [111] direction. On this surface the [111] direction lies at a polar angle of 35.2°. However, the spectrum recorded in this direction looks quite different from that from the (111) surface in the [111] direction. This is because

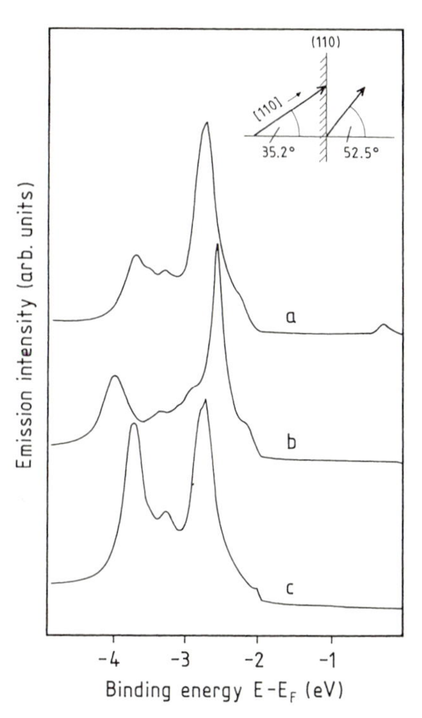

Fig. V.2. Angle-resolved photoemission spectra from copper surfaces. Spectrum (a) was recorded from a Cu(111) surface in the [111] direction. The spectra (b) and (c) stem from a Cu(110) surface and were measured at polar angles of 35.2° and 52.5°, respectively. The magnitude of the ***k***-vector of the initial state corresponding to spectrum (a) is determined from the projection of the ***k***-vector for which the same spectrum appears from the (110) surface. (After [V.2])

the electrons are refracted as they leave the surface. This refraction means that a spectrum like that from the (111) surface now occurs at a different angle, namely, at 52.5°. Using conservation of parallel momentum we can now establish the full ***k***-vector of the initial state (see insert in Fig. V.2). The construction gives the magnitude of the ***k***-vector as

$$k_{111} = k_{\text{ext}} \sin 52.5^\circ / \sin 35.2^\circ \, . \qquad \text{(V.2)}$$

As an exercise you may like to calculate which data points in Fig. 7.12 can be taken from the spectra.

With the help of this and other methods, one can determine both the binding energy of occupied electron states and their corresponding wave vectors. Angle-resolved photoemission is the most important method for the experimental determination of electronic bandstructures $E(\boldsymbol{k})$. For the sake of completeness, let us mention that it is also possible to establish the bandstructure of the unoccupied states. This is done by means of inverse photoemission: the surface of the solid is bombarded with electrons of a known momentum and one records the UV light that is emitted.

By integration the photoemission spectra over all angles comprising the halfspace, or by measuring with an angle-integrated analyzer, one can obtain a qualitative picture of the density of occupied states.

As an example, Fig. V.3 shows the UPS spectra measured for ZnO $(10\bar{1}0)$ surfaces with the two different spectral lines of He. The spectra are

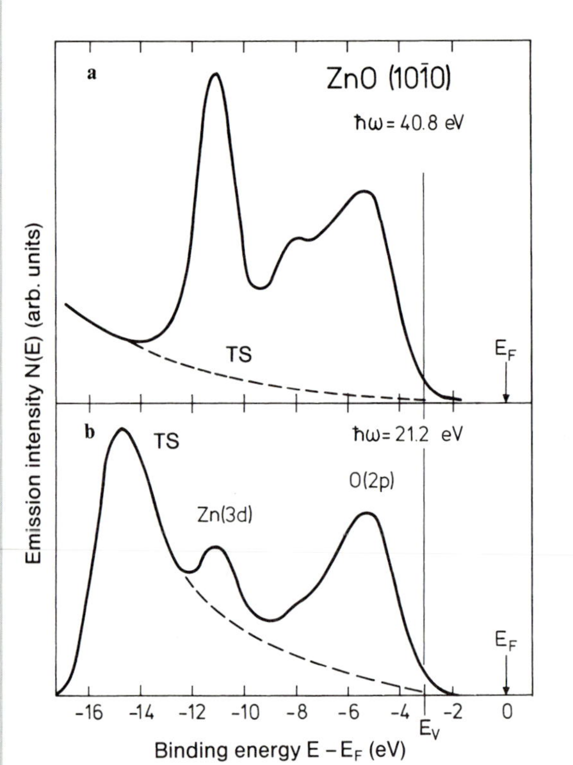

Fig. V.3 a, b. UPS spectra of a ZnO (10$\bar{1}$0) surface that was cleaned in UHV by annealing. The excitation was provided by the He II (**a**) and He I (**b**) lines. The background due to true secondary (TS) electrons is indicated by the dashed line [V.3]

plotted not against the kinetic energy, but as a function of the binding energy with respect to the Fermi level E_F. As a result of matrix element effects (Sect. 11.10), the Zn $(3d)$ and O $(2p)$ bands have different intensities in the two spectra. In spectrum (b) one sees the entire background due to the true secondary electrons (TS) (interpolated qualitatively by the dashed line). The Fermi level of ZnO lies in the forbidden energy gap just below the conduction band edge and at about 3.2 eV above the energy E_V at which the onset of emission is observed.

References

V.1 M. Cardona, L. Ley (eds.): *Photoemission in Solids I, II;* Topics Appl. Phys., Vols. 26, 27 (Springer, Berlin, Heidelberg 1979)
B. Feuerbacher, B. Fitton, R.F. Willis (eds.): *Photoemission and the Electronic Properties of Surfaces* (Wiley, New York 1978)
H. Lüth: *Solid Surfaces, Interfaces and Thin Films* (Springer, Berlin, Heidelberg 2001) 4th. edn.
S. Hüfner: *Photoelectron Spectroscopy*, 2nd edn., Springer Ser. Solid-State Sci., Vol. 82 (Springer, Berlin, Heidelberg 1996)
V.2 R. Courths, S. Hüfner: Physics Reports **112**, 55 (1984)
V.3 H. Lüth, G.W. Rubloff, W.D. Grobmann: Solid State Commun. **18**, 1427 (1975)

8 Magnetism

In our previous discussions of the electronic structure of materials we have assumed the one-electron approximation. The energy levels and the bandstructure were calculated for an electron in an effective potential consisting of the potential of the ion cores and an average potential due to the other electrons. Within this model quite acceptable bandstructures can be calculated. However, another aspect of the one-electron model is more important than the qualitative agreement with experiment and the (in principle) simple calculational method: Within the one-electron model it is also possible to understand conceptually the excited states of the electronic system, resulting for example from the interaction with photons and other particles or from thermal excitation. Just as the energy levels of the hydrogen atom serve as a model for describing the energy levels of all elements, so the one-electron model is the basic model for our understanding of the solid state. Furthermore, there are phenomena associated with the collective behavior of the electrons which can nonetheless be tackled within this framework; for example, Thomas-Fermi screening (Sect. 6.5) or the excitation of charge density waves (Sect. 11.9).

In the case of magnetic phenomena in solids, in particular ferromagnetism and antiferromagnetism, the one-electron and many-electron aspects are mixed in such a way that it is difficult to formulate a simple basic model. We will consider, for instance, excited states in which one electron spin is flipped but in which all valence electrons take part (spin wave). In addition, the electronic theory of magnetism is complicated by the fact that it contains both collective and local aspects. A particularly important topic in this chapter is the ferromagnetism of the $3d$ metals Ni, Co and Fe, which exists due to the exchange interaction between the largely delocalized $3d$ electrons. A local description is suitable for most magnetic compounds and particularly for the $4f$ transition metals and their compounds. Antiferromagnetism and spin waves can also be described relatively easily in terms of an exchange interaction between localized electrons.

8.1 Diamagnetism and Paramagnetism

The physical quantities magnetic field strength, $\boldsymbol{H}$, and magnetic induction, $\boldsymbol{B}$, in vacuum, are related by the equation

$$\boldsymbol{B} = \mu_0 \boldsymbol{H} \tag{8.1}$$

where $\mu_0 = 4\pi \times 10^{-7}$ Vs/Am is the permeability of free space. The magnetic state of the system (Sommerfeld system) will be specified by the magnetization $\boldsymbol{M}$, which is related to $\boldsymbol{B}$ and $\boldsymbol{H}$ by

$$\boldsymbol{B} = \mu_0 (\boldsymbol{H} + \boldsymbol{M}) . \tag{8.2}$$

The magnetization $\boldsymbol{M}$ is equal to density of magnetic dipole moments $\boldsymbol{m}$:

$$\boldsymbol{M} = \boldsymbol{m} \frac{N}{V} . \tag{8.3}$$

For the following discussion it is convenient to introduce, instead of the external field $\boldsymbol{H}$, an external induction $\boldsymbol{B}_0 = \mu_0 \boldsymbol{H}$, and, in order to keep the language straightforward, we will call the quantity $\boldsymbol{B}_0$ simply the "magnetic field strength". In most cases there is a linear relation between the "field" $\boldsymbol{B}_0$ and the magnetization $\boldsymbol{M}$:

$$\mu_0 \boldsymbol{M} = \chi \boldsymbol{B}_0 \tag{8.4}$$

where χ is the magnetic susceptibility. If χ is negative, then the induced magnetic polarization is opposite in sign to the applied field. Such behavior is denoted diamagnetic, while the reverse behavior is termed paramagnetic, and is characterized by $\chi > 0$. In general, the susceptibility of atoms, and therefore of solids, comprises a dia- and a paramagnetic component, which we will denote as χ_d and χ_p. The paramagnetic component is related to the orientation of intrinsic magnetic moments, which originate from the angular momentum and the spin of the electrons. For instance, the magnetic dipole moment of an electron due to its angular momentum is

$$\boldsymbol{m} = -\frac{e}{2m} \sum_i \boldsymbol{r}_i \times \boldsymbol{p}_i = -\mu_B \boldsymbol{L} \tag{8.5}$$

with $\hbar \boldsymbol{L} = \sum_i \boldsymbol{r}_i \times \boldsymbol{p}_i$ and the Bohr magneton $\mu_B = (e\hbar/2m)$ $(=5.7884 \cdot 10^{-5}$ eV/T $= 9.2741 \cdot 10^{-24}$ J/T; 1 T = 1 Tesla = 1 Vs/m^2).

The negative sign in (8.5) follows from the fact that the electric current has the opposite sense to that of the particle current because of the negative charge of the electron. (The elementary charge e is treated throughout this book as a positive number.) Besides a magnetic moment due to angular momentum, the electrons also possess a magnetic moment due to spin, and these add up to give the spin moment of the whole atom

$$\boldsymbol{m} = \mu_B g_0 \sum_i \boldsymbol{s}_i = \mu_B g_0 \boldsymbol{S} . \tag{8.6}$$

Here g_0 is the electronic g factor ($g_0 = 2.0023$) and $\boldsymbol{s}_i$ are the (negative) electron spins. As already stated (8.5, 6), $\boldsymbol{L}$ and $\boldsymbol{S}$ can, without further ado, be treated as operators. The choice of sign of the spin operator is best made so that the spin operator and the magnetic moment have the same sign. By evaluating the expectation values of the operators $\boldsymbol{L}$ and $\boldsymbol{S}$ for atoms, it can be seen that a non-vanishing expectation value results only for open shells. For closed shells the sum of the angular momentum and spins is 0. In solids, we have open shells for transition metals and rare earths. Paramagnetic behavior is thus expected for both.

Besides this paramagnetism due to the electrons, one must also consider diamagnetism. The latter results from the induction of eddy currents by an external magnetic field. According to Lenz's rule, the magnetic moment of these induced currents is opposed to the applied field. The susceptibility thereby acquires a negative, diamagnetic contribution. To calculate this diamagnetic contribution, we must replace the momentum operator in the Schrödinger equation, $\boldsymbol{p}$, by $\boldsymbol{p}+e\boldsymbol{A}$. Here $\boldsymbol{A}$ is the vector potential, which is related to the field $\boldsymbol{B}_0$ by

$$\boldsymbol{B}_0 = \mathrm{rot}\boldsymbol{A} \quad \text{and} \quad \mathrm{div}\boldsymbol{A} = 0 \; . \tag{8.7}$$

For a homogeneous field $\boldsymbol{B}_0$, a possible choice of vector potential is

$$\boldsymbol{A} = -\tfrac{1}{2}\boldsymbol{r} \times \boldsymbol{B}_0 \; . \tag{8.8}$$

It is easy to show that (8.8) fulfills the conditions (8.7). We can now write the kinetic component of the Hamiltonian as

$$\begin{aligned}\mathscr{H}_{\text{kin}} &= \frac{1}{2m}\sum_i (\boldsymbol{p}_i + e\boldsymbol{A}_i)^2 = \frac{1}{2m}\sum_i \left(\boldsymbol{p}_i - \frac{e}{2}\boldsymbol{r}_i \times \boldsymbol{B}_0\right)^2 \\ &= \frac{1}{2m}\sum_i \boldsymbol{p}_i^2 + \frac{e}{2m}\sum_i (\boldsymbol{r}_i \times \boldsymbol{p}_i)_z \cdot \boldsymbol{B}_{0_z} + \frac{e^2 \boldsymbol{B}_{0_z}^2}{8m}\sum_i (x_i^2 + y_i^2) \; . \end{aligned} \tag{8.9}$$

In the second step of the calculation we have assumed that $\boldsymbol{B}_0$ is parallel to the z axis, and have used the commutation rule to exchange terms in the triple product. The sum index runs over all electrons. The second term in the above expression is nothing other than the paramagnetism due to the angular momentum, which was discussed above.

By comparing (8.9) with (8.5), one sees that the expectation value of the magnetic moment in a state φ is

$$\boldsymbol{m} = -\frac{\partial\langle\varphi|\mathscr{H}|\varphi\rangle}{\partial \boldsymbol{B}_{0_z}} = -\mu_{\mathrm{B}}\langle\varphi|\boldsymbol{L}_z|\varphi\rangle - \frac{e^2}{4m}\boldsymbol{B}_{0_z}\langle\varphi|\sum_i (x_i^2 + y_i^2)|\varphi\rangle \; . \tag{8.10}$$

The first term in (8.10) represents a magnetic moment that exists even in the absence of a magnetic field. This term, together with the occupation statistics of the energy levels for different orientations of the magnetic moment in an external magnetic field, yields the temperature dependence of the paramagnetism.

The second term is responsible for diamagnetism. Due to the spherically symmetric charge distribution of atoms, we can set

$$\langle\varphi|x_i^2|\varphi\rangle = \langle\varphi|y_i^2|\varphi\rangle = \tfrac{1}{3}\langle\varphi|r_i^2|\varphi\rangle \; , \tag{8.11}$$

and therefore for the susceptibility we obtain (in the SI system)

$$\chi = -\frac{e^2 n}{6m}\mu_0 \sum_i \langle\varphi|r_i^2|\varphi\rangle \,, \tag{8.12}$$

where n is the number of atoms per unit volume. In the sum over the matrix elements the electrons in the outer shells are naturally of greatest importance because their mean square distance from the nucleus is largest. If the number of outer electrons is Z_a, and we insert in place of $\overline{r_i^2}$ the square of the ionic or atomic radius r_a, then we obtain

$$\chi \sim -\frac{e^2}{6m}\mu_0\, n\, Z_a\, r_a^2 \,. \tag{8.13}$$

The measured values of the diamagnetic susceptibility for atoms and ions with closed shells are indeed found to be in good agreement with $Z_a r_a^2$ (Fig. 8.1). However, the values indicate that the above estimate should be multiplied by a prefactor of approximately 0.35. From Fig. 8.1 we can see that for typical solid state densities of 0.2 mol/cm^3, the susceptibility is about 10^{-4} (SI), i.e. small compared with 1. A similar order of magnitude results for paramagnetic contributions. We therefore find that, apart from

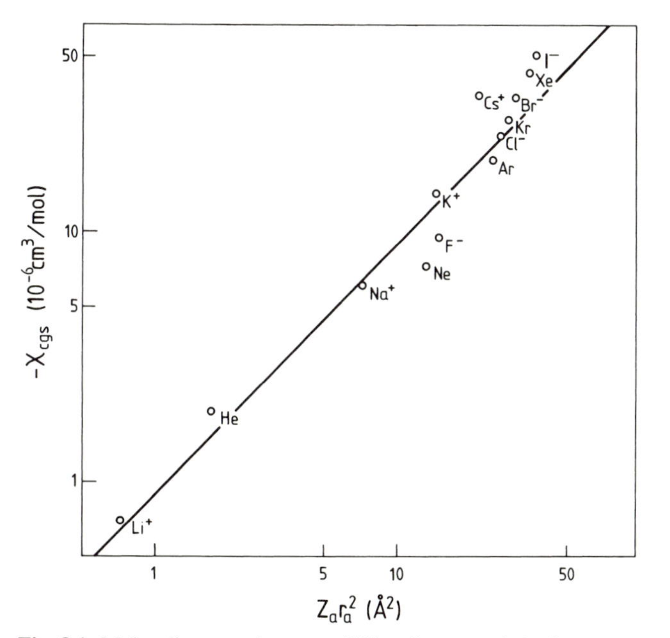

Fig. 8.1. Molar diamagnetic susceptibility (in cgs units) of atoms and ions with closed shells as function of $Z_a r_a^2$. In order to obtain the susceptibility of a material, e.g. a gas composed of these atoms or ions, one has to multiply by the density in mol cm^{-3}. If one inserts the value of the ionic radius r_a in [Å], then the value of $Z_a r_a^2$ immediately provides an estimate for χ in units of 10^{-6} cm^3/mol. To convert χ to SI units one must simply multiply by 4π

the case of ferromagnetism, to be discussed below, the magnetic susceptibility of solids is small. In contrast, the electric susceptibility is of the order of one or larger. This explains why, in solid state spectroscopy with electromagnetic radiation, which is one of the most important experimental methods, one usually considers only electric effects (Chap. 11).

So far we have treated only electrons that are bound to atoms. For free electrons in a metal, (8.10) is not applicable. To calculate the diamagnetism of free electrons, one must solve the Schrödinger equation for free electrons in a magnetic field (Panel VIII), and from the energy levels, one can then calculate the free energy in a magnetic field, and from that the susceptibility. This last part, however, is mathematically rather tedious and brings few new insights, and in any case the free electron gas model is only a very crude approximation. It should nonetheless be remarked that the diamagnetism of free electrons does represent a genuine quantum effect. For a classical gas of free electrons, the free energy does not depend on the magnetic field, and thus the diamagnetic susceptibility vanishes. This is already evident in (8.9): the magnetic field shifts the momentum by $e\boldsymbol{A}$. If one then integrates over all states, and thus over all momenta, the result does not depend on $\boldsymbol{A}$, and is therefore independent of the magnetic field.

Besides diamagnetism due to their angular momentum, free electrons also exhibit paramagnetism (Pauli paramagnetism). This part is easy to calculate and we do not even need to involve the free energy. In the absence of a magnetic field, states with different spin quantum numbers have the same energy (they are degenerate). In a magnetic field the spins adopt one of two alignments. Electrons with spins parallel to the field lines of $\boldsymbol{B}_0$ are in states whose energy is lowered by $\frac{1}{2}g_0\mu_B B_0$ with respect to the field-free condition. Electrons with spins antiparallel to the field lines have an energy raised by $\frac{1}{2}g_0\mu_B B_0$. The energy parabola $D(E)$ (Fig. 6.4) splits into two parabolas (Fig. 8.2), which are separated on the energy axis by $g_0\mu_B B_0$. From Fig. 8.2 it follows that, in the approximation $\not{k}T \ll E_F$, the volume density of electrons with uncompensated spins is approximately $\frac{1}{2}D(E_F)g_0\mu_B B_0$ (cross-hatched region in Fig. 8.2). Each of these electrons contributes a magnetic moment of $\frac{1}{2}g_0\mu_B$, and the resulting magnetization is

$$\boldsymbol{M} = \tfrac{1}{2}D(E_F)g_0\mu_B \boldsymbol{B}_0 \tfrac{1}{2}g_0\mu_B \ . \tag{8.14}$$

We thereby obtain a temperature-independent paramagnetic susceptibility χ_p given by

$$\chi_p = \mu_0 \frac{g_0^2}{4}\mu_B^2 D(E_F) \sim \mu_0\mu_B^2 D(E_F) \ . \tag{8.15}$$

If we include the diamagnetic component, which we have not calculated here, we find

$$\chi = \mu_0\mu_B^2 D(E_F)\left[1 - \frac{1}{3}\left(\frac{m}{m^*}\right)^2\right] \ . \tag{8.16}$$

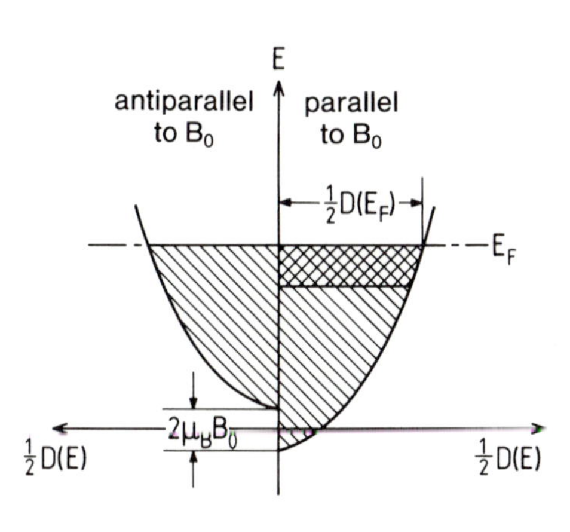

Fig. 8.2. Paramagnetism of free electrons. The density of states $D(E)$ splits in a magnetic field $\boldsymbol{B}_0$ into two parabolas which are shifted with respect to one another. This leads to a net a magnetic moment due to the excess (uncompensated) electron spins parallel to $\boldsymbol{B}_0$ (*cross-hatched area*)

Here m^* is the so-called effective mass of the charge carriers, which takes account of the fact that the electrons are moving in a crystal lattice and not in vacuum (Sect. 9.1). According to the value of the effective mass, the charge carriers may show paramagnetic or diamagnetic behavior.

We can also estimate the order of magnitude of the susceptibility of the conduction electrons. Values of the density of states at the Fermi level can be obtained from the electronic specific heat (Table 6.2). If $m^*=m$, then we obtain, e.g. for the molar susceptibility of sodium, $\chi_m=1.96 \times 10^{-4}\,\mathrm{cm^3/mol}$ or $\chi=8.6\times10^{-6}$ when one multiplies by the density of metallic sodium. Thus even Pauli paramagnetism does not lead to large values of the susceptibility. It is of the same order of magnitude as the diamagnetism of closed shells. One could now close the chapter on magnetic effects in solids and "file it away", were it not for the phenomenon of the collective coupling of electron spins which will now be addressed.

8.2 The Exchange Interaction

To describe the exchange interaction between localized electrons we return to the hydrogen molecule as the prototype of covalent bonding. The one-electron approximation has already been discussed in Sect. 1.2. Considering both electrons (labelled 1 and 2), the Hamiltonian operator $\mathcal{H}\,(1,2)$ can be decomposed into three parts

$$\mathcal{H}(1,2) = \mathcal{H}\,(1) + \mathcal{H}\,(2) + \mathcal{H}_{\mathrm{w}}\,(1,2)\,. \tag{8.17}$$

Here $\mathcal{H}\,(1)$ and $\mathcal{H}\,(2)$ are the Hamiltonian operators as in (1.2) expressed in terms of the coordinates of electrons 1 and 2 respectively, and $\mathcal{H}_{\mathrm{int}}(1,2)$ describes the residual interaction between the electrons. We can recover the one-electron approximation from (8.17) if we neglect $\mathcal{H}_{\mathrm{int}}\,(1,2)$ and set the total wavefunction (without spin) equal to the product of the one-electron solutions. For the ground state this would be

$$\Psi(1,2) = [\varphi_A(1) + \varphi_B(1)][\varphi_A(2) + \varphi_B(2)] \tag{8.18}$$

where φ_A and φ_B are the atomic wavefunctions. It is immediately clear that the eigenvalue problem is separable and the calculation proceeds as in Sect. 1.2. Carrying out the multiplication, (8.18) reads

$$\Psi(1,2) = \varphi_A(1)\varphi_B(2) + \varphi_B(1)\varphi_A(2) + \varphi_A(1)\varphi_A(2) + \varphi_B(1)\varphi_B(2) \ . \tag{8.19}$$

One sees that states in which both electrons are located on one atom ("ionic states") are equally represented. This is quite acceptable provided we neglect the repulsive Coulomb interaction or at least absorb it into an effective potential of the ions. For a Hamiltonian with electron-electron interactions, (8.19) is a poor postulate, particularly for nuclei which are far from one another. It is then better to omit the ionic states completely. This leads to the Heitler-London approximation

$$\Psi(1,2) = \varphi_A(1)\varphi_B(2) + \varphi_B(1)\varphi_A(2) \ . \tag{8.20}$$

This formula is symmetric with respect to the coordinates of the electrons. Because the total wavefunction, including the spin function, must be antisymmetric (generalized Pauli principle), the spins in this state must be antiparallel (singlet state). A triplet state with parallel spin orientation is described in space coordinates by the antisymmetric wavefunction

$$\Psi(1,2) = \varphi_A(1)\varphi_B(2) - \varphi_B(1)\varphi_A(2) \ . \tag{8.21}$$

With these two electron wavefunctions we can calculate the expectation value of the energy and, after a few intermediate steps, we obtain:

$$E = \frac{\langle \Psi(1,2)|\mathscr{H}|\Psi(1,2)\rangle}{\langle \Psi(1,2)|\Psi(1,2)\rangle} = 2\,E_{\mathrm{I}} + \frac{C \pm A}{1 \pm S} \tag{8.22}$$

where E_{I} is the ionization energy of the hydrogen atom, C the so-called Coulomb integral, A the exchange integral, and S the overlap integral. The + sign in (8.22) corresponds to the singlet state.

$$E_{\mathrm{I}} = \int \varphi_A^*(1)\left(-\frac{\hbar^2}{2m}\Delta_1 - \frac{e^2}{4\pi\,\varepsilon_0 r_{A1}}\right)\varphi_A(1)d\boldsymbol{r}_1 \ , \tag{8.23}$$

$$C = \frac{e^2}{4\pi\,\varepsilon_0}\int\left(\frac{1}{R_{AB}} + \frac{1}{r_{12}} - \frac{1}{r_{A2}} - \frac{1}{r_{B1}}\right)|\varphi_A(1)|^2|\varphi_B(2)|^2 d\boldsymbol{r}_1 d\boldsymbol{r}_2 \ , \tag{8.24}$$

$$A = \frac{e^2}{4\pi\,\varepsilon_0}\int\left(\frac{1}{R_{AB}} + \frac{1}{r_{12}} - \frac{1}{r_{A1}} - \frac{1}{r_{B2}}\right)\varphi_A^*(1)\varphi_A(2)\varphi_B(1)\varphi_B^*(2) d\boldsymbol{r}_1 d\boldsymbol{r}_2 \ , \tag{8.25}$$

$$S = \int \varphi_A^*(1)\varphi_A(2)\varphi_B(1)\varphi_B^*(2)d\boldsymbol{r}_1 d\boldsymbol{r}_2 \, . \tag{8.26}$$

The result for the energy levels in the two-electron model (8.22) is quite different from that in the one-electron model. The interpretation of the energy levels (8.22) is also different. We shall examine these differences with the aid of Fig. 8.3. In the one-electron model, the energy levels can be occupied by either two or one electron(s). Within this model, an excited state corresponds, for instance, to the occupation of the lowest level with one electron and the next higher level with the second electron. The excited state can be a singlet or a triplet state, and the energies are necessarily degenerate. The degeneracy of these two states is easy to see in the formula by neglecting the electron-electron interaction and forming the expectation value of the Hamiltonian (8.17). In the two-electron picture, however, the energy of a single electron is not defined, only a total energy for both. The ground state is, as in the one-electron model, a singlet state but the first excited state is necessarily a triplet state. We see from this example that the energy levels of the calculated band structures in Chaps. 6 and 7 only have meaning within the one-electron picture. In a many-body model the energy of the ground state can only by symbolized by a single total energy, or single eigenvalue. This difference between single-body and many-body models will be met again in Chap. 10.

The energy difference between the triplet and singlet state is obtained from (8.22) as

$$E_\mathrm{t} - E_\mathrm{s} = -J = 2\frac{CS - A}{1 - S^2} \, . \tag{8.27}$$

The quantity J gives the separation of the energy levels for parallel and antiparallel spins and is called the exchange constant. For the hydrogen molecule it is always negative and the singlet state is therefore the lower energy state. Using the exchange constant, a model Hamiltonian can be introduced which only affects the spin functions and produces the same split-

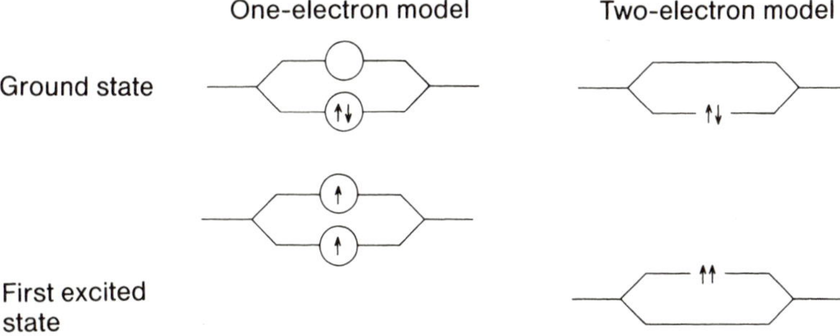

Fig. 8.3. Energy levels for a pair of electrons in the one-electron and two-electron models

ting between the energy levels for parallel and antiparallel orientations (for proof see e.g. [8.1], or textbooks on quantum mechanics)

$$\mathscr{H}_{\text{spin}} = -2J\sigma_1 \cdot \sigma_2 \ . \tag{8.28}$$

The operators σ can be represented by the Pauli spin matrices. Ferromagnetic coupling between the electrons is obtained for $J>0$. This Heisenberg Hamiltonian is the starting point of many modern theories of magnetism, in so far as the magnetism can be understood within a model that treats only pairwise coupling between electrons. This is not always the case, however, and particularly for typical ferromagnetics such as the $3d$ transition metals Ni, Co and Fe, this model is insufficient. Besides the Heisenberg Hamiltonian, we require a description of the collective exchange interaction. As an example we will study the free electron gas.

8.3 Exchange Interaction Between Free Electrons

Although the exchange interaction for an electron pair in a bond is negative, for free electrons it is positive. This can be shown by considering two free electrons i and j and their pair wavefunction Ψ_{ij}. For electrons with the same spin, the pair wavefunction must be antisymmetric in space coordinates. From this requirement we obtain

$$\begin{aligned} \Psi_{ij} &= \frac{1}{\sqrt{2}V}\left(e^{i\boldsymbol{k}_i\cdot\boldsymbol{r}_i}e^{i\boldsymbol{k}_j\cdot\boldsymbol{r}_j} - e^{i\boldsymbol{k}_i\cdot\boldsymbol{r}_j}e^{i\boldsymbol{k}_j\cdot\boldsymbol{r}_i}\right) \\ &= \frac{1}{\sqrt{2}V}e^{i(\boldsymbol{k}_i\cdot\boldsymbol{r}_i+\boldsymbol{k}_j\cdot\boldsymbol{r}_j)}\left(1 - e^{-i(\boldsymbol{k}_i-\boldsymbol{k}_j)\cdot(\boldsymbol{r}_i-\boldsymbol{r}_j)}\right) \ . \end{aligned} \tag{8.29}$$

The probability that electron i is to be found in volume element $d\boldsymbol{r}_i$ and that electron j is to be found in volume element $d\boldsymbol{r}_j$ is then equal to $|\psi_{ij}|^2 d\boldsymbol{r}_i d\boldsymbol{r}_j$:

$$|\Psi_{ij}|^2 d\boldsymbol{r}_i d\boldsymbol{r}_j = \frac{1}{V^2}\left[1 - \cos(\boldsymbol{k}_i - \boldsymbol{k}_j)\cdot(\boldsymbol{r}_i - \boldsymbol{r}_j)\right]d\boldsymbol{r}_i d\boldsymbol{r}_j \ . \tag{8.30}$$

This expression shows all of the crucial features: the probability of finding two electrons with the same spin at the same place vanishes for every $\boldsymbol{k}_i$ and $\boldsymbol{k}_j$. As a result, for a particular spin-up electron, the other electrons with the same spin cannot screen the Coulomb potential of the ion cores so well locally, which leads to a reduction of the energy of the spin-up electron. This energy reduction is reinforced if the highest possible percentage of all the electrons have the same spin as the spin-up electron. The net effect is thus a gain in the electronic energy for parallel spins and a collective exchange interaction with positive sign.

Before we expand these ideas into a model description of ferromagnetism, it is quite useful to consider the spatial correlation (8.30) further. From (8.30), we can obtain a k-averaged probability if we integrate over the Fermi sphere. We introduce relative coordinates between the electrons i and j with $\boldsymbol{r}=\boldsymbol{r}_i-\boldsymbol{r}_j$. We then ask what is the probability that a second spin-up electron is at a distance $\boldsymbol{r}$ in a volume element $d\boldsymbol{r}$. This probability is then

$$P(r)_{\uparrow\uparrow}d\boldsymbol{r} = n_\uparrow d\boldsymbol{r}\overline{[1-\cos(\boldsymbol{k}_i-\boldsymbol{k}_j)\cdot\boldsymbol{r}]} \tag{8.31}$$

with $n_\uparrow$ the concentration of electrons with the same spin, which is half as large as the total concentration n of electrons. Instead of the probability we can also speak of an effective electron density acting on the spin-up electron, we denote this by $\varrho_{\text{ex}}(\boldsymbol{r})$, because of its origin in the exchange interaction

$$\varrho_{\text{ex}}(\boldsymbol{r}) = \frac{e\,n}{2}\overline{[1-\cos(\boldsymbol{k}_i-\boldsymbol{k}_j)\cdot\boldsymbol{r}]} \quad \text{with} \quad n_\uparrow = n/2\,. \tag{8.32}$$

If we now take the average over the Fermi sphere, we obtain

$$\begin{aligned}\varrho_{\text{ex}}(\boldsymbol{r}) &= \frac{e\,n}{2}\overline{[1-\cos(\boldsymbol{k}_i-\boldsymbol{k}_j)\cdot\boldsymbol{r}]}\\ &= \frac{e\,n}{2}\left[1-\frac{1}{\left(\frac{4\pi}{3}k_{\text{F}}^3\right)^2}\int d\boldsymbol{k}_i\int d\boldsymbol{k}_j\frac{1}{2}\left(e^{i(\boldsymbol{k}_i-\boldsymbol{k}_j)\cdot\boldsymbol{r}}+e^{-i(\boldsymbol{k}_i-\boldsymbol{k}_j)\cdot\boldsymbol{r}}\right)\right].\end{aligned} \tag{8.33 a}$$

$$\varrho_{\text{ex}}(\boldsymbol{r}) = \frac{e\,n}{2}\left(1-\frac{1}{\left(\frac{4\pi}{3}k_{\text{F}}^3\right)^2}\int d\boldsymbol{k}_i\,e^{i\boldsymbol{k}_i\cdot\boldsymbol{r}}\int d\boldsymbol{k}_j\,e^{i\boldsymbol{k}_j\cdot\boldsymbol{r}}\right). \tag{8.33 b}$$

These integrals may be solved analogously to (3.40):

$$\varrho_{\text{ex}}(\boldsymbol{r}) = \frac{e\,n}{2}\left(1-9\,\frac{(\sin k_{\text{F}}r-k_{\text{F}}r\cos k_{\text{F}}r)^2}{(k_{\text{F}}r)^6}\right). \tag{8.34}$$

The total charge density seen by a free electron is the sum of the charge density of electrons with the same spin and the homogeneous charge density $e\,n/2$ of the electrons with opposite spin, for which the spatial part of the wavefunction remains symmetric and thus unaltered,

$$\varrho_{\text{eff}}(\boldsymbol{r}) = e\,n\left(1-\frac{9}{2}\,\frac{(\sin k_{\text{F}}r-k_{\text{F}}r\cos k_{\text{F}}r)^2}{(k_{\text{F}}r)^6}\right). \tag{8.35}$$

This charge density is plotted in Fig. 8.4. The charge density at $\boldsymbol{r}=0$ is reduced as a result of the exchange interaction. This creates an "exchange hole", whose size is approximately equal to twice the reciprocal Fermi vector. According to Table 6.1 the radius is about 1–2 Å. We can use the effec-

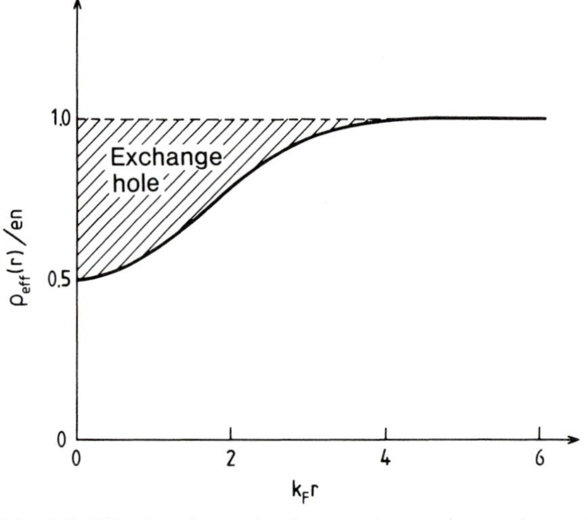

Fig. 8.4. Effective charge density seen by an electron in an electron gas. Due to the exchange interaction, the density of electrons with the same spin orientation in the neighborhood of a particular electron is reduced ("exchange hole"). If this electron moves, it must drag the exchange hole with it and thus its effective mass is raised. Furthermore, the existence of an exchange hole implies a positive exchange coupling. From this we can obtain a model for the ferromagnetism of band electrons

tive charge density $\varrho_{\text{eff}}(\boldsymbol{r})$ to form a new ("renormalized") Schrödinger equation for the free electron gas and this leads us to the Hartree-Fock approximation. We note that (8.35) does not properly reproduce the electron-electron correlation because the Coulomb interaction in fact forbids the presence of two electrons with the same spin at the same place. Furthermore, the correlation (8.30) between two electrons that are arbitrarily far apart is an unrealistic result of the assumption of plane waves.

8.4 The Band Model of Ferromagnetism

We now wish to construct a simple band model of ferromagnetism using the qualitatively derived renormalization of the one-electron levels due to correlation of electrons bearing the same spin. The model is due to Stoner and Wohlfarth. For the renormalized electron energies we make the ansatz

$$\begin{aligned} E_\uparrow(\boldsymbol{k}) &= E(\boldsymbol{k}) - I\,n_\uparrow/N\,, \\ E_\downarrow(\boldsymbol{k}) &= E(\boldsymbol{k}) - I\,n_\downarrow/N\,, \end{aligned} \tag{8.36}$$

$E(\boldsymbol{k})$ are the energies in a normal one-electron bandstructure, $n_\uparrow$ and $n_\downarrow$ are the number of electrons with corresponding spin, and N is the number of atoms. The Stoner parameter I describes the energy reduction due to the electron correlation. Its dependence on the wave vector will be neglected in this model. We now introduce the relative excess of electrons of one spin type

$$R = \frac{n_\uparrow - n_\downarrow}{N} \ . \tag{8.37}$$

This quantity is, apart from a factor $\mu_B(N/V)$, equal to the magnetization M (8.3). Furthermore, to simplify the formula, we subtract $I(n_\uparrow + n_\downarrow)/2N$ from the one-electron energies and obtain, in place of (8.36),

$$\left.\begin{aligned} E_\uparrow(\boldsymbol{k}) &= \tilde{E}(\boldsymbol{k}) - IR/2 \\ E_\downarrow(\boldsymbol{k}) &= \tilde{E}(\boldsymbol{k}) + IR/2 \end{aligned}\right\} \text{ with } \quad \tilde{E}(\boldsymbol{k}) = E(\boldsymbol{k}) - I(n_\uparrow + n_\downarrow)/2N \ . \tag{8.38}$$

The pair of equations (8.38) corresponds to a $\boldsymbol{k}$-independent splitting of the energy bands with different spin. The $\boldsymbol{k}$-independence of the exchange splitting is, of course, only an approximation. Newer theories nevertheless indicate that it holds to within a factor of about two. The value of the splitting depends on R, that is on the relative occupation of the sub-bands, which in turn is given by Fermi statistics. We therefore arrive at the self-consistency condition:

$$\begin{aligned} R &= \frac{1}{N}\sum_k f_\uparrow(\boldsymbol{k}) - f_\downarrow(\boldsymbol{k}) \\ &= \frac{1}{N}\sum_k \frac{1}{\mathrm{e}^{[\tilde{E}(\boldsymbol{k}) - IR/2 - E_F]/kT} + 1} - \frac{1}{\mathrm{e}^{[\tilde{E}(\boldsymbol{k}) + IR/2 - E_F]/kT} + 1} \ . \end{aligned} \tag{8.39}$$

Under certain conditions this equation has a non-zero solution for R, that is to say, a magnetic moment exists even in the absence of an external field, thus leading to ferromagnetism. It is possible to find a criterion for the appearance of ferromagnetism. For this purpose we expand the right-hand side of the equation for small R. Making use of the relation

$$f\left(x - \frac{\Delta x}{2}\right) - f\left(x + \frac{\Delta x}{2}\right) = -f'(x)\Delta x - \frac{2}{3!}\left(\frac{\Delta x}{2}\right)^3 f'''(x) \tag{8.40}$$

we obtain

$$R = -\frac{1}{N}\sum_k \frac{\partial f(\boldsymbol{k})}{\partial \tilde{E}(\boldsymbol{k})} IR - \frac{1}{24}\frac{1}{N}\sum_k \frac{\partial^3 f(\boldsymbol{k})}{\partial \tilde{E}(\boldsymbol{k})^3}(IR)^3 \ . \tag{8.41}$$

In this formula, the first derivative of the Fermi function is negative, while the third derivative is positive. The condition for ferromagnetism ($R>0$) is therefore

$$-1 - \frac{I}{N}\sum_k \frac{\partial f(\boldsymbol{k})}{\partial \tilde{E}(\boldsymbol{k})} > 0 \ . \tag{8.42}$$

The derivative of the Fermi function $-\partial f/\partial \tilde{E}$ obviously has its largest value for $T \to 0$ (Fig. 6.6). If the condition (8.42) is to be fulfilled at all, then it will be easiest at $T=0$. For the case of $T=0$ it is a straightforward matter to perform the summation over all $\boldsymbol{k}$ values:

$$-\frac{1}{N}\sum_{k}\frac{\partial f(\boldsymbol{k})}{\partial \tilde{E}(\boldsymbol{k})}=\frac{V}{(2\pi)^3 N}\int d\boldsymbol{k}\left(-\frac{\partial f}{\partial \tilde{E}}\right)$$
$$=\frac{V}{(2\pi)^3}\frac{1}{N}\int d\boldsymbol{k}\,\delta(\tilde{E}-E_{\mathrm{F}})=\frac{1}{2}\frac{V}{N}D(E_{\mathrm{F}})\,. \tag{8.43}$$

Here we have taken account of the fact that at $T=0$, the Fermi function is a step function and the first derivative $-\partial f/\partial \tilde{E}$ is equal to the δ-function $\delta(\tilde{E}-E_{\mathrm{F}})$. The factor 1/2 originates from the fact that, according to (8.39) and (8.41), the sum over $\boldsymbol{k}$ and also the integral over the electrons is taken over only *one* spin type, while the usual definition of the density of states considers the number of electrons with positive *and* negative spins per unit volume. The sum is thus equal to half the density of states for electrons at the Fermi level, in relation, however, not to the volume but to the number of atoms. We introduce the density of states per atom and spin as

$$\tilde{D}(E_{\mathrm{F}})=\frac{V}{2N}D(E_{\mathrm{F}})\,. \tag{8.44}$$

The condition for ferromagnetism to occur at all is then simply

$$I\tilde{D}(E_{\mathrm{F}})>1\,. \tag{8.45}$$

This is the so-called Stoner criterion for the existence of ferromagnetism. Under the assumption that this criterion is fulfilled, (8.42) also yields the temperature at which the magnetic moment disappears (Curie temperature). The Curie temperature is the point at which (8.42) becomes an equality instead of an inequality. This will be considered in the next section. Figure 8.5 shows the Stoner parameter, the density of states and their product from a theoretical treatment by Janak [8.2]. The theory correctly predicts that ferromagnetism exists only for the elements Fe, Co, and Ni. For the elements of the $4d$ series, the density of states and the Stoner parameter are too small to achieve the ferromagnetic state. Nevertheless, there is a considerable enhancement of the magnetic susceptibility due to the positive exchange interaction of the band electrons. For an external magnetic field B_0, (8.39) contains, in addition to the exchange splitting of $IR/2$, a splitting of $\mu_{\mathrm{B}}B_0$. In a first approximation for R at $T=0$, (8.41) then becomes

$$R=\tilde{D}(E_{\mathrm{F}})(IR+2\mu_{\mathrm{B}}B_0)\,. \tag{8.46}$$

For the magnetization M one thus obtains

$$M=\mu_{\mathrm{B}}\frac{N}{V}R=\tilde{D}(E_{\mathrm{F}})\left(IM+2\mu_{\mathrm{B}}^2\frac{N}{V}B_0\right)$$
$$M=2\mu_{\mathrm{B}}^2\frac{N}{V}\frac{\tilde{D}(E_{\mathrm{F}})}{1-I\tilde{D}(E_{\mathrm{F}})}B_0\,. \tag{8.47}$$

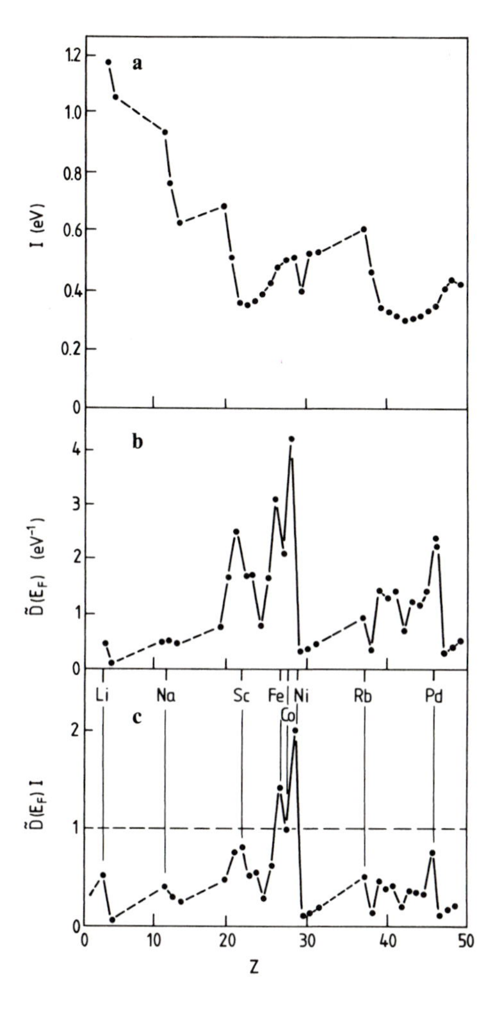

Fig. 8.5. (**a**) Integral of the exchange correlation (Stoner parameter) I as a function of the atomic number (after [8.2]). (**b**) Density of states per atom $\tilde{D}(E_F)$; (**c**) the product of the density of states $\tilde{D}(E_F)$ and the Stoner parameter I. The elements Fe, Co and Ni with values of $I\tilde{D}(E_F)>1$ display ferromagnetism. The elements Ca, Sc and Pd come very close to achieving ferromagnetic coupling

The numerator is just the normal Pauli susceptibility of band electrons (8.15), which is now considerably enhanced, however, by the denominator. If we denote the Pauli susceptibility by χ_0, we obtain

$$\chi = \frac{\chi_0}{1 - I\tilde{D}(E_F)} . \tag{8.48}$$

Janak [8.2] has calculated the factor χ/χ_0 and obtained values as large as 4.5 (Ca), 6.1 (Sc) and 4.5 (Pd). Thus, together with the rather high density of states, one also finds relatively large values for the susceptibilities of these elements. A direct comparison with experiment, however, must also take into account the magnetism due to angular momentum. It turns out that the previous assertion that $\chi \ll 1$ always remains valid.

8.5 The Temperature Behavior of a Ferromagnet in the Band Model

We now turn to the temperature dependence of the saturation magnetization of a ferromagnet. For this purpose one could, in principle, evaluate (8.39) with the aid of a one-electron bandstructure calculation. However, the associated mathematical effort would not be rewarded: the $\boldsymbol{k}$-independent and delocalized treatment of the exchange interaction does not provide any quantitatively significant results. A qualitative, but for our purposes quite sufficient, picture of the temperature behavior can be obtained from a highly simplified density of states model, which keeps the amount of mathematics to a minimum. Let us look at the density of states of Ni (Fig. 8.6 a), taken from a bandstructure calculation by Callaway and Wang

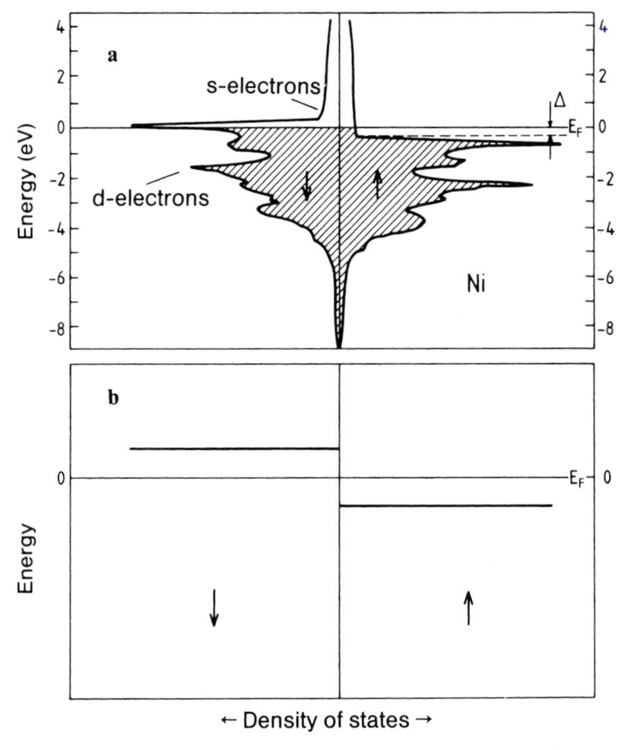

Fig. 8.6. (**a**) Calculated density of states of nickel (after [8.3]). The exchange splitting is calculated to be 0.6 eV. From photoelectron spectroscopy a value of about 0.3 eV is obtained. However the values cannot be directly compared, because a photoemitted electron leaves a hole behind, so that the solid remains in an excited state. The distance Δ between the upper edge of the d-band of majority spin electrons and the Fermi energy is known as the Stoner gap. In the bandstructure picture, this is the minimum energy for a spin flip process (the s-electrons are not considered in this treatment). (**b**) A model density of states to describe the thermal behavior of a ferromagnet

[8.3]. The largest contribution to the density of states at the Fermi level is provided by the d electrons, firstly because of their high number (9 per atom, or, more precisely, 9.46 per atom; see below), and secondly because the d band is only 4 eV wide (in contrast to the s band). In addition, the exchange splitting for s electrons is small. It is thus the differing occupation of the d bands for majority and minority spins that leads to the magnetization. For nickel at $T=0$, this is given simply by the number of unoccupied d states in the minority band. From the measured magnetization at $T=0$, the number of d holes is found to be 0.54 per atom in the case of nickel, i.e., the effective magnetic moment is $\mu_{B_{eff}} \sim 0.54\,\mu_{B_{eff}}$ per atom. The variation of the magnetization with temperature and the existence of the Curie point at which the magnetization vanishes, result from the interplay of the exchange splitting, the Fermi statistics and the density of states near the Fermi level, according to (8.39). For a qualitative discussion of (8.39), we do not need the actual functional behavior of the density of states, and it is sufficient to replace the sharp peak in the density of states at the upper edge of the d band with a δ-function in energy (Fig. 8.6b). If, in addition to the exchange splitting, there is also a field splitting as in (8.46), our model density of states would be

$$\tilde{D}(E) = \frac{\mu_{B_{eff}}}{\mu_B}\left[\delta(E - E_F - \mu_B B_0 - IR/2) + \delta(E - E_F + \mu_B B_0 + IR/2)\right] . \tag{8.49}$$

Since the states for majority and minority spin have equal weight in this model, the Fermi level always lies midway between these two levels, a feature that is already included in the above expression. In place of (8.39) we obtain

$$R = \frac{\mu_{B_{eff}}}{\mu_B}\left(\frac{1}{e^{(-\mu_B B_0 - IR/2)/kT} + 1} - \frac{1}{e^{(\mu_B B_0 + IR/2)/kT} + 1}\right) . \tag{8.50}$$

We now look for ferromagnetic solutions to this equation, i.e., solutions with $R \geq 0$ at $B_0=0$. With the abbreviations $T_c = I\mu_{B_{eff}}/\mu_B 4k$ and $\tilde{R} = \mu_B/\mu_{B_{eff}} R$, equation (8.50) becomes

$$\tilde{R} = \frac{1}{e^{-2\tilde{R}T_c/T} + 1} - \frac{1}{e^{+2\tilde{R}T_c/T} + 1} = \tanh\frac{\tilde{R}T_c}{T} . \tag{8.51}$$

In limiting cases this equation has the solutions $\tilde{R}=1$ for $T=0$ and $\tilde{R}=0$ for $T=T_c$. Thus T_c is identical to the previously introduced Curie temperature, above which the spontaneous magnetization vanishes. The behavior over the whole temperature range is shown in Fig. 8.7. Because the magnetization M is proportional to R, Fig. 8.7 should reproduce the temperature dependence of the spontaneous magnetization of a ferromagnet. The agreement with the experimental observations is quite acceptable (Fig. 8.7). For the limiting cases $T \ll T_c$ and $T \sim T_c$, the right-hand side of (8.51) can be expanded to give

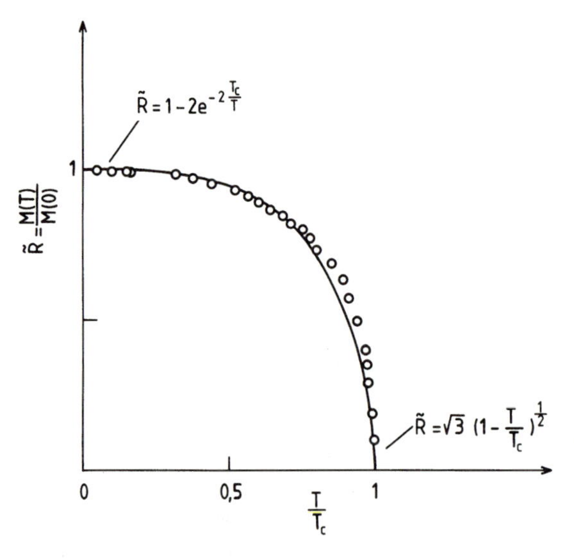

Fig. 8.7. Magnetization of a ferromagnet below the Curie temperature T_c. Experimental values for nickel from [8.4, 8.5]

$$\tilde{R} = 1 - 2e^{-2T_c/T} \qquad T \ll T_c\,, \tag{8.52}$$

$$\tilde{R} = \sqrt{3}\left(1 - \frac{T}{T_c}\right)^{1/2} T \sim T_c\,,\ \tilde{R} \ll 1 \quad \text{with} \quad \tanh x \approx x - \frac{1}{3}x^3 \ldots\,. \tag{8.53}$$

However, *neither* of these equations is confirmed by experiment. The critical exponent in the vicinity of the Curie point is $\frac{1}{3}$ (Fig. 8.8) and not $\frac{1}{2}$. The behavior at low temperature is likewise incorrectly described by (8.52). This is due to the fact that, in addition to the spin-flip accompanying the excitation from one band to another, other elementary excitations with smaller quantum energy are possible in a ferromagnet, and these can also cause spin-flip (Sect. 8.7).

Above the Curie temperature, (8.50) yields a magnetization only when the field B_0 is non-zero. We can expand the Fermi function for small R and B_0 to give

$$\tilde{R} = \frac{\mu_B}{kT} B_0 + \frac{T_c}{T}\tilde{R} \quad \text{or} \tag{8.54}$$

$$\tilde{R} = \frac{\mu_B}{k}\frac{1}{T - T_c} B_0\,. \tag{8.55}$$

As T approaches T_c from above the paramagnetic susceptibility should thus diverge according to the law

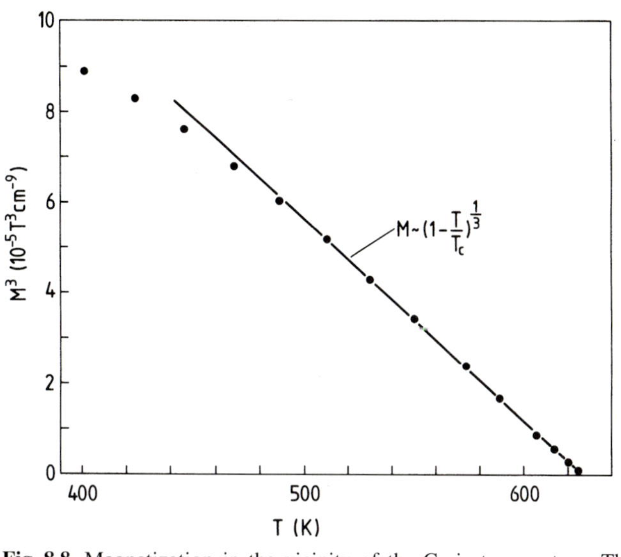

Fig. 8.8. Magnetization in the vicinity of the Curie temperature. The experimental values for nickel are from [8.4, 8.5]. The critical exponent in the vicinity of the transition to the paramagnetic phase above the Curie temperature is 1/3 and not 1/2 as predicted by the simple model

$$\chi = \frac{C}{T - T_c} . \tag{8.56}$$

This is the so-called Curie-Weiss law in which C is the Curie-Weiss constant. The Curie-Weiss law is experimentally fulfilled for $T \gg T_c$. As T_c is approached, one observes deviations from this law, and a better description is a decrease proportional to $(T-T_c)^{-4/3}$. According to our model the Curie-Weiss constant C is related to the saturation magnetization at $T=0$. This interrelation, however, gives a value for C which is too small. We could also try to use the relationship between the Stoner parameter I and the Curie temperature T_c (in this model $T_c = I\mu_{B_{\text{eff}}}/4\mu_B k$) to estimate a value of T_c from the measured values of the exchange splitting. This approach yields values of T_c that are much too large. The failure of our simple model in these respects does *not*, however, lie in the simple bandstructure. It is due to the fact that the model does not correctly treat excited states.

The Curie-Weiss law and the temperature dependence of the spontaneous magnetization are often derived in a mean field approximation, in which the exchange interaction is replaced by an average "internal" field. The derivation chosen here, whose principles are due to E.C. Stoner, is equivalent to the mean field approximation in that it allows no spatial variation of the spin distribution. Our treatment illustrates the role of the bandstructure more clearly than the conventional mean field approximation.

8.6 Ferromagnetic Coupling for Localized Electrons

Whereas the magnetic behavior of d-band transition metals is well described in the band model, the approach developed in Sect. 8.2 based on the exchange interaction between two localized electrons is particularly well suited to the rare earths with their partly filled f shells and to many ionic compounds of the d and f transition metals. The starting point is the Heisenberg Hamiltonian introduced in (8.28) for the exchange interaction between two electrons. In the following model treatment of ferromagnetism, we assume a primitive lattice of atoms each having one unpaired electron with zero angular momentum. This model of a spin lattice illustrates the essential consequences of the exchange interaction in a lattice. Taking into account an additional external magnetic field B_0, we obtain the Hamiltonian

$$\mathscr{H} = -\sum_i \sum_\delta J_{i\delta} \boldsymbol{S}_i \cdot \boldsymbol{S}_{i\delta} - g\,\mu_{\mathrm{B}} \boldsymbol{B}_0 \sum_i \boldsymbol{S}_i \,. \tag{8.57}$$

The index i runs over all atoms and the index δ over all the neighbors of an atom that participate in the exchange interaction. Unfortunately, the Heisenberg operator is a non-linear operator. Solutions can only be found in special cases or by introducing a linearizing approximation. One such approximation is the mean field approximation which will be discussed now.

In the mean field approximation the operator product in (8.57) is replaced by the product of the spin operator $\boldsymbol{S}_i$ and the expectation value of the spin operators of the neighbors $\langle \boldsymbol{S}_{i\delta} \rangle$. In the mean field approximation the Hamiltonian therefore becomes

$$\mathscr{H}_{\mathrm{MF}} = -\sum_i \boldsymbol{S}_i \cdot \left(\sum_\delta J_{i\delta} \langle \boldsymbol{S}_{i\delta} \rangle + g\,\mu_{\mathrm{B}} \boldsymbol{B}_0 \right) . \tag{8.58}$$

The exchange interaction thus acquires the character of an internal field

$$\boldsymbol{B}_{\mathrm{MF}} = \frac{1}{g\,\mu_{\mathrm{B}}} \sum_\delta J_{i\delta} \langle \boldsymbol{S}_{i\delta} \rangle \,. \tag{8.59}$$

For homogeneous systems (without a surface), $\langle \boldsymbol{S}_{i\delta} \rangle$ is the same for all atoms. The average value $\langle \boldsymbol{S}_{i\delta} \rangle = \langle \boldsymbol{S} \rangle$ can be expressed in terms of the magnetization

$$\boldsymbol{M} = g\,\mu_{\mathrm{B}} \frac{N}{V} \langle \boldsymbol{S} \rangle \tag{8.60}$$

with N/V the number of atoms per unit volume. We therefore obtain for the mean field $\boldsymbol{B}_{\mathrm{MF}}$

$$\boldsymbol{B}_{\mathrm{MF}} = \frac{V}{N g^2 \mu_{\mathrm{B}}^2} \nu J \boldsymbol{M} \,, \tag{8.61}$$

where the exchange interaction is restricted to the ν nearest neighbors. The Hamiltonian in the mean field approximation (8.58) is now mathematically identical to the Hamiltonian of N independent spins in an effective magnetic field $\boldsymbol{B}_{\text{eff}}=\boldsymbol{B}_{\text{MF}}+\boldsymbol{B}_0$. Its eigenvalues are

$$E = \pm\tfrac{1}{2} g\,\mu_{\text{B}} B_{\text{eff}} \tag{8.62}$$

for each electron spin. We denote the number of electrons in states with spin parallel and antiparallel to the B-field by $N_\uparrow$ and $N_\downarrow$.

In thermal equilibrium one has

$$\frac{N\downarrow}{N\uparrow} = \mathrm{e}^{-g\,\mu_{\text{B}} B_{\text{eff}}/kT} \tag{8.63}$$

and the magnetization is thus

$$M = \frac{1}{2} g\,\mu_{\text{B}} \frac{N\uparrow - N\downarrow}{V} = \frac{1}{2} g\,\mu_{\text{B}} \frac{N}{V} \tanh\left(\frac{1}{2} g\,\mu_{\text{B}} B_{\text{eff}}/kT\right) . \tag{8.64}$$

This equation together with (8.61) has non-zero solutions for the magnetization (even without an external magnetic field) provided $J>0$, i.e., whenever there is ferromagnetic coupling of the spins. With the abbreviations

$$M_{\text{s}} = \frac{N}{V}\frac{1}{2} g\,\mu_{\text{B}} \quad \text{and} \tag{8.65}$$

$$T_{\text{c}} = \tfrac{1}{4}\nu J/k \tag{8.66}$$

we obtain from (8.61, 8.64), and with no external magnetic field B_0

$$M(T)/M_{\text{s}} = \tanh\left(\frac{T_{\text{c}}}{T}\frac{M}{M_{\text{s}}}\right) . \tag{8.67}$$

This equation is equivalent to (8.51). It means that the temperature behavior of the magnetization in the band model and in the model of localized electrons is the same. Therefore, for $T \to 0$ and $T \approx T_{\text{c}}$ one finds

$$M(T) = M_{\text{s}}(1 - 2\mathrm{e}^{-2T_{\text{c}}/T}) , \qquad T \ll T_{\text{c}} \tag{8.68}$$

$$M(T)/M_{\text{s}} \approx \sqrt{3}\left(1 - \frac{T}{T_{\text{c}}}\right)^{1/2} , \quad T \sim T_{\text{c}} . \tag{8.69}$$

T_{c} is therefore the critical temperature at which the spontaneous magnetization vanishes. It depends on the strength of the exchange coupling and on the number of nearest neighbors. Interestingly, the critical exponent in (8.69) does not depend on the dimensionality of the system. A planar lattice has the same critical exponent in the mean field approximation as a $3d$ lattice, but its T_{c} is smaller because of the reduced number of nearest neighbors. This leads to an interesting behavior for the magnetization near the surface of a ferromagnet (Panel VII).

In the mean field approximation not only the magnetization, but also the short-range order of the magnetic moments vanishes at T_c. In reality a certain degree of short-range order survives. The Curie temperature is merely the temperature at which the magnetic order vanishes on a large length scale.

For temperatures above T_c, we can once again derive the Curie-Weiss law for the susceptibility. With an external field B_0, and using the series expansion (8.68), we obtain from (8.64)

$$M(T) = \frac{g^2 \mu_B^2 N}{4 V k} \frac{1}{T - T_c} B_0 \ . \tag{8.70}$$

8.7 Antiferromagnetism

Up until now we have assumed ferromagnetic coupling of the electron spins, i.e., $J > 0$. A number of compounds, for example, the oxides of Fe, Co and Ni, display antiferromagnetic coupling between the transition metal d electrons. They possess the NaCl lattice structure, i.e., the lattices of the paramagnetic d metal ions and of the O^{2-} ions each form a face-centered cubic sub-lattice. In the antiferromagnetically ordered structures, the metal ions form a magnetic elementary cell which is no longer face centered, but possesses a complicated magnetic superstructure. The magnetic behavior of an antiferromagnet will be treated within the mean field approximation for a simple magnetic superstructure. This superstructure should be constructed in such a way that all nearest neighbors have antiparallel spins (Fig. 8.9). In a model treatment with antiferromagnetic coupling ($J<0$) we can now apply (8.60–8.64) to each magnetic sub-lattice separately. We note that the

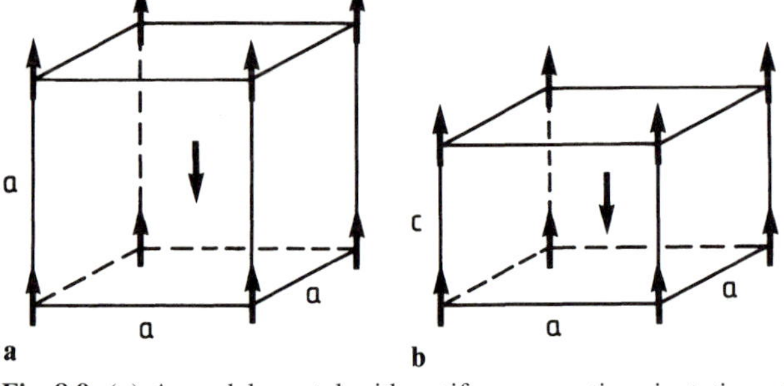

Fig. 8.9. (**a**) A model crystal with antiferromagnetic orientation of the nearest neighbor spins. (**b**) An equally simple spin structure, but with a tetragonal lattice, is observed for the compounds MnF_2, FeF_2 and CoF_2. In this case the atoms along the c axis are the nearest neighbors. If the transition metal ions form a face-centered cubic lattice, it is topologically impossible to have only antiferromagnetic orientation between nearest neighbors. The magnetic superstructures become correspondingly more complex

mean field for the sub-lattice with positive spin orientations is created by the sub-lattice with negative spins, and vice versa. For the antiferromagnetically ordered state we therefore obtain the pair of equations

$$M^+ = \frac{1}{2} g\,\mu_B \frac{N^+}{V} \tanh\left(\frac{V}{2 k T N^- g\,\mu_B} \nu J M^-\right) , \tag{8.71}$$

$$M^- = \frac{1}{2} g\,\mu_B \frac{N^-}{V} \tanh\left(\frac{V}{2 k T N^+ g\,\mu_B} \nu J M^+\right) , \tag{8.72}$$

where M^+ and M^- are the magnetizations of the two spin sub-lattices and $N^+ = N^-$ is the number of metal ions in each of the sub-lattices. In the antiferromagnetic state $M^+ = -M^-$ and we obtain, in analogy with (8.67),

$$M^+ = \frac{1}{2} g\,\mu_B \frac{N^+}{V} \tanh\left(-\frac{V}{2 k T N^+ g\,\mu_B} \nu J M^+\right) \tag{8.73}$$

and a corresponding equation for M^-. The magnetization of the sub-lattice vanishes above a critical temperature which is called the Néel temperature. It is analogous to the ferromagnetic case

$$T_N = -\frac{1}{4} \frac{\nu J}{k} . \tag{8.74}$$

The Néel temperature is positive since we now have $J < 0$.

In calculating the susceptibility we must differentiate between the cases of parallel and perpendicular orientation of the external field relative to the direction of the spins, at least for temperatures smaller than T_N. We first treat the case in which the external field B_0 is parallel or antiparallel to the spins. The external field causes only minor changes in the magnetization of the two spin sub-lattices; these we denote by ΔM^+ and ΔM^-. Instead of (8.71) and (8.72), we now obtain, with an additional field B_0

$$\begin{aligned} M^+ + \Delta M^+ &= \frac{1}{2} g\,\mu_B \frac{N^+}{V} \tanh\left\{\frac{1}{2} g\,\mu_B \frac{1}{k T}\right. \\ &\quad \times \left.\left[\frac{V \nu J}{N^- g^2 \mu_B^2} (M^- + \Delta M^-) + B_0\right]\right\} , \end{aligned} \tag{8.75}$$

$$\begin{aligned} M^- + \Delta M^- &= \frac{1}{2} g\,\mu_B \frac{N^-}{V} \tanh\left\{\frac{1}{2} g\,\mu_B \frac{1}{k T}\right. \\ &\quad \times \left.\left[\frac{V \nu J}{N^+ g^2 \mu_B^2} (M^+ + \Delta M^+) + B_0\right]\right\} . \end{aligned} \tag{8.76}$$

Taking into account that $M^+ = -M^-$ and $N^+ = N^- \equiv N/2$, and expanding (8.75) and (8.76) as series for small $\Delta M^\pm, B_0$, we find

$$\Delta M = \Delta M^+ + \Delta M^- = \frac{1}{\cosh^2(\zeta)} \left(\frac{\mu_B^2 g^2 N}{4 k T V} B_0 - \frac{T_N}{T} \Delta M \right) \quad \text{with} \tag{8.77}$$

$$\zeta = \frac{T_N}{T} \frac{M^+(T)}{M_s^+} \quad \text{and} \tag{8.78}$$

$$M_s^+ = \frac{1}{2} g \mu_B \frac{N^+}{V} \tag{8.79}$$

where $M_s^\pm$ is the saturation magnetization of one spin sub-lattice.

For temperatures above the Néel temperature, the magnetization of the sub-lattices vanishes and ζ is equal to zero. No particular direction in the crystal is any longer distinguishable and the susceptibility is isotropic:

$$\chi(T) = \mu_0 \frac{g^2 \mu_B^2 N}{4 V k} \frac{1}{T + T_N} . \tag{8.80}$$

We thus obtain a temperature dependence similar to the Curie-Weiss law (8.70), but the critical temperature now appears with a reversed sign. At the Néel temperature itself, the susceptibility remains finite. For temperatures far enough below the Néel temperature $M^+(T) = M_s^+$ and we obtain

$$\chi_\parallel(T) \approx \mu_0 \frac{g^2 \mu_B^2 N}{4 V k} \frac{1}{T \cosh^2(T_N/T) + T_N} , \tag{8.81}$$

which for low temperatures can further be approximated by

$$\chi_\parallel(T) \approx \mu_0 \frac{g^2 \mu_B^2 N}{V k T} \mathrm{e}^{-2T_N/T} \quad T \ll T_N . \tag{8.82}$$

This expression for the susceptibility and the equations (8.75, 8.76) are valid only for an external field oriented parallel to the polarization of the spin sub-lattice. For the direction perpendicular to the spin orientation, the Hamiltonian (8.58) should be interpreted as a classical energy equation. In an external field, each spin sub-lattice rotates its magnetic moment by an angle α in the direction of the field B_0. The energy of an elementary magnet in the field B_0 is then

$$E_r = -\tfrac{1}{2} g \mu_B B_0 \sin\alpha + \tfrac{1}{2} \nu J \cos\alpha . \tag{8.83}$$

The magnitude of the second term can be derived by considering that the energy needed to reverse an elementary magnet is νJ (8.27). The equilibrium condition

$$\partial E_r / \partial \alpha = 0 \tag{8.84}$$

leads, for small angles α, to

$$\alpha = -\frac{g\,\mu_B B_0}{\nu J} \ . \tag{8.85}$$

With the magnetization

$$M = M^+ + M^- = \tfrac{1}{2} g\,\mu_B \alpha N/V \tag{8.86}$$

one obtains for the susceptibility below T_N the (approximately) temperature-independent value

$$\chi_\perp = -\frac{g^2\mu_B^2 N}{2\,\nu J V} = \frac{g^2\mu_B^2 N}{2\,\nu |J| V} \ , \tag{8.87}$$

which is equal to the value of χ at the Néel temperature. The overall behavior is sketched in Fig. 8.10. It should be noted that the difference between $\chi_\parallel$ and $\chi_\perp$ is only measurable experimentally if a *single* magnetic domain is present. This is not the case, for instance, in lattices with many equivalent crystallographic directions for the possible magnetic orientation. Furthermore, the characteristic temperature in the equation for the susceptibility, (8.80), is only equal to the Néel temperature if the exchange interaction is confined to nearest neighbor metal ions. If an additional exchange coupling J_2 between next nearest neighbors exists, then T_N in (8.80) is replaced by a characteristic temperature $\theta \neq T_N$. For a simple rock salt type lattice one obtains

$$\theta = T_N \frac{J_1 + J_2}{J_1 - J_2} \ . \tag{8.88}$$

Thus θ is larger than T_N when the coupling between next nearest neighbors is antiferromagnetic and smaller than T_N if the coupling is ferromagnetic. Both cases are actually observed. It should also be noted that in lattices containing several different kinds of transition metal ions, or transition me-

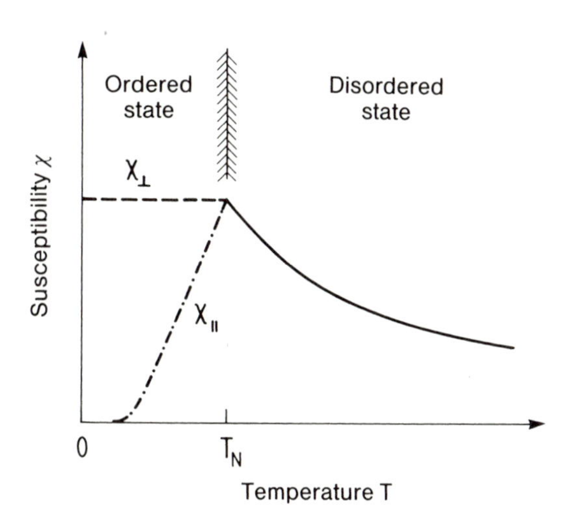

Fig. 8.10. Schematic representation of the magnetic susceptibility of an antiferromagnet. Below the Néel temperature T_N (i.e., in the antiferromagnetically ordered state) the susceptibility differs for parallel and perpendicular orientation of the magnetic field relative to the spin axis

tal ions in different valence states, the magnetic moments are not equal. Even if the coupling between the spins is antiferromagnetic, there will be a residual magnetization. This kind of magnetism is called ferrimagnetism, because is was first identified in ferrites.

8.8 Spin Waves

The energy necessary to reverse the spin of a particular electron is given by the exchange interaction. This is true in both the localized and band models. In the band model, the reversal of the spin of an electron means an interband transition of an electron into the corresponding exchange-shifted band. The minimum energy necessary to flip a spin in the band model is the energy separation between the upper edge of the majority spin band and the Fermi level, the so-called Stoner gap (Δ in Fig. 8.6). We will now learn about another excitation state, in which one spin is also reversed, but only as an average over the whole crystal. In other words, we are dealing with a collective excitation of all spins. The necessary energy is considerably smaller and may even become zero. To derive this excitation state we begin again with the spin Hamiltonian (8.57), but we must now consider the properties of the spin operator in (8.57) explicitly. The x-, y-, and z-components of the spin operators can be represented by the Pauli spin matrices

$$\boldsymbol{S}^z = \frac{1}{2}\begin{pmatrix} 1 & 0 \\ 0 & -1 \end{pmatrix}, \quad \boldsymbol{S}^x = \frac{1}{2}\begin{pmatrix} 0 & 1 \\ 1 & 0 \end{pmatrix}, \quad \boldsymbol{S}^y = \frac{1}{2}\begin{pmatrix} 0 & -\mathrm{i} \\ \mathrm{i} & 0 \end{pmatrix}. \tag{8.89}$$

Instead of the cartesian components $\boldsymbol{S}^x$ and $\boldsymbol{S}^y$, it is more convenient to use the spin reversal operators

$$\boldsymbol{S}^+ = \boldsymbol{S}^x + \mathrm{i}\,\boldsymbol{S}^y = \begin{pmatrix} 0 & 1 \\ 0 & 0 \end{pmatrix} \quad \text{and} \tag{8.90}$$

$$\boldsymbol{S}^- = \boldsymbol{S}^x - \mathrm{i}\,\boldsymbol{S}^y = \begin{pmatrix} 0 & 0 \\ 1 & 0 \end{pmatrix}. \tag{8.91}$$

The effect of these operators on the spin states

$$|\alpha\rangle = \begin{pmatrix} 1 \\ 0 \end{pmatrix} \quad \text{and} \quad |\beta\rangle = \begin{pmatrix} 0 \\ 1 \end{pmatrix} \quad \text{is} \tag{8.92}$$

$$\boldsymbol{S}^+|\alpha\rangle = 0\,, \ \boldsymbol{S}^+|\beta\rangle = |\alpha\rangle\,, \ \boldsymbol{S}^-|\beta\rangle = 0\,, \ \boldsymbol{S}^-|\alpha\rangle = |\beta\rangle\,. \tag{8.93}$$

The operators $\boldsymbol{S}^+$ and $\boldsymbol{S}^-$ therefore switch the spin to "+" or "–", and give zero if the spin is already in the state "+" or "–". The operator $\boldsymbol{S}^z$ "prepares" the eigenvalues in the usual way:

$$\boldsymbol{S}^z|\alpha\rangle = +\tfrac{1}{2}|\alpha\rangle\,, \ \boldsymbol{S}^z|\beta\rangle = -\tfrac{1}{2}|\beta\rangle\,. \tag{8.94}$$

Armed with these equations, we can begin to rewrite the Hamiltonian of a spin lattice in terms of the new operators. For a zero external field and an exchange coupling J between nearest neighbors, substitution of (8.90, 8.91) into (8.57) yields

$$\mathscr{H} = -J \sum_i \sum_\delta \boldsymbol{S}_i^z \cdot \boldsymbol{S}_{i+\delta}^z + \tfrac{1}{2}(\boldsymbol{S}_i^+ \boldsymbol{S}_{i+\delta}^- + \boldsymbol{S}_i^- \boldsymbol{S}_{i+\delta}^+) \,. \tag{8.95}$$

We assume ferromagnetic coupling between the electron spins ($J>0$). In the ground state all spins are therefore oriented. Such a state is described by the product of the spin states of all atoms

$$|0\rangle = \prod_i |\alpha\rangle_i \,. \tag{8.96}$$

It is immediately clear that this state is an eigenstate of the Hamiltonian (8.95), since the operators $\boldsymbol{S}^+\boldsymbol{S}^-$ and $\boldsymbol{S}^-\boldsymbol{S}^+$ yield zero and the $\boldsymbol{S}^z$ components leave the state unaltered with the corresponding eigenvalue of $\boldsymbol{S}^z$ as a prefactor

$$\mathscr{H}|0\rangle = -\tfrac{1}{4}J|0\rangle \sum_i \sum_\delta 1 = -\tfrac{1}{4}\nu J N|0\rangle \,. \tag{8.97}$$

A state with a reversed spin on atom j can be obtained by applying $\boldsymbol{S}_j^-$ to the ground state

$$|{\downarrow_j}\rangle \equiv \boldsymbol{S}_j^- \prod_n |\alpha\rangle_n \,. \tag{8.98}$$

This state, however, is not an eigenstate of $\mathscr{H}$ because applying the operators $\boldsymbol{S}_j^+ \boldsymbol{S}_{j+\delta}^-$ inside $\mathscr{H}$ would shift the reversed spin to the atom $j+\delta$ and create a different state. On the other hand, the linear combination

$$|\boldsymbol{k}\rangle = \frac{1}{\sqrt{N}} \sum_j \mathrm{e}^{\mathrm{i}\boldsymbol{k}\cdot\boldsymbol{r}_j} |{\downarrow_j}\rangle \tag{8.99}$$

is an eigenstate. This state represents a spin wave. The eigenvalues of $\boldsymbol{S}_i^z$ and $(\boldsymbol{S}_i^x)^2 + (\boldsymbol{S}_i^y)^2$ are conserved quantities with expectation values independent of the atom i. On the other hand, the expectation values of $\boldsymbol{S}_i^x$ and $\boldsymbol{S}_i^y$ vanish. The spin therefore precesses around the z axis with a phase shift between atoms that is determined by the wave vector $\boldsymbol{k}$ (Fig. 8.11). We now apply $\mathscr{H}$ to the spin wave state and obtain

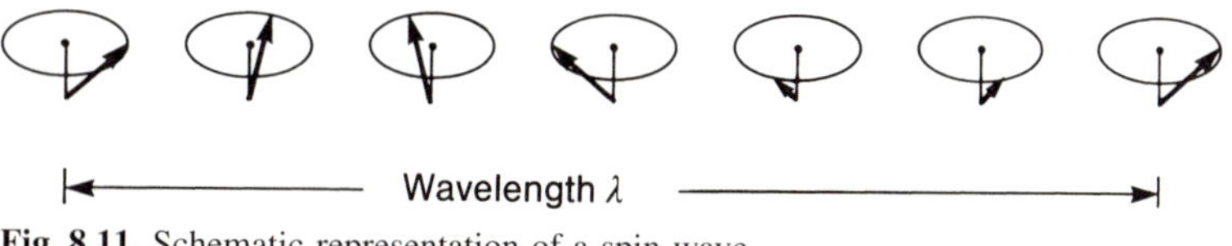

Fig. 8.11. Schematic representation of a spin wave

$$\mathcal{H}|\boldsymbol{k}\rangle = \frac{1}{\sqrt{N}}\sum_j \mathrm{e}^{\mathrm{i}\boldsymbol{k}\cdot\boldsymbol{r}_j}\left[-\frac{1}{4}\nu J(N-2)|\downarrow_j\rangle + \frac{1}{2}\nu J|\downarrow_j\rangle \right.$$
$$\left. -\frac{1}{2}J\sum_\delta |\downarrow_{j+\delta}\rangle + |\downarrow_{j-\delta}\rangle\right] . \quad (8.100)$$

By shifting the index j in the last two terms we can also express this result as

$$\mathcal{H}|\boldsymbol{k}\rangle = \left[-\frac{1}{4}\nu J N + J\nu - \frac{1}{2}J\sum_\delta(\mathrm{e}^{-\mathrm{i}\boldsymbol{k}\cdot\boldsymbol{r}_\delta} + \mathrm{e}^{\mathrm{i}\boldsymbol{k}\cdot\boldsymbol{r}_\delta})\right]$$
$$\times \frac{1}{\sqrt{N}}\sum_j \mathrm{e}^{\mathrm{i}\boldsymbol{k}\cdot\boldsymbol{r}_j}|\downarrow_j\rangle . \quad (8.101)$$

The state $|\boldsymbol{k}\rangle$ is therefore an eigenstate with the eigenvalue

$$E = E_0 + J\left(\nu - \frac{1}{2}\sum_\delta \mathrm{e}^{-\mathrm{i}\boldsymbol{k}\cdot\boldsymbol{r}_\delta} + \mathrm{e}^{\mathrm{i}\boldsymbol{k}\cdot\boldsymbol{r}_\delta}\right) , \quad (8.102)$$

where E_0 is the energy in the ferromagnetic ground state. As we will immediately see, this result is particularly significant for small k values. When k is small, (8.102) can be approximated by

$$E \approx E_0 + \tfrac{1}{2}J\textstyle\sum_\delta(\boldsymbol{k}\cdot\boldsymbol{r}_\delta)^2 . \quad (8.103)$$

This is the characteristic dispersion relation for ferromagnetic spin waves. According to this relation, the energy required to flip a spin vanishes for small k. In Fig. 8.12a, the dispersion relation is plotted together with the spectrum of one-electron excitations which also reverse a spin. These so-called Stoner excitations require an energy νJ for $k=0$. For $k \neq 0$ there is a whole spectrum of possibilities resulting from the dispersion of the one-electron states (Fig. 8.12b). In the region of the one-electron excitations, spin waves can decay into electron excitations. This reduces the lifetime of the spin waves and also affects the dispersion. Spin waves can be excited thermally and also by energy and momentum exchange with neutrons. Neutron scattering therefore allows an experimental determination of the dispersion curve of spin waves. The results for nickel are displayed in Fig. 8.13.

The thermal excitation of spin waves has an influence on the behavior of the magnetization at low temperature. We recall that the excitation of a spin wave flips on average one spin and hence reduces the magnetic moment. The magnetization of a spin lattice thus becomes

$$M = M_\mathrm{s} - \frac{1}{2}g\,\mu_\mathrm{B}\frac{1}{V}\sum_{\boldsymbol{k}} n(\boldsymbol{k}) , \quad (8.104)$$

where $n(k)$ is the number of excited spin waves with wave vector $\boldsymbol{k}$. If we neglect the fact that the Heisenberg operator is actually non-linear and pro-

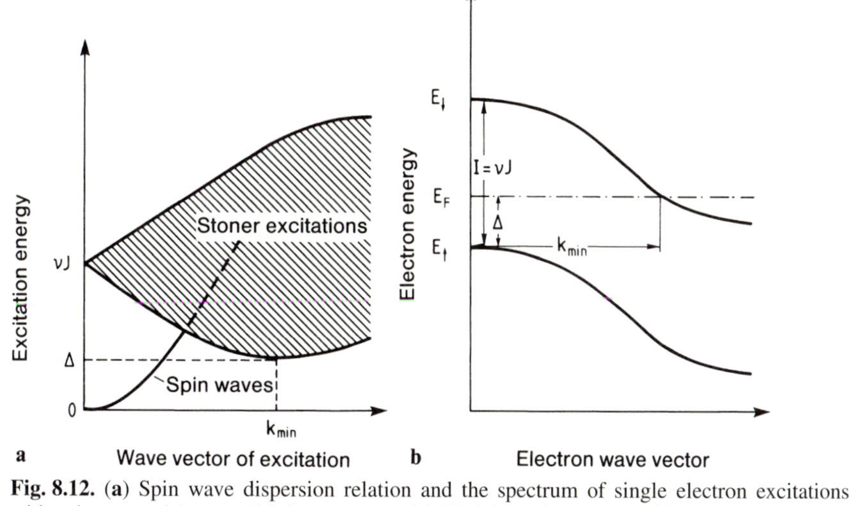

Fig. 8.12. (**a**) Spin wave dispersion relation and the spectrum of single electron excitations with spin reversal in a model ferromagnet. (**b**) Model band structure with an exchange splitting $I = \nu J$ and a Stoner gap Δ

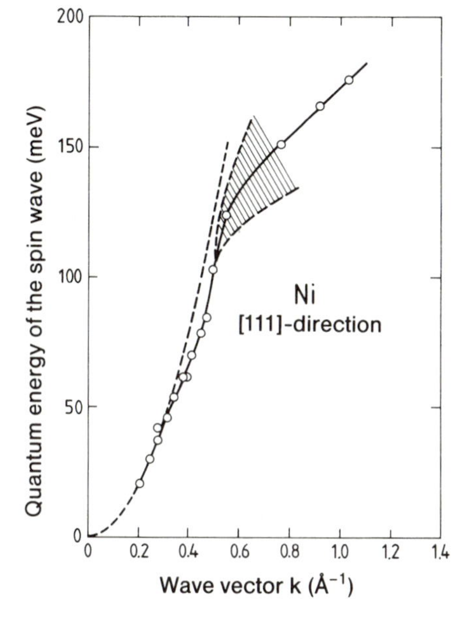

Fig. 8.13. Experimental dispersion relation for spin waves in nickel along the [111] direction [8.6]. The measurements were made at $T = 295$ K. The dashed line shows a dependence of the quantum energy proportional to k^2. Deviations from this line are due firstly to the exchange interaction between more distant neighbors, and secondly to the onset of one-electron excitations. The latter cause a reduction in the lifetime of the spin waves, leading to a lifetime broadening of the spectra (*shaded region*)

ceed as though the spin waves could be superposed, then the energy eigenvalues would be the same as for a harmonic oscillator

$$E(k) = n(k) \cdot \tfrac{1}{2} J \textstyle\sum_\delta (\boldsymbol{k} \cdot \boldsymbol{r}_\delta)^2 \tag{8.105}$$

and the occupation statistics would likewise correspond to the harmonic oscillator. The temperature dependence of the magnetization at low temperature can be calculated as in Sect. 5.3 and we obtain

$$M(T) - M(0) \sim -T^{3/2} \;. \tag{8.106}$$

This is the $T^{3/2}$ Bloch law, which replaces the exponential dependence (8.52), appropriate for a ferromagnet without spin wave excitations. The spin wave excitation is also observable in the specific heat where a $T^{3/2}$ term appears in addition to the T^3 term due to phonon excitation.

Problems

8.1 Calculate the diamagnetic susceptibility of the hydrogen atom using (8.12) and the ground state wavefunction $\psi = (a_0^3\pi)^{-1/2} e^{-r/a_0}$, where a_0 is the Bohr radius, $a_0 = 4\pi\hbar^2\varepsilon_0/(me^2) = 0.529$ Å.

8.2 Calculate the entropy of N spins in a magnetic field using the principles of statistical thermodynamics. Describe the process of adiabatic demagnetization and discuss this method for cooling a paramagnetic salt. How much heat can one extract from the lattice in the process of adiabatic demagnetization (cf. Panel IV)?

8.3 In crystalline ferromagnets the magnetization is preferentially oriented along a particular crystallographic direction (the "easy direction"). It costs energy to reorient the magnetization. This energy is called the anisotropy energy. It plays an essential part in determining the magnetic properties of materials, e.g., in technological applications. For cobalt the easy direction is along the c-axis of the hexagonal structure. Express the anisotropy energy by the two lowest order terms in the angle Θ which describes the orientation of the magnetization with respect to the c-axis. Determine also the form of the anisotropy energy for a cubic crystal in terms of the cosines of the angles Θ_1, Θ_2, Θ_3 of the magnetization with respect to the cubic axes. Note that none of the cubic axes is polar. The expression must also be invariant under interchange of the cubic axes, and finally must satisfy $\cos^2\Theta_1 + \cos^2\Theta_2 + \cos^2\Theta_3 \equiv 1$. Why is it also necessary to consider a sixth order term in $\cos\Theta$?

8.4 (To be tackled after Problem 8.3). It is energetically favorable for a ferromagnet of finite size to break up the magnetic order into domains of different orientation of the magnetization, despite the fact that the formation of domain walls costs energy. Why? Consider a wall parallel to the (100) plane between two domains in which the magnetization is parallel and antiparallel to the easy [001] axis. For the sake of simplicity, assume that the [001] axis is the only easy axis (as is the case, e.g., in a tetragonal structure). The wall between the two domains of opposite magnetization has a finite thickness, i.e., the direction of the magnetization rotates gradually. Throughout the wall the magnetization vector remains in the (100) plane. Such a wall is known as a Bloch wall. Calculate the total exchange energy and the total anisotropy energy in a Bloch wall as a function of the wall thickness. The exchange energy may be calculated classically according to (8.83). Show that the equilibrium between exchange and anisotropy energy determines the thickness of the wall.

Panel VI
Magnetostatic Spin Waves

In the absence of an external magnetic field, the quantum energy for spin waves in a ferromagnet vanishes for small k. This is because the difference in spin orientation from atom to atom becomes smaller and smaller for increasing wavelength, and the exchange interaction therefore provides an ever smaller contribution to the energy. The exchange coupling eventually becomes comparable to the energy of magnetic dipoles in an external field, i.e. the dispersion of the spin waves is dependent on the magnetic field. In the following we treat such spin waves with small k values in an external field. The k value is assumed to be small compared to a reciprocal lattice vector but large compared to ω/c, where ω is the spin wave frequency. This condition allows us to neglect the explicit interaction with the electromagnetic field (Sect. 11.4) and to assume that

$$\nabla \times \boldsymbol{H} \equiv \boldsymbol{0} \tag{VI.1}$$

and also of course

$$\nabla \cdot (\boldsymbol{H} + \boldsymbol{M}) \equiv 0 \tag{VI.2}$$

Spin waves of this kind are called magnetostatic spin waves. They can be handled with the classical equations of motion, which relate the rate of change of the angular momentum of an electron to the torque. If the magnetic dipole moment of the electron is $\boldsymbol{p}_i$, then

$$\frac{d\boldsymbol{p}_i}{dt} = \gamma(\boldsymbol{p}_i \times \boldsymbol{B}) \quad \text{with} \quad \gamma = \frac{g\,\mu_\mathrm{B}}{h} \; . \tag{VI.3}$$

In the following, instead of the dipole moments of single electrons, it is convenient to introduce a space- and time-dependent magnetization $\boldsymbol{M}(\boldsymbol{r},t)$, which can be constructed from a (local) average over the dipole moments per unit volume. In (VI.3) $\boldsymbol{p}_i$ can be replaced by the local magnetization $\boldsymbol{M}(\boldsymbol{r},t)$. The magnetization is essentially the saturation magnetization $\boldsymbol{M}_\mathrm{s}$ of the ferromagnet, which is oriented by the external field $\boldsymbol{H}_0$ along the z axis, with small space- and time-dependent deviations m_x and m_y in the x and y components. We make the ansatz

$$\boldsymbol{M}(\boldsymbol{r},t) = \begin{pmatrix} m_x(\boldsymbol{r})\mathrm{e}^{-\mathrm{i}\omega t} \\ m_y(\boldsymbol{r})\mathrm{e}^{-\mathrm{i}\omega t} \\ M_\mathrm{s} \end{pmatrix} ,$$

$$\boldsymbol{B}(\boldsymbol{r},t) = \mu_0 \begin{pmatrix} (h_x + m_x)\mathrm{e}^{-\mathrm{i}\omega t} \\ (h_y + m_y)\mathrm{e}^{-\mathrm{i}\omega t} \\ (H_0 + M_\mathrm{s}) \end{pmatrix} \tag{VI.4}$$

and by noting that $m_x, m_y \ll M$ and $h_x, h_y \ll H_0$, we obtain from (VI.3) the pair of equations

$$\begin{pmatrix} m_x \\ m_y \end{pmatrix} = \begin{pmatrix} \kappa & \mathrm{i}\,\nu \\ -\mathrm{i}\,\nu & \kappa \end{pmatrix} \begin{pmatrix} h_x \\ h_y \end{pmatrix} \tag{VI.5}$$

with

$$\kappa = \mu_0 \gamma^2 M_\mathrm{s} B_0 / (\gamma^2 B_0^2 - \omega^2) \, ,$$
$$\nu = \mu_0 \omega \gamma M_\mathrm{s} / (\gamma^2 B_0^2 - \omega^2) \, , \quad B_0 = \mu_0 H_0 \, .$$

According to (VI.1), the two component vector $\boldsymbol{h}$ can be written as the gradient of a potential

$$\boldsymbol{h} = \nabla \varphi \, , \tag{VI.6}$$

where for φ we insert as a trial solution a wave propagating in the y direction, i.e.,

$$\varphi = \psi(x) \mathrm{e}^{\mathrm{i}\, k_y y} \, . \tag{VI.7}$$

Application of (VI.2) then leads us to

$$\nabla \cdot (\boldsymbol{h} + \boldsymbol{m}) = \Delta \varphi + \nabla \cdot \boldsymbol{m} = 0 \, , \tag{VI.8}$$

from which, using (VI.5, 7), we obtain an equation resembling a wave equation for the potential

$$(1 + \kappa) \left(\frac{\partial^2}{\partial x^2} - k_y^2 \right) \varphi = 0 \, . \tag{VI.9}$$

A possible solution of this equation is obviously obtained when $\kappa = -1$. This condition yields a spin-wave frequency of

$$\omega = \gamma \sqrt{B_0^2 + \mu_0 M_\mathrm{s} B_0} \, . \tag{VI.10}$$

The frequency is thus independent of k if the exchange coupling is completely neglected, as is the case here. It is interesting to look for solutions for special geometries of the sample. The expression (VI.7) can also be applied to the case of a slab in a $\boldsymbol{B}$ field or, simpler still, to the half space $x < 0$. Equation (VI.9) then has further solutions with $\kappa \neq -1$ when

$$\psi''(x) = k_y^2 \psi(x) \tag{VI.11}$$

i.e., for $\psi(x)$ of the form

$$\psi(x) = A\, \mathrm{e}^{\pm |k_y| x} \, . \tag{VI.12}$$

This solution is evidently localized at the surface of the half-space and is a surface spin wave. The condition that the normal component of $\boldsymbol{B}$ is continuous across the surface yields the eigenfrequency of this mode

$$h_x + m_x|_{x<0} = h_x|_{x>0} \,, \tag{VI.13}$$

$$(1+\kappa)|k_y| - \nu\, k_y = -|k_y| \,. \tag{VI.14}$$

We can differentiate between the two cases $k_y = \pm|k_y|$, that is, the two different directions of propagation:

Case 1:	*Case 2*:	
$k_y = +\lvert k_y\rvert$	$k_y = -\lvert k_y\rvert$	
$\nu = \kappa + 2$,	$\nu = -(\kappa + 2)$	(VI.15)
$\omega_1 = -\gamma(\frac{1}{2}\mu_0 M_s + B_0)$,	$\omega_2 = \gamma(\frac{1}{2}\mu_0 M_s + B_0)$.	(VI.16)

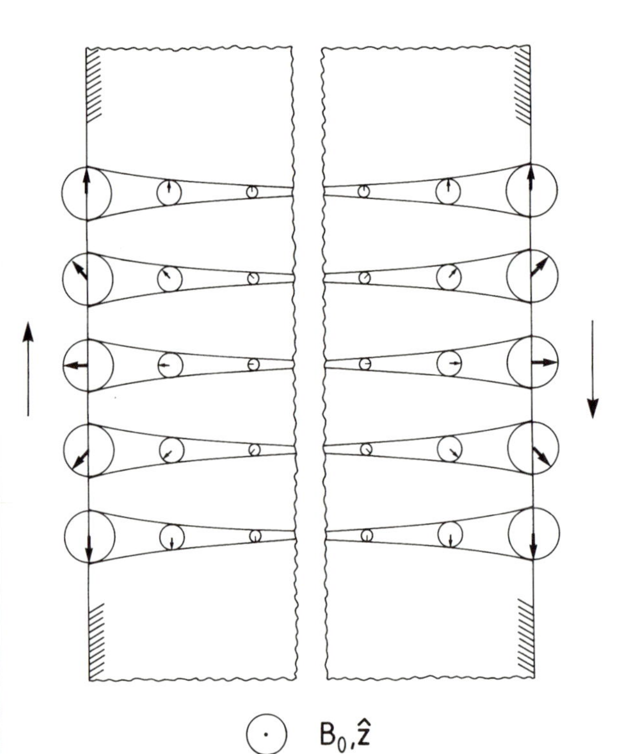

Fig. VI.1. Snapshot of the magnetization in a Damon-Eshbach spin wave on the two surfaces of a (thick) slab. The external magnetic field $\boldsymbol{B}_0$ points out of the page. The Damon-Eshbach waves then propagate in a clockwise direction. For smaller values of the slab thickness d, or larger wavelengths, the waves on the two surfaces interact and a coupling dispersion is obtained $\omega = \omega(k_y d)$

Clearly, only the second case yields a positive frequency. We thus have a curious situation in which a wave exists in one direction but not in the opposite direction. This is a particularly interesting demonstration of the break-down of time-reversal invariance by a magnetic field. Corresponding solutions are obtained if the sample has the form of a thin plate with the $\boldsymbol{B}_0$ field in the plane of the plate. The spin waves then propagate in opposite directions on the two surfaces (Fig. VI.1). Because of the coupling between the two surfaces the frequency becomes dependent on k_y when $k_y d \lesssim 1$. These magnetic surface waves are known as "Damon-Eshbach" waves after their discoverers [VI.1]. They were first identified in the absorption of microwaves [VI.2].

A very nice demonstration of the unidirectional nature of Damon-Eshbach waves, and at the same time of the conservation of wave vector (including its sign) during scattering (4.28), is provided by the Raman effect (Panel III). Coupling of the light occurs in this case via the (weak) magnetooptic effect. Figure VI.2 shows the experimental setup of Grünberg and Zinn [VI.3]. The sample is illuminated with a laser and the Raman effect is observed in backscattering. The frequency of Damon-Eshbach waves lies in the GHz region. A spectrometer with high resolution is required, for example, a Fabry-Pérot interferometer. To reduce the background due to elastic diffuse scattering at the sample, the beam passes through the interferometer several times (twice in Fig. VI.2). A frequency spectrum is obtained by moving the Fabry-Pérot mirrors relative to one another.

Figure VI.3 shows a Raman spectrum for two different positions of a EuO sample relative to the light beam. The surface spin wave can only travel in the direction labelled $q_{\|}$. In the geometry shown above, one thus has

$$(\boldsymbol{k}_0 - \boldsymbol{k})_{\|} = \boldsymbol{q}_{\|} \ .$$

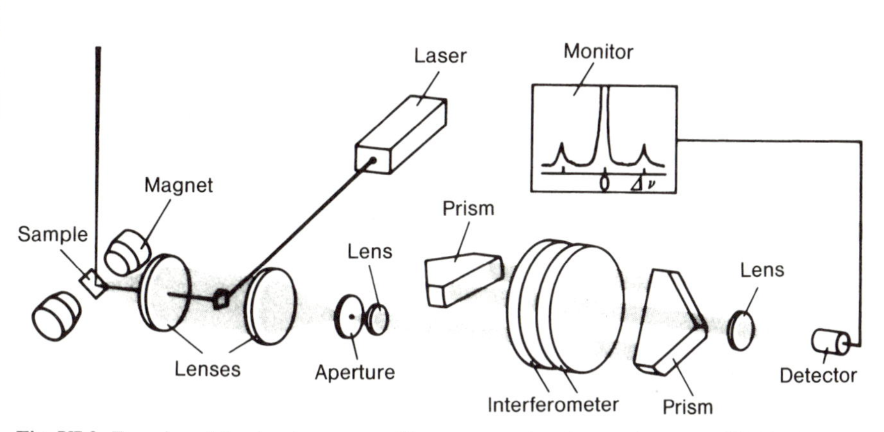

Fig. VI.2. Experimental setup to measure Raman scattering from spin waves [VI.2]. The multiple passes of the light through the Fabry-Pérot spectrometer reduces the background so that very weak inelastic signals can be observed

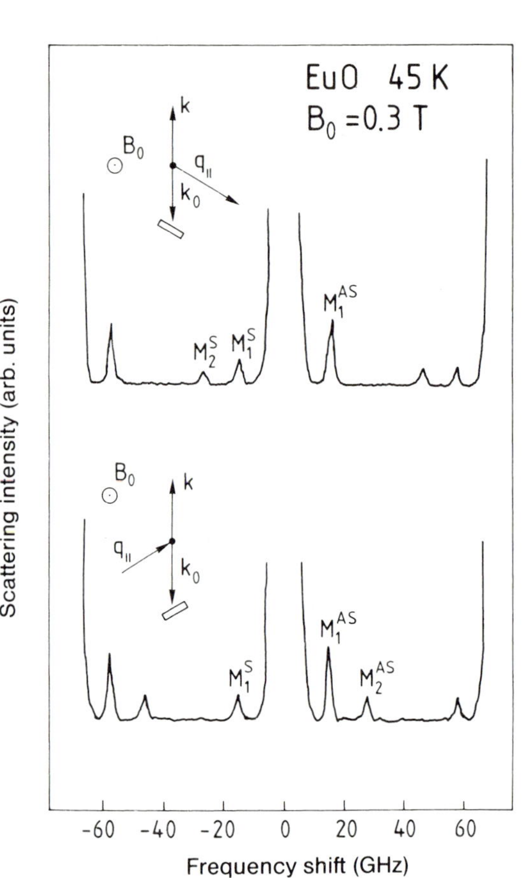

Fig. VI.3. Raman spectrum from EuO [VI.2]. According to the orientation of the sample one observes the Damon-Eshbach spin waves (labelled as M_2) as a Stokes line (*above*) or as an anti-Stokes line (*below*), while the volume spin waves appear with equal intensity in both geometries, although higher intensity is observed for the anti-Stokes line [VI.3]

Energy conservation (4.28) demands

$$\omega_0 - \omega = \omega(q_\parallel) .$$

The associated Raman line is shifted to smaller ω. We thus obtain the Stokes line, whereas the anti-Stokes line for the surface wave does not exist. For the scattering geometry sketched at the bottom of Fig. VI.3, we have

$$(\boldsymbol{k}_0 - \boldsymbol{k})_\parallel = -\boldsymbol{q}_\parallel$$

and therefore

$$\omega_0 - \omega = -\omega(q_\parallel) .$$

Correspondingly, only the anti-Stokes line is observed for surface waves. For the bulk wave, one observes both the Stokes and anti-Stokes lines, but curiously the anti-Stokes line has higher intensity: the number of contributing bulk waves with the same frequency is different for the two scattering geometries.

References

VI.1 R.W. Damon, J.R. Eshbach: J. Phys. Chem. Solids **19**, 308 (1961)
VI.2 P. Grünberg, W. Zinn: IFF Bulletin **22** (KFA, Jülich 1983) p. 3
VI.3 R.E. Camley, P. Grünberg, C.M. Mayer: Phys. Rev. B. **26**, 2609 (1982)

Panel VII
Surface Magnetism

In our discussion of the ferromagnetic behavior of solids with exchange coupling between localized electrons, it was found that the Curie temperature is usually dependent on the strength of the exchange coupling and on the number of neighboring atoms (8.66). At the surface of a solid the coordination number is considerably reduced. One would therefore expect a strongly reduced Curie temperature for the surface atoms. In fact, the magnetic order within the volume forces the surface into a magnetically ordered state, but the critical behavior near T_c is nonetheless strongly influenced by the surface. It is instructive to study this within a simple model for the exchange coupling in the framework of mean field theory. If the atomic layers parallel to the surface are labelled l, then the average magnetic moment of an atom in the lth layer $\langle m_l \rangle$ couples to the average moment of the neighboring atom δ according to (8.58),

$$\langle m_l \rangle / m_s = \tanh \left(\frac{T_c}{T} \sum_\delta \langle m_\delta \rangle J_{l\delta} \Big/ \sum_\delta m_s \right) . \qquad \text{(VII.1)}$$

Instead of the average magnetic moment we can also introduce a magnetization of each layer

$$M_l(T) = \frac{1}{A} \sum_{\text{layer}} \langle m_l \rangle . \qquad \text{(VII.2)}$$

For a finite number of layers, this set of equations can be solved numerically and self-consistently with relatively fast convergence. As an example, we show the result for 110 layers of a face centered cubic crystal with a (110) surface, isotropic exchange coupling, and interaction between nearest neighbors only. The coordination number of the surface atoms is 7. If the surface were not coupled to the bulk, the magnetization would behave as in the bulk but with the Curie temperature reduced by a factor of $\frac{7}{12}$ (dashed curve in Fig. VII.1). In fact, the Curie temperature is the same as in the bulk, but the behavior of the magnetization is quite different. Only with increasing depth into the solid does the magnetization approach the behavior of the bulk magnetization. This is further illustrated in Fig. VII.2 where the relative layer magnetization is plotted as a function of reduced temperature. For the surface, the critical exponent in the vicinity of T_c is equal to 1 (within the mean field approximation). This is valid also for the deeper

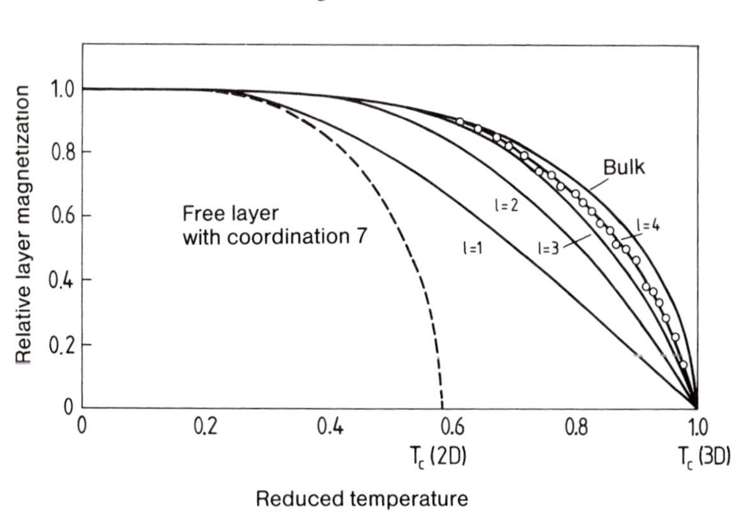

Fig. VII.1. Relative magnetization per layer $M_l(T)/M_s$ near a surface (isotropic exchange coupling, mean field approximation). The index $l=1$ labels the surface atoms, $l=2$ the atoms of the next layer, and so on. In this approximation a two-dimensional lattice has the same magnetization curve as in the bulk of a three-dimensional crystal, but with a smaller T_c due to the reduced coordination number. The data points show the spin polarization of secondary electrons [VII.1] as a function of temperature

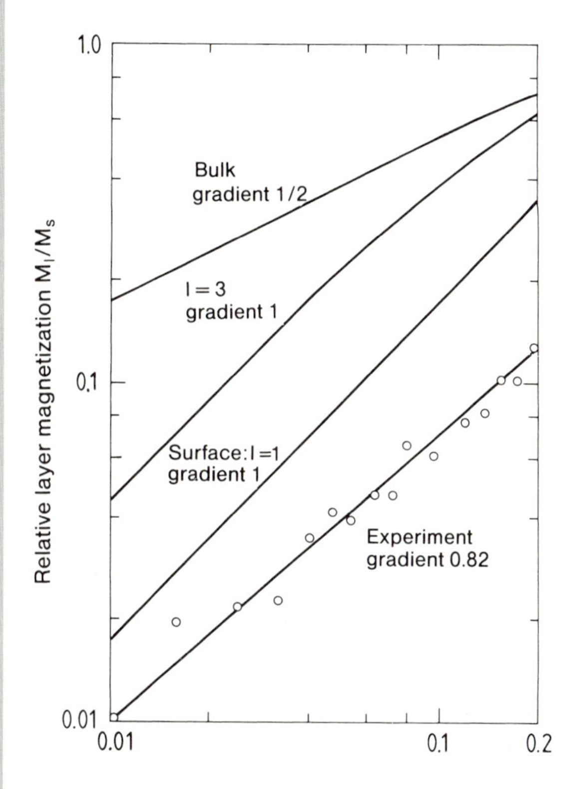

Fig. VII.2. Relative magnetization per layer $M_l(T)/M_s$ vs. the reduced temperature $t=1-(T/T_c)$ in a double logarithmic plot. For the surface, the gradient is 1 at small t. This is also valid for deeper layers, but one must go to ever smaller t in order to see this behavior. For the bulk or for deeper layers where t is not too small, the gradient is 0.5. Experimental values for the spin asymmetry in the scattering of electrons [VII.2], which is proportional to the surface magnetism, show a smaller slope than that predicted by the mean field model

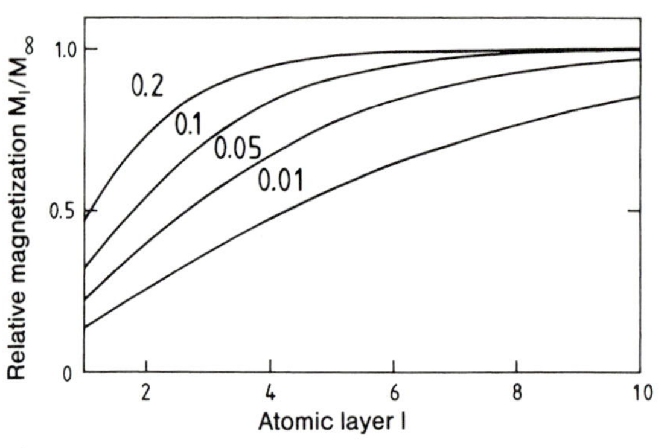

Fig. VII.3. Magnetization in the near-surface layers l divided by the magnetization in a deep layer ($l=50$) as a function of l. The variable parameter is the reduced temperature $t=1-(T/T_c)$. On approaching T_c, the reduced surface magnetism influences ever deeper regions of the bulk thereby reducing their magnetization

layers, but one must approach T_c ever more closely, in order to see the "true" critical exponent. In this connection it is useful to consider the magnetic moment as a function of the depth at fixed temperature (Fig. VII.3). The closer the temperature is to T_c, the farther the reduced magnetization at the surface penetrates into the bulk.

It is possible to observe surface magnetism experimentally. A sensitive method is based on a spin analysis of secondary electrons or the scattering of spin-polarized electrons. To avoid the deflection of electrons in the stray field of the sample, the sample has the form of a closed magnetic loop (Fig. VII.4). The spin polarization of the secondary electrons can be determined with a so-called Mott detector. In the Mott detector, electrons with high energy are scattered from a foil of a material with high atomic number. The spin-orbit interaction then converts the spin polarization of the electrons into an intensity asymmetry between the electrons scattered to the left and the right. The result of the polarization analysis of secondary electrons from a Ni(110) surface as a function of temperature is shown in Fig. - VII.1 [VII.1]. The temperature dependence corresponds approximately to the magnetization of the fourth layer. From this it is concluded that the secondary electrons come from an average depth corresponding to the fourth layer. The more exact analysis of Abraham and Hopster [VII.1] gives an average escape depth of about 9 Å, equivalent to seven layers.

Magnetic behavior can also be examined via the scattering of already polarized electrons. Commonly used sources of spin-polarized electrons are GaAs and $Al_xGa_{1-x}As$. Electrons are excited from the valence band to the conduction band at the Γ point with right or left circularly polarized light. If the surface is suitably prepared, the spin-polarized electrons are then emitted as photoelectrons. The intensity asymmetry for the scattering of electrons with opposite polarizations is then a measure of the magnetization. Alvarado et al. [VII.2] were thereby able to determine the critical exponent for surface magnetization (Fig. VII.2). With a value of 0.8, it deviates signifi-

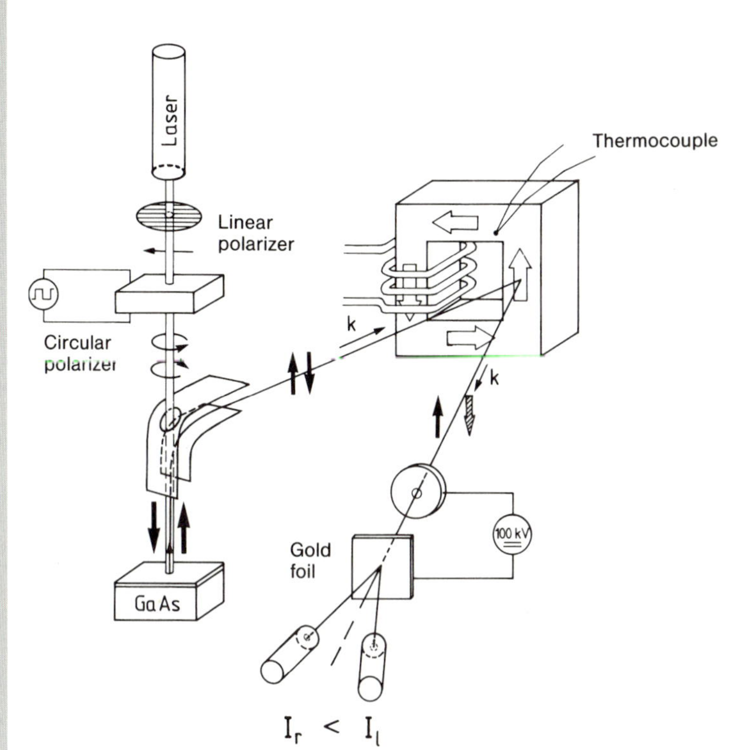

Fig. VII.4. Sketch of a scattering experiment with polarized electrons and spin analysis of the scattered electrons. Spin-polarized electrons are created by photoemission from GaAs with right or left circularly polarized light. By deflecting the electron beam through 90°, a process in which the spin direction is conserved, the spin becomes oriented perpendicular to the scattering plane. The exchange interaction with the electrons in a ferromagnet makes the scattering intensity dependent on the spin orientation of the incident beam. The spin polarization of the scattered beam can be determined from the asymmetry in the scattering intensity measured in a Mott detector

cantly from the mean field value. We have already seen in the discussion of bulk magnetization that the mean field approximation does not yield the correct critical exponent. The surface Curie temperature, in the mean field model, is just as large as the bulk Curie temperature. For very strong exchange coupling between the surface atoms, that is, when the total exchange coupling in the first layer (VII.1) is larger than in the bulk, the Curie temperature of the surface is larger than that of the bulk. Magnetization persists on the surface even above the bulk Curie temperature. This phenomenon has been experimentally observed for gadolinium [VII.3].

References

VII.1 D.L. Abraham, H. Hopster: Phys. Rev. Lett. **58**, 1352 (1987)
VII.2 S.F. Alvarado, M. Campagna, F. Ciccacci, H. Hopster: J. Appl. Phys. **53**, 7920 (1982)
VII.3 D. Weller, S.F. Alvarado, W. Gudat, K. Schröder, M. Campagna: Phys. Rev. Lett. **54**, 1555 (1985)

9 Motion of Electrons and Transport Phenomena

A number of important phenomena such as electrical and thermal conductivity are based on the motion of electrons in solids. The description of electron motion goes beyond our previous considerations since it involves a time-dependent Schrödinger equation; previously we have discussed only the time-independent Schrödinger equation and results for thermodynamic equilibrium (Fermi statistics, etc.). The present chapter deals with the question of how electrons in bands behave, if, for example, an external electric field is applied, so that thermodynamic equilibrium is disturbed. The simplest case is that of the stationary state, where the external forces, an electric field or temperature gradient for example, are independent of time.

9.1 Motion of Electrons in Bands and the Effective Mass

In describing the motion of an electron in a solid, one is confronted with the same problem that appears when the motion of a more-or-less localized free particle is to be described from a wave theory point of view. The motion of a free electron with fixed momentum $\boldsymbol{p}$ can be described by an infinitely extended plane wave. An exactly defined value of the wave vector $\boldsymbol{k}$, however, implies complete uncertainty about the electron's position in real space, since the plane wave extends along the entire x-axis. If, on the other hand, the electron is localized to within an interval Δx, e.g., due to a measurement of its position, then the momentum, i.e., wave vector, becomes uncertain. Mathematically, localization can be described by expressing the state of the electron as a wave packet, i.e., a linear superposition of waves with wave vector in the interval $\{k–\Delta k/2, k+\Delta k/2\}$:

$$\psi(x,t) \sim \int\limits_{k-\Delta k/2}^{k+\Delta k/2} a(k)\mathrm{e}^{\mathrm{i}[kx-\omega(k)t]}dk \;, \qquad (9.1)$$

where $\omega(k)$ is determined by a special dispersion relation. As an example, Fig. 9.1 shows a wave packet in a real space representation.

From the Fourier representation (9.1) one obtains the uncertainty relation of wave mechanics:

$$\Delta p \cdot \varDelta x = \hbar \Delta k \cdot \Delta x \sim \hbar \;. \qquad (9.2)$$

The translational motion of a wave packet may be described by the group velocity:

$$v = \frac{\partial \omega}{\partial k} \ . \tag{9.3}$$

The group velocity is the velocity of the center of gravity of the spatially localized wave packet (which is to be distinguished from the phase velocity $c = \omega/k$ of a plane wave). We note that Schrödinger's wave mechanics already contains a dispersion $\omega = c(k)k$ for a free electron, as a result of which the wave packet "spreads out" as a function of time, and also changes its shape, as shown in Fig. 9.1.

In a crystal, electrons are described by Bloch waves, which represent spatially modulated, infinitely extended waves with wavevector $\boldsymbol{k}$. To describe localized crystal electrons it is therefore appropriate to introduce wave packets of Bloch waves (also known as "Wannier functions"). The localization in real space implies, via the uncertainty relation, an associated uncertainty in the momentum or $\boldsymbol{k}$-vector. The velocity of a crystal electron in this semiclassical picture is given by the group velocity of the Bloch wave packet:

$$\boldsymbol{v} = \nabla_k \, \omega(\boldsymbol{k}) = \frac{1}{\hbar} \nabla_k \, E(\boldsymbol{k}) \ . \tag{9.4}$$

Here, $E(\boldsymbol{k})$ is the wave-vector dependence of the energy in the band from which the electron originates. This description naturally includes the case

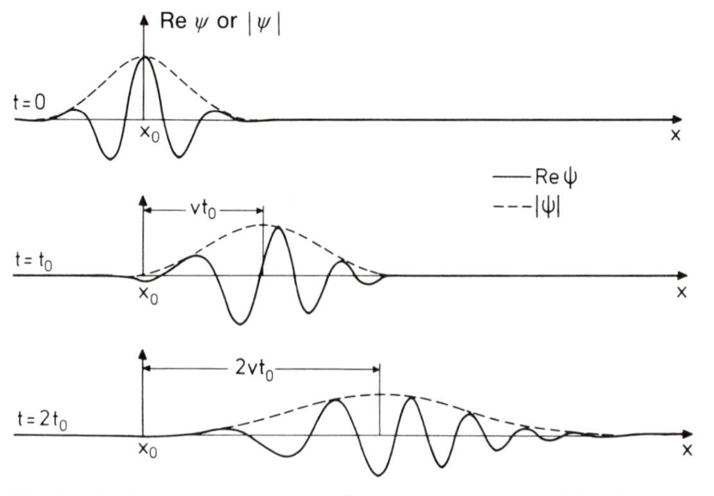

Fig. 9.1. Real space representation of the wave packet describing the motion of a spatially localized free electron at times $t=0, t_0, 2t_0 \ldots$ (Re$\{\psi$: ——; $|\psi|$: – – –). The center of the wave packet, i.e., in the particle picture the electron itself, moves with the group velocity $v = \partial\omega/\partial k$. The halfwidth of the envelope increases with time. As the wave packet spreads, the wavelength of the oscillations of Re$\{\psi\}$ becomes smaller at the front and larger at the rear

of the free electron, for which $E=\hbar^2 k^2/2m$. From (9.4) it then follows that $\boldsymbol{v}=\boldsymbol{k}\hbar/m=\boldsymbol{p}/m$.

Following the Correspondence Principle we are led to the semiclassical equations of motion for a crystal electron. A crystal electron described by a wave packet with average wave vector $\boldsymbol{k}$ in the presence of an external electric field gains, in a time, δt, an additional energy

$$\delta E = -e\mathscr{E} \cdot \boldsymbol{v}\delta t \,, \tag{9.5}$$

where $\boldsymbol{v}$ represents the group velocity (9.4), of the wave packet. From (9.4, 9.5) it follows that:

$$\delta E = \nabla_k E(\boldsymbol{k}) \cdot \delta\boldsymbol{k} = \hbar\boldsymbol{v} \cdot \delta\boldsymbol{k} \,, \tag{9.6a}$$

$$\hbar\,\delta\boldsymbol{k} = -e\mathscr{E}\delta t \,, \tag{9.6b}$$

$$\hbar\dot{\boldsymbol{k}} = -e\mathscr{E} \,. \tag{9.6c}$$

This equation of motion, which for free electrons can be deduced immediately from the Correspondence Principle, states that the wave vector $\boldsymbol{k}$ of an electron in the crystal changes according to (9.6c) in an external electric field. From the time-dependent Schrödinger equation, it can be shown that (9.6c) applies quite generally to wave packets of Bloch states, provided certain conditions hold: Namely, the electromagnetic fields must not be too large compared with atomic fields, and they must be slowly varying on an atomic length and time scale.

The equations (9.6) enable us to write down a semiclassical equation of motion for crystal electrons in the presence of an external field, with the influence of the atomic crystal field appearing only phenomenologically in the form of the bandstructure $E(\boldsymbol{k})$. From (9.4) and (9.6), it follows that the rate of change of the group velocity component v_i of an electron is:

$$\dot{v}_i = \frac{1}{\hbar}\frac{d}{dt}(\nabla_k E)_i = \frac{1}{\hbar}\sum_j \frac{\partial^2 E}{\partial k_i \partial k_j}\dot{k}_j \,, \tag{9.7a}$$

$$\dot{v}_i = \frac{1}{\hbar^2}\sum_j \frac{\partial^2 E}{\partial k_i \partial k_j}(-e\mathscr{E}_j) \,. \tag{9.7b}$$

This equation is completely analogous to the classical equation of motion $\dot{\boldsymbol{v}} = m^{-1}(-e\mathscr{E})$ of a point charge $(-e)$ in a field $\mathscr{E}$, if the scalar mass m is formally replaced by the so-called *effective-mass tensor* m^*_{ij}. The inverse of this mass tensor

$$\left(\frac{1}{m^*}\right)_{ij} = \frac{1}{\hbar^2}\frac{\partial^2 E(\boldsymbol{k})}{\partial k_i \partial k_j} \tag{9.8}$$

is simply given by the curvature of $E(\boldsymbol{k})$. Because the mass tensor m^*_{ij} and also its inverse $(m^*_{ij})^{-1}$ are symmetric, they can be transformed to principal axes.

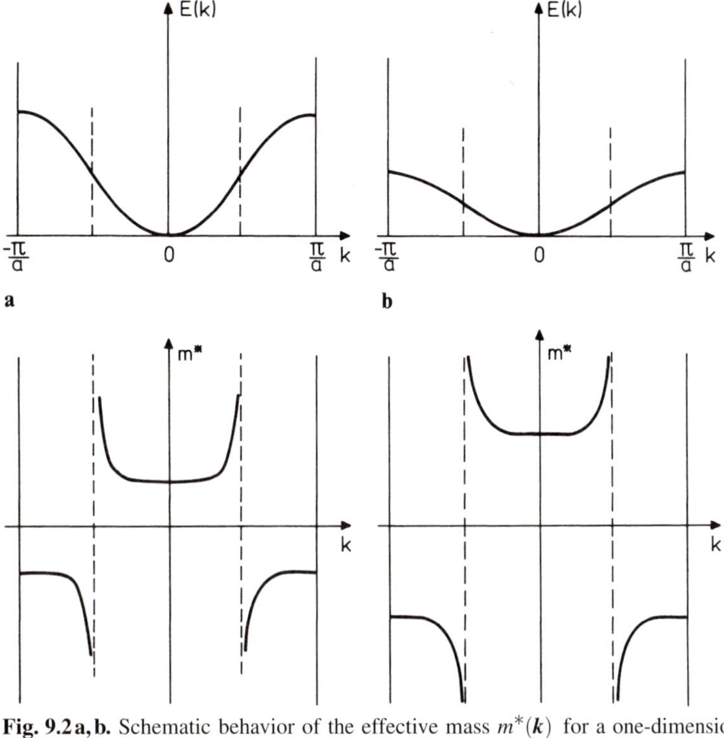

Fig. 9.2a,b. Schematic behavior of the effective mass $m^*(\boldsymbol{k})$ for a one-dimensional bandstructure $E(\boldsymbol{k})$: (**a**) for strong curvature of the bands, i.e., small effective masses; (**b**) for weak curvature, i.e., large effective masses. The dashed lines denote the points of inflection of $E(\boldsymbol{k})$

In the simplest case in which the three effective masses in the principal axes are equal to m^*, we have

$$m^* = \frac{\hbar^2}{d^2E/dk^2} \ . \tag{9.9}$$

This is the case at the minimum or maximum of a "parabolic" band where the dependence of E on $\boldsymbol{k}$ can be well approximated by

$$E(\boldsymbol{k}) = E_0 \pm \frac{\hbar^2}{2m^*}(k_x^2 + k_y^2 + k_z^2) \ . \tag{9.10}$$

In the vicinity of such a critical point, the so-called *effective-mass approximation* is particularly useful, because here m^* is a constant. On moving away from this point along the band, the deviation of the $E(\boldsymbol{k})$ surface from the parabolic form (9.10) means that m^* depends on $\boldsymbol{k}$.

Figure 9.2 shows two one-dimensional bands $E(\boldsymbol{k})$ with strong (a) and weak (b) curvature at the top and bottom of the bands. The effective mass is correspondingly small (a) and large (b) at these points. At the Brillouin zone boundary (top of the band), where the curvature is negative, the effective mass is also negative. Here it can clearly be seen that the effective-mass concept conveniently describes the effect of the periodic potential in terms of the $\boldsymbol{k}$-dependent quantity m^*.

It should be remarked with respect to the above description that one can still retain all the results derived in Chap. 6 for the free electron gas in metals. This is true so long as we consider only electrons in bands $E(\boldsymbol{k})$ that can be described by the parabolic approximation (9.10). Near to the band extrema the free electron mass m is simply replaced by the effective mass m^* which is a constant.

9.2 Currents in Bands and Holes

Since the effective mass in a band can vary over a wide range of values, such that electrons at the upper band edge move in the opposite direction to those at the bottom, the question arises of how electrons with different $\boldsymbol{k}$-vectors, that is in different band states, contribute to an electrical current. A volume element of $d\boldsymbol{k}$ at $\boldsymbol{k}$ contributes to the particle current density $\boldsymbol{j}_n$

$$d\boldsymbol{j}_n = \boldsymbol{v}(\boldsymbol{k})\frac{d\boldsymbol{k}}{8\pi^3} = \frac{1}{8\pi^3\hbar}\nabla_k E(\boldsymbol{k})\,d\boldsymbol{k} \tag{9.11}$$

since the density of states in $\boldsymbol{k}$-space is $[V/(2\pi)^3]$ (or $[1/(2\pi)^3]$ when related to the volume V of the crystal). We have taken into account that spin-degenerate states must be counted twice.

Taken together, the electrons in a fully occupied band therefore make the following contribution to the electrical current density $\boldsymbol{j}$:

$$\boldsymbol{j} = \frac{-e}{8\pi^3\hbar}\int\limits_{\text{1st Br.z.}} \nabla_k E(\boldsymbol{k})\,d\boldsymbol{k} \ . \tag{9.12}$$

Because the band is fully occupied, the integral extends over the whole of the first Brillouin zone. Thus, for each velocity $\boldsymbol{v}(\boldsymbol{k}) = \nabla_k E(\boldsymbol{k})/\hbar$ there is also a contribution from $\boldsymbol{v}(-\boldsymbol{k})$. Since the reciprocal lattice possesses the point symmetry of the real lattice (Sect. 3.2), for crystal structures with inversion symmetry one has

$$E(\boldsymbol{k}) = E(-\boldsymbol{k}) \ . \tag{9.13}$$

If we denote the two possible spin states of an electron by two arrows ($\uparrow$ and $\downarrow$), then for crystal structures without an inversion center (9.13) can be generalized to

$$E(\boldsymbol{k}\uparrow) = E(-\boldsymbol{k}\downarrow) \ . \tag{9.14}$$

The proof of (9.14) follows from the time-reversal invariance of the Schrödinger equation when the spin variable is taken into account. For simplicity, we assume spin degeneracy here, and obtain for the electron velocity associated with $\boldsymbol{k}$

$$v(-\boldsymbol{k}) = \frac{1}{\hbar}\nabla_{-k}E(-\boldsymbol{k}) = -\frac{1}{\hbar}\nabla_k E(\boldsymbol{k}) = -v(\boldsymbol{k})\ . \tag{9.15}$$

It thus follows, as a general consequence of (9.12), that the current density carried by a full band is zero:

$$\boldsymbol{j}\ (\text{full band}) \equiv \boldsymbol{0}\ . \tag{9.16}$$

If we now imagine a band which is only partially filled with electrons, then an external electric field $\mathscr{E}$ will, according to (9.6c), redistribute the electrons from states symmetric about $\boldsymbol{k}=\boldsymbol{0}$ to states which are no longer symmetric about $\boldsymbol{k}=\boldsymbol{0}$. The distribution of the occupied states is then no longer symmetric under inversion because $\mathscr{E}$ distinguishes a direction, and thus the current is different from zero

$$\boldsymbol{j}\ (\text{partially filled band}) \neq \boldsymbol{0}\ . \tag{9.17}$$

The integral now extends only over the occupied states and not over the whole Brillouin zone, so that using (9.12) and (9.16) we obtain:

$$\begin{aligned} \boldsymbol{j} &= \frac{-e}{8\pi^3} \int\limits_{\boldsymbol{k}\ \text{occupied}} v(\boldsymbol{k})\, d\boldsymbol{k} \\ &= \frac{-e}{8\pi^3} \int\limits_{\text{1st Br.z.}} v(\boldsymbol{k})\, d\boldsymbol{k} - \frac{-e}{8\pi^3} \int\limits_{\boldsymbol{k}\ \text{empty}} v(\boldsymbol{k})\, d\boldsymbol{k}\ , \\ \boldsymbol{j} &= \frac{+e}{8\pi^3} \int\limits_{\boldsymbol{k}\ \text{empty}} v(\boldsymbol{k})\, d\boldsymbol{k}\ . \end{aligned} \tag{9.18}$$

The total current calculated as an integral over the occupied states of a partially filled band may thus be formally described as a current of positive particles, assigned to the unoccupied states of the band (empty $\boldsymbol{k}$). These quasiparticles are known as *holes*, and they can be shown to obey equations of motion analogous to those derived in Sect. 9.1.

Holes also behave like positive particles with regard to their dynamics in an external field. If a band is almost completely filled, then only the highest energy part in the vicinity of the maximum contains unoccupied states. In thermodynamic equilibrium electrons always adopt the lowest energy states, so that holes are found at the upper edge of the band. Near to the maximum, the parabolic approximation for $E(\boldsymbol{k})$ applies

$$E(\boldsymbol{k}) = E_0 - \frac{\hbar^2 k^2}{2|m^*_\wedge|}\ . \tag{9.19}$$

$m^*_\wedge$ indicates that we are concerned with the effective mass at the top of the band, which is negative. Using (9.6c), the acceleration of a hole in one of these states under the influence of an electric field is

$$\dot{v} = \frac{1}{\hbar}\frac{d}{dt}[\nabla_k E(\boldsymbol{k})] = -\frac{1}{|m^*_\wedge|}\hbar\dot{\boldsymbol{k}} = \frac{e}{|m^*_\wedge|}\mathscr{E} \, . \tag{9.20}$$

Here, according to (9.18), $\nabla_k E$ and $\dot{\boldsymbol{k}}$ are assigned to the unoccupied electronic states. The equation of motion (9.20) is that of a positively charged particle with a positive effective mass, i.e., *holes at the top of a band* have a *positive effective mass.*

From the fact that a fully occupied band cannot conduct current, it immediately follows that a crystal with an absolute gap between its highest occupied and lowest unoccupied bands is an insulator. This is true, however, only for a temperature of absolute zero. Because the Fermi function is finite for $E \gg E_F$, even at very low temperatures, there are always at least a few thermally excited electrons in the lowest almost unoccupied band (the "conduction band"), which produce a current flow when an electric field is applied. Thermal excitation also creates holes in the highest occupied band, the so-called "valence band", and these can likewise produce a current. At non-zero temperature, the current is therefore carried by both electrons and holes. This behavior is typical of semiconductors and insulators (note that the distinction between a semiconductor and an insulator is not necessarily clear-cut). How well a material conducts at room temperature, for example, depends mainly on the energetic width of the band gap over which the electrons must be thermally excited (Chap. 12).

From the above, it is clear that a material with a partially filled electron band is a metal, having an essentially constant number of free carriers at all temperatures. The electrons in a partially filled band have already been treated in the framework of the potential well model (Chap. 6).

9.3 Scattering of Electrons in Bands

In our treatment so far, we have neglected the important fact that electrons moving under the influence of an external field in a crystal undergo collisions, which restrict their movement. If this were not the case, there would be no electrical resistance; once created by a temporarily applied electric field, a current would continue to flow indefinitely according to the semiclassical equations of motion (9.7b, 9.20). This phenomenon of "superconductivity" has indeed been observed for many materials (Chap. 10). Normal conductors, however, have a finite, often high electrical resistance. What, therefore, are the important scattering processes for electrons accelerated in external fields?

Drude (1900) [9.1] assumed that the electrons scatter from the positive cores that form the periodic lattice. This model implies a mean free path between collisions of 1–5 Å. In obvious contradiction to this prediction, most metals are found to have a mean free path at room temperature which is about two orders of magnitude higher (Sect. 9.5).

The explanation of this discrepancy came with the recognition that an exactly periodic lattice of positive cores does not cause scattering. This is immediately apparent within the *one-electron approximation*, because Bloch waves travelling through the lattice are stationary solutions of the Schrödinger equation. Since $\psi^*\psi$ is time independent, these solutions describe the unperturbed propagation of electron waves. These results naturally apply also to packets of Bloch waves which describe localized electrons. Deviations from this undisturbed propagation, i.e., perturbations of the stationary Bloch states can only occur in two ways:

I) Within the one-electron approximation, where interactions between electrons are neglected, the only sources of electron scattering are *deviations from strict periodicity* in the lattice. These may be:
   a) defects in the lattice that are fixed in time and space (vacancies, dislocations, impurities, etc.),
   b) deviations from periodicity that vary in time, i.e., lattice vibrations.

II) The one-electron approximation neglects interactions between electrons. Electron-electron collisions, which are not contained in the concept of a non-interacting Fermi gas, can in fact perturb the stationary Bloch states. As we will see, this effect is usually much less significant than those noted in (I).

The decisive quantity for the description of an electron scattering process is the probability $w_{k'k}$ that the electron will be scattered from a Bloch state $\psi_k(\boldsymbol{r})$ to a state $\psi_{k'}(\boldsymbol{r})$ under the influence of one of the previously described imperfections. According to quantum mechanical perturbation theory, this transition probability is

$$w_{k'k} \sim |\langle \boldsymbol{k}'|\mathcal{H}'|\boldsymbol{k}\rangle|^2 = |\int d\boldsymbol{r}\psi^*_{k'}(\boldsymbol{r})\mathcal{H}'\psi_k(\boldsymbol{r})|^2 , \tag{9.21}$$

where $\mathcal{H}'$ is the perturbation to the Hamiltonian. Because of the "Bloch character" of $\psi_k(\boldsymbol{r})$ we have

$$\langle \boldsymbol{k}'|\mathcal{H}'|\boldsymbol{k}\rangle = \int d\boldsymbol{r} u^*_{k'} \mathrm{e}^{-\mathrm{i}\boldsymbol{k}'\cdot\boldsymbol{r}}\mathcal{H}' u_k \mathrm{e}^{\mathrm{i}\boldsymbol{k}\cdot\boldsymbol{r}} . \tag{9.22}$$

If the perturbing potential $\mathcal{H}'$ can be written as a function of real-space coordinates, then (9.22) becomes comparable to the Fourier integral (3.6) describing diffraction amplitudes from periodic structures, when $(u^*_{k'}\mathcal{H}'u_k)$ is identified with the scattering electron density $\varrho(\boldsymbol{r})$ in (3.6). By comparing the calculations for diffraction from static and moving scattering centers (Chap. 3 and Sect. 4.4) we can immediately draw the conclusion: If $\mathcal{H}'(\boldsymbol{r})$ is a potential that is constant in time, such as that of a static defect, then we expect only *elastic scattering* of the Bloch waves with conservation of energy.

If, on the other hand, $\mathcal{H}'(\boldsymbol{r},t)$ is a potential that varies in time, as appropriate for the perturbation due to a *lattice wave* (phonon), then the scattering is *inelastic* (Sect. 4.4). In analogy to the results of Sect. 4.4, energy conservation also applies to the scattering of conduction electrons by phonons:

$$E(\boldsymbol{k}') - E(\boldsymbol{k}) = \hbar\omega(\boldsymbol{q}) \ . \tag{9.23}$$

For scattering by a phonon with wave vector $\boldsymbol{q}$, the perturbing potential $\mathcal{H}'$ naturally has a spatial dependence $\exp(\mathrm{i}\boldsymbol{q}\cdot\boldsymbol{r})$. This means that the scattering probability (9.21) contains a matrix element (9.22) of the form:

$$\langle \boldsymbol{k}'|\mathrm{e}^{\mathrm{i}\boldsymbol{q}\cdot\boldsymbol{r}}|\boldsymbol{k}\rangle = \int d\boldsymbol{r} u_{\boldsymbol{k}'}^{*} u_{\boldsymbol{k}} \mathrm{e}^{\mathrm{i}(\boldsymbol{k}-\boldsymbol{k}'+\boldsymbol{q})\cdot\boldsymbol{r}} \ . \tag{9.24}$$

Because $(u_{\boldsymbol{k}'}^{*} u_{\boldsymbol{k}})$ has the periodicity of the lattice and can be expanded as a Fourier series in terms of reciprocal lattice vectors, the matrix element in (9.24) is non-zero (as in Sect. 4.3) only when

$$\boldsymbol{k}' - \boldsymbol{k} = \boldsymbol{q} + \boldsymbol{G} \ . \tag{9.25}$$

To within a reciprocal lattice vector $\boldsymbol{G}$, this relation resembles a momentum conservation law. Note that $\boldsymbol{k}$ is nothing other than the wavevector of a Bloch state, i.e. a quantum number. Only for a free electron is $\hbar\boldsymbol{k}$ the true momentum. If we take *energy conservation* (9.23) and $\boldsymbol{k}$*-conservation* (9.25) together, then, just as in Sect. 4.3, we see that scattering of Bloch-state electrons can be well described in the particle picture.

Let us leave the one-electron approximation and proceed to the subject of *electron-electron scattering*. It can be shown, also in a many-body description, that energy and wavevector (quasi-momentum) conservation also apply here. For a collision between two electrons (1)+(2) → (3)+(4), we must have

$$E_1 + E_2 = E_3 + E_4 \ , \tag{9.26}$$

where $E_i = E(\boldsymbol{k}_i)$ denotes the one-particle energy of an electron in a non-interacting Fermi gas. Furthermore for the corresponding $\boldsymbol{k}$-vectors:

$$\boldsymbol{k}_1 + \boldsymbol{k}_2 = \boldsymbol{k}_3 + \boldsymbol{k}_4 + \boldsymbol{G} \ . \tag{9.27}$$

Naively, one would expect a high probability of scattering since the packing density is of the order of one electron per elementary cell and because of the strength of the Coulomb repulsion. The Pauli exclusion principle, however, to a large extent forbids scattering of this type, for the reasons detailed below.

Let us assume that one electron occupies the state $E_1 > E_F$, an excited state just above the Fermi level; the second electron involved in the collision is inside the Fermi sphere with $E_2 < E_F$. For scattering to states E_3 and E_4, the Pauli principle demands that E_3 and E_4 must be unoccupied. Thus, the necessary conditions for scattering, besides (9.26), are:

$$E_1 > E_F \ , \ E_2 < E_F \ , \ E_3 > E_F \ , \ E_4 > E_F \ . \tag{9.28 a}$$

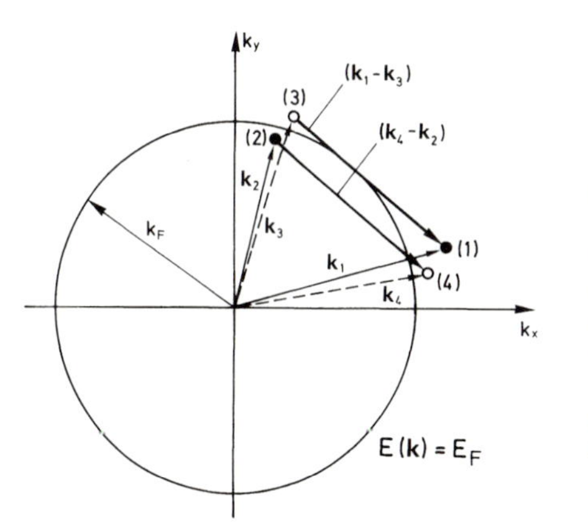

Fig. 9.3. Illustration in $\boldsymbol{k}$-space of a two-electron collision (E_F is the radius of the Fermi sphere): Two electrons (1) and (2), with initial wave vectors $\boldsymbol{k}_1$ and $\boldsymbol{k}_2$, scatter from one another to final states with wave vectors $\boldsymbol{k}_3$ and $\boldsymbol{k}_4$. Their total energy and total momentum are conserved in the process. In addition, the Pauli principle demands that the electrons scatter to final states $\boldsymbol{k}_3, \boldsymbol{k}_4$, that were initially unoccupied ($E_3 > E_F, E_4 > E_F$)

From energy conservation (9.26), it then follows that

$$E_1 + E_2 = E_3 + E_4 > 2E_F \tag{9.28b}$$

and

$$(E_1 - E_F) + (E_2 - E_F) > 0 \,. \tag{9.28c}$$

If $(E_1 - E_F) < \varepsilon_1 \, (\varepsilon_1 \ll E_F)$, i.e. E_1 is just outside the Fermi surface (Fig. 9.3), then $|E_2 - E_F| = |\varepsilon_2| < \varepsilon_1$ from (9.28c), i.e. E_2 must be only a little ($\sim \varepsilon_1$) below E_F. Therefore only the fraction $\sim \varepsilon_1/E_F$ of all electrons may scatter with the electron in the state E_1. If E_1 and E_2 are in the shell $\pm\varepsilon_1$ around E_F, then because of $\boldsymbol{k}$ conservation (9.27) and (9.28b,c), E_3 and E_4 must also lie in the shell $\pm\varepsilon_1$ around E_F. $\boldsymbol{k}$ conservation in the form $\boldsymbol{k}_1 - \boldsymbol{k}_3 = \boldsymbol{k}_4 - \boldsymbol{k}_2$ means that the connecting lines (1)–(3) and (2)–(4) in Fig. 9.3 must be equal. Because only a fraction $\sim \varepsilon_1/E_F$ of all occupied states are allowed final states, the Pauli principle further reduces the scattering probability by a factor ε_1/E_F.

The thermal broadening of the Fermi function is of the order kT, so that the final state E_1 must lie within this energy of E_F, i.e. $\varepsilon_1 \sim kT$. The temperature-dependent reduction in the cross-section for electron-electron scattering which stems from the Pauli principle can be estimated as follows:

$$\Sigma \propto \left(\frac{kT}{E_F}\right)^2 \Sigma_0 \,. \tag{9.29}$$

Here Σ_0 is the cross-section which would apply without the Pauli principle for a classical gas of screened point charges.

Let us assume that the cross-section for the scattering of an electron from a defect in the lattice (e.g. the charged core of an impurity) is of the order Σ_0. For both scattering partners of the electron we are concerned with screened Coulomb fields. The probability of electron-electron scattering at a temperature of 1 K is about a factor $\sim 10^{-10}$ smaller than that of electron-defect scattering (E_F/k is assumed to be about 10^5 K, Sect. 6.2, Table 6.1).

These simple arguments again illustrate, but from a different viewpoint, that the Pauli principle allows electrons in solids to be treated, to a good approximation, as non-interacting, despite their high density. Therefore, in the following treatment of electrical conduction in solids we need to consider only the scattering of electrons from defects and phonons.

9.4 The Boltzmann Equation and Relaxation Time

Transport phenomena, such as the flow of electric current in solids, involve two characteristic mechanisms with opposite effects: the driving force of the external fields and the dissipative effect of the scattering of the carriers by phonons and defects. The interplay between the two mechanisms is described by the Boltzmann equation. With the help of this equation one may investigate how the distribution of carriers in thermal equilibrium is altered in the presence of external forces and as a result of electron scattering processes. In thermal equilibrium, i.e. under homogeneous temperature conditions and with no external fields, this distribution function is simply the Fermi distribution derived in Sect. 6.3:

$$f_0[E(\boldsymbol{k})] = \frac{1}{e^{[E(\boldsymbol{k})-E_F]/kT} + 1} . \tag{9.30}$$

This equilibrium distribution f_0 is independent of $\boldsymbol{r}$ because of the assumed homogeneity. Away from equilibrium, where we merely assume local equilibrium over regions large compared with atomic dimensions, the required distribution $f(\boldsymbol{r},\boldsymbol{k},t)$ can be both space and time dependent: the $\boldsymbol{r}$ and $\boldsymbol{k}$ values of an electron are altered by external fields and collisions. To derive f, we first consider the effect of external fields, and then introduce the scattering as a correction. We consider the change in f during the time from t–dt to t. Under an applied external field $\mathscr{E}$, an electron that is at $\boldsymbol{r}$ and $\boldsymbol{k}$ at time t, will have had the coordinates

$$\boldsymbol{r} - \boldsymbol{v}(\boldsymbol{k})dt \quad \text{and} \quad \boldsymbol{k} - (-e)\mathscr{E}dt/\hbar \tag{9.31}$$

at time t–dt (for e positive).

In the absence of collisions, each electron with coordinates $\boldsymbol{r}-\boldsymbol{v}dt$ and $\boldsymbol{k}+e\mathscr{E}\,dt/\hbar$ at t–dt must arrive at $\boldsymbol{r},\boldsymbol{k}$ at time t, i.e.,

$$f(\boldsymbol{r},\boldsymbol{k},t) = f(\boldsymbol{r} - \boldsymbol{v}\,dt, \boldsymbol{k} + e\mathscr{E}dt/\hbar, t - dt) . \tag{9.32}$$

An additional term must be added to this equation because along with the changes described by (9.31), electrons may also be scattered to or from $\boldsymbol{r}$ and $\boldsymbol{k}$ in time dt. If we express the change in f due to scattering by the term $(\partial f/\partial t)_s$, then the correct equation is

$$f(\boldsymbol{r},\boldsymbol{k},t) = f(\boldsymbol{r} - \boldsymbol{v}\,dt, \boldsymbol{k} + e\mathscr{E}dt/\hbar, t - dt) + \left(\frac{\partial f}{\partial t}\right)_{\mathrm{s}} dt \; . \tag{9.33}$$

Expanding this up to terms linear in dt gives

$$\frac{\partial f}{\partial t} + \boldsymbol{v} \cdot \nabla_r f - \frac{e}{\hbar}\mathscr{E} \cdot \nabla_k f = \left(\frac{\partial f}{\partial t}\right)_{\mathrm{s}} \; . \tag{9.34}$$

This is the Boltzmann equation, and represents the starting point for the treatment of transport problems in solids. The terms on the left-hand side are called the drift terms, and the as yet unspecified term on the right is the scattering term. This contains all of the atomic aspects of the scattering mechanisms introduced briefly in Sect. 9.3.

The generalized quantum mechanical transition probability discussed in Sect. 9.3, $w_{k'k} \propto |\langle \boldsymbol{k}'|\mathscr{H}'|\boldsymbol{k}\rangle|^2$ for transitions from the Bloch state ψ_k to $\psi_{k'}$, immediately enables us to write the scattering term in its most general form:

$$\left(\frac{\partial f(\boldsymbol{k})}{\partial t}\right) = \frac{V}{(2\pi)^3}\int d\boldsymbol{k}'\{[1 - f(\boldsymbol{k})]w_{kk'}f(\boldsymbol{k}') - [1 - f(\boldsymbol{k}')]w_{k'k}f(\boldsymbol{k})\} \; . \tag{9.35}$$

In this equation the first term in the integrand takes account of all scattering events from occupied states $\boldsymbol{k}'$ to unoccupied states $\boldsymbol{k}$, while the second term includes all scattering out of $\boldsymbol{k}$ into some other state $\boldsymbol{k}'$. If (9.35) is inserted in (9.34), it can be seen that the Boltzmann equation, in its most general form, is a complicated integro-differential equation for determining the non-equilibrium distribution $f(\boldsymbol{r},\boldsymbol{k},t)$.

Because of this, the scattering term (9.35) is often modelled in the so-called *relaxation time ansatz*, which is a plausible postulate for many problems. It assumes that the rate at which f returns to the equilibrium distribution f_0 due to scattering is proportional to the deviation of f from f_0, i.e.,

$$\left(\frac{\partial f}{\partial t}\right)_{\mathrm{s}} = -\frac{f(\boldsymbol{k}) - f_0(\boldsymbol{k})}{\tau(\boldsymbol{k})} \; . \tag{9.36}$$

The so-called relaxation time $\tau(\boldsymbol{k})$ depends only on the position of the state in $\boldsymbol{k}$-space, which in a localized electron picture corresponds to the mean $\boldsymbol{k}$-value of the partial waves in the wave packet. For a spatially inhomogeneous distribution, the relaxation time $\tau(\boldsymbol{k},\boldsymbol{r})$ in (9.36) will also depend on the position $\boldsymbol{r}$. The distribution $f(\boldsymbol{r},\boldsymbol{k},t)$ is then defined for volume elements $d\boldsymbol{r}$ which are large compared to atomic dimensions, but small compared with the macroscopic distances over which current, heat flux, etc., change significantly.

The assumption behind the relaxation time model is that scattering merely serves to drive a non-equilibrium distribution back towards thermal equilibrium. The significance of the relaxation time becomes clearer if we consider the case in which an external field is suddenly switched off: If the external field creates a stationary non-equilibrium distribution $f_{\rm stat}(\boldsymbol{k})$, and the field is then switched off, then from the moment t of switching off:

$$\frac{\partial f}{\partial t} = -\frac{f - f_0}{\tau} \ . \tag{9.37}$$

With the initial condition $f(t=0,\boldsymbol{k})=f_{\rm stat}$, the Boltzmann equation (9.37) has the solution

$$f - f_0 = (f_{\rm stat} - f_0)\mathrm{e}^{-t/\tau} \ . \tag{9.38}$$

The deviation of the distribution f from the equilibrium distribution $f_0(\boldsymbol{k})$ decays exponentially with decay constant τ (relaxation time). The relaxation time τ is therefore the time constant with which the non-equilibrium distribution relaxes via scattering to the equilibrium state when the external perturbation is switched off.

Furthermore, the Boltzmann equation allows an approximate calculation of the stationary non-equilibrium distribution which exists under the influence of an external field, e.g. an electric field $\mathscr{E}$. If f does not depend on position (i.e., $\nabla_r f=\boldsymbol{0}$), then from (9.34, 9.36) it follows that the *stationary state* (i.e., $\partial f/\partial t=0$) is given by

$$-\frac{e}{\hbar}\mathscr{E} \cdot \nabla_k f = -[f(\boldsymbol{k}) - f_0(\boldsymbol{k})]/\tau(\boldsymbol{k}) \ , \tag{9.39}$$

$$f(\boldsymbol{k}) = f_0(\boldsymbol{k}) + \frac{e}{\hbar}\tau(\boldsymbol{k})\mathscr{E} \cdot \nabla_k f(\boldsymbol{k}) \ . \tag{9.40}$$

This differential equation for $f(\boldsymbol{k})$ can be solved iteratively by initially approximating f in the $(\nabla_k f)$-term on the right-hand side by the equilibrium distribution f_0. This gives a solution of f linear in $\mathscr{E}$, which can again be inserted into (9.40) to give a solution quadratic in $\mathscr{E}$. Successive insertion of the solutions leads to an expression for the non-equilibrium distribution $f(\boldsymbol{k})$ in terms of powers of the external field $\mathscr{E}$. If we are only interested in phenomena that depend linearly on the external field, e.g. Ohmic conduction in solids, then we can restrict the calculation to the first approximation where f depends linearly on $\mathscr{E}$, i.e. where only the equilibrium distribution f_0 is considered in $\nabla_k f$. This gives the so-called linearized Boltzmann equation for the determination of the non-equilibrium distribution

$$f(\boldsymbol{k}) \simeq f_0(\boldsymbol{k}) + \frac{e}{\hbar}\tau(\boldsymbol{k})\mathscr{E} \cdot \nabla_k f_0(\boldsymbol{k}) \ . \tag{9.41 a}$$

In the present approximation for small electric fields, i.e., for small deviations from thermal equilibrium (the linearized problem), (9.41 a) may be regarded as an expansion of $f_0(\boldsymbol{k})$ about the point $\boldsymbol{k}$:

$$f(\boldsymbol{k}) \simeq f_0\Big(\boldsymbol{k} + \frac{e}{\hbar}\tau(\boldsymbol{k})\mathscr{E}\Big) \ . \tag{9.41 b}$$

The stationary distribution resulting from an external field $\mathscr{E}$ and including the effects of scattering (described by τ) can therefore be represented by a

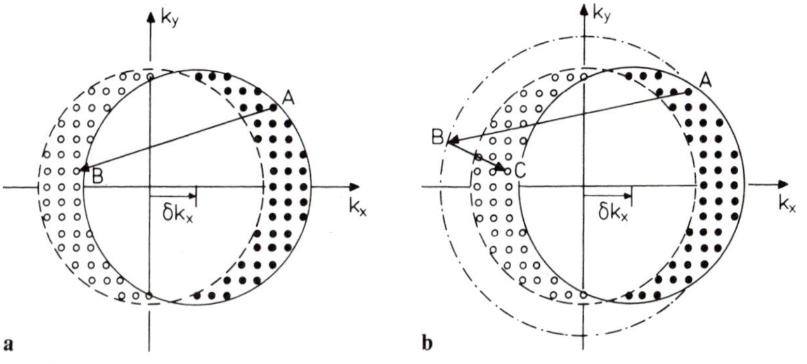

Fig. 9.4 a, b. The effect of a constant electric field $\mathscr{E}_x$ on the $\boldsymbol{k}$-space distribution of quasi-free electrons: (**a**) The Fermi sphere of the equilibrium distribution [– – –, centered at (0,0,0)] is displaced in the stationary state by an amount $\delta k_x = -e\tau\mathscr{E}_x/\hbar$. (**b**) The new Fermi distribution $f(E(k))$ only differs significantly from the equilibrium distribution f_0 (– – –) in the vicinity of the Fermi energy (Fermi radius)

Fermi distribution shifted by $e\tau\mathscr{E}/\hbar$ from the equilibrium position, as shown in Fig. 9.4.

It is interesting to consider the effect of elastic and inelastic scattering on the approach to equilibrium in $\boldsymbol{k}$-space. The stationary state of the distribution is represented as a displaced Fermi sphere in Fig. 9.5 (full line). If the external field is switched off, the displaced sphere relaxes back to the

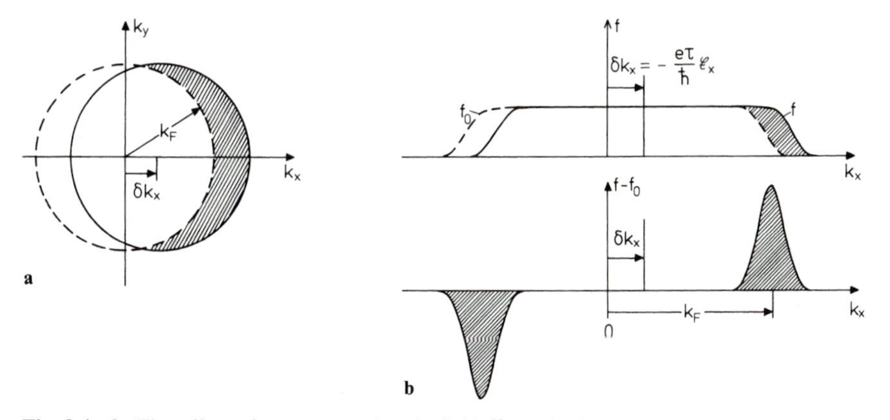

Fig. 9.5 a, b. Electron scattering processes in k-space. The dashed circle represents the Fermi surface in thermodynamic equilibrium ($\mathscr{E} = 0$). Under the influence of an electric field $\mathscr{E}_x$ and for a constant current, the Fermi surface is displaced as shown by the full circle. (**a**) When the electric field is switched off, the displaced Fermi surface relaxes back to the equilibrium distribution by means of electron scattering from occupied states (●) to unoccupied states (○). Since the states A and B are at different distances from the $\boldsymbol{k}$-space origin (i.e., have different energies), the relaxation back to equilibrium must involve inelastic scattering events (e.g., phonon scattering). (**b**) For purely elastic scattering (from states A to B), the Fermi sphere would simply expand. When the field is switched off, equilibrium can only be achieved by inelastic scattering into states C within the dashed (equilibrium) Fermi sphere

equilibrium distribution (dashed line). *Only inelastic scattering* (Fig. 9.5 a) can cause this return to equilibrium. Elastic collisions alone, e.g. from defects, cannot facilitate the return to equilibrium. Without inelastic collisions the Fermi sphere would merely expand due to the effect of elastic collisions (Fig. 9.5 b).

9.5 The Electrical Conductivity of Metals

In about 1900, long before an exact theory of the solid state was available, Drude [9.1] described metallic conductivity using the assumption of an ideal electron gas in the solid. For an ideal electron gas in an external field $\mathscr{E}$, the dynamics of the electrons is described by the classical equation of motion

$$m\dot{v} + \frac{m}{\tau} v_{\mathrm{D}} = -e\mathscr{E} \ . \tag{9.42}$$

The dissipative effect of scattering is accounted for by the friction term mv_{D}/τ where $v_{\mathrm{D}} = v - v_{\mathrm{therm}}$ is the so-called drift velocity, i.e., the additional velocity due to the field, over and above the thermal velocity v_{therm}. Since v relaxes exponentially to the thermal velocity with a time constant τ after switching off the field, τ also has the meaning of a relaxation time. For the stationary case ($\dot{v}=0$) one has

$$v_{\mathrm{D}} = -\frac{e\tau}{m}\mathscr{E} \tag{9.43}$$

and hence the current density j in the direction of the field is

$$j = -env_{\mathrm{D}} = ne\mu\mathscr{E} = \frac{e^2\tau n}{m}\mathscr{E} \ . \tag{9.44}$$

Here n is the volume density of all free electrons and the mobility μ is defined as the proportionality constant between the drift velocity and the external field. In the Drude model, the electrical conductivity is therefore

$$\sigma = j/\mathscr{E} = \frac{e^2 n\tau}{m} \tag{9.45}$$

and the electron mobility becomes

$$\mu = \frac{e\tau}{m} \ . \tag{9.46}$$

Note that, in this simple model, all free electrons contribute to the current. This view is in contradiction to the Pauli principle. For the Fermi gas this forbids electrons well below the Fermi level from acquiring small amounts of energy, since all neighboring higher energy states are occupied.

To relate the material-specific conductivity to atomically-determined features of the band structure, we consider, as in (9.2), the contribution to the current of electrons in the volume element $d\boldsymbol{k}$. In the expressions (9.11,

9.12) for the particle and electrical current densities, only the occupied states in $\boldsymbol{k}$-space are summed. This summation, i.e., integration, over occupied states can be replaced by a much tidier sum over all states in the first Brillouin zone if one introduces the occupation probability $f(\boldsymbol{k})$ for the states $\boldsymbol{k}$. In analogy to (9.11), the particle current density $\boldsymbol{j}_n$ is given by

$$\boldsymbol{j}_n = \frac{1}{8\pi^3} \int\limits_{\text{1st Br.z.}} \boldsymbol{v}(\boldsymbol{k}) f(\boldsymbol{k}) d\boldsymbol{k} \; . \tag{9.47}$$

Because we are confining our attention to linear effects in an external field, i.e. Ohm's law, it is sufficient to insert $f(\boldsymbol{k})$ in the linearized form (9.41 a). For an electric field $\mathscr{E}_x$ in the x-direction, the electrical current density is

$$\begin{aligned} \boldsymbol{j} &= -\frac{e}{8\pi^3} \int d\boldsymbol{k}\, \boldsymbol{v}(\boldsymbol{k}) f(\boldsymbol{k}) \; , \\ &= -\frac{e}{8\pi^3} \int d\boldsymbol{k}\, \boldsymbol{v}(\boldsymbol{k}) \left[f_0(\boldsymbol{k}) + \frac{e\tau(\boldsymbol{k})}{\hbar} \mathscr{E}_x \frac{\partial f_0}{\partial k_x} \right] \; . \end{aligned} \tag{9.48}$$

For an isotropic medium and for cubic lattices, the y and z components of $\boldsymbol{j}$ vanish when the electric field points in the x-direction. In other words, the tensor relationship between current density $\boldsymbol{j}$ and electric field $\mathscr{E}$ reduces to a scalar relationship,

$$j_y = j_z = 0 \; . \tag{9.49}$$

Since the integral is over the whole Brillouin zone and $f_0(\boldsymbol{k})$ has inversion symmetry about $\boldsymbol{k} = \boldsymbol{0}$, the integral over $v_x f_0$ vanishes. Furthermore, since

$$\frac{\partial f_0}{\partial k_x} = \frac{\partial f_0}{\partial E} \hbar v_x \; , \tag{9.50}$$

we have

$$j_x = -\frac{e^2}{8\pi^3} \mathscr{E}_x \int d\boldsymbol{k}\, v_x^2(\boldsymbol{k}) \tau(\boldsymbol{k}) \frac{\partial f_0}{\partial E} \; . \tag{9.51}$$

The specific electrical conductivity is therefore

$$\sigma = j_x / \mathscr{E}_x = -\frac{e^2}{8\pi^3} \int d\boldsymbol{k}\, v_x^2(\boldsymbol{k}) \tau(\boldsymbol{k}) \frac{\partial f_0}{\partial E} \; . \tag{9.52}$$

The energy region over which the Fermi function $f_0(E)$ changes rapidly has a width of about $4 k T$ (Sect. 6.3). It also has inversion symmetry about the point $(E_\mathrm{F}, f_0(E_\mathrm{F}) = 1/2)$. Thus, to a good approximation,

$$\frac{\partial f_0}{\partial E} \approx -\delta(E - E_\mathrm{F}) \; . \tag{9.53}$$

This approximate solution is largely self-evident, and the derivation will not be given here.

Using (9.53) with

$$d\boldsymbol{k} = df_E dk_\perp = df_E \frac{dE}{|\nabla_k E|} = df_E \frac{dE}{\hbar v(\boldsymbol{k})} , \tag{9.54}$$

it follows immediately from (9.52) that

$$\sigma \simeq \frac{e^2}{8\pi^3\hbar} \int df_E dE \frac{v_x^2(\boldsymbol{k})}{v(\boldsymbol{k})} \tau(\boldsymbol{k}) \delta(E - E_F) , \tag{9.55}$$

and because of the property of the δ-function in the integral

$$\sigma \simeq \frac{e^2}{8\pi^3\hbar} \int\limits_{E=E_F} \frac{v_x^2(\boldsymbol{k})}{v(\boldsymbol{k})} \tau(\boldsymbol{k}) df_E . \tag{9.56}$$

In the general case, $v(\boldsymbol{k})$ and $\tau(\boldsymbol{k})$ vary over the Fermi surface. In (9.56) however, an average value $\langle v_x^2(\boldsymbol{k})\tau(\boldsymbol{k})/v(\boldsymbol{k})\rangle_{E_F}$ over the Fermi surface can be taken outside the integral. For a "Fermi sphere", i.e., for nearly free electrons with a constant effective mass m^* (see below), this average value is simply equal to $v(E_F)\tau(E_F)/3$.

The *electrical conductivity* σ of a metal can thus be expressed as a *surface integral over the Fermi surface* $E(\boldsymbol{k}) = E_F$ in $\boldsymbol{k}$-space. Only the velocity $v(E_F)$ and the relaxation time $\tau(E_F)$ of the electrons at the Fermi surface appear in the microscopic description. Equation (9.56) expresses precisely the fact that only electrons in the vicinity of the Fermi energy are relevant for current transport in a metal, as expected from the Pauli exclusion principle. This conclusion can also be drawn from the diagrams in Figs. 9.4 and 9.5: Electrons well below the Fermi energy are not affected by the slight stationary shift, $-(e\tau/\hbar)\mathscr{E}_x$, of the Fermi sphere in $\boldsymbol{k}$-space.

For the simple case of a conduction band of a metal (with the approximation $kT \ll E_F$) containing only electrons in an energy region for which the parabolic approximation is valid with a constant effective mass m^*, (9.56) can be evaluated further. For electrons in an exactly parabolic band (quasi-free electrons) we have

$$v(E_F) = \hbar k_F/m^* \tag{9.57 a}$$

and

$$\int\limits_{E_F} df_E = 2(4\pi k_F^2) . \tag{9.57 b}$$

Furthermore, since $kT \ll E_F$,

$$n = \frac{2(4/3)\pi k_F^3}{8\pi^3} , \text{ i.e.,} \quad k_F^3 = 3\pi^2 n . \tag{9.57 c}$$

It therefore follows that the electrical conductivity σ and mobility μ are given by

$$\sigma \simeq \frac{e^2\tau(E_\mathrm{F})}{m^*} n \tag{9.58 a}$$

and

$$\mu \simeq \frac{e\tau(E_\mathrm{F})}{m^*} . \tag{9.58 b}$$

These relationships are formally equivalent to those of the Drude model (9.45, 9.46). The relaxation time, which was not exactly specified in that model, is that of electrons at the Fermi level, and the effective mass m^* replaces the free electron mass m. The total concentration of electrons in the conduction band, which appears in the Drude model, also appears in the correct formulas (9.58 a, b). However, this is because of the formal integration over $\boldsymbol{k}$-space (9.52) and *not*, as assumed in the Drude model, because all electrons contribute to charge transport. The similarity between (9.58 a, b) and (9.45, 9.46) explains why the Drude model yields satisfactory results for many purposes.

It should also be remarked that for semiconductors, in which the carrier concentration n is strongly temperature dependent (Sect. 9.2, Chap. 12), the evaluation of (9.52) naturally leads to complicated averages over $\tau(\boldsymbol{k})$ rather than to the expressions (9.58 a, b).

To understand the temperature dependence of the resistance of metals, it suffices to consider the temperature dependence of $\tau(E_\mathrm{F})$ or μ, because the electron concentration n is independent of temperature. Instead of a rigorous quantum mechanical calculation of the scattering probability $w_{\boldsymbol{k}'\boldsymbol{k}}$ (Sect. 9.3), we give here a qualitative discussion of both scattering mechanisms: *phonon* and *defect* scattering. Assuming that the two mechanisms are independent of one another, the total scattering probability is the sum of the individual scattering probabilities. The scattering probability is inversely proportional to the average free time of flight τ_TF of a carrier, and therefore inversely proportional to the relaxation time. It therefore follows that

$$\frac{1}{\tau} = \frac{1}{\tau_\mathrm{ph}} + \frac{1}{\tau_\mathrm{def}} \tag{9.59}$$

where τ_ph and τ_def are the average times between scattering from phonons and from defects, respectively.

This treatment directly provides the important features of the temperature behavior of the resistivity of a metal. The number of collisions per unit time is proportional to the scattering cross-section Σ and to the velocity v of the particle: $1/\tau \propto \Sigma v$. For metals, v is simply the velocity at the Fermi surface $v(E_\mathrm{F})$ and is therefore temperature independent. For *defects* the cross-section Σ is also temperature independent. Defect scattering leads therefore to a temperature-independent component of the resistivity, ϱ_def.

For *phonon scattering*, the scattering cross-section can be set proportional to the mean square vibrational amplitude $\langle s^2(\boldsymbol{q})\rangle$ of the appropriate phonon (wavevector $\boldsymbol{q}$, frequency $\omega_{\boldsymbol{q}}$) (Chap. 5). In the classical, limiting case of high temperature, it follows from the equipartition theorem that

$$M\omega_{\boldsymbol{q}}^2\langle s^2(\boldsymbol{q})\rangle = k T\,, \quad (T \gg \Theta) \tag{9.60}$$

where M is the mass of the heavy ion cores. We thus obtain

$$\frac{1}{\tau_{\mathrm{ph}}} \sim \langle s^2(\boldsymbol{q})\rangle \sim \frac{k T}{M\omega_{\boldsymbol{q}}^2}\,. \tag{9.61 a}$$

The phonon frequency $\omega_{\boldsymbol{q}}$ contains information about the elastic properties of the material. A rough estimate is obtained by replacing $\omega_{\boldsymbol{q}}$ by the Debye cutoff frequency $\omega_{\mathrm{D}} = k\,\Theta/\hbar$ (Sect. 5.3), where Θ is the Debye temperature. This yields

$$\tau_{\mathrm{ph}} \sim \frac{M\Theta^2}{T} \quad \text{for} \quad T \gg \Theta\,. \tag{9.61 b}$$

At temperatures $T < \Theta$, the phonon excitation probability decreases rapidly and, together with the decreasing phonon energy, this leads to predominantly small angle scattering. Grüneisen [9.2] has developed an exact theory for $T < \Theta$, and gave the following universal expression for the phonon contribution $\varrho_{\mathrm{ph}}(T) \propto 1/\sigma_{\mathrm{ph}}$ to the resistivity of metals:

$$\varrho_{\mathrm{ph}}(T) = A(T/\Theta)^5 \int_0^{\Theta/T} \frac{x^5 dx}{(\mathrm{e}^x - 1)(1 - \mathrm{e}^{-x})} \tag{9.62}$$

At low temperatures, this varies as T^5.

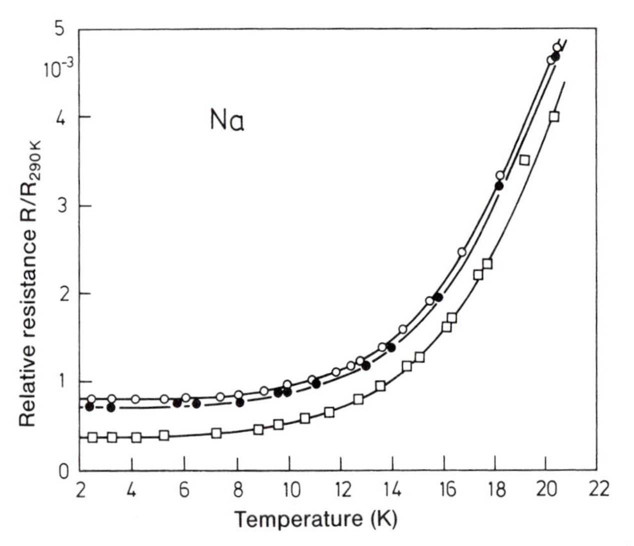

Fig. 9.6. Electrical resistance of sodium compared to the value at 290 K as a function of temperature. The data points (○, ●, □) were measured for three different samples with differing defect concentrations. (After [9.3])

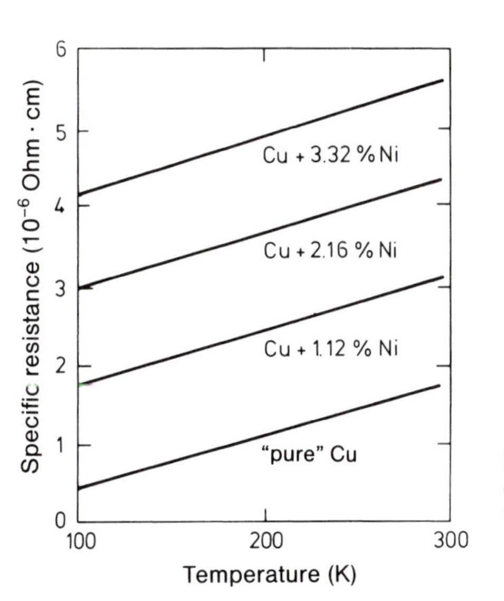

Fig. 9.7. Resistivity ϱ of pure copper and of copper-nickel alloys of various compositions. (After [9.4])

According to (9.59) we can write the resistivity $\varrho = 1/\sigma \sim 1/\tau$ of a metal as the sum of a temperature-independent residual resistivity ϱ_{def} (due to defects) and a part due to phonon scattering $\varrho_{\text{ph}}(T)$ which is linear in temperature at high temperature:

$$\varrho = \varrho_{\text{ph}}(T) + \varrho_{\text{def}} \; . \tag{9.63}$$

This behavior, which was first identified experimentally, is known as *Matthiesen's Rule.* Note that (9.63) is only valid approximately.

Figure 9.6 shows the experimentally measured electrical resistance of Na at low temperature. Below about 8 K, a temperature-independent residual resistance is observed, which depends on the defect concentration of the sample. At higher temperatures, the component described by the Grüneisen formula (9.62) becomes evident, and above 18 K ϱ_{ph} displays the linear dependence $\varrho_{\text{ph}} \sim T$.

The influence of the defect concentration (nickel impurities) on the resistance of copper at moderately low temperatures is shown in Fig. 9.7.

Figure 9.8 shows how the universal Grüneisen formula (9.62) applies to a series of metals, when the resistance is plotted against reduced temperature T/Θ. At higher temperatures there is clearly a linear increase in resistance with temperature.

For the sake of completeness, it should also be mentioned that magnetic defects can cause a quite special kind of scattering in metals that results from spin interactions. At low temperatures ($T < 20$ K) this leads to a minimum in $\varrho(T)$ instead of a constant residual resistance. This phenomenon is known as the *Kondo Effect.*

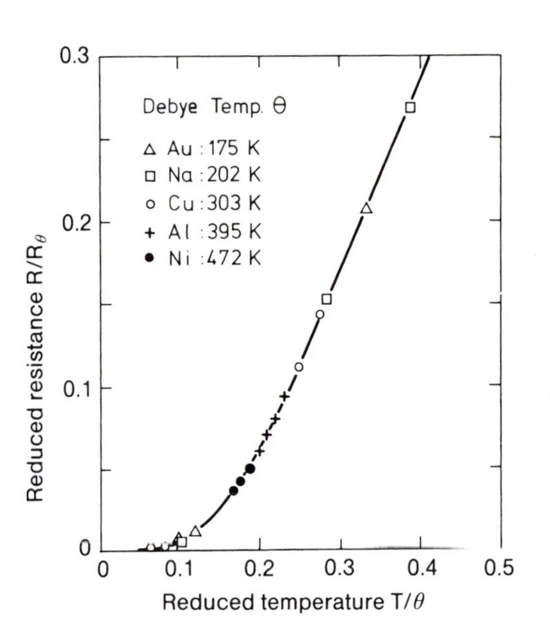

Fig. 9.8. The universal curve for the reduced resistance (R/R_Θ) as a function of reduced temperature (T/Θ, where Θ is the Debye temperature). Data points are plotted for a number of different metals

9.6 Thermoelectric Effects

In calculating the electrical conductivity (9.56) using the Boltzmann equation (9.34) and the relaxation time ansatz (9.36), we assumed for simplicity a homogeneous temperature over the conductor (i.e., $\nabla_r T = \mathbf{0}$). If we relax this assumption, the distribution function $f(\boldsymbol{k})$ is affected not only by the electric field $\mathscr{E}$ but also by the temperature gradient $\nabla_r T \neq \mathbf{0}$. The adjustment to the equilibrium distribution $f_0(\boldsymbol{k})$ via scattering will again be described by the relaxation time postulate (9.36). For the stationary state $[\dot{\mathscr{E}} = \mathbf{0},\ \partial(\nabla_r T)/\partial t = \mathbf{0},\ \partial f/\partial t = 0]$, the non-equilibrium distribution follows from (9.34) and (9.36):

$$f(\boldsymbol{k}) = f_0(\boldsymbol{k}) + \frac{e}{\hbar}\tau\mathscr{E}\cdot\nabla_k f - \tau\boldsymbol{v}\cdot\nabla_r f\ . \tag{9.64}$$

Since $\nabla_r f[\boldsymbol{k}, T(\boldsymbol{r})] = (\partial f/\partial T)\nabla_r T$, and assuming only a small perturbation, (9.64) can be linearized in a similar way to (9.41 a), and one finds

$$f(\boldsymbol{k}) \approx f_0(\boldsymbol{k}) + \frac{e}{\hbar}\tau\mathscr{E}\cdot\nabla_k f_0 - \tau\frac{\partial f_0}{\partial T}\boldsymbol{v}\cdot\nabla_r T\ . \tag{9.65}$$

To calculate the resulting electric current density in the x-direction, we insert the expression (9.65) into the relation for the current density (9.48). The first two terms in (9.65), particularly the one linear in $\mathscr{E}_x$, give Ohm's law, as in the calculation in Sect. 9.5 [(9.49–9.52)], so that the current density can be written as

$$j_x = \sigma \mathscr{E}_x + \frac{e}{8\pi^3} \int d\boldsymbol{k}\, \tau\, v_x^2 \frac{\partial f_0}{\partial T} \frac{\partial T}{\partial x} \,. \tag{9.66}$$

It is instructive to evaluate the integral with the assumptions of a spherical Fermi surface and a relaxation time that depends on energy only. Using the definition of the density of states (7.41), one obtains

$$j_x = \sigma \mathscr{E}_x + \frac{1}{3} e \int dE \tau(E) v^2(E) D(E) \frac{\partial f_0}{\partial T} \frac{\partial T}{\partial x} \,. \tag{9.67}$$

Since the derivative $\partial f_0/\partial T$ is non-zero only in the immediate vicinity of the Fermi energy the integrands $\tau(E)$ and $v^2(E)$ can be replaced by their average values τ_F and v_F^2. By expanding $D(E)$ into a Taylor series around E_F and by using (6.43–6.45) it can be shown that the remaining integral is proportional to the derivative of the density of states $D'(E_F)$

$$j_x = \sigma \mathscr{E}_x + \frac{\pi^2}{9} e v_F^2 \tau_F k^2 T D'(E_F) \frac{\partial T}{\partial x} \,. \tag{9.68}$$

In deriving (9.66) it was tacitly assumed that the Fermi energy E_F can be treated as a constant. This is a good approximation for metals (Chap. 6). For semiconductors, however (Chap. 12), a temperature gradient $\partial T/\partial x$ is associated with a considerable spatial dependence of the Fermi energy, $E_F(\boldsymbol{r})$. The derivative of the distribution f in (9.34) also brings into play the derivative of the Fermi energy $\partial E_F/\partial x$ and (9.66) must now be expressed in the general form

$$j_x = \sigma \mathscr{E}'_x + \mathscr{L}_{xx}^{12} \left(-\frac{\partial T}{\partial x} \right) , \tag{9.69}$$

where $\mathscr{E}' = \mathscr{E} + e^{-1} \nabla_r E_F(\boldsymbol{r})$ is a generalized electric field strength, which includes the effects of the space-dependent Fermi energy. $\mathscr{L}_{xx}^{12}$ is a so-called *transport coefficient* [defined by (9.66)] responsible for the fact that a temperature gradient is generally associated with an electrical current.

The Boltzmann equation can be used to calculate not only electric current but also the heat flux due to electrons in an external field $\mathscr{E}$ and for a temperature gradient $\nabla_r T$. The quantity of heat transported dQ is associated with an entropy change dS, or a change in internal energy dU and particle number dn via the following thermodynamic relation (E_F is the chemical potential of the electrons, i.e., Fermi energy):

$$dQ = TdS = dU - E_F dn \,. \tag{9.70}$$

With $\boldsymbol{j}_E$ as the energy flux density and $\boldsymbol{j}_n$ as the particle flux density, the heat flux density follows as

$$\boldsymbol{j}_Q = \boldsymbol{j}_E - E_F \boldsymbol{j}_n \,, \tag{9.71}$$

where $\boldsymbol{j}_E = (8\pi^3)^{-1} \int d\boldsymbol{k}\, E(\boldsymbol{k})\, \boldsymbol{v}(\boldsymbol{k})\, f(\boldsymbol{k},\boldsymbol{r})$. As above, the insertion of the linearized non-equilibrium distribution (9.65) leads to a linear relation between $\boldsymbol{j}_Q$, $\mathscr{E}'$ (or $\mathscr{E}$) and $\nabla_r T$, as in (9.69). In general, one thus has the following equations for the electrical and thermal currents as a function of their origins, generalized electric field $\mathscr{E}'$ and temperature gradient $\nabla_r T$:

$$j = \mathscr{L}^{11}\,\mathscr{E}' + \mathscr{L}^{12}\,(-\nabla_r T)\,, \tag{9.72 a}$$

$$j_Q = \mathscr{L}^{21}\,\mathscr{E}' + \mathscr{L}^{22}\,(-\nabla_r T)\,. \tag{9.72 b}$$

For non-cubic crystals the transport coefficients $\mathscr{L}^{ij}$ are of course tensors. Their general representation [e.g., (9.66) for $\mathscr{L}^{12}_{xx}$] leads to interesting relations between the coefficients, which, however, will not be pursued here.

We now wish to summarize the physical behavior expressed by (9.72 a, b). An electric field or a temperature gradient causes both a thermal and an electric current; in particular, an electric field can be created by a thermal gradient. Consider a conducting loop consisting of two different metals A and B, which is either open, or connected only by a high resistance voltmeter (Fig. 9.9 a). The two junctions between metals A and B are at different temperatures $T_1 \neq T_2 \neq T_0$. Because $j=0$, we obtain from (9.72 a) (with $\mathscr{E}' = \mathscr{E}$ inside the metal):

$$\mathscr{E}_x = (\mathscr{L}^{11})^{-1}(\mathscr{L}^{12})\partial T/\partial x = K\partial T/\partial x\,. \tag{9.73}$$

The quantity K is called the absolute thermopower and is a material-specific property. We consider a one-dimensional problem in which x is the distance along the wire. The loop voltage measured by the voltmeter in Fig. 9.9 a is

$$\begin{aligned} U &= \int_0^1 \mathscr{E}_B\,dx + \int_1^2 \mathscr{E}_A\,dx + \int_2^0 \mathscr{E}_B\,dx \\ &= \int_2^1 K_B \frac{\partial T}{\partial x}dx + \int_1^2 K_A \frac{\partial T}{\partial x}dx = \int_{T_1}^{T_2} (K_A - K_B)dT\,. \end{aligned} \tag{9.74}$$

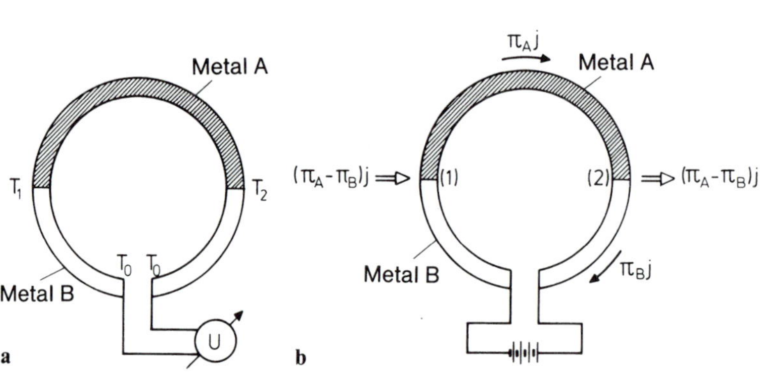

Fig. 9.9. (**a**) Schematic arrangement for measuring the Seebeck effect. When the two junctions between metals A and B are at different temperatures ($T_1 \neq T_2 \neq T_0$) a thermovoltage appears across the ends of the loop. (**b**) Adaptation for measuring the Peltier effect. A current is passed through the loop (now at constant temperature) and leads to heat transport between junctions (1) and (2)

This so-called thermopotential produced by the thermoelectric, or *Seebeck effect*, depends on the temperature difference T_2–T_1 and the difference in the absolute thermopowers K_A and K_B. This effect is exploited in thermocouples for measuring temperatures using essentially the same scheme as shown in Fig. 9.9a. From (9.72b) one can also derive the inverse effect, known as the *Peltier effect*. In Fig. 9.9b the temperature is constant around the loop, i.e., $\partial T/\partial x=0$. We then have

$$j_Q = \mathscr{L}^{12}\mathscr{E} \quad \text{and} \quad j = \mathscr{L}^{11}\mathscr{E} , \tag{9.75}$$

and thus

$$j_Q = \mathscr{L}^{21}\left(\mathscr{L}^{11}\right)^{-1} j = \Pi j , \tag{9.76}$$

where Π is the Peltier coefficient. If we attach a battery to the conducting loop in Fig. 9.9b, then the resulting electric current density j will be accompanied by a thermal current according to (9.76). In A the thermal current is $\Pi_A j$, while in B, it is $\Pi_B j$. Thus there must be a net heat flow of $(\Pi_A - \Pi_B)j$ into junction (2), which must be extracted at junction (1). Point (1) becomes cooler and point (2) warmer, if $\Pi_A > \Pi_B$. The Peltier effect is thus a convenient way to obtain simple and easily controlled cooling.

9.7 The Wiedemann-Franz Law

As mentioned in passing in Sect. 9.6, relationships between the transport coefficients $\mathscr{L}^{ij}$ can be derived from their explicit representations as integrals over the linearized non-equilibrium distribution $f(\boldsymbol{k},\boldsymbol{r})$. In particular, one can derive a relationship between the electrical conductivity σ and the conduction electron contribution to the thermal conductivity λ_{E}. The conductivity σ is essentially equal to $\mathscr{L}^{11}$, and λ_{E} to $\mathscr{L}^{22}$. Making use of (9.65, 9.71), the thermal current density is found to be

$$j_{Qx} = \int d\boldsymbol{k}\,(E - E_{\mathrm{F}})v_x^2(\boldsymbol{k})\tau(k)\frac{\partial f_0}{\partial T}\left(\frac{-\partial T}{\partial x}\right) . \tag{9.77}$$

For a spherical Fermi surface, and under the assumption that τ depends only on energy, we can relate this integral to the electronic specific heat, as in (9.66–9.68), and, for the thermal conductivity λ_{E} due to conduction electrons, we obtain

$$\lambda_{\mathrm{E}} = \tfrac{1}{3}\, v_{\mathrm{F}}^2 \tau(E_{\mathrm{F}}) c_v . \tag{9.78}$$

For a free electron gas, using (6.50), this becomes

$$\lambda_{\mathrm{E}} = \frac{\pi^2}{3}\tau(E_{\mathrm{F}}) n \kappa^2 T/m^* . \tag{9.79}$$

From (9.58a), the electrical conductivity is $\sigma = e^2\tau(E_{\mathrm{F}})n/m^*$, so that the quotient

Table 9.1. Experimentally derived values of the Lorentz number at 0 °C, $L = \lambda_E/\sigma T$, deduced from published data for electrical and thermal conductivity

Metal	$L(10^{-8}\ \mathrm{W\Omega/K^2})$
Na	2.10
Ag	2.31
Au	2.35
Cu	2.23
Pb	2.47
Pt	2.51

$$\lambda_E/\sigma = \frac{\pi^2}{3}\left(\frac{k}{e}\right)^2 T \tag{9.80}$$

depends linearly on temperature. This is known as the Wiedemann-Franz law, and the factor $\pi^2 k^2/3e^2$ is the *Lorentz number L*.

The theoretical value of the Lorentz number is, according to (9.80),

$$L = 2.45 \times 10^{-8}\,\mathrm{W\,\Omega\,K^{-2}}\,. \tag{9.81}$$

Table 9.1 shows a few experimental values for the Lorentz number at 0 °C; they show relatively good agreement with the value in (9.81) even for transition metals. At low temperatures ($T \ll \Theta$, where Θ is the Debye temperature) the value of L generally decreases strongly. For Cu at 15 K, for example, L is an order of magnitude smaller than the value in Table 9.1 at 0 °C. The reason for this deviation is that the electrical and thermal relaxation times are no longer the same at low temperatures.

The difference between the two relaxation times can be explained within a simple picture: An electric current arises due to a displacement of the Fermi sphere by an amount δk (Fig. 9.4). A thermal current, on the other hand, stems from the fact that electrons with a wave vector $+k_F$ have a different average temperature to electrons with $-k_F$. The relaxation of an electric current thus demands scattering processes with wave vector of about $q = 2k_F$, whereas relaxation of a thermal current requires only $\boldsymbol{q}$-vectors of the order of $q \approx kT/\hbar v_F$ to change the energy levels of electrons near the Fermi level. At sufficiently high temperatures, the difference between the scattering processes is not so important. At low temperatures, the phonons available for scattering are mostly those with small q. As a result, the thermal conductivity becomes smaller relative to the electrical conductivity and the Lorentz number L decreases with temperature.

9.8 Electrical Conductivity of Localized Electrons

So far we have considered only transport phenomena associated with delocalized band electrons. The delocalization of band electrons is a consequence of the translational invariance of crystalline solids. In an amorphous solid, on the other hand, electrons may be either delocalized or localized. In Sect. 7.6 we have argued why electrons in the tails of the energy band extending into the forbidden gap of amorphous Si and Ge are localized, whereas the electrons in states further away from the gap remain delocalized as in an periodic solid. In this section we discuss the contribution of localized electrons to the electrical conductivity. We consider a simple model for the density of states with localized and delocalized electrons below and above a "mobility edge" E_m, respectively (Fig. 9.10). The lower the electron energy, the stronger is the localization and the larger also is the mean spatial distance between localized electron states. In such a system the electrical conductivity depends on the relative positions of the Fermi energy E_F and mobility edge E_m. If the Fermi energy is above the mobility edge, then the conductivity is as in a crystalline metal with a high concentration of defects. If the Fermi energy is a few $k T$ below the mobility edge then the conductivity is dominated by the thermally excited electrons in the states above the mobility edge. The conductivity is proportional to the concentration of electrons in the states above the mobility edge E_m, hence

$$\sigma = \sigma_0 e^{-(E_m - E_F)/k T} . \qquad (9.82)$$

If the Fermi energy is below the mobility edge by more than a few $k T$ then the electrical transport is carried by localized electrons that hop from one localized state to another via a tunneling process. This transport process is therefore known as "hopping conduction". Tunneling from one localized state to another requires an overlap of the wave function. Since the

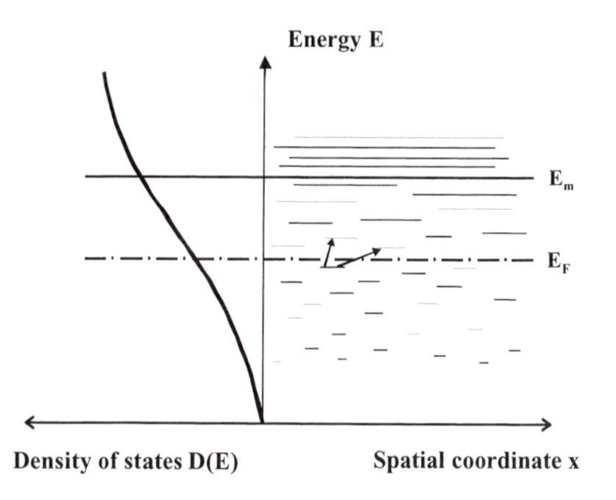

Fig. 9.10. Electronic states of an amorphous solid (schematic). The density of states tails into the energy gap. The states are localized below the mobility edge E_m and delocalized above E_m. The extension of the wave function increases as the mobility edge is approached

amplitudes of the wave functions decay exponentially the probability for a hopping process from an atom i to an atom j depends exponentially on the distance r_{ij} between the atoms.

$$p(r_{ij}) \propto \mathrm{e}^{-\alpha r_{ij}} \ . \tag{9.83}$$

The constant α characterizes the decay of the wave function. In a hopping process, the initial state must be occupied, the final state unoccupied. Hence, the energy of the initial state must lie below the Fermi level and the energy of the final state above, so that one has $E_j > E_i$. The jump frequency therefore contains a Boltzmann factor, with the energy difference between E_j and E_i in the numerator of the exponent. Thus, the probability for a jump from i to j becomes

$$p(r_{ij}, E_j, E_i) \propto \mathrm{e}^{-\alpha r_{ij}} \mathrm{e}^{-(E_j - E_i)/k T} \ . \tag{9.84}$$

The conductivity is proportional to the sum over all jump probabilities $p(r_{ij}, E_j, E_i)$. We now consider the possible jumps into a fixed final state E_j out of a continuous distribution of initial states E_i that lie in a sphere with a radius $r_{ij} < R$ around the final state j. A jump is possible if there is at least one state with the initial energy $E_i < E_j$ within the sphere of radius R. In order to simplify the calculation we assume that the density of states is constant around E_F. In reality this assumption is not well fulfilled (Fig. 9.10). The condition for the existence of an initial state within the sphere of radius R out of which a jump can take place is then

$$\frac{4\pi}{3} R^3 D(E_F)(E_j - E_i) > 1 \ . \tag{9.85}$$

Hence, all jumps are allowed for which

$$E < E_{\max} = E_j - \frac{3}{4\pi R^3 D(E_F)} \ . \tag{9.86}$$

The total jump rate and thus the conductivity is therefore proportional to

$$\sigma \propto \mathrm{e}^{-\alpha R} \int_{-\infty}^{E_{\max}} \mathrm{e}^{-(E_j - E)/k T} dE \propto \mathrm{e}^{-\alpha R} \, \mathrm{e}^{-3/(4\pi R^3 D(E_F) k T)} \ . \tag{9.87}$$

The jump rate vanishes for large R because of the first term describing the overlap of the wave functions. The jump rate also vanishes for small R, since there are no energy terms within the sphere of radius R. In between, the jump rate has a maximum where the exponent has a minimum

$$\frac{d}{dR}\left(\alpha R + \frac{3}{4\pi R^3 D(E_F) k T}\right) = 0 \ . \tag{9.88}$$

The minimum is for a radius

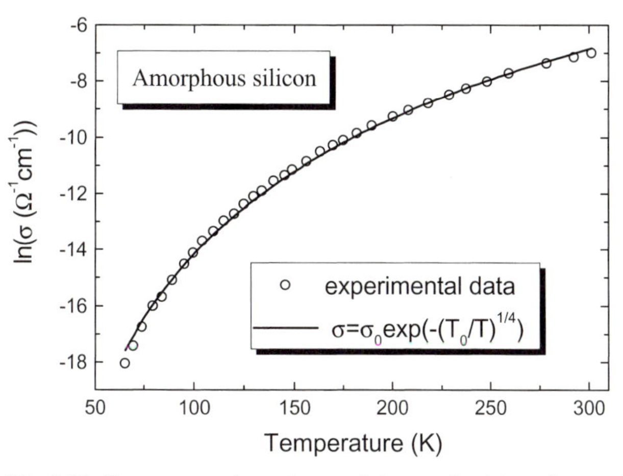

Fig. 9.11. Temperature dependence of the conductivity of amorphous Si (open circles, after [9.7]). The experimental data are reasonably well represented by Mott's law with an exponent of 1/4. The data would be fitted even better with an exponent of 1/2

$$R_\mathrm{m} = \left(\frac{9}{4\pi\alpha D(E_\mathrm{F})\,kT}\right)^{1/4} . \tag{9.89}$$

For $T=300$ K, e.g., a density of states of 5×10^5 cm^{-3} and an α of 10^7 cm^{-1} one calculates an optimal jump distance of 5×10^{-6} cm. After inserting the optimum jump distance R_m into (9.87) one obtains the temperature dependence of the conductivity

$$\sigma \propto \mathrm{e}^{-(T_0/T)^{1/4}} \quad \text{with} \quad T_0 = \frac{(4\alpha)^3}{9\pi D(E_\mathrm{F})\,k} . \tag{9.90}$$

This temperature dependence of the hopping conductivity was first derived by Mott [9.6]. It describes the temperature dependence of the conductivity of amorphous Si quite well (Fig. 9.11). An even better fit to the experimental data is obtained for an exponent 1/2 instead of 1/4. The main reason is that the density of states around the Fermi level is not constant, as assumed in the model, but increases with energy (Figs. 9.10, 7.15).

Problems

9.1 Discuss the difference between relaxation time, τ, and free flight time, τ_c, using the example of the electrical conductivity of a classical gas. Assume that at time $t=0$ the charged particles are moving with a velocity v_{0x} in the x-direction. After switching off the electric current the particles are elastically scattered after a time τ_c with a non-isotropic scattering probability $p(\Theta)$, where Θ is the angle with the x-axis. Show that the mean velocity of the particles in the x-direction decreases exponentially with a relaxation time $\tau = \frac{\tau_c}{1 - \langle \cos \Theta \rangle}$ where $\langle \cos \Theta \rangle$ is the average value of $\cos \Theta$ over the distribution $p(\Theta)$. Compare the free flight time and the relaxation time. Under what conditions are they identical? In a solid, electrons are scattered by defects. Discuss the various types of defect with regard to the expected relation between τ and τ_c.

9.2 Using the stationary Boltzmann equation, show that the mobility of charged particles in a classical gas is given by

$$\mu = \frac{e\langle v^2 \tau(\boldsymbol{k}) \rangle}{m\langle v^2 \rangle} \quad .$$

9.3 Calculate the electric field dependence of the conductivity $\sigma(\mathscr{E})$ using the second iteration for the solution of the Boltzmann equation. In the second iteration step, as in the first, use a field-independent distribution f_0. Discuss interesting applications of a material with a highly field-dependent conductivity.

9.4 Consider a semi-infinite metallic half-space with two point contacts on the surface. A magnetic field $\boldsymbol{B}$ is applied parallel to the surface and perpendicular to the line joining the point contacts. One point contact is used as a source of electrons injected into the metal, the second as a detector. Draw electron trajectories for the case where the transport is ballistic.
– At what values of $\boldsymbol{B}$ do you expect a maximum signal at the detector?
– Calculate the signal as a function of $\boldsymbol{B}$.

9.5 Consider a semiclassical model of a cyclotron resonance experiment, where the motion of a quasi-free particle in a solid is described by the effective mass m^* and relaxation time τ.
a) Derive the cyclotron frequency ω_c.
b) Derive the differential equations of motion for the electron drift velocity in a static magnetic field $\boldsymbol{B}=(0,0,B)$ and a simultaneous high-frequency (rf) electric field $\mathscr{E} = [A \exp(\mathrm{i}\omega t),\ A \exp \mathrm{i}(\omega t + \pi/2),\ 0]$, where the associated rf magnetic field is neglected. Scattering processes are described by a frictional term $m^* v/\tau$.
c) Calculate from the current density in the x-direction the conductivity $\sigma = j_x/\mathscr{E}_x$, whereby $\mathrm{Re}\{\sigma\}$ gives the damping of the electric field.
d) Plot the real part of the conductivity $\mathrm{Re}\{\sigma\}$ as a function of ω_c/ω for the values $\omega_c\tau = 0.2$, 1, 3 and discuss the curves.
e) How can electron resonances be distinguished from hole resonances?
f) A clystron yields rf radiation at a frequency of 2.4×10^{10} Hz. Maximum absorption is observed for a magnetic field $B = 8.6 \times 10^{-2}$ Tesla. Determine the mass ratio m^*/m_0 of the carriers. Over what range of relaxation time and of mobility μ is a resonance observable?

Panel VIII
Quantum Oscillations and the Topology of Fermi Surfaces

The Fermi surface of a metal is a surface defined in k-space by the relation $E(\boldsymbol{k})=E_F$. Electrons at the Fermi surface are particularly important because only they are capable of absorbing energy in infinitesimal quantities. It is these electrons, therefore, that determine all transport properties, such as electrical conductivity (Sect. 9.5), which, as we have seen, depends on an integral over the Fermi surface.

Fermi surfaces can be calculated from a known bandstructure by applying the condition $E(\boldsymbol{k})=E_F$. Experimental methods are also available for the direct and very exact measurement of Fermi surfaces. These methods are based on the observation of quantized oscillations in a magnetic field. The origins and manifestations of quantum oscillations are not easy to understand, and their explanation will take us somewhat further afield.

Free electrons in a magnetic field experience an additional force, the so-called Lorenz force. This acts perpendicular to the magnetic field and the velocity and causes a rate of change of momentum:

$$m\dot{\boldsymbol{v}} = e(\boldsymbol{v} \times \boldsymbol{B}) \ . \tag{VIII.1}$$

For electrons in the solid, we can use (9.4) to set $m=\hbar\boldsymbol{k}$ and $\boldsymbol{v}=\hbar^{-1}\nabla_k E(\boldsymbol{k})$. From the time-dependent variation of the momentum, we obtain the corresponding rate of change of the wave vector

$$\frac{d\boldsymbol{k}}{dt} = \frac{e}{\hbar^2}[\nabla_k E(\boldsymbol{k}) \times \boldsymbol{B}] \ . \tag{VIII.2}$$

Thus the electrons move in a plane perpendicular to the magnetic field and tangential to the surfaces of constant energy in $\boldsymbol{k}$-space. We wish now to calculate the period of this orbit:

$$T = \int dt = \frac{\hbar^2}{e}\frac{1}{B}\oint \frac{d\boldsymbol{k}}{[\nabla_k E(\boldsymbol{k})]_\perp} \ . \tag{VIII.3}$$

Here we have introduced $[\nabla_k E(\boldsymbol{k})]_\perp$, which is the component of the gradient perpendicular to the $\boldsymbol{B}$-field. We can also write this component as $dE/dk_\perp$ with $k_\perp$ being the component of $\boldsymbol{k}$ perpendicular both to the $\boldsymbol{B}$-field and to the contour line on the energy surface along which the electron moves (Fig. VIII.1). From Fig. VIII.1, we can also see that the path integral is none other than the derivative of the cross-sectional area with respect to energy

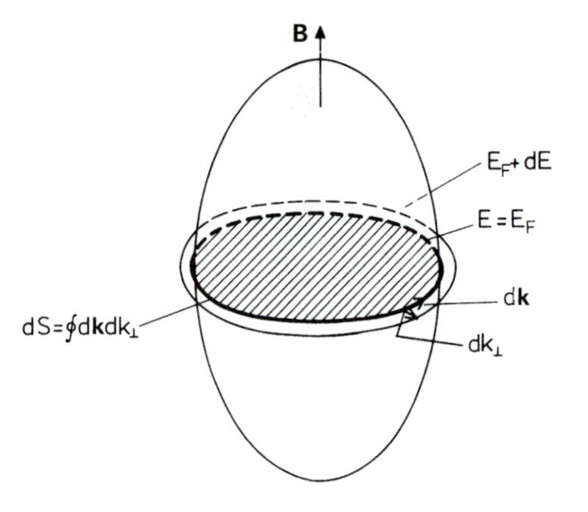

Fig. VIII.1. The Fermi-surface orbit of an electron in a solid under the influence of an applied magnetic field. The period of the orbit is proportional to the derivative of the cross-sectional area with respect to energy

$$\oint \frac{dk_\perp}{dE} d\boldsymbol{k} - \frac{dS}{dE} . \tag{VIII.4}$$

The period

$$T = \frac{\hbar^2}{eB} \frac{dS}{dE} \tag{VIII.5}$$

is thereby connected to the energy dependence of the area enclosed by the electron orbit in $\boldsymbol{k}$-space. In general, the orbital periods are different for different electrons in $\boldsymbol{k}$-space; however, for a free electron gas they are all equal. For this case, one can readily convince oneself that, since $S = \pi k^2$,

$$\frac{dS}{dE} = 2\pi \frac{m^*}{\hbar^2} \tag{VIII.6}$$

and thus

$$T = \frac{2\pi m^*}{eB} . \tag{VIII.7}$$

Here m^* is the effective mass (Sect. 9.1).

For a magnetic field of 1 T (10^4 G), we obtain, for example, a period of $T = 3.6 \times 10^{-11}$ s. This period in $\boldsymbol{k}$-space is also the period in real space, and we could also have derived this result for free electrons from the equality of the centrifugal and Lorentz forces ($e\,vB = m^*\,v\omega_c$, see also Panel XV). The electron orbit is then a closed path, and the frequency $\omega_c = 2\pi\,T$ is also called the cyclotron frequency.

According to our derivation, electrons under the influence of a magnetic field move on *continuous* paths. How can this be made consistent with the discrete $\boldsymbol{k}$-vectors, which we have assumed up until now? The discrete values for the components of the wave vector and their uses as quantum numbers arise from the three-dimensional translational invariance. In a magnetic field, however, the Schrödinger equation is no longer translationally invariant in the directions perpendicular to the magnetic field. In the Schrödinger equation, $\boldsymbol{p}$ must thus be replaced by $\boldsymbol{p}+e\boldsymbol{A}$, where $\boldsymbol{B}=\text{curl}\,\boldsymbol{A}$. The components of the $\boldsymbol{k}$-vector perpendicular to the magnetic field axis (k_x, k_y) are therefore no longer good quantum numbers and the whole band model collapses. Fortunately, we can still make some predictions about the energy levels, provided the magnetic fields are not too large. According to the correspondence principle, the energetic splitting of two neighboring levels, $E_{n+1}-E_n$, is

$$E_{n+1} - E_n = \frac{2\pi\hbar}{T} = \hbar\omega_c \ , \qquad \text{(VIII.8)}$$

where T is the period of the orbit around the cross-sectional area, and ω_c is the frequency [VIII.1]. The estimate of 3.6×10^{-11} s for the period allows us to estimate a difference between energy levels of 1.1×10^{-4} eV for a magnetic field of 1 T. The existence of such discrete levels can be demonstrated via the absorption of electromagnetic radiation, clearly transitions can occur only from occupied states below the Fermi level to unoccupied states above the Fermi level (*cyclotron resonance*, cf. Panel XV). The interpretation of cyclotron resonances in metals is rather difficult though, because the period T depends, among other things, on the position of the cross-sectional area S which the electron path circumscribes in $\boldsymbol{k}$-space.

We now tackle the question of how $\boldsymbol{k}$-space can be quantized although k_x, k_y are no longer good quantum numbers. Inserting (VIII.5) in (VIII.8), and replacing the differential quotient dS/dE by the difference quotient $(S_{n+1}-S_n)/(E_{n+1}-E_n)$, we obtain

$$S_{n+1} - S_n = 2\pi\frac{eB}{\hbar} \ . \qquad \text{(VIII.9)}$$

Hence, in a magnetic field, the areas enclosed by the electron path are quantized in $\boldsymbol{k}$-space. Against this, the states on the orbital path are degenerate.

Interestingly, the difference between the areas $S_{n+1}-S_n$ is constant (for large n) and independent of k_z. The electrons are located on a concentric set of so-called *Landau tubes* (Fig. VIII.2). The allowed states in $\boldsymbol{k}$-space "condense" onto these tubes as soon as even a minimal field is switched on. It must be remembered, however, that the distance between tubes is very small for normal magnetic fields. The total number of states, of course, does not change on "condensation". For each particular k_z, all states which were between the tubes prior to condensation must be on the tubes after condensation.

So far we have considered only a constant magnetic field. What happens when we slowly increase $\boldsymbol{B}$? Equation (VIII.9) tells us that the Landau tubes slowly expand, and thus successively leave the Fermi surface $E(\boldsymbol{k})=E_F$ (Fig. VIII.2) passing through a contour of contact. If the contact contour of the Fermi level is exactly between two Landau tubes, the highest occupied states lie below the Fermi level. The average energy is then

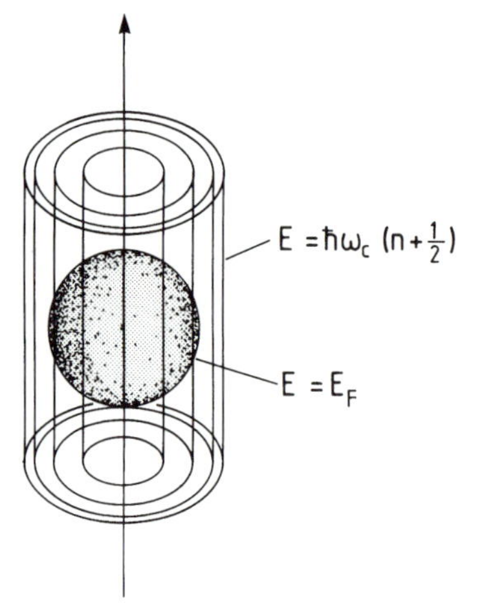

Fig. VIII.2. In a magnetic field, the allowed states of the electrons lie on so-called Landau tubes. These tubes are characterized by two properties: (1) The orbit of an electron in the plane perpendicular to B is around a surface of constant energy. (2) The magnetic flux enclosed between two successive tubes is constant. The axis of the tubes is only parallel to the magnetic field when the latter is directed along a principle symmetry axis of the solid. For the sake of simplicity, we show here only the tubes of circular cross section that would be obtained in the case of a free electron gas with a spherical Fermi surface at zero field. In a magnetic field, the occupied states lie on those parts of the Landau tubes that are within the sphere

lower than when the Landau tube is just leaving the Fermi surface at the contact contour. The average energy therefore oscillates. These periodic quantum oscillations are in principle observable in all solid state properties, and have in fact been observed in many. According to (VIII.9), the period of this oscillation in $\boldsymbol{B}$ is

$$\Delta B = \frac{2\pi e}{\hbar S_{\mathrm{F}}} B^2 \,. \tag{VIII.10}$$

The period ΔB directly measures cross-sectional areas of the Fermi surface. It is for this reason that quantum oscillations are so significant in understanding the topology of Fermi surfaces.

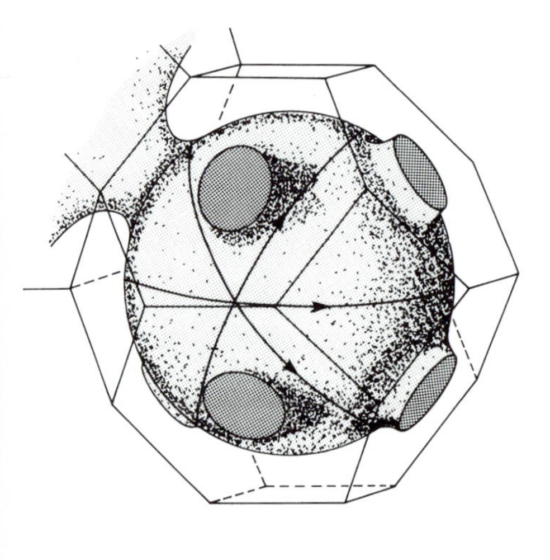

Fig. VIII.3. The Fermi surface of the noble metals Cu, Ag and Au make contact with the Brillouin zone boundary around the point L. Thus, perpendicular to the $\langle 111\rangle$ directions for example, there are two orbits with extremal area. Perpendicular to the $\langle 110\rangle$ and $\langle 100\rangle$ directions there are also closed orbits that lie between the Fermi surfaces of neighboring Brillouin zones and thus enclose empty states (hole orbits). In addition, one finds open trajectories in which the electrons travel through the extended zone scheme of k-space. Such open trajectories clearly have no orbital period and it is not possible to apply the arguments derived for closed orbits

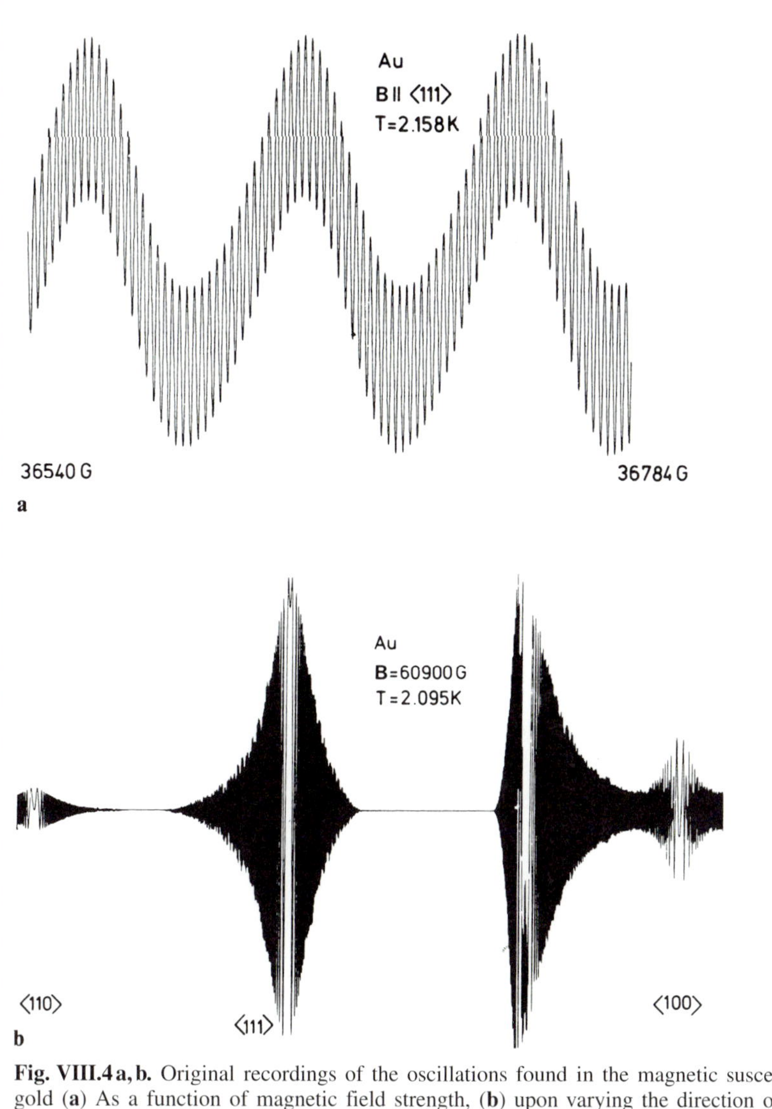

Fig. VIII.4 a, b. Original recordings of the oscillations found in the magnetic susceptibility of gold (**a**) As a function of magnetic field strength, (**b**) upon varying the direction of the magnetic field (after [VIII.3]). Such quantum oscillations can only be observed of course for a sufficiently sharp Fermi distribution ($kT < \hbar\omega_c$)

The periodic oscillations of the magnetic susceptibility known as the de Haas-van Alphen effect [VIII.2] have gained particular practical importance. As an example we show some measurements on gold.

In gold (as in Cu, cf. Fig. 7.12) the *d*-states are occupied and the Fermi level falls in the region of the *sp*-band. Hence the Fermi surface is approximately spherical, although it is deformed somewhat to make contact with the zone boundaries along the [111] directions. This distortion of the sphere means that the Fermi surface is joined by a neck to the next Brillouin zone (Fig. VIII.3).

For a magnetic field along the [111] direction, there are therefore two types of closed orbits along which the Landau tubes can peel off from the Fermi surface with increasing magnetic field. These are the so-called neck and belly orbits. Their cross-sectional areas determine the period of the de Haas-van Alphen oscillations. Figure VIII.4 a shows an original spectrum for this situation. The ratio of the cross-sectional areas can be read off directly by comparing the periods (1 : 29). In addition, using (VIII.9), we can calculate from the data that the cross-sectional areas are 1.5×10^{15} cm^{-2} and 4.3×10^{16} cm^{-2}, respectively. With this information, one can construct an image of the Fermi surface such as that shown in Fig. VIII.3.

If the angle of the magnetic field axis is varied at constant field strength, one again observes oscillations. This is because at some angles the Landau tubes touch the Fermi surface along the contact contour, while at other angles they do not (Fig. VIII.4 b). The oscillations disappear at angles where the orbit is not closed. Closed orbits with definite periods were prerequisites for the derivation of the relations (VIII.3–10).

References

VIII.1 L. Onsager: Philos. Mag. **43**, 1006 (1952)
VIII.2 W.J. de Haas, P.M. van Alphen: Leiden Comm. **208** d, **212** a (1930)
VIII.3 B. Lengeler: *Springer Tracts Mod. Phys.* **82**, 1 (Springer, Berlin, Heidelberg 1978)

10 Superconductivity

In Chap. 9, the basic aspects of the phenomenon of electrical conductivity in metals were introduced and explained. The finite resistance of these materials originates from the fact that a real crystal always exhibits deviations from perfect lattice periodicity: phonons and defects. An infinitely high electrical conductivity is unthinkable in this description, because (1) a crystal without a certain degree of disorder is inconceivable according to the second law of thermodynamics, and (2) even in the absence of phonon and defect scattering, electron-electron scattering will still cause resistance (Sect. 9.3). However, in the year 1911, Onnes [10.1] discovered that the electrical resistance of mercury approaches an unmeasurably small value when it is cooled below 4.2 K. This phenomenon is called "superconductivity", and in the subsequent years many more materials were found to be superconducting when cooled to below a certain critical temperature T_c. A microscopic explanation, which, as we know from Chap. 9, cannot be obtained within the framework of a one-electron approximation, was not found for nearly half a century. It was not until shortly before 1960 that Bardeen, Cooper and Schrieffer [10.2] achieved the decisive break-through and proposed the theory now named after them (BCS theory). After a short introduction to the basic phenomena of superconductivity, this chapter will give a simplified illustration of how the important properties of superconductors can be understood within the framework of BCS theory.

10.1 Some Fundamental Phenomena Associated with Superconductivity

The property that gives superconductivity its name was discovered by Onnes [10.1] in mercury. Figure 10.1 shows the classical experimental results: on cooling below about 4.2 K the electrical resistance of the mercury sample collapses. It was assumed at that time to be below $10^{-5}\,\Omega$. In fact, it is fundamentally impossible to test experimentally whether the resistance is zero; one can only set an upper limit to the value of the electrical resistance, and this of course still applies to today's much improved measurement methods. The best experimental method of measuring very small resistances, which Onnes also used in addition to current-voltage measurements, is to measure the decrease of a current in a closed conducting loop.

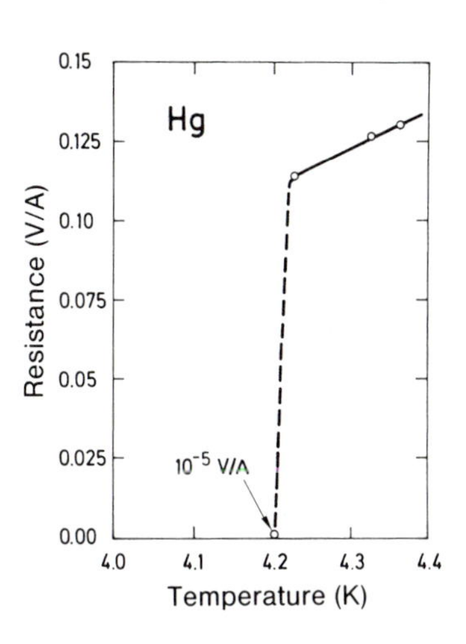

Fig. 10.1. Original measurement curve for Hg with which, in 1911, the phenomenon of superconductivity was first discovered. On cooling to below 4.2 K, the electrical resistance falls below the experimental detection limit (at that time $\sim 10^{-5}\,\Omega$). (After [10.1])

A magnetic flux through the loop is created with a magnet, and the material is then cooled to below the critical temperature T_c, so that it becomes superconducting. Removal of the magnet alters the flux through the loop, thereby inducing a current. If the ring had a finite resistance R, the current would decrease according to

$$I(t) = I_0 \exp(-Rt/L) , \tag{10.1}$$

where L is the self-induction of the loop. With this and similar arrangements it is possible to show that the resistance of a metal decreases by at least 14 orders of magnitude when it enters the superconducting state. Superconductivity is not confined to just a few chemical elements; the periodic table in Fig. 10.2 shows that this property is found in many elements. Both non-transition metals such as Be and Al as well as transition metals such as Nb, Mo and Zn show the effect. Semiconductors like Si, Ge, Se and Te transform to a metallic phase under high pressure and then become superconducting at low temperatures. Superconductivity may depend on the crystal structure. This is clearly seen in the case of Bi, where several of the modifications are superconducting with different transition temperatures, while one modification does not show the effect down to 10^{-2} K. A crystalline structure is not a necessary condition for superconductivity. Amorphous and polycrystalline superconducting samples, which are produced by rapid condensation onto cooled substrates, play an important role in modern research. In addition, a large number of alloys display superconductivity, with $Nb_3Al_{0.75}Ge_{0.25}$, for example, having a relatively high transition temperature of 20.7 K. At present, the most fascinating superconductors are the ceramic materials, consisting, for example, of the elements Ba-Y-Cu-O. These materials can have an extremely high transition temperature in the vicinity of 100 K (Sect. 10.11). It is noticeable that the ferromagnetic materials like Fe, Co, Ni (Fig. 10.2) do not display superconductivity. It is now believed that their strong ferromagnetism is the reason for this and we

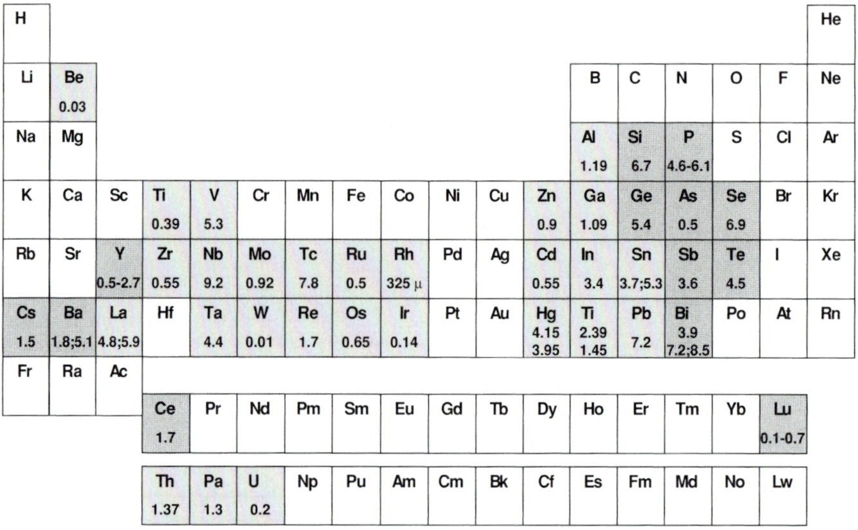

H																	He
Li	Be 0.03											B	C	N	O	F	Ne
Na	Mg											Al 1.19	Si 6.7	P 4.6-6.1	S	Cl	Ar
K	Ca	Sc	Ti 0.39	V 5.3	Cr	Mn	Fe	Co	Ni	Cu	Zn 0.9	Ga 1.09	Ge 5.4	As 0.5	Se 6.9	Br	Kr
Rb	Sr	Y 0.5-2.7	Zr 0.55	Nb 9.2	Mo 0.92	Tc 7.8	Ru 0.5	Rh 325 μ	Pd	Ag	Cd 0.55	In 3.4	Sn 3.7;5.3	Sb 3.6	Te 4.5	I	Xe
Cs 1.5	Ba 1.8;5.1	La 4.8;5.9	Hf	Ta 4.4	W 0.01	Re 1.7	Os 0.65	Ir 0.14	Pt	Au	Hg 4.15 3.95	Tl 2.39 1.45	Pb 7.2	Bi 3.9 7.2;8.5	Po	At	Rn
Fr	Ra	Ac															
			Ce 1.7	Pr	Nd	Pm	Sm	Eu	Gd	Tb	Dy	Ho	Er	Tm	Yb	Lu 0.1-0.7	
			Th 1.37	Pa 1.3	U 0.2	Np	Pu	Am	Cm	Bk	Cf	Es	Fm	Md	No	Lw	

Fig. 10.2. The periodic table of the elements showing known superconductors (*shaded*) and their transition temperatures T_c (in [K]). Dark shading means that only a high pressure phase of the element is superconducting

will see in Sect. 10.6 that strong magnetic fields have the effect of suppressing superconductivity.

It should be emphasized that, in general, the transition of a metal from its normal state to a superconducting state has nothing to do with a change of crystallographic structure; furthermore, no ferromagnetic, ferrimagnetic or antiferromagnetic transition is involved. This can be demonstrated by scattering experiments with X-rays, electrons and neutrons (Panel I). What actually occurs in the transition from the normal to the superconducting state is a thermodynamic change of state, or phase transition, which is clearly manifest in other physical quantities. The specific heat as a function of temperature, for example, changes discontinuously at the transition temperature T_c. Figure 10.3 shows the example of Al with a transition temperature of 1.19 K. The specific heat c_n of a normally conducting metal is composed of a lattice-dynamical part c_{nl} and an electronic part c_{ne}. From (6.51) we have

$$c_n = c_{ne} + c_{nl} = \gamma T + \beta T^3 \ . \tag{10.2}$$

In the normal conducting state $(n) c_n$ varies smoothly at low temperatures, similar to Fig. 6.8. At the transition to the superconducting state (at T_c) the specific heat jumps and then at very low temperatures sinks to below the value of the normal phase. Normally one assumes that at the transition to the superconducting state the lattice dynamical part c_{nl} remains constant. A more exact analysis then shows that, for a superconductor, the electronic part ($c_{ne} \sim T$ for a normal conductor) must be replaced by a component which, well below the critical temperature, decreases exponentially

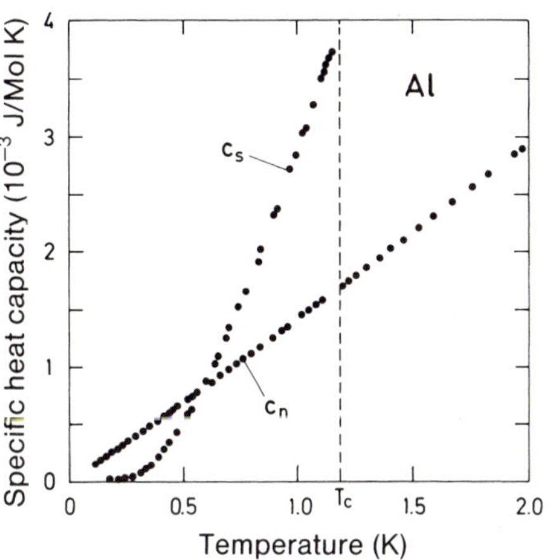

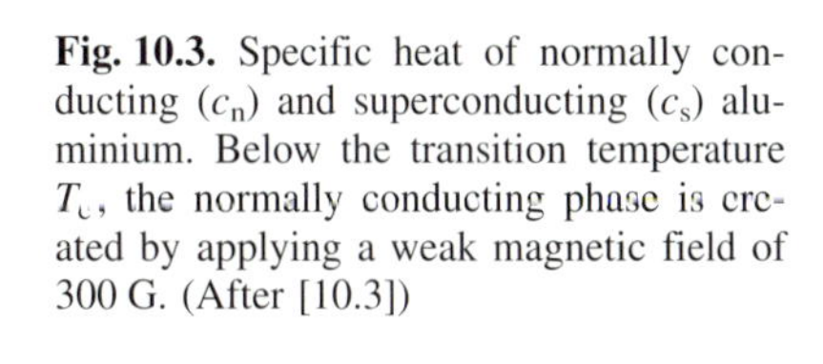

Fig. 10.3. Specific heat of normally conducting (c_n) and superconducting (c_s) aluminium. Below the transition temperature T_c, the normally conducting phase is created by applying a weak magnetic field of 300 G. (After [10.3])

$$c_{se} \sim \exp(-A/k T) . \tag{10.3}$$

A further distinctive property of a superconductor is its magnetic behavior. The magnetic properties of a material that transforms to the superconducting state cannot simply be described by the Maxwell equations with the additional requirement of a vanishing electrical resistance ($R \rightarrow 0$). Figure 10.4 A shows schematically how an ideal conductor, whose resistance is zero below a critical temperature, would behave, if it were cooled in an external magnetic field $\boldsymbol{B}_{ext}$, i.e. if it loses its resistance in the magnetic field. For any closed path (enclosing area $\boldsymbol{S}$) in a material, it must be true that

$$IR = U = \int_F \operatorname{curl} \mathscr{E} \cdot d\boldsymbol{S} = -\dot{\boldsymbol{B}} \cdot \boldsymbol{S} . \tag{10.4}$$

Vanishing resistance implies that the magnetic flux $\boldsymbol{B} \cdot \boldsymbol{S}$ through the closed loop may not alter, that is, that the magnetic field inside the material must be maintained both after cooling and after switching off the external field $\boldsymbol{B}_{ext}$. On switching off $\boldsymbol{B}_{ext}$ in the cooled state, this requirement is satisfied because the process of switching off the field induces persistent currents inside the material, which maintain the magnetic field in the interior. If such a conductor were cooled to below T_c in a field-free space ($\boldsymbol{B}_{ext} = \boldsymbol{0}$), and thereafter the external field $\boldsymbol{B}_{ext}$ was switched on, then because $R = 0$ the interior of the material must remain field-free, again due to the effect of the induced persistent current. After switching off $\boldsymbol{B}_{ext}$ in the cooled state the interior of the material now remains field-free. An ideal conductor in the sense of (10.4) could thus, for $\boldsymbol{B}_{ext} = \boldsymbol{0}$ and $T < T_c$, adopt two different states: with and without an internal field, depending on the order of events leading to this state. Hence, if a superconductor were merely such an "ideal conductor", the superconducting state would not be a state in the thermodynamic sense. In fact, in the superconductor not only is $\dot{\boldsymbol{B}} = \boldsymbol{0}$, but also $\boldsymbol{B} = \boldsymbol{0}$, independent of the path by which the state is reached. A superconductor

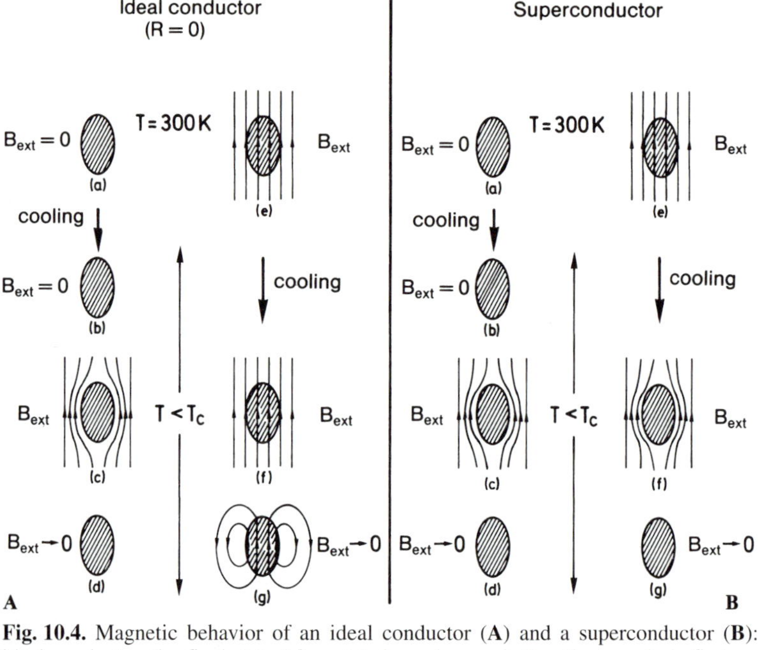

Fig. 10.4. Magnetic behavior of an ideal conductor (**A**) and a superconductor (**B**): (**A**) In an ideal conductor, the final state (*d*) or (*g*) depends on whether the sample is first cooled to below T_c before applying the magnetic field $\boldsymbol{B}_{ext}$, or alternatively, cooled in the presence of the field. ($a \rightarrow b$) The sample loses its resistance when cooled in a field-free region. (*c*) Application of $\boldsymbol{B}_{ext}$ to sample with zero resistance. (*d*) Magnetic field $\boldsymbol{B}_{ext}$ switched off. ($e \rightarrow f$) Sample loses its resistance in the magnetic field. (*g*) Magnetic field $\boldsymbol{B}_{ext}$ switched off. (**B**) For a superconductor, the final states (*d*) and (*g*) are identical, regardless of whether $\boldsymbol{B}_{ext}$ is switched on before or after cooling the sample: ($a \rightarrow b$) sample loses its resistance upon cooling in the absence of a magnetic field. (*c*) Application of the field $\boldsymbol{B}_{ext}$ to the superconducting sample. (*d*) $\boldsymbol{B}_{ext}$ switched off. ($e \rightarrow f$) Sample becomes superconducting in the applied magnetic field $\boldsymbol{B}_{ext}$. (*g*) Magnetic field $\boldsymbol{B}_{ext}$ switched off

behaves as sketched in Fig. 10.4B. This effect, which represents a property independent of the vanishing resistance ($R=0$), is called, after its discoverers, the "Meissner-Ochsenfeld effect" [10.4].

Because of the Meissner-Ochsenfeld effect, the magnetic state of a superconductor can be described as ideal diamagnetism. Persistent surface currents maintain a magnetisation $\boldsymbol{M}=-\boldsymbol{H}_{ext}$ in the interior, and this magnetisation is exactly opposite to the applied magnetic field $\boldsymbol{H}_{ext}$. If an external magnetic field is applied to a superconductor at a temperature $T<T_c$, then a certain amount of energy is consumed to induce the supercurrents which maintain the interior of the sample magnetically field-free. If the magnetic field strength $\boldsymbol{H}_{ext}$ is increased, then at a critical field strength $\boldsymbol{H}_c$ it is energetically more favorable for the material to convert to the normally conducting phase, in which the magnetic field penetrates the material.

This transition is admittedly associated with an increase of the free energy (see the specific heat in Fig. 10.3), but a strong increase of the field

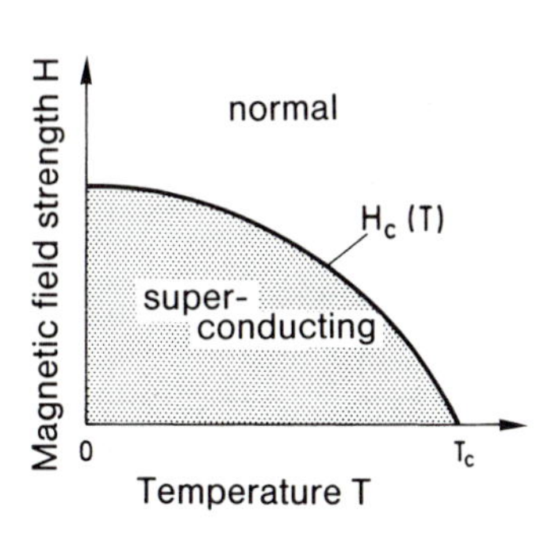

Fig. 10.5. Schematic phase diagram of a superconductor. The phase boundary between superconducting and normally conducting states corresponds to the critical magnetic field $H_c(T)$

$\boldsymbol{H}_{ext}$ would lead to an even higher energy expenditure in inducing the screening currents. There thus exists an upper critical magnetic field strength $\boldsymbol{H}_c$ (which is also temperature dependent), above which the superconducting phase no longer exists. A qualitative phase diagram of a superconductor as a function of temperature T and external magnetic field is shown in Fig. 10.5.

10.2 Phenomenological Description by Means of the London Equations

A purely phenomenological description of superconductivity, particularly the effects described in Sect. 10.1, requires a modification of the conventional equations of electrodynamics. This description, which is not an atomistic explanation of superconductivity, was first suggested by London and London [10.5]. The vanishing resistivity, $\varrho=0$, is included in the classical equations for the motion of an electron in an external field $\mathscr{E}$ (9.42) by neglecting the "friction" term $(m v_D/\tau)$. Thus, for a superconducting electron, one has

$$m\dot{\boldsymbol{v}} = -e\mathscr{E} \ . \tag{10.5}$$

For the current density $\boldsymbol{j}_s = -e n_s \boldsymbol{v}$ of the superconducting electrons (density n_s) the first London equation follows

$$\dot{\boldsymbol{j}}_s = \frac{n_s e^2}{m}\mathscr{E} \ , \tag{10.6}$$

with this, and the Maxwell equation curl $\mathscr{E} = -\dot{\boldsymbol{B}}$, it follows that

$$\frac{\partial}{\partial t}\left(\frac{m}{n_s e^2}\operatorname{curl}\boldsymbol{j}_s + \boldsymbol{B}\right) = \boldsymbol{0} \ . \tag{10.7}$$

This equation describes the behavior of an ideal conductor ($\varrho = 0$), but not the ideal diamagnetism associated with the Meissner-Ochsenfeld effect, i.e. the "expulsion" of a magnetic field. From (10.7), as from (10.4), it merely follows that the magnetic flux through a conducting loop remains constant. Integrating (10.7) gives an integration constant, which, if set to zero, yields the second London equation in which the Meissner-Ochsenfeld effect is correctly described

$$\mathrm{curl}\, \boldsymbol{j}_\mathrm{s} = -\frac{n_\mathrm{s} e^2}{m} \boldsymbol{B} \,. \tag{10.8}$$

It should again be emphasised that (10.8) was derived under two assumptions: (1) that ϱ vanishes and (2) that the integration constant in (10.7) vanishes.

To describe a superconductor in a magnetic field we now have, with the abbreviation,

$$\lambda_\mathrm{L} = \frac{m}{n_\mathrm{s} e^2} \tag{10.9}$$

the following system of equations at our disposal:

the two London equations

$$\mathscr{E} = \lambda_\mathrm{L} \dot{\boldsymbol{j}}_\mathrm{s} \,, \tag{10.10 a}$$

$$\boldsymbol{B} = -\lambda_\mathrm{L}\, \mathrm{curl}\, \boldsymbol{j}_\mathrm{s} \,, \tag{10.10 b}$$

and, in addition, the Maxwell equation

$$\mathrm{curl}\, \boldsymbol{H} = \boldsymbol{j}_\mathrm{s} \quad \text{or} \quad \mathrm{curl}\, \boldsymbol{B} = \mu_0 \boldsymbol{j}_\mathrm{s} \,. \tag{10.11}$$

Equation (10.10) is the superconducting equivalent of Ohm's law for a normal conductor. Combining (10.11) with (10.10) yields

$$\mathrm{curl}\,\mathrm{curl}\, \boldsymbol{B} = \mu_0\, \mathrm{curl}\, \boldsymbol{j}_\mathrm{s} = -\frac{\mu_0}{\lambda_\mathrm{L}} \boldsymbol{B} \tag{10.12 a}$$

$$\mathrm{curl}\,\mathrm{curl}\, \boldsymbol{j}_\mathrm{s} = -\frac{1}{\lambda_\mathrm{L}}\, \mathrm{curl}\, \boldsymbol{B} = -\frac{\mu_0}{\lambda_\mathrm{L}} \boldsymbol{j}_\mathrm{s} \,, \quad \text{or} \tag{10.12 b}$$

$$\Delta \boldsymbol{B} - \frac{\mu_0}{\lambda_\mathrm{L}} \boldsymbol{B} = 0 \,, \tag{10.13 a}$$

$$\Delta \boldsymbol{j}_\mathrm{s} - \frac{\mu_0}{\lambda_\mathrm{L}} \boldsymbol{j}_\mathrm{s} = 0 \,. \tag{10.13 b}$$

For the two-dimensional problem of a semi-infinite superconductor in a vacuum under a homogeneous magnetic field $\boldsymbol{B} = (B_x, 0, 0)$, as sketched in Fig. 10.6, one obtains

$$\frac{\partial^2 B_x}{\partial z^2} - \frac{\mu_0}{\lambda_\mathrm{L}} B_x = 0 \,, \tag{10.14 a}$$

$$\frac{\partial^2 j_{\mathrm{s}y}}{\partial z^2} - \frac{\mu_0}{\lambda_\mathrm{L}} j_{\mathrm{s}y} = 0 \,. \tag{10.14 b}$$

From (10.11) it follows that $\boldsymbol{j}_\mathrm{s}$ has the form $(0, j_{\mathrm{s}y}, 0)$.

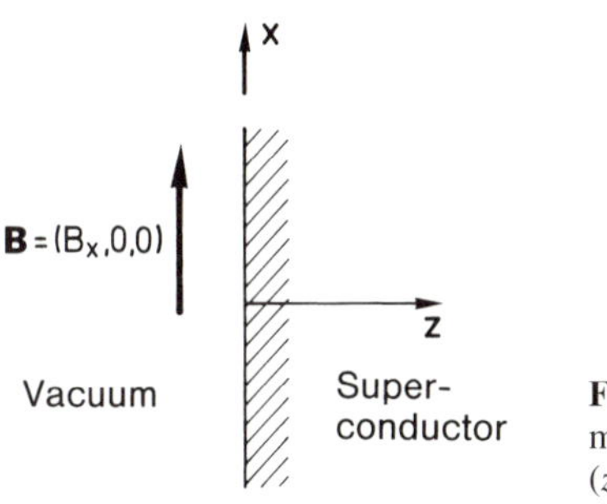

Fig. 10.6. Semi-infinite superconductor ($z>0$) in a magnetic field that is homogeneous in the vacuum ($z<0$)

The solutions of these equations,

$$B_x = B_x^0 \exp\left(-\sqrt{\mu_0/\lambda_L} z\right) = B_x^0 \exp(-z/\Lambda_L) \, , \tag{10.15 a}$$

$$j_{sy} = j_{sy}^0 \exp\left(-\sqrt{\mu_0/\lambda_L} z\right) = j_{sy}^0 \exp(-z/\Lambda_L) \tag{10.15 b}$$

show that the magnetic field actually penetrates the superconductor. However, the field decays exponentially into the interior over the so-called London penetration depth

$$\Lambda_L = \sqrt{m/\mu_0 n_s e^2} \, . \tag{10.16}$$

The solution (10.15b) for j_{sy} indicates that the superconducting currents, which screen the interior of the superconductor against external fields, also decay exponentially with distance into the solid. For an order of magnitude estimate of the London penetration depth, we set m equal to the electron mass and take n_s to be the atomic density, i.e., we assume that each atom provides a superconducting electron. For Sn, for example one obtains $\Lambda_L = 260$ Å. An important element of the microscopic theory of superconductivity is that it is not electrons but electron pairs, so-called Cooper pairs, which are the current carriers. This can be included in the London equations if, instead of n_s, one uses half of the electron density $n_s/2$. Because the London equations describe the Meissner-Ochsenfeld effect particularly well, a microscopic theory of superconductivity should be able to provide, among other things, an equation of the type (10.10b). This, and an explanation of the vanishing resistance at sufficiently low temperature ($T<T_c$), is the central concern of the BCS theory mentioned above. Because superconductivity is evidently a widespread phenomenon, the theory must also be sufficiently general; it should not hinge on one special metallic property. In the following sections the basic aspects of the microscopic theory of superconductivity will be treated in a somewhat simplified form.

10.3 Instability of the "Fermi Sea" and Cooper Pairs

From the previously mentioned fundamental experiments on the phenomenon of superconductivity, it is clear that we are dealing with a new phase of the electron gas in a metal which displays the unusual property of "infinitely high" conductivity. An important contribution to our understanding of this new phase was made by *Cooper* [10.6] who, in 1956, recognized that the ground state (T=0 K) of an electron gas (Sect. 6.2) is unstable if one adds a weak attractive interaction between each pair of electrons. Such an interaction had already been discussed by Fröhlich [10.7] in the form of the phonon-mediated interaction. As it passes through the solid, an electron, on account of its negative charge, leaves behind a deformation trail affecting the positions of the ion cores. This trail is associated with an increased density of positive charge due to the ion cores, and thus has an attractive effect on a second electron. The lattice deformation therefore causes a weak attraction between pairs of electrons (Fig. 10.7a). This attractive electron-electron interaction is retarded because of the slow motion of the ions in comparison with the almost instantaneous Coulomb repulsion between electrons; at the instant when an electron passes, the ions receive a pull which, only after the electron has passed, leads to a displacement and thereby to a polarization of the lattice (Fig. 10.7b). The lattice deformation reaches its maximum at a distance from the first electron which can

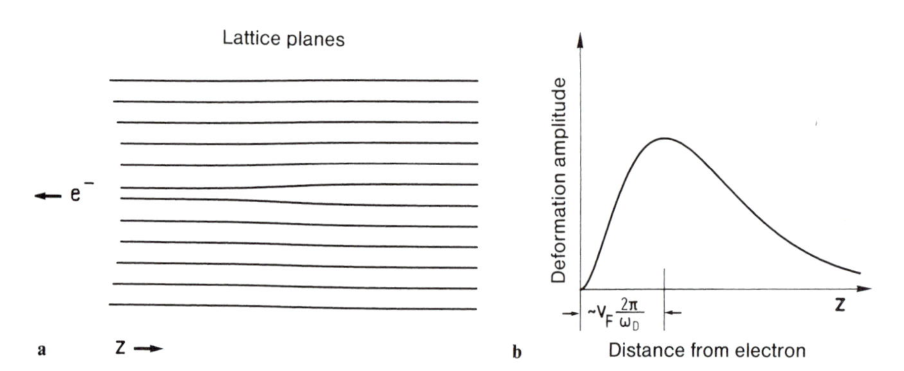

Fig. 10.7a,b. Schematic representation of the phonon-induced electron–electron interaction that leads to the formation of Cooper pairs. (**a**) An electron (e^-) travelling through the crystal lattice leaves behind a deformation trail, which can be regarded as an accumulation of the positively charged ion cores, i.e., as a compression of the lattice planes (exaggerated here for clarity). This means that an area of enhanced positive charge (compared to the otherwise neutral crystal) is created behind the electron, and exerts an attractive force on a second electron. (**b**) Qualitative plot of the displacement of the ion cores as a function of their distance behind the first electron. Compared with the high electron velocity v_F ($\sim 10^8$ cm/s), the lattice follows only very slowly and has its maximum deformation at a distance $v_F 2\pi/\omega_D$ behind the electron, as determined by the typical phonon frequency ω_D (Debye frequency). Thus the coupling of the two electrons into a Cooper pair occurs over distances of more than 1000 Å, at which their Coulomb repulsion is completely screened

be estimated from the electron velocity (Fermi velocity $v_F \sim 10^8$ cm/s) and the phonon vibration period ($2\pi/\omega_D \sim 10^{-13}$ s). The two electrons correlated by the lattice deformation thus have an approximate separation of about 1000 Å. This then corresponds to the "size" of a Cooper pair (which is also estimated in Sect. 10.7 using a different method). The extremely long interaction range of the two electrons correlated by a lattice deformation explains why the Coulomb repulsion is insignificant; it is completely screened out over distances of just a few Ångstroms.

Quantum mechanically, the lattice deformation can be understood as the superposition of phonons which the electron, due to its interaction with the lattice, continuously emits and absorbs. To comply with energy conservation, the phonons constituting the lattice deformation may only exist for a time $\tau = 2\pi/\omega$ determined by the uncertainty relation; thereafter, they must be absorbed. One thus speaks of "virtual" phonons.

The ground state of a non-interacting Fermi gas of electrons in a potential well (Chap. 6) corresponds to the situation where all electron states with wave vector $\boldsymbol{k}$ within the Fermi sphere $[E_F^0(T=0\,\mathrm{K}) = \hbar^2 k_F^2/2m]$ are filled and all states with $E > E_F^0$ are unoccupied. We now perform a *Gedankenexperiment* and add to this system two electrons $[\boldsymbol{k}_1, E(\boldsymbol{k}_1)]$ and $[\boldsymbol{k}_2, E(\boldsymbol{k}_2)]$ in states just above E_F^0. A weak attractive interaction between these two electrons is switched on in the form of phonon exchange. All other electrons in the Fermi sea are assumed to be non-interacting, and, on account of the Pauli exclusion principle, they exclude a further occupation of states with $|\boldsymbol{k}| < k_F$. Due to phonon exchange the two additional electrons continually change their wave vector, whereby, however, momentum must be conserved:

$$\boldsymbol{k}_1 + \boldsymbol{k}_2 = \boldsymbol{k}_1' + \boldsymbol{k}_2' = \boldsymbol{K} \ . \tag{10.17}$$

Since the interaction in $\boldsymbol{k}$-space is restricted to a shell with an energy thickness of $\hbar\omega_D$ (with ω_D=Debye frequency) above E_F^0 the possible $\boldsymbol{k}$-states are given by the shaded area in Fig. 10.8. This area and therefore the number of energy-reducing phonon exchange processes – i.e. the strength of the attractive interaction – is maximum for $\boldsymbol{K} = \boldsymbol{0}$. It is therefore sufficient in what follows to consider the case $\boldsymbol{k}_1 = -\boldsymbol{k}_2 = \boldsymbol{k}$, i.e. electron pairs with equal and opposite wave vectors. The associated two particle wave function $\psi(\boldsymbol{r}_1, \boldsymbol{r}_2)$ must obey the Schrödinger equation

$$\begin{aligned} -\frac{\hbar^2}{2m}(\Delta_1 + \Delta_2)\psi(\boldsymbol{r}_1,\boldsymbol{r}_2) + V(\boldsymbol{r}_1,\boldsymbol{r}_2)\psi(\boldsymbol{r}_1,\boldsymbol{r}_2) \\ = E\psi(\boldsymbol{r}_1,\boldsymbol{r}_2) = (\varepsilon + 2E_F^0)\psi(\boldsymbol{r}_1,\boldsymbol{r}_2) \ . \end{aligned} \tag{10.18}$$

ε is the energy of the electron pair relative to the interaction-free state $(V = 0)$, in which each of the two electrons at the Fermi level would possess an energy $E_F^0 = \hbar^2 k_F^2/2m$. The two-particle function in this case consists of two plane waves

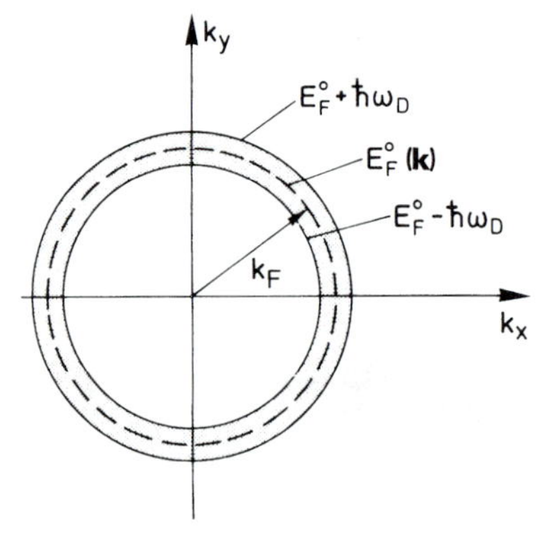

Fig. 10.8. Representation (in reciprocal space) of electron pair collisions for which $\boldsymbol{k}_1+\boldsymbol{k}_2=\boldsymbol{k}_1'+\boldsymbol{k}_2'=\boldsymbol{K}$ remains constant. Two spherical shells with Fermi radius k_F and thickness Δk describe the pairs of wave vectors $\boldsymbol{k}_1$ and $\boldsymbol{k}_2$. All pairs for which $\boldsymbol{k}_1+\boldsymbol{k}_2=\boldsymbol{K}$ end in the shaded volume (rotationally symmetric about $\boldsymbol{K}$). The number of pairs $\boldsymbol{k}_1,\boldsymbol{k}_2$ is proportional to this volume in $\boldsymbol{k}$ space and is maximum for $\boldsymbol{K}=0$

$$\left(\frac{1}{\sqrt{L^3}}e^{i\boldsymbol{k}_1\cdot\boldsymbol{r}_1}\right)\left(\frac{1}{\sqrt{L^3}}e^{i\boldsymbol{k}_2\cdot\boldsymbol{r}_2}\right)=\frac{1}{L^3}e^{i\boldsymbol{k}\cdot(\boldsymbol{r}_1-\boldsymbol{r}_2)}\ . \tag{10.19}$$

We note that (10.19) implies that the two electrons have opposite spin (compare Sect. 8.3). The most general representation of a two-particle state for the case of a non-vanishing interaction ($V\neq 0$) is given by the series

$$\psi(\boldsymbol{r}_1-\boldsymbol{r}_2)=\frac{1}{L^3}\sum_{\boldsymbol{k}} g(\boldsymbol{k})e^{i\boldsymbol{k}\cdot(\boldsymbol{r}_1-\boldsymbol{r}_2)}\ , \tag{10.20}$$

which depends only on the relative coordinate $\boldsymbol{r}=\boldsymbol{r}_1-\boldsymbol{r}_2$. The summation is confined to pairs with $\boldsymbol{k}=\boldsymbol{k}_1=-\boldsymbol{k}_2$, which, because the interaction is restricted to the region $\hbar\omega_D$ (Fig. 10.9), must obey the condition

$$E_F^0<\frac{\hbar^2k^2}{2m}<E_F^0+\hbar\omega_D\ . \tag{10.21 a}$$

In Fig. 10.9, unlike in the restriction (10.21 a) used here, interaction is assumed to take place in a shell of thickness $2\hbar\omega_D$ positioned symmetrically about E_F, as is required in the many-body description of the BCS ground state (Sect. 10.4).

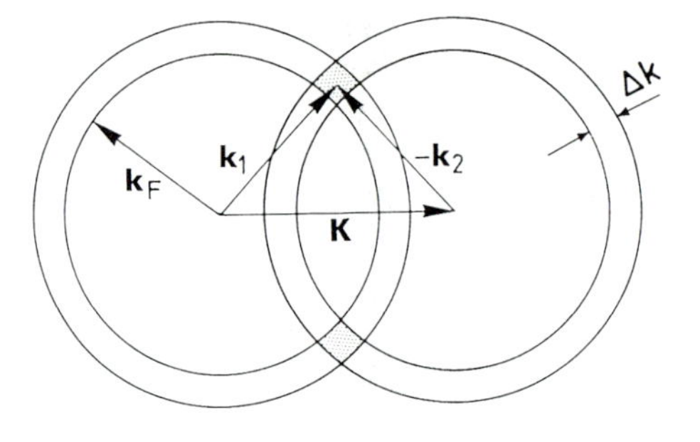

Fig. 10.9. Diagram to illustrate the simplest attractive interaction between two electrons which leads to Cooper pairing within BCS theory. The interaction potential is assumed to be constant ($=-V_0$) in the dashed region of $\boldsymbol{k}$ space, i.e., between the energy shells $E_F^0+\hbar\omega_D$ and $E_F^0-\hbar\omega_D$ (ω_D is the Debye frequency of the material). For the case of one extra Cooper pair considered here only the k-space with $E>E_F$ is of interest

The quantity $|g(\boldsymbol{k})|^2$ is the probability of finding one electron in state $\boldsymbol{k}$ and the other in $-\boldsymbol{k}$, that is, the electron pair in $(\boldsymbol{k},-\boldsymbol{k})$. Due to the Pauli principle and the condition (10.21 a) we have

$$g(\boldsymbol{k}) = 0 \quad \text{for} \quad \begin{cases} k < k_\mathrm{F} \\ k > \sqrt{2m(E_\mathrm{F}^0 + \hbar\omega_\mathrm{D})/\hbar^2} \, . \end{cases} \tag{10.21 b}$$

Inserting (10.20) in (10.18), multiplying by $\exp(-\mathrm{i}\boldsymbol{k}'\cdot\boldsymbol{r})$ and integrating over the normalization volume yields

$$\frac{\hbar^2 k^2}{m} g(\boldsymbol{k}) + \frac{1}{L^3}\sum_{\boldsymbol{k}'} g(\boldsymbol{k}')V_{\boldsymbol{k}\boldsymbol{k}'} = (\varepsilon + 2E_\mathrm{F}^0)g(\boldsymbol{k}) \, . \tag{10.22}$$

The interaction matrix element

$$V_{\boldsymbol{k}\boldsymbol{k}'} = \int V(\boldsymbol{r})\mathrm{e}^{-\mathrm{i}(\boldsymbol{k}-\boldsymbol{k}')\cdot\boldsymbol{r}} d\boldsymbol{r} \tag{10.23}$$

describes scattering of the electron pair from $(\boldsymbol{k},-\boldsymbol{k})$ to $(\boldsymbol{k}',-\boldsymbol{k}')$ and vice versa. In the simplest model this matrix element $V_{\boldsymbol{k}\boldsymbol{k}'}$ is assumed to be independent of $\boldsymbol{k}$ and attractive, that is, $V_{\boldsymbol{k}\boldsymbol{k}'} < 0$:

$$V_{\boldsymbol{k}\boldsymbol{k}'} = \begin{cases} -V_0 \, (V_0 > 0) & \text{for} \quad E_\mathrm{F}^0 < \left(\dfrac{\hbar^2 k^2}{2m}, \dfrac{\hbar^2 k'^2}{2m}\right) < E_\mathrm{F}^0 + \hbar\omega_\mathrm{D} \\ 0 \quad \text{otherwise} \, . \end{cases} \tag{10.24}$$

It thus follows from (10.22) that

$$\left(-\frac{\hbar^2 k^2}{m} + \varepsilon + 2E_\mathrm{F}^0\right) g(\boldsymbol{k}) = -A \, , \quad \text{where} \tag{10.25 a}$$

$$A = \frac{V_0}{L^3}\sum_{\boldsymbol{k}'} g(\boldsymbol{k}') \tag{10.25 b}$$

is independent of $\boldsymbol{k}$.

After summing (10.25 a) over $\boldsymbol{k}$ and comparing with (10.25 b), consistency demands that

$$1 = \frac{V_0}{L^3}\sum_{\boldsymbol{k}} \frac{1}{-\varepsilon + \hbar^2 k^2/m - 2E_\mathrm{F}^0} \, . \tag{10.26}$$

We denote $\xi = \hbar^2 k^2/2m - E_\mathrm{F}^0$ and replace the sum over pair states $\boldsymbol{k}$ by the integral over the $\boldsymbol{k}$ space $(L^{-3}\sum_k \Rightarrow \int d\boldsymbol{k}/4\pi^3)$ and keep in mind that the sum as well as the integral extends only over the manifold of states of *one* spin type. We have already encountered such summations in the context of exchange interaction between electrons (Sect. 8.4). With reference to Sect. 7.5 we split the integral over the entire $\boldsymbol{k}$-space into an integral over the Fermi sphere and the energy (using (7.41) as an expression for the density of states) and obtain from (10.26)

$$1 = V_0 \frac{1}{(2\pi)^3} \iint \frac{dS_E}{|\mathrm{grad}_k E(\boldsymbol{k})|} \frac{dE}{2E - \varepsilon - 2E_F^0} \quad \text{with } E = \frac{\hbar^2 k^2}{2m} . \tag{10.27a}$$

Since the integral over the energy extends only over the narrow interval between E_F^0 and $E_F^0 + \hbar\omega_D$, the density of the free electron gas (7.41)

$$D(E_F^0 + \xi) = \frac{(2m)^{3/2}}{2\pi^2\hbar^3} (E_F^0 + \xi)^{1/2} \approx D(E_F^0) \tag{10.27b}$$

can be considered as a constant $[\approx D(Ea_F^0)]$. Since we integrate over pair states $(\boldsymbol{k}, -\boldsymbol{k})$, half the density of states $Z(E_F^0) = D(E_F^0)/2$ at the Fermi energy has to be taken. From (10.26) and (10.27) one then obtains:

$$1 = V_0 Z(E_F^0) \int_0^{\hbar\omega_D} \frac{1}{2\xi - \varepsilon} d\xi . \tag{10.28}$$

By performing the integration we obtain

$$1 = \frac{1}{2} V_0 Z(E_F^0) \ln \frac{\varepsilon - 2\hbar\omega_D}{\varepsilon} \quad \text{or} \tag{10.29a}$$

$$\varepsilon = \frac{2\hbar\omega_D}{1 - \exp[2/V_0 Z(E_F^0)]} . \tag{10.29b}$$

For the case of a weak interaction, $V_0 Z(E_F^0) \ll 1$, it follows that

$$\varepsilon \approx -2\hbar\omega_D \mathrm{e}^{-2/V_0 Z(E_F^0)} . \tag{10.30}$$

There thus exists a two-electron bound state, whose energy is lower than that of the fully occupied Fermi sea ($T=0$) by an amount $\varepsilon = E - 2E_F^0 < 0$. The ground state of the non-interacting free electron gas, as treated in Sect. 6.2, becomes unstable when a minute attractive interaction between electrons is "switched on". It should be noted that the energy reduction ε (10.30) results from a *Gedankenexperiment* in which the Fermi-sea for states with $\hbar^2 k^2/2m < E_F^0$ is assumed to be fixed and only the effect of the attraction between two additional electrons in the presence of the Fermi-sea is treated. In reality the instability leads to the formation of a high density of such electron pairs, so called *Cooper pairs* $(\boldsymbol{k}, -\boldsymbol{k})$, via which the system tries to achieve a new lower-energy ground state. This new ground state is identical to the superconducting phase, as we shall see.

In the aforegoing treatment it was important that the Pauli principle applies to both electrons. The two-particle wavefunction (10.20) was symmetric in spatial coordinates $(\boldsymbol{r}_1, \boldsymbol{r}_2)$ under exchange of electrons 1 and 2, but the whole wavefunction including spins must be antisymmetric (the most general formulation of the Pauli principle). The spin part of the wave-function, not indicated in (10.20), must therefore be antisymmetric. The Cooper pair therefore comprises two electrons with opposite wave vectors and opposite spins $(\boldsymbol{k}\uparrow, -\boldsymbol{k}\downarrow)$. In this connection one often speaks of singlet pairs. It should be noted that a more complicated electron-electron coupling could lead to parallel spin pairs, so-called triplets. Models of this type have been discussed, but experimental proof of the existence of such states has not yet been found. Triplet pairs have however been found in liquid ^{3}He. At low temperatures this system behaves like a degenerate Fermi gas.

10.4 The BCS Ground State

In Sect. 10.3 we saw how a weak attractive interaction, resulting from electron-phonon interaction, leads to the formation of "Cooper pairs". The energy reduction of the Fermi sea due to a single pair was calculated in (10.29). Such a Cooper pair must be imagined as an electron pair in which the two electrons always occupy states $(\boldsymbol{k}\uparrow, -\boldsymbol{k}\downarrow)$, $(\boldsymbol{k}'\uparrow, -\boldsymbol{k}'\downarrow)$, and so on, with opposed $\boldsymbol{k}$-vectors and spin. The scattering of the pair from $(\boldsymbol{k}\uparrow, -\boldsymbol{k}\downarrow)$ to $(\boldsymbol{k}'\uparrow, -\boldsymbol{k}'\downarrow)$ mediated by $V_{\boldsymbol{k}\boldsymbol{k}'}$ leads to an energy reduction on the formation of a Cooper pair (Fig. 10.10). Due to this energy reduction ever more Cooper pairs are formed. The new ground state of the Fermi sea after pair formation is achieved through a complicated interaction between the electrons. The total energy reduction is not obtained by simply summing the contributions (10.29) of single Cooper pairs. The effect of each single Cooper pair depends on those already present. One must thus seek the minimum total energy of the whole system for all possible pair configurations, taking into account the kinetic one-electron component and the energy reduction due to "pair collisions", i.e., the electron-phonon interaction. Since an excitation above E_{F}^0 is necessary, the pairing is associated with an increase in kinetic energy. The kinetic component can be given immediately: If $w_{\boldsymbol{k}}$ is the probability that the pair state $(\boldsymbol{k}\uparrow, -\boldsymbol{k}\downarrow)$ is occupied, the kinetic component E_{kin} is

$$E_{\mathrm{kin}} = 2\sum_{\boldsymbol{k}} w_{\boldsymbol{k}}\xi_{\boldsymbol{k}} , \quad \text{with} \quad \xi_{\boldsymbol{k}} = E(\boldsymbol{k}) - E_{\mathrm{F}}^0 . \tag{10.31}$$

The total energy reduction due to the pair collisions $(\boldsymbol{k}\uparrow, -\boldsymbol{k}\downarrow) \rightleftarrows (\boldsymbol{k}'\uparrow, -\boldsymbol{k}'\downarrow)$ can be most easily calculated via the Hamiltonian $\mathscr{H}$, which explicitly takes account of the fact that the "annihilation" of a pair $(\boldsymbol{k}\uparrow, -\boldsymbol{k}\downarrow)$ and the "simultaneous creation" of a pair $(\boldsymbol{k}'\uparrow, -\boldsymbol{k}'\downarrow)$, i.e. a

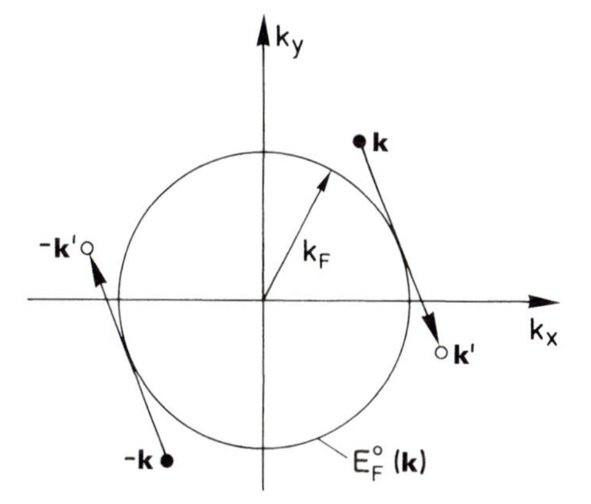

Fig. 10.10. Representation in $\boldsymbol{k}$ space of the scattering of an electron pair with wave vectors $(\boldsymbol{k}, -\boldsymbol{k})$ to the state $(\boldsymbol{k}', -\boldsymbol{k}')$

scattering from $(\boldsymbol{k}\uparrow, -\boldsymbol{k}\downarrow)$ to $(\boldsymbol{k}'\uparrow, -\boldsymbol{k}'\downarrow)$, leads to an energy reduction of $V_{kk'}$ (Sect. 10.3 and Fig. 10.10). Since a pair state $\boldsymbol{k}$ can be either occupied or unoccupied, we choose a representation consisting of two orthogonal states $|1\rangle_k$ and $|0\rangle_k$, where $|1\rangle_k$ is the state in which $(\boldsymbol{k}\uparrow, -\boldsymbol{k}\downarrow)$ is occupied, and $|0\rangle_k$ is the corresponding unoccupied state. The most general state of the pair $(\boldsymbol{k}\uparrow, -\boldsymbol{k}\downarrow)$ is thus given by

$$|\psi_k\rangle = u_k|0\rangle_k + v_k|1\rangle_k \; . \tag{10.32}$$

This is an alternative representation of the Cooper pair wavefunction (10.19). Hence $w_k = v_k^2$ and $1-w_k = u_k^2$ are the probabilities that the pair state is occupied and unoccupied, respectively. We shall assume that the probability amplitudes v_k and u_k are real. It can be shown in a more rigorous theoretical treatment that this restriction is not important. In the representation (10.32) the ground state of the many-body system of all Cooper pairs can be approximated by the product of the state vectors of the single pairs

$$|\phi_{\mathrm{BCS}}\rangle \simeq \prod_k (u_k|0\rangle_k + v_k|1\rangle_k) \; . \tag{10.33}$$

This approximation amounts to a description of the many-body state in terms of non-interacting pairs, i.e., interactions between the pairs are neglected in the state vector. In the two-dimensional representation

$$|1\rangle_k = \begin{pmatrix} 1 \\ 0 \end{pmatrix}_k , \quad |0\rangle_k = \begin{pmatrix} 0 \\ 1 \end{pmatrix}_k \tag{10.34}$$

one can use the Pauli matrices

$$\sigma_k^{(1)} = \begin{pmatrix} 0 & 1 \\ 1 & 0 \end{pmatrix}_k , \quad \sigma_k^{(2)} = \begin{pmatrix} 0 & -\mathrm{i} \\ \mathrm{i} & 0 \end{pmatrix}_k \tag{10.35}$$

to describe the "creation" or "annihilation" of a Cooper pair:

The operator

$$\sigma_k^+ = \tfrac{1}{2}(\sigma_k^{(1)} + \mathrm{i}\,\sigma_k^{(2)}) \tag{10.36 a}$$

transforms the unoccupied state $|0\rangle_k$. into the occupied state $|1\rangle_k$, while

$$\sigma_k^- = \tfrac{1}{2}(\sigma_k^{(1)} - \mathrm{i}\,\sigma_k^{(2)}) \tag{10.36 b}$$

transforms the state $|1\rangle_k$ into $|0\rangle_k$ From the representations (10.34–10.36) one can deduce the following properties:

$$\sigma_k^+|1\rangle_k = 0 \, , \quad \sigma_k^+|0\rangle_k = |1\rangle_k \, , \tag{10.37 a}$$

$$\sigma_k^-|1\rangle_k = |0\rangle_k \, , \quad \sigma_k^-|0\rangle_k = 0 \; . \tag{10.37 b}$$

The matrices σ_k^+ and σ_k^- are formally identical to the spin operators introduced in Sect. 8.7. Their physical interpretation as "creator" and "annihilator" of Cooper pairs is however completely different from that in Sect. 8.7, where the reversal of a spin was described.

Scattering from $(\boldsymbol{k}\uparrow, -\boldsymbol{k}\downarrow)$ to $(\boldsymbol{k}'\uparrow, -\boldsymbol{k}'\downarrow)$ is associated with an energy reduction by an amount $V_{\boldsymbol{k}\boldsymbol{k}'}$. In the simple BCS model of superconductivity (as in Sect. 10.3) this interaction matrix element $V_{\boldsymbol{k}\boldsymbol{k}'}$ is assumed to be independent of $\boldsymbol{k},\boldsymbol{k}'$, i.e. constant. We relate this to the normalization volume of the crystal, L^3, by setting it equal to V_0/L^3. The scattering process is described in the two-dimensional representation as annihilation of $\boldsymbol{k}$ and creation of $\boldsymbol{k}'$. The operator that describes the corresponding energy reduction is thus found to be $-(V_0/L^3)\sigma_{k'}^+\sigma_k^-$. The total energy reduction due to pair collisions $\boldsymbol{k} \to \boldsymbol{k}'$ and $\boldsymbol{k}' \to \boldsymbol{k}$ is given by summing over all collisions and may be expressed in operator terminology as

$$\mathscr{H} = -\frac{1}{L^3} V_0 \sum_{kk'} \frac{1}{2}(\sigma_{k'}^+\sigma_k^- + \sigma_k^+\sigma_{k'}^-) = -\frac{V_0}{L^3}\sum_{kk'} \sigma_k^+\sigma_{k'}^- \ . \qquad (10.38)$$

Since V_0 is restricted to the shell $\pm\hbar\omega_{\mathrm{D}}$ around E_{F}^0, the sum over $\boldsymbol{k},\boldsymbol{k}'$ likewise includes only pair states in this shell. In the sum, scattering in both directions is considered; the right-hand side of (10.38) follows when one exchanges the indices.

The energy reduction due to collisions is given from perturbation theory as the expectation value of the operator $\mathscr{H}$ (10.38) in the state $|\phi_{\mathrm{BCS}}\rangle$ (10.33) of the many-body system

$$\langle\phi_{\mathrm{BCS}}|\mathscr{H}|\phi_{\mathrm{BCS}}\rangle = -\frac{V_0}{L^3}\left[\prod_p (u_{pp}\langle 0| + v_{pp}\langle 1|) \sum_{kk'} \sigma_k^+\sigma_{k'}^- \prod_q (u_q|0\rangle_q + v_q|1\rangle_q)\right] . \qquad (10.39)$$

In evaluating (10.39), one should note that the operator σ_k^+ (σ_k^-) acts only on the state $|1\rangle_k$ $(|0\rangle_k)$. The explicit rules are given in (10.37). Furthermore, from (10.34) we have the orthonormality relations

$${}_k\langle 1|1\rangle_k = 1 \ , \quad {}_k\langle 0|0\rangle_k = 1 \ , \quad {}_k\langle 1|0\rangle_k = 0 \ . \qquad (10.40)$$

We thus obtain

$$\langle\phi_{\mathrm{BCS}}|\mathscr{H}|\phi_{\mathrm{BCS}}\rangle = -\frac{V_0}{L^3}\sum_{kk'} v_k u_{k'} u_k v_{k'} \ . \qquad (10.41)$$

According to (10.31, 10.41) the total energy of the system of Cooper pairs can therefore be represented as

$$W_{\mathrm{BCS}} = 2\sum_k v_k^2 \xi_k - \frac{V_0}{L^3}\sum_{kk'} v_k u_k v_{k'} u_{k'} \ . \qquad (10.42)$$

The BCS ground state at $T=0\,\mathrm{K}$ of the system of Cooper pairs is given by the minimum, W_{BCS}^0, of the energy density W_{BCS}. By minimizing (10.42) as a function of the probability amplitudes u_k and v_k we obtain the energy of the ground state W_{BCS}^0 and the occupation and non-occupation probabilities $w_k = v_k^2$ and $(1-w_k) = u_k^2$. Because of the relationship between v_k and u_k, the calculation is greatly simplified by setting

$$v_k = \sqrt{w_k} = \cos\theta_k \ , \tag{10.43 a}$$

$$u_k = \sqrt{1 - w_k} = \sin\theta_k \ , \tag{10.43 b}$$

which guarantees that

$$u_k^2 + v_k^2 = \cos^2\theta_k + \sin^2\theta_k = 1 \ . \tag{10.43 c}$$

The minimizing is then with respect to θ_k

The quantity to be minimized can be written

$$W_{\mathrm{BCS}} = \sum_k 2\xi_k \cos^2\theta_k - \frac{V_0}{L^3}\sum_{kk'} \cos\theta_k \sin\theta_{k'} \cos\theta_{k'} \sin\theta_k$$

$$= \sum_k 2\xi_k \cos^2\theta_k - \frac{1}{4}\frac{V_0}{L^3}\sum_{kk'} \sin 2\theta_k \sin 2\theta_{k'} \ . \tag{10.44}$$

The condition for the minimum of W_{BCS} then reads

$$\frac{\partial W_{\mathrm{BCS}}}{\partial\theta_k} = -2\xi_k \sin 2\theta_k - \frac{V_0}{L^3}\sum_{k'} \cos 2\theta_k \sin 2\theta_{k'} = 0 \ , \quad \text{or} \tag{10.45 a}$$

$$\xi_k \tan 2\theta_k = -\frac{1}{2}\frac{V_0}{L^3}\sum_{k'} \sin 2\theta_{k'} \ . \tag{10.45 b}$$

We let

$$\Delta = \frac{V_0}{L^3}\sum_{k'} u_{k'} v_{k'} = \frac{V_0}{L^3}\sum_{k'} \sin\theta_{k'} \cos\theta_{k'} \ , \tag{10.46}$$

$$E_k = \sqrt{\xi_k^2 + \Delta^2} \tag{10.47}$$

and obtain from standard trigonometry

$$\frac{\sin 2\theta_k}{\cos 2\theta_k} = \tan 2\theta_k = -\Delta/\xi_k \ , \tag{10.48}$$

$$2u_k v_k = \sin 2\theta_k = \Delta/E_k \ , \tag{10.49}$$

$$v_k^2 - u_k^2 = -\xi_k/E_k \ . \tag{10.50}$$

Thus the occupation probability, $w_k = v_k^2$, of a pair state $(\boldsymbol{k}\uparrow, -\boldsymbol{k}\downarrow)$ in the BCS ground state at $T=0$ K is given by

$$w_k = v_k^2 = \frac{1}{2}\left(1 - \frac{\xi_k}{E_k}\right) = \frac{1}{2}\left(1 - \frac{\xi_k}{\sqrt{\xi_k^2 + \Delta^2}}\right) \ . \tag{10.51}$$

This function is plotted in Fig. 10.11. At $T=0$ K(!) it has a form similar to the Fermi function at finite temperature and a more exact analysis shows that it is like the Fermi distribution at the finite critical temperature T_{c}. It should be

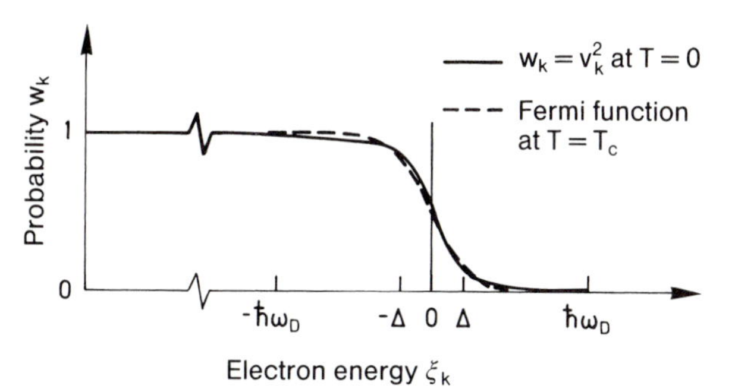

Fig. 10.11. The BCS occupation probability v_k^2 for Cooper pairs in the vicinity of the Fermi energy E_F^0. The energy is given as $\xi_k = E(k) - E_F^0$, i.e., the Fermi energy ($\xi_k = 0$) serves as a reference point. Also shown for comparison is the Fermi-Dirac distribution function for normally conducting electrons at the critical temperature T_c (*dashed line*). The curves are related to one another by the BCS relationship between $\Delta(0)$ and T_c (10.67)

noted that this form of v_k^2 results from the representation in terms of single-particle states with well-defined quantum numbers $\boldsymbol{k}$. This representation is not particularly appropriate for the many-body problem; for example, one does not recognize the energy gap in the excitation spectrum of the superconductor (see below). On the other hand, it can be seen in the behavior of v_k^2 that the Cooper pairs which contribute to the energy reduction of the ground state are constructed from one-particle wavefunctions from a particular $\boldsymbol{k}$-region, corresponding to an energy shell of $\pm\Delta$ around the Fermi surface.

The energy of the superconducting BCS ground state W_{BCS}^0 is obtained by inserting in W_{BCS} – (10.42 or 10.44) – relationships (10.48–10.51) which follow from the minimization. The result is

$$W_{\mathrm{BCS}}^0 = \sum_k \xi_k (1 - \xi_k / E_k) - L^3 \frac{\Delta^2}{V_0} \,. \tag{10.52 a}$$

The condensation energy of the superconducting phase is obtained by subtracting from W_{BCS}^0 the energy of the normal conducting phase, i.e., the energy of the Fermi-sea without the attractive interaction $W_{\mathrm{n}}^0 = \sum_{|k|<k_{\mathrm{F}}} 2\xi_k$. The calculation can be carried out by using (10.51, 10.52 a) to express the energy of the superconducting ground state as

$$\begin{aligned} W_{\mathrm{BCS}}^0 &= 2\sum_{k<k_{\mathrm{F}}} (1 - u_k^2)\xi_k + 2\sum_{k>k_{\mathrm{F}}} v_k^2 \xi_k - L^3 \frac{\Delta^2}{V_0} \\ &= 2\sum_{k<k_{\mathrm{F}}} \xi_k - 2\sum_{k<k_{\mathrm{F}}} u_k^2 \xi_k + 2\sum_{k>k_{\mathrm{F}}} v_k^2 \xi_k - L^3 \frac{\Delta^2}{V_0} \,. \end{aligned} \tag{10.52 b}$$

Here the first term is exactly $\boldsymbol{W}_{\mathrm{n}}^0$. Going from the sum in $\boldsymbol{k}$-space to an integral $(L^{-3}\sum_k \Rightarrow \int d\boldsymbol{k}/4\pi^3)$, after integration and using (10.46–10.50), one obtains, with some calculation (Problem 10.6),

$$(W^0_{\rm BCS} - W^0_{\rm n})/L^3 = -\tfrac{1}{2}Z(E^0_{\rm F})\Delta^2 \ . \tag{10.53}$$

Thus, for finite Δ, there is always a reduction in energy for the superconducting state, whereby Δ is a measure of the size of the reduction. One can visualize (10.53) by imagining that $Z(E^0_{\rm F})\Delta$ electron pairs per unit volume from the energy region Δ below the Fermi level all "condense" into a state at exactly Δ below $E^0_{\rm F}$. Their average gain in energy is thus $\Delta/2$.

The decisive role of the parameter Δ is clear from the following: The first excitation state above the BCS ground state involves the breaking up of a Cooper pair due to an external influence. Here an electron is scattered out of $(\boldsymbol{k}\uparrow)$ leaving behind an unpaired electron in $(-\boldsymbol{k}\downarrow)$. In order to calculate the necessary excitation energy, we rewrite the ground state energy $W^0_{\rm BCS}$ (10.52) as follows

$$\begin{aligned} W^0_{\rm BCS} &= \sum_{\boldsymbol{k}} \xi_{\boldsymbol{k}}(1-\xi_{\boldsymbol{k}}/E_{\boldsymbol{k}}) - \frac{L^3\Delta^2}{V_0} \\ &= \sum_{\boldsymbol{k}} E_{\boldsymbol{k}}(u^2_{\boldsymbol{k}} - v^2_{\boldsymbol{k}}) - \sum_{\boldsymbol{k}} E_{\boldsymbol{k}}(u^2_{\boldsymbol{k}} - v^2_{\boldsymbol{k}})^2 - \frac{L^3\Delta^2}{V_0} \\ &= 2\sum_{\boldsymbol{k}} E_{\boldsymbol{k}}u^2_{\boldsymbol{k}}v^2_{\boldsymbol{k}} + \sum_{\boldsymbol{k}} E_{\boldsymbol{k}}[u^2_{\boldsymbol{k}}(1-u^2_{\boldsymbol{k}}) - v^2_{\boldsymbol{k}}(1+v^2_{\boldsymbol{k}})] - \frac{L^3\Delta^2}{V_0} \\ &= \Delta\sum_{\boldsymbol{k}} u_{\boldsymbol{k}}v_{\boldsymbol{k}} - \frac{L^3\Delta^2}{V_0} + \sum_{\boldsymbol{k}} E_{\boldsymbol{k}}v^2_{\boldsymbol{k}}(u^2_{\boldsymbol{k}} - 1 - v^2_{\boldsymbol{k}}) \\ &= -2\sum_{\boldsymbol{k}} E_{\boldsymbol{k}}v^4_{\boldsymbol{k}} \ . \end{aligned} \tag{10.54}$$

If $(\boldsymbol{k}'\uparrow, -\boldsymbol{k}'\downarrow)$ is occupied, i.e. $v^2_{\boldsymbol{k}'}=1$, then the first excited state $W^1_{\rm BCS}$ is achieved by breaking up the pair, i.e., $v^2_{\boldsymbol{k}'}=0$, and therefore

$$W^1_{\rm BCS} = -2\sum_{\boldsymbol{k}\neq\boldsymbol{k}'} E_{\boldsymbol{k}}v^4_{\boldsymbol{k}} \ . \tag{10.55}$$

The necessary excitation energy is the difference between the energies of the initial and final states

$$\Delta E = W^1_{\rm BCS} - W^0_{\rm BCS} = 2E_{\boldsymbol{k}'} = 2\sqrt{\xi^2_{\boldsymbol{k}'} + \Delta^2} \ . \tag{10.56}$$

The first term in the square-root, $\xi^2_{\boldsymbol{k}'}$, describes the kinetic energy of the two electrons "scattered" out of the Cooper pair. Since $\xi_{\boldsymbol{k}'}=\hbar^2k'^2/2m-E^0_{\rm F}$, this can be arbitrarily small, i.e., the excitation requires a minimum finite energy

$$\Delta E_{\rm min} = 2\Delta \ . \tag{10.57}$$

The excitation spectrum of the superconducting state contains a gap of 2Δ, which corresponds to the energy required to break up a Cooper pair. Equation (10.56) describes the excitation energy of the two electrons that result from the destruction of a Cooper pair. If we imagine that a single electron is added to the BCS ground state, then it can naturally find no partner for Cooper pairing. Which energy state can this electron occupy? From (10.56) we conclude that the possible states of this excited system are given by $E_k=(\xi_k^2+\Delta^2)^{1/2}$. If the unpaired electron is at $\xi_k=0$ it thus has an energy which lies at least Δ above the BCS ground state (Fig. 10.12). However it can also occupy states with finite ξ_k; for $\xi_k^2 \gg \Delta^2$ the one-electron energy levels

$$E_k = \sqrt{\xi_k^2 + \Delta^2} \approx \xi_k = \frac{\hbar^2 k^2}{2m} - E_F^0 \tag{10.58}$$

become occupied. These are exactly the levels of the free electron gas (of a normal conductor). Thus, for energies well above the Fermi energy ($\xi_k^2 \gg \Delta^2$), the continuum of states of a normal conductor results. To compare the density of states in the energy range Δ about the Fermi level for excited electrons $D_s(E_k)$ in a superconductor with that of a normal conductor $D_n(\xi_k)$ (Sect. 6.1), we note that in the phase transition no states are lost, i.e.

$$D_s(E_k)dE_k = D_n(\xi_k)d\xi_k \,. \tag{10.59 a}$$

Because we are only interested in the immediate region Δ around E_F^0, it is sufficient to assume that $D_n(\xi_k) \approx D_n(E_F^0)$=const. According to (10.56) it then follows that

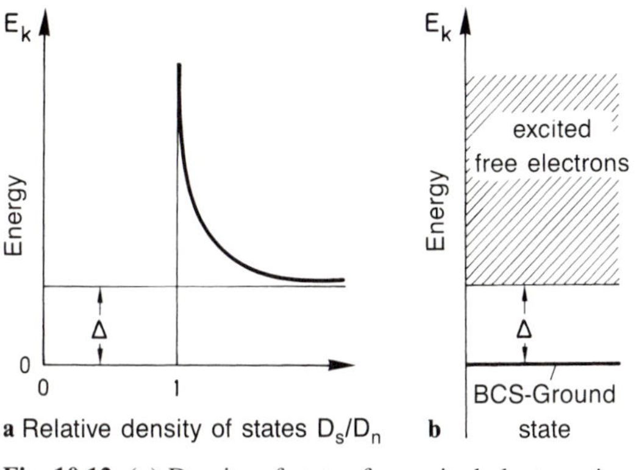

Fig. 10.12. (**a**) Density of states for excited electrons in a superconductor D_s relative to that of a normal conductor. $E_k=0$ corresponds to the Fermi energy E_F^0. (**b**) Simplified representation of the excitation spectrum of a superconductor on the basis of one-electron energies E_k. In the BCS ground state, the one-electron picture collapses: At $T=0$ K all Cooper pairs occupy one and the same ground state (like Bosons). This state is therefore energetically identical to the chemical potential, i.e., the Fermi energy E_F^0. The energy level drawn here can thus be formally interpreted as the "many-body energy of one electron" (total energy of all particles divided by the number of electrons). Note that a minimum energy of 2Δ is necessary to split up a Cooper pair

$$D_s(E_k)/D_n(E_F^0) = \frac{d\xi_k}{dE_k} = \begin{cases} \dfrac{E_k}{\sqrt{E_k^2 - \Delta^2}} & \text{for} \quad E_k > \Delta \\ 0 & \text{for} \quad E_k < \Delta \,. \end{cases} \tag{10.59 b}$$

This function possesses a pole at Δ and for $E_k \gg \Delta$ converts, as expected, to the density of states of a normal conductor; it is depicted in Fig. 10.12b.

It should again be emphasized that the illustration in Fig. 10.12 does not express the fact that the breaking up of a Cooper pair requires a minimum energy 2Δ. It says only that the "addition" of an unpaired electron to the BCS ground state makes possible the occupation of one-particle states which lie at least Δ above the BCS ground state energy (for one electron). The density of states in the vicinity of the minimum energy single-particle states is singular (Fig. 10.12b).

The "addition" of electrons to the BCS ground state can be realized in an experiment by the injection of electrons via an insulating tunnel barrier (Panel IX). Such tunnel experiments are today very common in superconductor research. They may be conveniently interpreted with reference to diagrams such as Fig. 10.12b.

We now wish to determine the gap Δ (or 2Δ) in the excitation spectrum. For this we combine (10.49) with (10.46, 10.47) and obtain

$$\Delta = \frac{1}{2}\frac{V_0}{L^3}\sum_k \frac{\Delta}{E_k} = \frac{1}{2}\frac{V_0}{L^3}\sum_k \frac{\Delta}{\sqrt{\xi_k^2 + \Delta^2}} \,. \tag{10.60}$$

As in (10.26, 10.28), the sum in $\boldsymbol{k}$-space is replaced by an integral ($L^{-3}\sum_k \Rightarrow \int d\boldsymbol{k}/4\pi^3$). We note that we are summing again over pair states, i.e., that instead of the one-particle density of states $D(E_F^0+\xi)$ we must take the pair density of states $Z(E_F^0 + \xi) = \frac{1}{2}D(E_F^0 + \xi)$.

Furthermore, in contrast to Sect. 10.3, the sum is taken over a spherical shell $\pm\hbar\omega_D$ located symmetrically around E_F^0. We then have

$$1 = \frac{V_0}{2}\int_{-\hbar\omega_D}^{\hbar\omega_D} \frac{Z(E_F^0 + \xi)}{\sqrt{\xi^2 + \Delta^2}}\,d\xi \,. \tag{10.61 a}$$

In the region $[E_F^0 - \hbar\omega_D, E_F^0 + \hbar\omega_D]$ where V_0 does not vanish, $Z(E_F^0+\xi)$ varies only slightly, and, due to the symmetry about E_F^0, it follows that

$$\frac{1}{V_0 Z(E_F^0)} = \int_0^{\hbar\omega_D} \frac{d\xi}{\sqrt{\xi^2 + \Delta^2}} \,, \quad \text{or} \tag{10.61 b}$$

$$\frac{1}{V_0 Z(E_F^0)} = \operatorname{arc\,sinh} \frac{\hbar\omega_D}{\Delta} \,. \tag{10.62}$$

In the case of a weak interaction, i.e., $V_0 Z(E_F^0) \ll 1$, the gap energy is thus

$$\Delta = \frac{\hbar\omega_D}{\sinh[1/V_0 Z(E_F^0)]} \approx 2\hbar\omega_D e^{-1/V_0 Z(E_F^0)} . \tag{10.63}$$

This result bears a noticeable similarity to (10.30), i.e., to the binding energy ε of two electrons in a Cooper pair in the presence of a fully occupied sea. As in (10.30), one sees that even a very small attractive interaction, i.e., a very small positive V_0, results in a finite gap energy Δ, but that Δ cannot be expanded in a series for small V_0. A perturbation calculation would thus be unable to provide the result (10.63). For the sake of completeness, it should also be mentioned that superconductors have now been discovered that have a vanishingly small gap energy.

10.5 The Excitation Spectrum of a Superconductor

The theoretical description of modern devices based on superconductivity requires an advanced level of understanding of the transport properties that goes beyond the heuristic London theory (Sect. 10.2) and the so-called two-fluid model in which a normal current is shunted by a supercurrent made up of Cooper pairs. The theoretical description of superconductor devices and electronics requires an understanding of non-equilibrium states. Because of the strong electron–electron correlation as the definitive property, the transport theory for a superconductor is significantly more complex than for a normal conductor. The single-electron approximation that proved so successful for the description of transport phenomena in normal conductors (Chap. 9) fails grossly for superconductors. The purpose of this section is to introduce the reader to an essential ingredient of any transport theory that is a correct description of the spectrum of excited states near the Fermi level: Transport involves a non-equilibrium state of matter and thus always requires excitations above the ground state (here, of the Cooper-pair condensate). The excitations of a superconductor can be described as "Quasi-particles", just as for other highly correlated electron systems such as "heavy Fermion materials" and "Fermi liquids".

By way of introduction we compare the occupation statistics of a normal conductor (metal) with that of a superconductor at low temperature $T \simeq 0$ (Fig. 10.13). In a normal conductor the Fermi occupation probability function $f(\boldsymbol{k})$ is a step function, which extends symmetrically along a k-axis in reciprocal space around zero (Γ-point) up to the Fermi radius k_F and $-k_F$, respectively. The probability $w_{\boldsymbol{k}} = v_{\boldsymbol{k}}^2$ for the occupation of a Cooper-pair state $(\boldsymbol{k}\uparrow, -\boldsymbol{k}\downarrow)$, on the other hand, is a function that is "washed out" at $\pm \boldsymbol{k}_F$, even at $T=0$. There are still unoccupied electronic states slightly below $|k_F|$ at $T=0$ since $w_{\boldsymbol{k}}(T=0)<1$ for $|\boldsymbol{k}|<k_F$ (Figs. 10.11, 10.13b). These states can be occupied both with Cooper pairs and with single electrons. Empty states slightly below $|k_F|$ at $T=0$ are necessary in order to allow electron-scattering processes for Cooper pairing that decreases the total energy of the system.

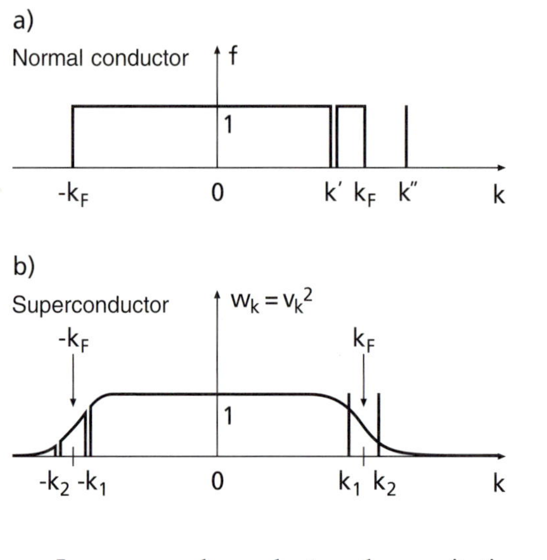

Fig. 10.13. (**a**) Occupation probability $f(k)$ for electrons in a normal conductor at $T=0$ K (Fermi distribution). Electron hole pair excitations are indicated at k' and k''; $\pm k_F$ are the Fermi wave vectors. (**b**) Occupation probability w_k for Cooper pairs ($\boldsymbol{k}\uparrow$, $-\boldsymbol{k}\downarrow$) in a superconductor at $T=0$ K. Excitations involving the occupation of empty states by electrons at k_1 and k_2 and the non-occupation at $-k_1$ and $-k_2$ are qualitatively indicated

In a normal conductor the excitation of an electron from a state $E'(k')<E_F$ into an unoccupied state $E''(k'')$ is described as (see Figs. 10.13 a, 10.14 a).

$$\Delta E = E'' - E' = (E'' - E_F) + (E_F - E') . \tag{10.64}$$

The excitation energy ΔE can formally be separated into a part $(E''-E_F)$, which corresponds to the occupation of E'' above E_F, and a part (E_F-E')

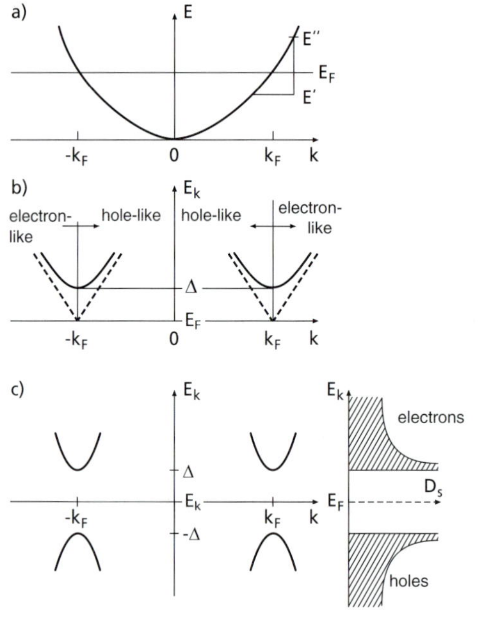

Fig. 10.14. (**a**) Dispersion of electron and hole states in a normal conductor. The excitation of an electron from E' to E'' can be described by the occupation of E'' and the formation of a hole at E'. (**b**) Dispersion of electron and hole states in a superconductor near the Fermi edge $\pm k_F$ (full line) in comparison with a normal conductor (*dashed line*). (**c**) Dispersion of electron and hole states in a superconductor in a graph where the hole states are plotted in mirror symmetry around the Fermi level. The resulting density of states D_s for quasi-electrons and quasi-holes is qualitatively shown in the right panel

describing the generation of an empty state E' below E_F. This is schematically shown in Fig. 10.13. The excitation spectrum of a normal conductor can thus be described by the generation of excited electrons (energy referred to E_F) and of so-called holes with the energy $-(E'-E_F)$. This description is analogous to that which was introduced by $\xi_k = (\hbar^2 k^2/2m) - E_F$ in the BCS theory. Holes, empty states in the ground state of a normal conductor (Fermi sea), are thus excited states. In the transport of electric current they behave as positively charged carriers. This is easily seen from the expressions for the current density: In contrast to the ground state, where

$$j = \frac{-e}{8\pi^3} \int v(k) f(k) dk = 0 \tag{10.65 a}$$

the excited state in Fig. 10.13 a carries a current density

$$j = -ev(k'') - \frac{e}{8\pi^3} \int_{1\text{st BZ}} v(k) f(k) dk - [-ev(k')] \, . \tag{10.65 b}$$

The last term results from the missing electron with $|k'| < k_F$; it contributes as a current of a positive carrier, the hole in the Fermi sea. The hole is defined here analogously to Sect. 9.2. Such hole states will be treated in further detail in the context of semiconductors (Chap. 12).

In superconductors we find a new interesting phenomenon, namely the coupling of excited states consisting of excited electrons and holes. Excitation of a single electron on a state $\boldsymbol{k}$ in a superconductor means that the Cooper pair ($\boldsymbol{k}\uparrow$, $-\boldsymbol{k}\downarrow$) no longer exists, the state $-\boldsymbol{k}\downarrow$ must be empty. An excitation process implies ($\boldsymbol{k}\uparrow$ occupied, $-\boldsymbol{k}\downarrow$ empty). The definition of an excitation thus makes a statement simultaneously about the occupation of $+\boldsymbol{k}$ and $-\boldsymbol{k}$; the plot of the pair-occupation probability $w_{\boldsymbol{k}}$ for positive and negative k values as in Fig. 10.13 b is thus very convenient. In contrast to a normal conductor (Fig. 10.13 a) the occupation probability function $w_{\boldsymbol{k}}$ of a superconductor allows the occupation of a state k_1 within the Fermi surface ($|k_1| < k_F$) by an electron. Excitation of the state k_1, i.e. occupation by an uncoupled electron, then means $w_{\boldsymbol{k}}(k = k_1) = 1$ (Fig. 10.13 b), but also the absence of an electron on the state $-k_1$. Analogously, excitation of an electron in $k_2 > k_F$ results in an empty state at $-k_2$. The excitation k_1 adds to the BCS ground state described by $w_{\boldsymbol{k}}(T=0)$ about 25% "electron character", while approximately 75% electron character is missing at $-k_1$. The excitation in total exhibits a "hole character". Correspondingly, the excitation $k_2 > k_F$ in Fig. 10.13 b has mainly electron character, the accompanying hole character at $-k_2$ is less important. The described excitations in a superconductor can thus not be described as entirely electron- or hole-like. That is why one speaks of "quasi-particles". Depending on their nature, the quasi-particles here are either called quasi-electrons or quasi-holes. From Fig. 10.13 b one easily derives that $w_{\boldsymbol{k}} = v_{\boldsymbol{k}}$ is the occupation probability for quasi-electrons, while $1 - w_{\boldsymbol{k}} = u_{\boldsymbol{k}}^2$ is the distribution function for quasi-holes.

Notice that the original definition of v_k^2 was that of the occupation probability for $\pm\boldsymbol{k}$ with a Cooper pair $(\boldsymbol{k}\uparrow, -\boldsymbol{k}\downarrow)$.

Furthermore it is evident from Fig. 10.13b that a charge can be attributed to the quasi-particle. By using (10.50) the charge in a state $\boldsymbol{k}$ is obtained as

$$q_k = -e(u_k^2 - v_k^2) = -e\xi_k/E_k \ . \tag{10.66}$$

As shown in the treatment of the BCS ground state (10.56, 10.58) the excitation energy E_k of an electron is

$$E_k^2 = \xi_k^2 + \Delta^2 \ , \quad \text{with}$$
$$\xi_k^2 = (\hbar^2k^2/2m - E_\mathrm{F}^0)^2 \ . \tag{10.67}$$

This equation already implicitly contains the sought after representation of excitations of a superconductor in terms of quasi-electrons and quasi-holes. The positive square root of ξ_k^2 describes electron-like excitations, while the negative square root corresponds to quasi-holes. Both types of excitations occur in the vicinity of $+k_\mathrm{F}$ and $-k_\mathrm{F}$ (Fig. 10.14). In a superconductor there is the gap Δ between the excitation spectrum and the Fermi energy E_F; this gap does not exist in a normal conductor (dashed in Fig. 10.14b). For many applications an extended representation of the excitation spectrum $E_k(\boldsymbol{k})$ (Fig. 10.14c) is useful in which the excitation of quasi-holes is plotted with negative energy values, since E_k (electron) $=-E_k$ (hole). This representation with a gap of 2Δ resembles the band scheme of a semiconductor (Fig. 7.13). The density of states for both quasi-electrons and for quasi-holes has a pole at the edges of the gap 2Δ (Fig. 10.14c). Since quasi-electrons and quasi-holes in a superconductor are coupled to each other with probability amplitudes u_k for electron-like and v_k for hole-like excitations, it is adequate to represent the coupled excitation consisting of both quasi-particles by one and the same state vector

$$\begin{pmatrix} \tilde{u}_k(\boldsymbol{r},t) \\ \tilde{v}_k(\boldsymbol{r},k) \end{pmatrix} = g(\boldsymbol{r},t) \begin{pmatrix} u_k \\ v_k \end{pmatrix} , \tag{10.68}$$

which has the same time and space dependences for both particles. In order to treat inhomogeneous systems with an interface between a normal and a superconductor it would be necessary to have a Schrödinger equation at hand, which describes the dynamics both of the coupled quasi-particles in the superconductor and the uncoupled free carriers in the normal conductor by one and the same scheme.

For a normal conductor this Schrödinger equation is simply obtained by using a standard Hamiltonian $\mathscr{H}$ with an external potential $V(\boldsymbol{r})$. For holes the corresponding negative operator $-\mathscr{H}$ must be used. In a superconductor an additional term must be introduced that couples the amplitudes u_k and v_k of both quasi-particles. In order to introduce this term we consider the total energy of the Cooper-pair condensate (10.42) and compare it with the expression for the superconductor gap energy Δ (10.46). In (10.42) the amplitudes u_k and v_k are indeed coupled by the following term:

$$\frac{V_0}{L^3}\sum_{kk'} v_k u_k v_{k'} u_{k'} = \Delta \sum_k u_k v_k \ . \tag{10.69}$$

In (10.69) the gap energy $\Delta = \Delta(\{u_k, v_k\})$ acts as a coupling potential; its value depends on the entire ensemble of probability amplitudes u_k and v_k. The following ansatz for the generalized Schrödinger equation seems therefore reasonable:

$$\mathrm{i}\hbar\frac{\partial}{\partial t}\tilde{u}_k = \left[-\frac{\hbar^2}{2m}\nabla^2 - E_\mathrm{F} + V(r)\right]\tilde{u}_k + \Delta\tilde{v}_k \ , \tag{10.70 a}$$

$$\mathrm{i}\hbar\frac{\partial}{\partial t}\tilde{v}_k = -\left[-\frac{\hbar^2}{2m}\nabla^2 - E_\mathrm{F} + V(r)\right]\tilde{v}_k + \Delta\tilde{u}_k \ . \tag{10.70 b}$$

Note that for the solution of a particular problem $\Delta(\{u_k, v_k\})$ has to be calculated self-consistently with the solutions $\tilde{u}_k$ and $\tilde{v}_k$ of (10.70). In contrast to classical BCS theory, (10.70) allows time- and spatial-dependent gap energies $\Delta(\boldsymbol{r}, t)$. In addition, Δ can have complex values that also gives physically reasonable solutions, as will be shown below. In order to retain the gap energy as a real value $(\Delta^*\Delta)^{1/2}\Delta$ has to be replaced by Δ^* in (10.70 b). The Schrödinger equation (10.70) then is written in matrix form as

$$\mathrm{i}\hbar\frac{\partial}{\partial t}\begin{pmatrix}\tilde{u}_k\\ \tilde{v}_k\end{pmatrix} = \begin{pmatrix}\mathscr{H} & \Delta\\ \Delta^* & -\mathscr{H}\end{pmatrix}\begin{pmatrix}\tilde{u}_k\\ \tilde{v}_k\end{pmatrix} . \tag{10.71}$$

In a normal conductor with vanishing gap energy Δ the amplitudes $\tilde{u}_k$ and $\tilde{v}_k$ in (10.70) decouple, as required for free electrons and holes. In the simple case of a homogeneous superconductor (Δ=const., E_F=const.) without external field (V=0) we may try an ansatz in terms of plane waves as a solution for (10.70):

$$\tilde{u}_k = u_k \mathrm{e}^{\mathrm{i}\boldsymbol{k}\cdot\boldsymbol{r} - \mathrm{i}Et/\hbar} \ , \tag{10.72 a}$$

$$\tilde{v}_k = v_k \mathrm{e}^{\mathrm{i}\boldsymbol{k}\cdot\boldsymbol{r} - \mathrm{i}Et/\hbar} \ . \tag{10.72 b}$$

With the single particle energies $\xi_k = \dfrac{\hbar^2 k^2}{2m} - E_\mathrm{F}$ one obtains

$$E\begin{pmatrix}u_k\\ v_k\end{pmatrix} = \begin{pmatrix}\xi_k & \Delta\\ \Delta^* & -\xi_k\end{pmatrix}\begin{pmatrix}u_k\\ v_k\end{pmatrix} . \tag{10.73}$$

Solving this set of equations yields the two energy eigen-values

$$E_k^\pm = \pm\sqrt{\xi_k^2 + \Delta^2} \ . \tag{10.74}$$

These energies are exactly the excitation energies (10.58) for electrons (+) and holes (–) in classical BCS theory. After inserting the eigen-values

(10.74) into (10.73) and by using the normalization condition for the amplitudes u_k, v_k

$$u_k^2 + v_k^2 = 1 \; , \tag{10.75}$$

one obtains directly the mathematical expressions (10.48–10.51) for the BCS probability amplitudes for electrons u_k and holes v_k. The Schrödinger equations (10.70, 10.71) thus correctly describe the dynamics of excited electrons and holes in normal conductors and of quasi-electrons and quasi-holes in superconductors. The equations (10.70, 10.71) were first derived in a general form by Bogoliubov [10.8] using field-theoretical methods; they are called Bogoliubov equations.

With (10.74), respectively (10.47), one obtains from (10.50)

$$u_k^2 = 1 - v_k^2 = \frac{1}{2}\left[1 + \frac{\sqrt{E_k^2 - \Delta^2}}{E_k}\right] \; . \tag{10.76}$$

The equation has complex solutions for the probability amplitudes u_k and v_k for single particle energies $|E_k|<\Delta$. In that case also $\Delta(\{u_k, v_k\})$ becomes complex. This is a realistic situation in inhomogeneous systems at the interface between a normal conductor and a superconductor when normal electrons with $E<\Delta$ are injected into the superconductor.

For further reading on transport in superconductors and devices the reader is referred to [10.9, 10.10, 11.11]. After this excursion we return to the classical BCS model of a superconductor.

10.6 Consequences of the BCS Theory and Comparison with Experimental Results

An important prediction of the BCS theory is the existence of a gap Δ (or 2Δ) in the excitation spectrum of a superconductor. The tunnel experiments described in Panel IX provide direct experimental evidence for the gap. Indications for the existence of a gap are also present in the behavior of the electronic specific heat capacity of a superconductor at very low temperature. The exponential behavior of this specific heat capacity (10.3) is readily understandable for a system whose excited states are reached via excitation across an energy gap. The probability that the excited state is occupied is then proportional to an exponential Boltzmann expression, which also appears in the specific heat capacity (temperature derivative of the internal energy) as the main factor determining its temperature dependence.

A further direct determination of the energy gap 2Δ is possible using spectroscopy with electromagnetic radiation (optical spectroscopy). Electromagnetic radiation is only absorbed when the photon energy $\hbar\omega$ exceeds the energy necessary to break up the Cooper pair, i.e., $\hbar\omega$ must be larger than the gap energy 2Δ. Typical gap energies for classical superconductors lie in the region of a few meV. The appropriate experiments must therefore be carried out with microwave radiation. The curves in Fig. 10.15 result from an experiment in which the microwave intensity I is measured by a bolometer after multiple reflection in a cavity made of the material to be examined. With an external magnetic field, the material can be driven from the superconducting state (intensity I_S) to the normal state (intensity I_N). This allows the measurement of a difference quantity $(I_S - I_N)/I_N$. At a photon energy corresponding to the gap energy 2Δ, this quantity drops suddenly. This corresponds to an abrupt decrease in the reflectivity of the superconducting material for $\hbar\omega > 2\Delta$, whereas for photon energies below 2Δ the superconductor reflects totally since there are no excitation mechanisms.

At all temperatures above $T = 0$ K there is a finite possibility of finding electrons in the normal state. As the temperature rises, more and more Cooper pairs break up; thus a temperature increase has a destructive effect on the superconducting phase. The critical temperature T_c (transition point) is defined as the temperature at which the superconductor transforms to the normal state, and Cooper pairs cease to exist. At this temperature the gap Δ (or 2Δ) must have closed because the normal conducting state has a continuous excitation spectrum (Chap. 6). The gap energy Δ must therefore be a function of temperature with $\Delta(T) = 0$ for $T = T_c$. Hence the gap Δ of a superconductor cannot be compared with the almost temperature-independent forbidden band of a semiconductor (Chap. 12). In the framework of BCS theory it is possible to calculate the temperature dependence of Δ. At finite temperature the occupation of the excited one-electron states $E_k = (\xi_k^2 + \Delta^2)^{1/2}$ (10.58) obeys Fermi statistics with the Fermi distribution $f(E_k, T)$ (Sect. 6.3). In the equation determining Δ (10.61) this fact is taken into account by including the non-occupation of the corresponding pair states. In place of (10.61) one thus has

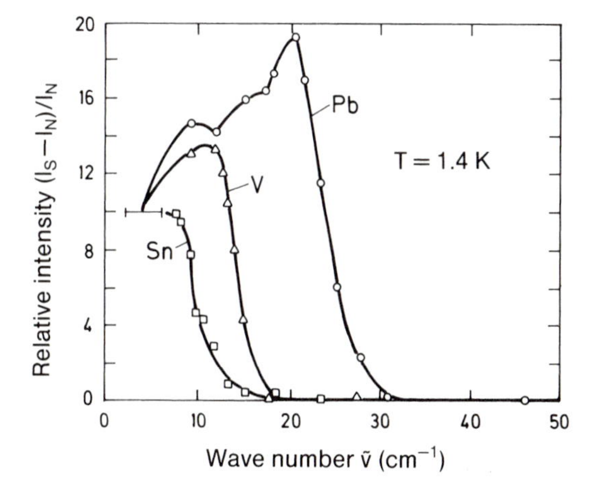

Fig. 10.15. Infrared reflectivity of various materials, determined from the intensity I of multiply reflected microwave radiation. The intensities I_S and I_N refer respectively to the superconducting and normal states of the material. The curves thus represent the difference between the infrared reflectivity of the superconducting and normally conducting states. (After [10.11])

$$\frac{1}{V_0 Z(E_F^0)} = \int_0^{\hbar\omega_D} \frac{d\xi}{\sqrt{\xi^2+\Delta^2}} [1 - 2f(\sqrt{\xi^2+\Delta^2} + E_F^0, T)] . \quad (10.77)$$

The factor of 2 multiplying the Fermi function appears because either one of the states $\boldsymbol{k}$ or $-\boldsymbol{k}$ may be occupied. Since $f(E_{\boldsymbol{k}}, T)$ vanishes for $T \to 0$, the more general formula (10.77) also contains the limiting case (10.61). Integration of (10.77) in analogy to (10.61–10.63) yields the energy gap Δ as a function of temperature T. A normalized plot of $\Delta(T)/\Delta(T{=}0)$ against T/T_c gives a universal curve for all superconductors. This curve is shown in Fig. 10.16 together with measured data for the three superconductors In, Sn and Pb. Deviations from the theoretical curve are due mostly to the fact that in BCS theory the assumption of a constant interaction matrix element V_0/L^3 is too simple. Because phonons are the origin of the coupling, the phonon structure of each material will also influence $V_{\boldsymbol{k}\boldsymbol{k}'}$. Improvements of the BCS theory in this direction provide a very good description of the simpler superconductors.

From (10.77) one can easily derive an equation for the critical temperature T_c. One simply sets Δ equal to zero to obtain

$$\frac{1}{V_0 Z(E_F^0)} = \int_0^{\hbar\omega_D} \frac{d\xi}{\xi} \tanh\frac{\xi}{2\ell T_c} . \quad (10.78)$$

A numerical treatment of the integral (10.78) yields

$$1 = V_0 Z(E_F^0) \ln\frac{1.14\hbar\omega_D}{\ell T_c} \quad \text{or} \quad (10.79\,a)$$

$$k T_c = 1.14\hbar\omega_D e^{-1/V_0 Z(E_F^0)} . \quad (10.79\,b)$$

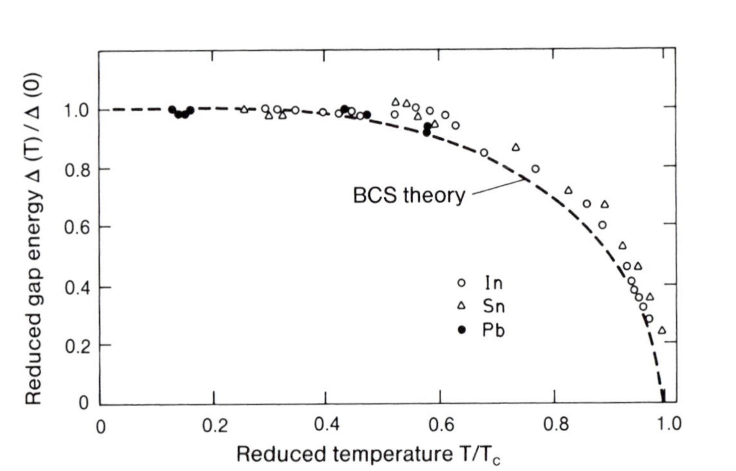

Fig. 10.16. Temperature dependence of the gap energy $\Delta(T)$ relative to the value $\Delta(0)$ at $T{=}0$ for In, Sn and Pb. Values determined from tunnel experiments (Panel IX) are compared with those predicted by BCS theory (*dashed*). (After [10.12])

Table 10.1. Debye temperature θ_D, transition temperature T_c, superconducting coupling constant $Z(E_F^0)\,V_0$ and gap energy Δ relative to $k\,T_c$ for several superconductors

Metal	θ_D [K]	T_c [K]	$Z(E_F^0)\,V_0$	Δ/kT_c
Zn	235	0.9	0.18	1.6
Cd	164	0.56	0.18	1.6
Hg	70	4.16	0.35	2.3
Al	375	1.2	0.18	1.7
In	109	3.4	0.29	1.8
Tl	100	2.4	0.27	1.8
Sn	195	3.75	0.25	1.75
Pb	96	7.22	0.39	2.15

This formula for the transition temperature T_c is, to within a constant factor, identical to that for the gap energy $\Delta(0)$ at $T=0$ K (10.63). A comparison of (10.79b) with (10.63) gives the BCS relationship between the gap energy $\Delta(0)$ and the transition temperature

$$\Delta(0)/k\,T_c = 2/1.14 = 1.764\,. \qquad (10.80)$$

Thus kT_c corresponds to about half of the gap energy at $T=0$ K. How well this relationship is fulfilled can be seen for a few superconductors in Table 10.1. An experimental measurement of T_c or $\Delta(0)$ also allows a determination of the so-called coupling constant $Z(E_F^0)\,V_0$ using (10.79b) or (10.63). As seen in Table 10.1, the values for common superconductors lie between 0.18 and 0.4.

According to (10.63, 10.79b) the gap energy $\Delta(0)$ and the transition temperature T_c are proportional to the phonon cut-off frequency ω_D (Debye frequency). We have seen in Sects. 4.3, 5.3 that the phonon frequency for a constant restoring force varies with the atomic mass M as $M^{-1/2}$. For two different isotopes of the same material, both the electronic properties and the chemical binding forces are the same. However, because of the different atomic masses, both T_c and $\Delta(0)$ should thus be proportional to $M^{-1/2}$. This is called the "isotope effect". Figure 10.17 shows experimental results for Sn [10.13] from a number of different authors.

The agreement between the expected $M^{-1/2}$ dependence and experiment is very good for the case of Sn. One finds, however, that particularly for the transition metals considerable deviations from 0.5 occur for the mass exponent, e.g., 0.33 for Mo and 0.2 for Os. These deviations are not surprising when one recalls the highly simplified assumptions made for the interaction matrix element V_0 (10.34, 10.28). Within the simple BCS theory, the Debye frequency ω_D is the only quantity that contains information about the material's phonon spectrum. On the other hand, the very existence of the isotope effect demonstrates the important influence of the phonons on the attractive electron interaction appearing in the BCS theory of superconductivity.

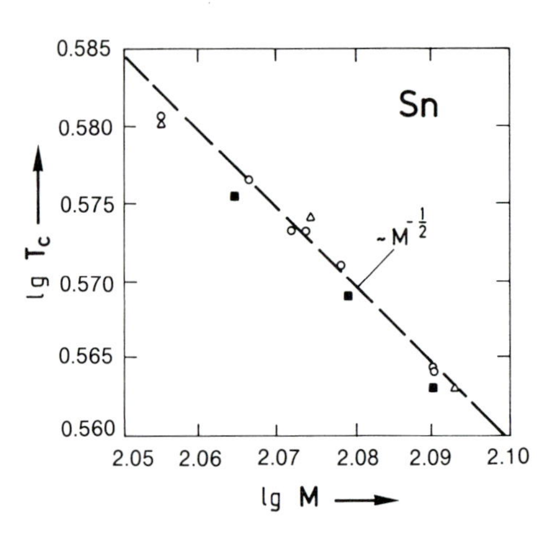

Fig. 10.17. Isotope effect for tin (Sn). The results of several authors are summarized [10.13]: Maxwell (○); Lock, Pippard, Shoenberg (■); Serin, Reynolds and Lohman (△)

10.7 Supercurrents and Critical Currents

The main goal of a theory of superconductivity is of course to explain the two fundamental properties of the superconducting phase: the disappearance of electrical resistance for $T<T_c$, and the ideal diamagnetism exhibited in the Meissner-Ochsenfeld effect. How do the existence of Cooper pairs and the properties of the BCS ground state yield a situation in which scattering processes do not lead to a finite resistance? Let us consider the influence of a current flux on the BCS ground state, and particularly on the existence of the gap 2Δ in the excitation spectrum. According to Sect. 9.5, a current density $\boldsymbol{j}$ can be described through an increase of the momentum or $\boldsymbol{k}$-vector of the current carriers. If n_s is the density of the individual electrons which, in the form of Cooper pairs, carry the supercurrent density $\boldsymbol{j}_s$, then

$$\boldsymbol{j}_s = -n_s e\,\boldsymbol{v} \quad \text{and} \tag{10.81}$$

$$m\,\boldsymbol{v} = \hbar \boldsymbol{k} \;. \tag{10.82}$$

When a current flows, each individual electron in a Cooper pair therefore experiences a change in $\boldsymbol{k}$-vector of

$$\frac{1}{2}\boldsymbol{K} = -\frac{m}{n_s e\,\hbar}\boldsymbol{j}_s \;. \tag{10.83}$$

Here we have introduced the change in the total wavevector of the Cooper pair as $\boldsymbol{K}$. The Cooper pair as a whole therefore has an additional momentum equal to $\boldsymbol{P}=\hbar\boldsymbol{K}$. When a current flows, the pair $(\boldsymbol{k}\uparrow, -\boldsymbol{k}\downarrow)$ must be described by the expression

$$(\boldsymbol{k}_1\uparrow, \boldsymbol{k}_2\downarrow) = (\boldsymbol{k} + \tfrac{1}{2}\,\boldsymbol{K}\uparrow, -\boldsymbol{k} + \tfrac{1}{2}\,\boldsymbol{K}\downarrow) \tag{10.84}$$

Neglecting the spin, the wavefunction of a Cooper pair (10.20) in the presence of a current ($\boldsymbol{K}\neq 0$) can thus be written

$$\psi(\boldsymbol{r}_1,\boldsymbol{r}_2) = \frac{1}{L^3}\sum_{\boldsymbol{k}} g(\boldsymbol{k})\mathrm{e}^{\mathrm{i}\boldsymbol{k}_1\cdot\boldsymbol{r}_1+\mathrm{i}\boldsymbol{k}_2\cdot\boldsymbol{r}_2}$$
$$= \frac{1}{L^3}\sum_{\boldsymbol{k}} g(\boldsymbol{k})\mathrm{e}^{\mathrm{i}\boldsymbol{K}\cdot(\boldsymbol{r}_1+\boldsymbol{r}_2)/2}\mathrm{e}^{\mathrm{i}\boldsymbol{k}\cdot(\boldsymbol{r}_1-\boldsymbol{r}_2)} . \qquad (10.85)$$

With $\boldsymbol{R}=(\boldsymbol{r}_1+\boldsymbol{r}_2)/2$ as the center of mass coordinate, and $\boldsymbol{r}=\boldsymbol{r}_1-\boldsymbol{r}_2$ as the relative coordinate of the Cooper pair, this wavefunction can be expressed as:

$$\psi(\boldsymbol{r}_1,\boldsymbol{r}_2)= \mathrm{e}^{\mathrm{i}\boldsymbol{K}\cdot\boldsymbol{R}}\frac{1}{L^3}\sum_{\boldsymbol{k}} g(\boldsymbol{k})\mathrm{e}^{\mathrm{i}\boldsymbol{k}\cdot\boldsymbol{r}} = \mathrm{e}^{\mathrm{i}\boldsymbol{K}\cdot\boldsymbol{R}}\psi(\boldsymbol{K}=\boldsymbol{0},\boldsymbol{r}_1-\boldsymbol{r}_2) . \qquad (10.86)$$

The flow of current therefore alters the Cooper pair wavefunction only by a phase factor, which does not affect the measurable probability density

$$|\psi(\boldsymbol{K}\neq\boldsymbol{0},\boldsymbol{r})|^2 = |\psi(\boldsymbol{K}=\boldsymbol{0},\boldsymbol{r})|^2 . \qquad (10.87)$$

Because the attractive interaction $V(\boldsymbol{r}_1-\boldsymbol{r}_2)$ depends only on the relative separation of the two electrons constituting the Cooper pair, it follows from (10.86) that the interaction matrix element (10.23) in the presence of a current is

$$V_{\boldsymbol{k}\boldsymbol{k}'}(\boldsymbol{K}\neq\boldsymbol{0}) = \int d\boldsymbol{r}\psi^*(\boldsymbol{K}\neq\boldsymbol{0},\boldsymbol{r})\, V(\boldsymbol{r})\psi(\boldsymbol{K}\neq\boldsymbol{0},\boldsymbol{r})$$
$$= \int d\boldsymbol{r}\psi^*(\boldsymbol{K}=\boldsymbol{0},\boldsymbol{r})\, V(\boldsymbol{r})\psi(\boldsymbol{K}=\boldsymbol{0},\boldsymbol{r}) = V_{\boldsymbol{k}\boldsymbol{k}'}(\boldsymbol{K}=\boldsymbol{0}) . \qquad (10.88)$$

The invariance of $V_{\boldsymbol{k}\boldsymbol{k}'}$ with respect to current flux means that in the BCS ground state energy (10.42) the same interaction parameter V_0/L^3 can be taken for both the current-carrying and non-current-carrying states.

Equation (10.84) implies, in the framework of the BCS theory, that a current flow causes only a shift of the $\boldsymbol{k}$-coordinate system by $\boldsymbol{K}/2$ (10.83). All equations from (10.42) onwards remain the same in the displaced reciprocal space; in particular, the gap energy Δ (10.60, 10.63) is unchanged in the presence of a current, because the integral (10.61) is evaluated over the same region in $\boldsymbol{k}$-space as in the current-free case; only the coordinate system is shifted. Correspondingly, the value of Δ (10.62) is also independent of $\boldsymbol{k}$. If a supercurrent is excited in a superconductor (e.g. by a changing magnetic flux), the gap in the excitation spectrum continues to exist. An alteration of the state, by inelastic electron scattering at least (e.g. phonons, Sects. 9.3–9.5), can only occur by excitation across the gap 2Δ, i.e. by breaking up at least one Cooper pair. Inelastic electron collisions can thus be ruled out as a source of charge carrier relaxation, i.e., current decay, provided that the total momentum of the Cooper pairs $\boldsymbol{P}$ is not associated with an energy increase that allows excitations of 2Δ or more. Elastic collisions, however, also alter the direction of the electron velocity, i.e. the current. In Sect. 10.9 we will see from the properties of the BCS ground state that in a current-carrying superconducting loop the magnetic flux may only change in fixed "quantum jumps". To contribute to the current relaxation, an elastic collision must therefore alter the current by exactly one such "flux quantum". The probability of such an event is vanishingly small. A current-carrying superconductor is therefore in a stable state.

If, due to the current flux itself, i.e., through an increase of the center of mass momentum $\boldsymbol{P}$ of the Cooper pairs, the energy 2Δ is achieved, then the Cooper pairs break up and the superconductivity collapses. For a single electron in a Cooper pair, the energy associated with an increase in wave vector $\boldsymbol{k}$ by $\boldsymbol{K}/2$ (10.83) is

$$E = \frac{(\boldsymbol{k}+\boldsymbol{K}/2)^2\hbar^2}{2m} = \frac{\hbar^2 k^2}{2m} + \frac{\hbar^2 \boldsymbol{k}\cdot\boldsymbol{K}/2}{m} + \frac{K^2\hbar^2}{8m} . \tag{10.89}$$

We consider the energy of a single electron since, on dissociating, the second electron of the Cooper pair does not retain coherence. In a linear approximation, and since $|\boldsymbol{K}| \ll k_\mathrm{F}$, the increase in energy per particle in the presence of a current flux compared with the current-free state is

$$\delta E \approx \frac{1}{2}\frac{\hbar^2 k_\mathrm{F} K}{m} . \tag{10.90}$$

It is assumed here that only electrons in the vicinity of the Fermi energy, i.e. with $k \simeq k_\mathrm{F}$, need to be treated. For the superconducting state to collapse the energy taken up by one Cooper pair $2\delta E$ must be larger than the energy to break up the pair 2Δ; from (10.83, 10.90) it must thus hold that

$$2\delta E \approx \frac{\hbar^2 k_\mathrm{F} K}{m} = \frac{2\hbar k_\mathrm{F}}{e\, n_\mathrm{s}} j_\mathrm{s} \geq 2\Delta . \tag{10.91}$$

From this we may estimate an upper critical current density j_c for the existence of the superconducting phase

$$j_\mathrm{c} \approx \frac{e\, n_\mathrm{s}\Delta}{\hbar k_\mathrm{F}} . \tag{10.92}$$

For Sn one finds, in the limiting case of low temperature $T \to 0$ K, an experimental value for the critical current density of $j_\mathrm{c} = 2\cdot 10^7$ A/cm^2, above which the superconductivity collapses. From the values in Table 10.1 and with an electron velocity at the Fermi edge of $v_\mathrm{F} = \hbar k_\mathrm{F}/m$ of $6.9\cdot 10^7$ cm/s one obtains a concentration n_s of superconducting current-carrying electrons of about $8\cdot 10^{21}$ cm^{-3}.

Via the Maxwell equation

$$\mathrm{curl}\,\boldsymbol{H} = \boldsymbol{j} \quad \text{or} \tag{10.93 a}$$

$$\int \mathrm{curl}\,\boldsymbol{H}\cdot d\boldsymbol{S} = \oint \boldsymbol{H}\cdot d\boldsymbol{l} = \int \boldsymbol{j}\cdot d\boldsymbol{S} \tag{10.93 b}$$

a magnetic field along a closed path is uniquely associated with the electrical current flowing through this path. This relation should also be valid for supercurrents in a long wire with a magnetic field on the surface. If we imagine a closed path around the circumference of a wire (radius r), then from (10.93) the magnetic field strength H on the surface of the wire is given by

$$2\pi r H = \int \boldsymbol{j} \cdot d\boldsymbol{S} \ . \tag{10.94}$$

As described in Sect. 10.2, the supercurrent in a "long, thick" wire is contained in a surface zone with a typical thickness of 100–1000 Å (London penetration depth Λ_L). If we describe the exponential decay of the current density into the interior of the wire by an equation of the form $j = j^0 \exp(-z/\Lambda_L)$, then from (10.94) we have

$$2\pi r H = 2\pi r \Lambda_L j^0 \ . \tag{10.95}$$

The critical current density j_c, (10.92), on the surface of a superconductor therefore corresponds to a critical magnetic field H_c on the surface, above which the superconductivity breaks down:

$$H_c = \Lambda_L j_c \approx \Lambda_L \frac{e\, n_s \Delta}{\hbar k_F} \ . \tag{10.96}$$

We therefore have demonstrated, on the basis of BCS theory, the existence of a critical magnetic field above which superconductivity can no longer exist.

It should also be remarked that the equation (10.96) for the critical magnetic field strength can also be derived by setting the condensation energy density for the superconducting phase (10.53) equal to the magnetic field density at the critical field $H_c B_c$. If the magnetic field energy becomes larger than the condensation energy for the superconducting phase, the Cooper pairs break up.

According to Sect. 10.6, and particularly (10.77), the gap energy Δ is a temperature-dependent quantity, which at the critical temperature T_c shrinks to zero. Therefore, the same must be true for the critical current density $j_c(T)$ and the critical magnetic field strength $H_c(T)$; see (10.92) and (10.96). The behavior of $H_c(T)$ described qualitatively in Fig. 10.5 can therefore be traced back to the temperature dependence of the gap energy $\Delta(T)$ (Fig. 10.16).

10.8 Coherence of the BCS Ground State and the Meissner-Ochsenfeld Effect

After successfully describing a state with vanishing electrical resistance, our next goal is to use the microscopic BCS theory to understand the Meissner-Ochsenfeld effect, i.e. ideal diamagnetism and the expulsion of magnetic flux from a superconductor. According to Sect. 10.2 it is sufficient in a first approximation to derive the second London equation (10.10b) from the properties of Cooper pairs or from the behavior of the

BCS ground state in the presence of a magnetic field. This material-specific equation, together with the Maxwell equations, constitutes a description of the behavior of a superconductor in a magnetic field.

Let us look more closely at the structure of Cooper pairs and the wavefunction of the ground state. Within the BCS approximation the wavefunction of the ground state is represented as a product of similar two-particle wavefunctions of Cooper pairs $\psi(\boldsymbol{r}_1 - \boldsymbol{r}_2, \uparrow\downarrow)$. According to (10.20) the pair wavefunctions can be represented by one-particle wavefunctions whose wave vector $\boldsymbol{k}$ and spins ($\uparrow\downarrow$) are opposite. The $\boldsymbol{k}$-space or energy region from which the one-particle states are taken to construct this wavefunction can be recognized from the behavior of the occupation probability $w_{\boldsymbol{k}}$ for the pair state $(\boldsymbol{k}\uparrow, -\boldsymbol{k}\downarrow)$ (Fig. 10.11). It is only in a region of approximately $\pm\Delta$ around the Fermi level E_F^0 that the one-particle occupation in the superconductor is modified compared to that of a normal conductor. The one-particle wavefunctions that constitute the Cooper pairs therefore originate from this region.

From the energy uncertainty 2Δ, one can calculate the magnitude of the momentum uncertainty δp for electrons in Cooper pairs:

$$2\Delta \sim \delta\left(\frac{p^2}{2m}\right) \simeq \frac{p_\mathrm{F}}{m}\delta p \,. \tag{10.97}$$

From the uncertainty relation this momentum distribution corresponds to a spatial extent of the Cooper pair wavefunction of

$$\xi_\mathrm{CP} = \delta x \sim \frac{\hbar}{\delta p} \approx \frac{\hbar p_\mathrm{F}}{m2\Delta} = \frac{\hbar^2 k_\mathrm{F}}{m2\Delta} \,. \tag{10.98}$$

Since $k_\mathrm{F} = 2mE_\mathrm{F}^0/\hbar^2 k_\mathrm{F}$, it follows that

$$\xi_\mathrm{CP} \sim \frac{E_\mathrm{F}^0}{k_\mathrm{F}\Delta} \,, \tag{10.99}$$

with E_F^0/Δ typically on the order 10^3 or 10^4 and k_F about $10^8\,\mathrm{cm}^{-1}$, the Cooper pair wavefunction usually extends over a spatial region of about 10^3 to 10^4 Å. We have already estimated the "size" of a Cooper pair to be about this order of magnitude from the space-time behavior of the lattice deformation associated with the Cooper pair (10.7). Spatial alterations of the superconducting state thus involve a region of at least 10^3 to 10^4 Å. If ξ_coh is the coherence length, that is, the distance from the interface between a normal conductor and a superconductor over which the density of Cooper pairs rises from zero to its maximum value, then it is always true that $\xi_\mathrm{coh} > \xi_\mathrm{CP}$. Since $E_\mathrm{F}^0/\Delta \simeq 10^4$, we can assume that out of the roughly 10^{23} electrons per cm^{-3} about 10^{19} are coupled in Cooper pairs. Thus, within the volume of about $10^{-12}\,\mathrm{cm}^3$ occupied by one Cooper pair, about another 10^6 or 10^7 other pairs also have their centers of gravity. The pairs are therefore not to be regarded as independent of each other; they are spatially "anchored" to one another. One is inclined to compare this high coherence of the many-body state with the photon coherence in a laser beam. It is this high coherence of the BCS ground state that is the root of its high stability. The collective behavior of so many Cooper pairs in the superconducting state leads one to expect that quantum mechanical effects will become macroscopically observable. This will be confirmed in our treatment of magnetic fields (Sects. 10.9, 10.10).

As we have seen, a Cooper pair consists of two electrons with opposite spins and thus a total spin of zero. This means that the statistics of Cooper pairs can be approximated by those of bosons (particles with integer spin). In other words, in a first approximation, the Pauli principle does not apply to Cooper pairs: Cooper pairs are all in the same BCS ground state, i.e., the time-dependent parts of their wavefunctions all contain one and the same energy in the exponent, namely, that of the BCS ground state. However, because of the high coherence of the ground state, one should exercise caution when treating the Cooper pairs as a non-interacting Bose gas.

A more adequate description to the BCS ground state is in terms of a "condensate" of Cooper pairs.

In the following we calculate the density of a supercurrent $\boldsymbol{j}_s$ carried by this condensate in the presence of a magnetic field $\boldsymbol{B} = \nabla \times \boldsymbol{A}$. Our starting point is the two-particle wave function of a Cooper pair (10.86). Neglecting Cooper-pair interaction, the many-particle wave function is formed as a product of the two-particle wave functions $\psi(\boldsymbol{r}_1, \boldsymbol{r}_2)$ (10.86). In any case, the exact many-particle wave function can be represented as an expansion in products of two-particle wave functions. Because of (10.86) the wave function of one single Cooper pair under current flow can be expressed in terms of the wave function of a Cooper pair without current flow ($\boldsymbol{K}=0$):

$$\psi(\boldsymbol{r}_1, \boldsymbol{r}_2) = \mathrm{e}^{\mathrm{i}\boldsymbol{K}\cdot\boldsymbol{R}} \psi(\boldsymbol{K} = \boldsymbol{0}, \boldsymbol{r}_1 - \boldsymbol{r}_2) \ . \tag{10.100}$$

Here, $\hbar\boldsymbol{K}$ is the additional momentum of the Cooper pair that is to be added to both electron momenta and $\boldsymbol{R}$ is the center of mass co-ordinate of the Cooper pair. The wave function without current $\psi(\boldsymbol{K}=\boldsymbol{0}, \boldsymbol{r}_1-\boldsymbol{r}_2)$ depends only on the "internal" co-ordinate $\boldsymbol{r} = \boldsymbol{r}_1 - \boldsymbol{r}_2$ of the two electrons that constitute the Cooper pair, not on the co-ordinate of the pair as such. The ground-state wave function of the entire Cooper pair condensate in space representation is therefore

$$\Phi_{\mathrm{BCS}} \simeq \mathscr{A}\, \mathrm{e}^{\mathrm{i}\boldsymbol{K}\cdot\boldsymbol{R}_1} \mathrm{e}^{\mathrm{i}\boldsymbol{K}\cdot\boldsymbol{R}_2} \ldots \mathrm{e}^{\mathrm{i}\boldsymbol{K}\cdot\boldsymbol{R}_\nu} \ldots \Phi(\boldsymbol{K} = \boldsymbol{0}, \ldots \boldsymbol{r}_\nu \ldots) \ . \tag{10.101}$$

All Cooper pairs with their individual center of mass co-ordinates $\boldsymbol{R}_1, \boldsymbol{R}_2, \ldots \boldsymbol{R}_\nu \ldots$ have gained the same change of center of mass momentum $\boldsymbol{K}$ due to the current flow. $\Phi(\boldsymbol{K}=\boldsymbol{0}, \ldots \boldsymbol{r}_\nu \ldots) = \Phi(0)$ is a product of wave functions $\psi(\boldsymbol{K}=\boldsymbol{0}, \boldsymbol{r}_\nu)$ of single Cooper pairs without current flow that depends only on the internal relative co-ordinates $\boldsymbol{r}_\nu$ (ν counts the pairs). $\mathscr{A}$ is a so-called anti-symmetrization operator, which anti-symmetrizes the entire wave function, similarly as the Slater determinant does in a many-electron wave function: it adds up expressions of the type (10.101) with varying signs, such that Φ_{BCS} is antisymmetric with respect to the exchange of single-particle states. In the specific form of the ground-state wave function (10.33) the effect of $\mathscr{A}$ is already explicitly contained because of the algebraic properties of the two-dimensional state vectors $|0\rangle$ and $|1\rangle$ and spin matrices, respectively. In (10.101) $\mathscr{A}$ essentially has the effect of adding up similar terms and can therefore be neglected for the subsequent considerations. We therefore work with the simplified representation for the wave function of the Cooper-pair condensate in the following, rather than using (10.101).

$$\Phi_{\mathrm{BCS}} \simeq \mathrm{e}^{\mathrm{i}\boldsymbol{K}\cdot(\boldsymbol{R}_1+\boldsymbol{R}_2+\ldots\boldsymbol{R}_\nu\ldots)}\Phi(0) . \tag{10.102}$$

Now a distinctive feature of the superconducting ground state Φ_{BCS} in relation to a common Fermi gas enters the consideration. Because of the high coherence of the many-particle superconducting ground state (10^6 to 10^7 Cooper pairs overlap) the wave function of the Cooper-pair condensate can be assumed as internally "rigid": One simple center-of-mass coordinate $\tilde{\mathbf{R}}$ can be attributed to the entire ensemble of Cooper pairs

$$\tilde{\mathbf{R}} = \sum_\nu 2m\boldsymbol{R}_\nu \Big/ \sum_\nu 2m = \frac{1}{N_{\mathrm{CP}}}\sum_\nu \boldsymbol{R}_\nu , \tag{10.103}$$

where N_{CP} is the number of Cooper pairs. This approximation is possible also because of the existence of the gap Δ in the excitation spectrum of the superconducting state, which forbids scattering processes of electrons by phonons and impurities for temperatures far below the critical temperature T_{c}. The part $\Phi(0)$ of the wave function (10.102) still contains the internal relative co-ordinates $\boldsymbol{r}_\nu$ of the two electrons within the single Cooper pairs (counted by ν), which denote the distances between the constituting electrons. Within the approximation of a rigid Cooper-pair condensate this dependence on the internal distances (10.102) is neglected. Consequently, all effects in which the Cooper-pair wave function varies over spatial distances comparable to or smaller than the coherence length ξ_{coh} ($\gtrsim 10^4$ Å for type I superconductors) can not be described in this particular approximation. For relatively homogeneous situations, as are found in the bulk of "simple" superconductors (type I superconductors, Sect. 10.9), the essential space dependence of Φ_{BCS} (10.102) is the phase dependence of the wave function of the Cooper-pair condensate on the center-of-mass coordinate $\tilde{\mathbf{R}}$ (10.103). Using (10.102) and (10.103) one derives the following expression for the wave function of the condensate.

$$\Phi_{\mathrm{BCS}} \simeq \mathrm{e}^{\mathrm{i}N_{\mathrm{CP}}\boldsymbol{K}\cdot\tilde{\mathbf{R}}}\Phi(0) = \mathrm{e}^{\mathrm{i}\tilde{\mathbf{K}}\cdot\tilde{\mathbf{R}}}\Phi(0) , \tag{10.104}$$

with $\tilde{\mathbf{K}} = N_{\mathrm{CP}}\boldsymbol{K}$ the total wave vector (quasi-momentum) transfer to the Cooper-pair condensate because of the current flow. In the model for the superconductivity presented here we describe the entire rigid Cooper-pair condensate as a large "hyper-particle" of mass $N_{\mathrm{CP}}m$ and charge $2eN_{\mathrm{CP}}$.

For calculating the supercurrent density we can therefore use the single-particle quantum-mechanical current density operator:

$$j = -\frac{2eN_{\mathrm{CP}}}{4mN_{\mathrm{CP}}}\left(\Phi_{\mathrm{BCS}}\, p^{*} \Phi_{\mathrm{BCS}}^{*} + \Phi_{\mathrm{BCS}}^{*}\, p\, \Phi_{\mathrm{BCS}}\right) . \tag{10.105}$$

The momentum operator $\tilde{p}$ acts on the entire Cooper-pair condensate with center-of-mass coordinate $\tilde{\mathbf{R}}$ and the total charge $2N_{\mathrm{CP}}e$:

$$p = \frac{\hbar}{\mathrm{i}}\nabla_{\tilde{\mathbf{R}}} + 2eN_{\mathrm{CP}}\boldsymbol{A} . \tag{10.106}$$

Since the Cooper-pair condensate is considered as one single "hyper-particle", we require the normalization condition:

$$\int |\Phi(0)|^2 dV = 1 , \tag{10.106 a}$$

i.e.

$$|\Phi(0)|^2 = 1/L^3 , \tag{10.106 b}$$

where L^3 is the volume of the superconducting sample. From (10.104, 10.105, 10.106) follows

$$\begin{aligned} \boldsymbol{j}_{\mathrm{s}} = \frac{-2eN_{\mathrm{CP}}}{4mN_{\mathrm{CP}}} \Bigg[& \Phi_{\mathrm{BCS}}\left(\frac{\hbar}{\mathrm{i}}\nabla_{\mathbf{R}} + 2eN_{\mathrm{CP}}\boldsymbol{A}\right)\Phi_{\mathrm{BCS}}^{*} \\ & + \Phi_{\mathrm{BCS}}^{*}\left(-\frac{\hbar}{\mathrm{i}}\nabla_{\mathbf{R}} + 2eN_{\mathrm{CP}}\boldsymbol{A}\right)\Phi_{\mathrm{BCS}}\Bigg] , \end{aligned} \tag{10.107 a}$$

$$\boldsymbol{j}_{\mathrm{s}} = -\frac{e}{2m}\left[4eN_{\mathrm{CP}}\boldsymbol{A}|\Phi(0)|^2 - 2\hbar\tilde{\mathbf{K}}|\Phi(0)|^2\right] . \tag{10.107 b}$$

Since $\tilde{\mathbf{K}}$ does not depend on a space co-ordinate curl $\tilde{\mathbf{K}}=\mathbf{0}$ holds, and the application of the curl operator on (10.107) yields

$$\mathrm{curl}\,\boldsymbol{j}_{\mathrm{s}} = -\frac{2\mathrm{e}^2}{\mathrm{m}}\mathrm{N}_{\mathrm{CP}}|\Phi(0)|^2\,\mathrm{curl}\,\boldsymbol{A} . \tag{10.108}$$

With the normalization condition (10.106 b), with $\boldsymbol{B}=\mathrm{curl}\,\boldsymbol{A}$ and with N_{CP}/L^3 as the density of Cooper pairs (i.e. $n_{\mathrm{s}}=2N_{\mathrm{CP}}/L^3$ as the density of superconducting electrons) one obtains

$$\mathrm{curl}\,\boldsymbol{j}_{\mathrm{s}} = -\frac{\mathrm{n_s e^2}}{\mathrm{m}}\boldsymbol{B} , \tag{10.109}$$

which, according to (10.8), is exactly the 2nd London equation. We have thereby derived the Meissner-Ochsenfeld effect within the approximation that the Cooper pair density does not vary strongly in space. One recognizes further that the London equation is *only* valid under this assumption. For strongly spatially varying Cooper pair densities, i.e. varying over dimensions comparable to the "size" of a Cooper pair (coherence length),

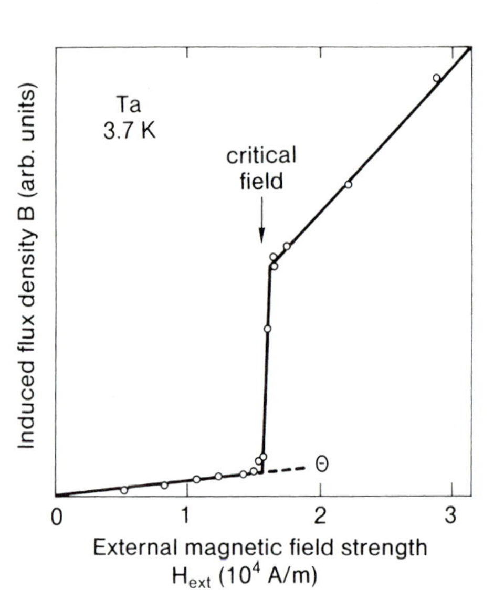

Fig. 10.18. Induced magnetic flux density B through a Ta sample, measured at 3.7 K as a function of external magnetic field H_{ext}. The sample is in a large coil which creates H_{ext}. The field B is measured with a small solenoid, which is located directly around the Ta sample. The weak background signal θ is due to flux induced in the windings of the measurement solenoid. (After [10.15])

non-local phenomenological extensions of the London theory were developed even before the conception of the BCS theory [10.14].

Experimentally, the Meissner-Ochsenfeld effect manifests itself most clearly in measurements of the internal magnetic field in a superconductor. Figure 10.18 shows the measured flux density B penetrating a Ta sample as a function of the external applied magnetic field H_{ext}. The external magnetic field H_{ext} is created by a long coil, in which the Ta sample is placed within a small solenoid. This solenoid measures the magnetic flux density B that penetrates the sample. One clearly sees that there is a critical magnetic field strength H_c (10.96), above which superconductivity ceases. Above H_c the flux density B created in the material grows in proportion to the applied field H_{ext}. Below H_c, i.e. in the superconducting phase, the magnetic flux is expelled due to the Meissner-Ochsenfeld effect. The low background signal (slope θ) is due to the magnetization of the windings of the solenoid.

10.9 Quantization of Magnetic Flux

We will now investigate further the influence of an external magnetic field $\boldsymbol{B}=\mathrm{curl}\,\boldsymbol{A}$ on a superconductor. As in Sect. 10.8, this will be done within the framework of a Cooper pair density that does not vary rapidly in space (type I superconductors). The wavefunction in the presence of a supercurrent density $\boldsymbol{j}_s$ should have the form (10.104), whereby the spatial dependence appears only in the phase $\varphi = \tilde{\mathbf{K}} \cdot \tilde{\mathbf{R}} = N_{CP}\boldsymbol{K} \cdot \mathbf{R}$ and not in the amplitude. It therefore follows from (10.106, 10.107) that the supercurrent density is

$$\boldsymbol{j}_{\mathrm{s}} = -\left[\frac{e^2 n_{\mathrm{s}}}{m}\boldsymbol{A} - \frac{e\hbar n_{\mathrm{s}}}{2m}\boldsymbol{K}\right] . \tag{10.110}$$

We consider a closed path around the superconducting ring, which is threaded by an external magnetic field (Fig. 10.19). The path integral over the supercurrent density (10.110) is

$$\oint \boldsymbol{j}_{\mathrm{s}} \cdot d\boldsymbol{l} = -\frac{n_{\mathrm{s}} e^2}{m}\oint \boldsymbol{A} \cdot d\boldsymbol{l} + \frac{e\hbar n_{\mathrm{s}}}{2m}\frac{1}{N_{\mathrm{CP}}}\oint \nabla_{\tilde{\mathbf{R}}}\varphi(\tilde{\mathbf{R}}) \cdot d\boldsymbol{l} . \tag{10.111}$$

The many-particle wavefunction of the Cooper pairs is now unambiguously defined along the whole loop in terms of the coordinates of the single pairs. Thus, for a stationary state the phase change around the closed loop must be zero or an integral multiple of 2π:

$$\oint \nabla_{\tilde{\mathbf{R}}}\varphi(\tilde{\mathbf{R}}) \cdot d\boldsymbol{l} = 2\pi N . \tag{10.112}$$

From (10.111) it then follows that

$$\frac{m}{n_{\mathrm{s}} e^2}\oint \boldsymbol{j}_{\mathrm{s}} \cdot d\boldsymbol{l} + \int \boldsymbol{B} \cdot d\boldsymbol{S} = N\frac{h}{2e} . \tag{10.113}$$

The second London equation (10.10b, 10.109) results as a special case for a closed path in a simply connected region, through which, for a superconductor, no magnetic flux may penetrate (N=0). The expression on the left-hand side of (10.113) is called the fluxoid; this fluxoid can only exist as an integral multiple of $h/2e$. Normally one would expect that with an appropriate choice of magnetic field any arbitrary supercurrent could be "switched on" through a closed loop. Condition (10.113) dictates however that the current density and magnetic flux through the ring must obey a "quantization rule". Noting that the supercurrent flows only inside a thin shell with a thickness (the London penetration depth) of a few hundred

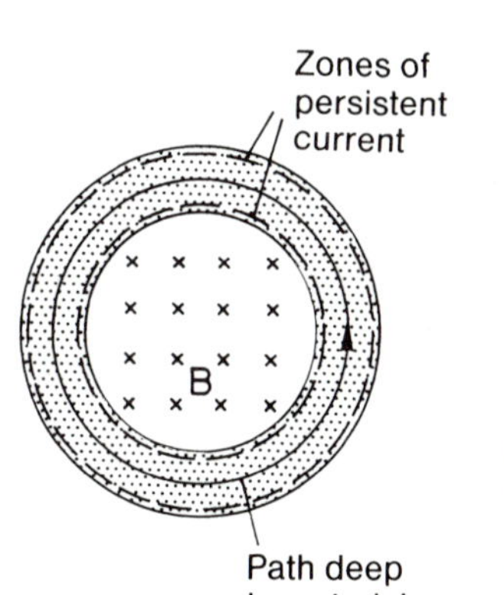

Fig. 10.19. A superconducting ring (*shaded*) which is threaded by a magnetic flux density B. The dashed lines indicate zones in which persistent supercurrent flows. The integration contour (*solid line*) of (10.111–10.113) thus passes through a virtually current-free region

Ångstroms we see that the closed path in (10.113) lies in a region where $\boldsymbol{j}_s$ is effectively zero (Fig. 10.19). Equation (10.113) thus simplifies to

$$\int \boldsymbol{B} \cdot d\boldsymbol{S} = N \frac{h}{2e} \ . \tag{10.114}$$

This condition says that a closed superconducting loop can only enclose magnetic flux in integral multiples of the so-called flux quantum

$$\phi_0 = \frac{h}{2e} = 2.0679 \times 10^{-7} \mathrm{G\,cm}^2 \approx 2 \times 10^{-7}\,\mathrm{G\,cm}^2 \tag{10.115}$$

($1\,\mathrm{G} = 10^{-4}\,\mathrm{V\,s/m}^2$). To provide a concrete example of the size of ϕ_0, one can imagine a small cylinder with a diameter of about 1/10 mm. One flux quantum ϕ_0 will penetrate this cylinder when the magnetic field in the middle has a strength of about 1% of the Earth's magnetic field.

The relations (10.113, 10.114) can also be derived formally by applying the Bohr-Sommerfeld quantization condition to the macroscopic supercurrent flowing in a ring. If we assume that the current is carried by particles with charge q and density n, then the momentum of one particle is

$$\boldsymbol{p} = m\boldsymbol{v} + q\boldsymbol{A} \ . \tag{10.116}$$

Since $\boldsymbol{j} = nq\boldsymbol{v}$ it follows that

$$\boldsymbol{p} = \frac{m}{nq}\boldsymbol{j} + q\boldsymbol{A} \ . \tag{10.117}$$

The quantization condition demands that the closed path integral over $\boldsymbol{p}$ is a multiple of the Planck quantum of action h:

$$\oint \boldsymbol{p} \cdot d\boldsymbol{l} = \frac{m}{nq}\oint \boldsymbol{j} \cdot d\boldsymbol{l} + q\oint \boldsymbol{A} \cdot d\boldsymbol{l} = Nh \ , \tag{10.118 a}$$

$$\frac{m}{nq^2}\oint \boldsymbol{j} \cdot d\boldsymbol{l} + \int \boldsymbol{B} \cdot d\boldsymbol{S} = \frac{Nh}{q} \ . \tag{10.118 b}$$

In this derivation we have treated the current-carrying circuit as a giant molecule, that is to say, we have applied quantum conditions to a macroscopic system. This again shows, from another point of view, that the superconducting phase must be described by a macroscopic many-body wavefunction due to the high coherence of the Cooper pairs. Typical quantum-mechanical microscopic properties are manifest in the superconductor as macroscopically observable quantum phenomena. In a non-superconducting ring, the quantization condition (10.118) may not be applied, because for electrons not in a Cooper pair there is no wavefunction that encloses the whole ring. Collision processes destroy the coherence over large distances.

By comparing (10.118 b) with (10.113), we observe that the BCS derivation of flux quantization also yields information about the charge of the superconducting particles: namely, that the particles have exactly twice the electronic charge ($q=2e$, Cooper pairs). The experimental detection of flux quantization and, in particular, the measurement of the flux quantum (10.115) is one of the most important confirmations of the BCS theory. These experiments were carried out at about the same time, and with the same result, by Doll and Näbauer [10.16] and Deaver and Fairbank [10.17]. Doll and Näbauer used tiny lead cylinders made by evaporating the metal onto quartz tubes with a diameter of about 10 µm. With this diameter a flux quantum corresponds to a magnetic flux density inside the lead cylinder of about 0.25 G. A permanent supercurrent was then created in the lead cylinders by cooling (below T_c) in a "freezing in" magnetic field B_f and then switching off B_f (Fig. 10.20, inset). In a measuring magnetic field B_m the lead cylinder acts as a magnetic dipole on which a torque acts. In principle, the torque could be measured statically by the deflection of a light beam, thereby allowing a determination of the magnetic flux through the cylinder. However, because of the small size of the effect, Doll and Näbauer used a dynamic method in which torsional oscillations of the system were excited. The amplitude at resonance is then proportional to

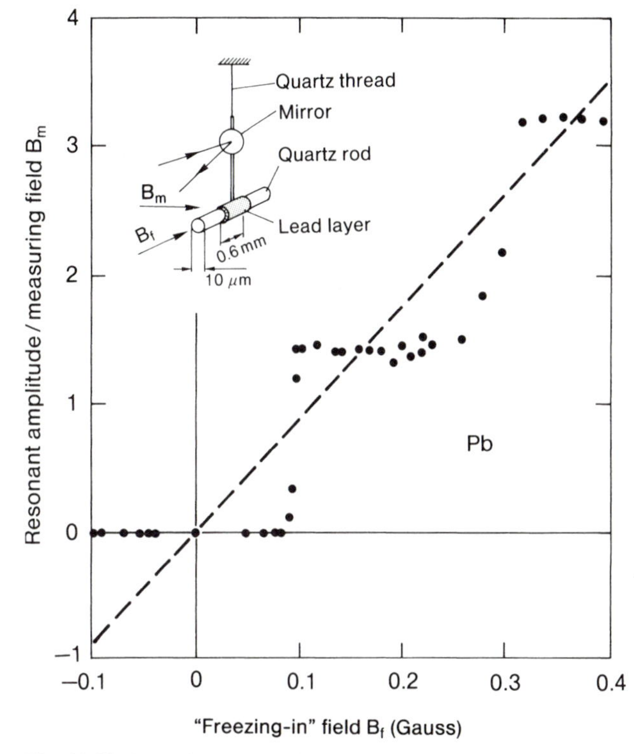

Fig. 10.20. Experimental results for flux quantization in a Pb cylinder. The flux through the small Pb cylinder, which is evaporated onto a quartz rod (*inset*), is determined from vibrations of the system in a measuring field B_m. (After [10.16])

the magnetic moment of the lead cylinder. Figure 10.20 shows a plot of the amplitude at resonance in relation to the field strength B_m (proportional to the flux through the cylinder) as a function of the "freezing in" field B_f. Without flux quantization one would expect the dashed straight line. The data points clearly show that only flux quanta of a particular size $\phi_0 N$ can be "frozen into" the lead cylinder. The experimental value of ϕ_0 corresponds to the value obtained in (10.115) on the basis of Cooper pairs. It has therefore been unambiguously shown that electron pairs and not single electrons are the carriers of the superconducting current.

10.10 Type II Superconductors

The behavior of superconductors in a magnetic field has been characterized by the fact that in an external magnetic field B^0 the supercurrent decays exponentially with the London penetration depth Λ_L (10.16) (Fig. 10.21 a). The supercurrent flows in this thin layer and keeps the interior of the superconductor field free (Sect. 10.2). Superconductivity collapses when the external magnetic field B^0 exceeds a critical value $B_c = \mu_0 \mu H_c$ (10.96), at which the magnetic energy density $\frac{1}{2} H_c B_c$ exceeds the condensation energy density for Cooper pairs (10.53). Above this value, the Cooper pairs are broken up and the normal conducting state sets in.

If instead of a superconducting half space (Fig. 10.21 a) we consider a layer of thickness d (Fig. 10.21 b), then for a sufficiently thin layer the magnetic field can no longer decay fully. Inside the layer a substantial field remains, since the screening effect of the superconducting currents cannot

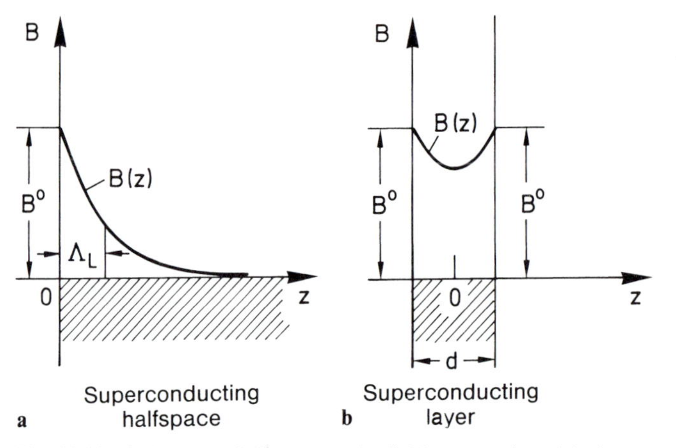

Fig. 10.21 a,b. Decay of the magnetic field penetrating. (**a**) A superconducting half-space ($z>0$). (**b**) A superconducting layer of thickness d. The field strength at the surface is in each case B_0. The London penetration depth Λ_L is the characteristic length for the exponential decay in the case of the half-space

develop fully in this thin layer. Mathematically one obtains this behavior by solving the differential equations (10.14) for the layer geometry. The general solution is a combination of the two partial solutions at the boundaries

$$B_1(z) = B_1 e^{-z/\Lambda_L} , \quad B_2(z) = B_2 e^{z/\Lambda_L} . \tag{10.119}$$

With the boundary condition

$$B_1 e^{d/2\Lambda_L} + B_2 e^{-d/2\Lambda_L} = B^0 \tag{10.120}$$

and from the symmetry of the problem ($B_1 = B_2 = \bar{B}$) it then follows that

$$\bar{B} = B^0/2\cosh\frac{d}{2\Lambda_L} . \tag{10.121}$$

From this the magnetic field strength in the superconducting layer (Fig. 10.21 b) is

$$B(z) = B^0 \frac{\cosh(z/\Lambda_L)}{\cosh(d/2\Lambda_L)} . \tag{10.122}$$

For small layer thicknesses the variation of the magnetic field over the layer is no longer significant; the field penetrates the superconducting layer almost completely. This emphasizes the fact that the screening of magnetic fields by the superconducting currents cannot be fully effective in the layer. In this situation, if the critical magnetic field B_c (10.96) is applied to the surface of the layer, the magnetic field energy would not be sufficient to break up all the Cooper pairs in the layer and destroy the superconductivity. Correspondingly, the critical magnetic field B_c for a sufficiently thin superconducting layer must be higher than for a half-space. The critical field increases continuously with decreasing layer thickness. For thicknesses $d \ll \Lambda_L$ it can be a factor 10 higher than the corresponding field for a superconducting half space in thermodynamic equilibrium. Thus, as the layer thickness decreases, the reaction of a superconductor to an applied magnetic field becomes ever smaller.

As a direct consequence of this result one would expect that the application of a magnetic field above the critical value ($>B_c$) to a massive superconductor would lead to its break up into alternate superconducting and normally conducting regions parallel to the magnetic field. If the superconducting regions are thin enough they can withstand a considerably higher magnetic field without becoming unstable. That this does not occur in most superconductors (type I superconductors) is related to the fact that the creation of interfaces between normal and superconducting phases costs energy. The situation in the vicinity of such an interface is qualitatively portrayed in Fig. 10.22. The normally conducting state transforms to the superconducting state at $z=0$. The critical magnetic field B_c being applied parallel to the interface ($\perp z$) falls off exponentially over the London penetration length Λ_L (10.16) in the superconductor. In the superconductor the density of Cooper pairs

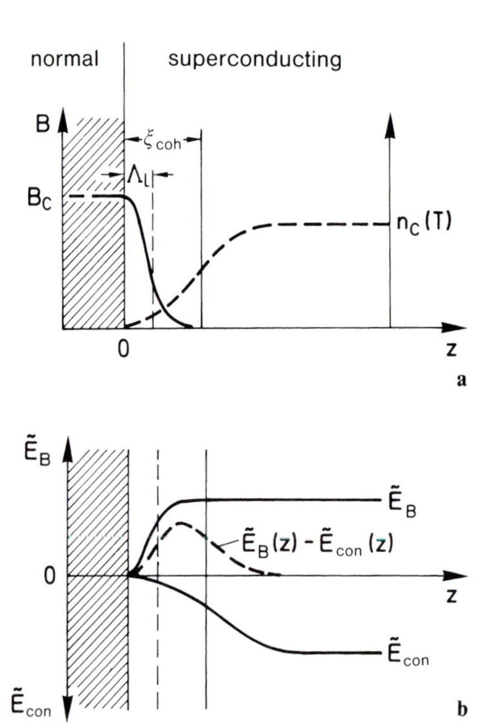

Fig. 10.22 a, b. Spatial variation of some important parameters of a superconductor near to the interface between a normal region (*shaded*) and a superconducting region ($z>0$). (**a**) The critical magnetic field B_c is applied parallel to the interface layer in the normal region; in the superconductor, $B(z)$ (*full line*) decreases exponentially over the London penetration depth Λ_L. The rise in the density of Cooper pairs n_c (*dashed line*) is determined by the coherence length ξ_{coh}. (**b**) Spatial variation of the energy density $\tilde{E}_B$ associated with the expulsion of the magnetic field, and of the energy density $\tilde{E}_{con}$ due to Cooper pair condensation (*full lines*). The difference (*dashed curves*) determines the interfacial energy between the normal and superconducting phases

$n_c = n_s/2$ rises within the coherence length ξ_{coh} from zero to its thermodynamical equilibrium value $n_c(T)$. Here ξ_{coh} is assumed to be larger than Λ_L: the coherence length must always be larger than the spatial extent of a Cooper pair ξ_{CP} (10.99), because the density of pairs can never change over a distance smaller than the "size" of a single pair. In the vicinity of $z=0$ two contributions to the energy density must be compared; namely a contribution associated with the expulsion of the magnetic field $\tilde{E}_B(z)$ and the energy density (10.53) due to the condensation of Cooper pairs $\tilde{E}_{con}(z)$. In a normal conductor both contributions vanish; deep inside the superconductor they compensate one another, i.e. for $z \to \infty$ and with a critical external field B_c

$$\tilde{E}_B(\infty) = -\tilde{E}_{con} = \frac{1}{2\mu_0} B_c^2 V \, . \tag{10.123}$$

Within the interfacial layer, however (Fig. 10.22 b), a finite difference remains because of the different characteristic decay lengths Λ_L and ξ_{coh}

$$\Delta\tilde{E}(z) = \tilde{E}_{con} - \tilde{E}_B = \frac{1}{2\mu_0} B_c^2 [(1 - e^{-z/\xi_{coh}}) - (1 - e^{-z/\Lambda_L})] \, . \tag{10.124}$$

Integrating across the whole boundary layer yields the interfacial energy (per unit area) which is necessary to create such an interface between the normal and superconducting phases

$$\gamma_{\mathrm{n/s}} = \int\limits_0^\infty \Delta \tilde{E}(z)dz = (\xi_{\mathrm{coh}} - \Lambda_{\mathrm{L}})\frac{1}{2\mu_0}B_{\mathrm{c}}^2 \,. \tag{10.125}$$

For the case $\xi_{\mathrm{coh}}>\Lambda_{\mathrm{L}}$ considered here, the loss of condensation energy is greater than the gain in energy due to expulsion of the magnetic field. To create such an interface a surface energy density of $\gamma_{\mathrm{n/s}}$ per unit area (10.125) must be added to the system. As one expects for a "normal" superconductor (type I superconductor), the creation of alternating normal and superconducting regions costs energy and is thus avoided.

The behavior is quite different in superconductors in which the magnetic field can penetrate deeper than the distance over which the superconducting state "switches on", i.e. superconductors for which $\xi_{\mathrm{coh}}<\Lambda_{\mathrm{L}}$. In such type II superconductors the creation of interfaces between normal and superconducting regions is energetically favorable, and for certain external magnetic fields a so-called "mixed state" exists, in which alternate normal and superconducting regions border on one another. The two lengths which are decisive for the existence of a type II superconductor are the decay length of the magnetic field Λ_{L} and the coherence length ξ_{coh} (larger than the "size" ξ_{CP} (10.99) of a Cooper pair), which determines the "switching on" of the superconducting state. As seen from (10.16, 10.99), these two lengths depend differently on the crucial superconducting parameters n_{s} (density of superconducting electrons) and Δ (gap energy). Both n_{s} and Δ are determined by, among other things, the strength of the electron-phonon interaction. This interaction also determines the mean free path $\Lambda(E_{\mathrm{F}})$ for electron-phonon scattering in the relevant material. It is clear therefore that a search for type II superconductors will be associated with attempts to vary the mean free paths for electron-phonon collisions. It is known that a reduction of the mean free path leads to the tendency $\xi_{\mathrm{coh}}<\Lambda_{\mathrm{L}}$. This result can also be obtained from the Ginzburg-Landau theory [10.14] for type II superconductors. A reduction of the mean free path is easy to achieve by alloying a certain amount of another metal with the superconductor. From experience, type II superconductors are in general metal alloys with relatively short mean free paths for electron-phonon scattering. Lead-bismuth alloys, for example, remain superconducting in magnetic fields up to 20 kG, that is, in fields twenty times as strong as the critical field for pure lead.

The difference between type I and type II superconductors is particularly evident in the magnetization curve $M(B)$ (Fig. 10.23). In a type I superconductor ($\xi_{\mathrm{coh}}>\Lambda_{\mathrm{L}}$), the magnetic field $\boldsymbol{B}=\mu_0\mu\boldsymbol{H}=\mu_0(\boldsymbol{H}+\boldsymbol{M})$ is completely expelled from the interior of the superconducting phase ($\boldsymbol{B}=\boldsymbol{0}$). The superconducting ring currents in the outer "shell" build up a magnetization $\boldsymbol{M}$ which is equal and opposite to the external field ($\boldsymbol{M}=-\boldsymbol{H}$). Up to the critical field strength $B_{\mathrm{c}}=\mu_0\mu H_{\mathrm{c}}$ the magnetization of the material grows in proportion to the external field and then breaks off at the transition to the normally conducting state (Fig. 10.23 a). For a type II superconductor – we

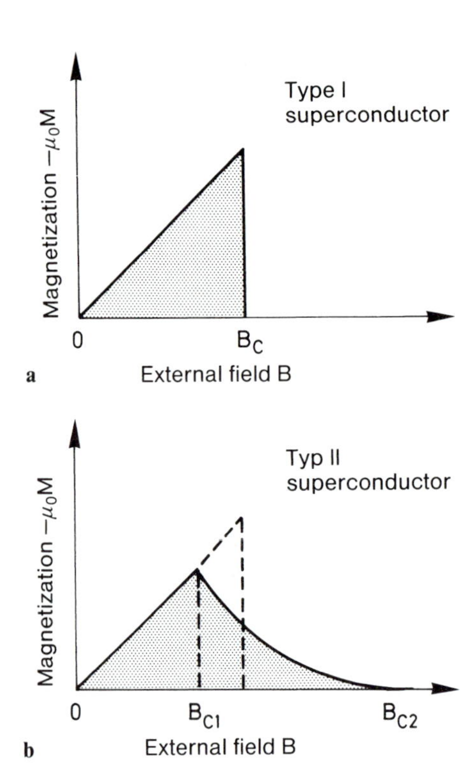

Fig. 10.23. Magnetization curve for a type I superconductor (**a**) and a type II superconductor (**b**). In a type I superconductor there is a single critical field B_c, whereas a type II superconductor has lower and upper critical fields, B_{c1} and B_{c2}

consider a sample in the form of a long rod in Fig. 10.23b – the gain in expulsion energy on penetration of the field is larger than the loss in condensation energy due to the spatial variation of the Cooper pair density. This penetration begins at a lower critical field B_{c1}; the magnetization M decreases monotonically with increasing magnetic field, until it vanishes completely at an upper field B_{c2}. The external field has then completely penetrated and destroyed the superconductivity. Analogous behavior is shown in measurements of the penetrating flux density B (mean internal magnetic field) in type II superconductors. In contrast to the behavior of a type I superconductor, shown in Fig. 10.18, the transition from a vanishing internal field ($<H_c$) to a linearly rising internal field ($>H_c$) is not sharp. The internal field already begins to build up slowly at a lower critical field strength $H_{c1}(<H_c)$ and then at an upper boundary value $H_{c2}(>H_c)$ takes on a linear behavior as in Fig. 10.18.

As in the case of type I superconductors, where the critical magnetic fields H_c and B_c are functions of temperature (Fig. 10.5), the upper and lower critical field strengths $B_{c1}(T)$ or $B_{c2}(T)$ also depend on the temperature. For type II superconductors there are thus three thermodynamically stable phases, which are usually called the superconducting ("Meissner") phase below $B_{c1}(T)$, the normally conducting phase above $B_{c2}(T)$ and the intermediate phase of the mixed state (also called the Shubnikov phase) be-

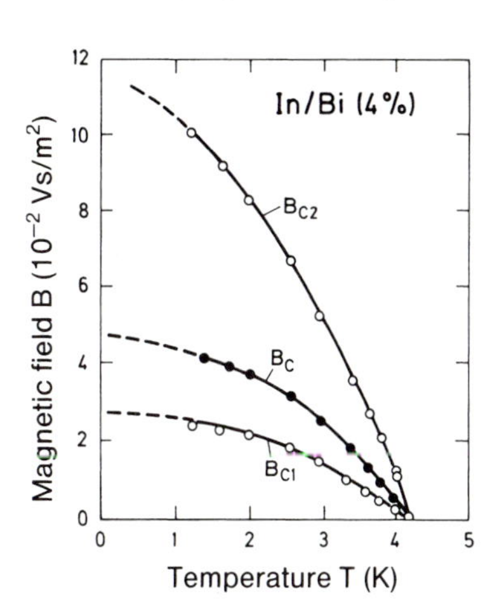

Fig. 10.24. Temperature dependence of the critical magnetic field of an indium-bismuth alloy (In+4 at. % Bi). (After [10.18])

tween the critical states at $B_{c1}(T)$ and $B_{c2}(T)$. An experimental example is shown in Fig. 10.24. Pure elemental In is a type I superconductor displaying only superconducting and normally conducting phases below and above $B_c(T)$ respectively. Alloying with 4% Bi creates a type II superconductor, whose upper and lower critical magnetic field curves $B_{c1}(T)$ and $B_{c2}(T)$ define the boundaries of the Meissner, Shubnikov and normally conducting phases.

While the Meissner phase consists of a homogeneous phase as in a type I superconductor, the Shubnikov phase consists of alternate superconducting and normally conducting regions. Although magnetic flux penetrates a type II superconductor, the superconducting regions locally expel the field, and are surrounded by screening supercurrents. The currents of course must have closed current paths, because only then can they be stationary. In the Shubnikov phase there is thus a stationary, spatially varying distribution of magnetic field strength and supercurrent density. It follows from the arguments presented in Sect. 10.9 that a closed supercurrent circuit can only enclose magnetic flux in integer multiples of the flux quantum $\phi_0 = h/2e \simeq 2\times10^{-7}\,\mathrm{G\,cm}^2$. One might therefore expect that in the Shubnikov phase the superconducting regions consist of superconducting ring currents which include exactly one flux quantum ϕ_0. Higher and lower densities of the superconducting electrons in this phase would then be reflected in a variation of the density of these individual "flux tubes". As a solution of the Ginzburg-Landau theory [10.14] shows, this is actually the case. The schematic representation in Fig. 10.25 shows that each flux quantum is associated with a system of ring currents that create the magnetic flux through the tube. With increasing external field B_{ext} the distance between the flux tubes (flux vortices) becomes smaller. Because there is a repulsive interaction between the flux tubes, at least at small distances, it is clear that the state of minimum enthalpy is given by an ordered array of flux vortices on a two dimensional hexagonal lattice. This arrangement has indeed been

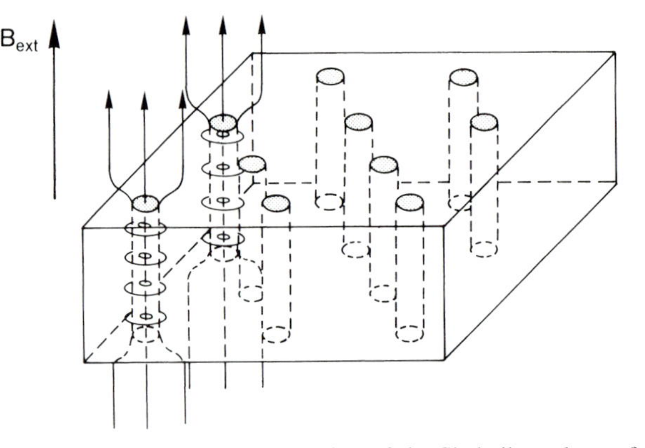

Fig. 10.25. Schematic representation of the Shubnikov phase of a type II superconductor. So-called flux tubes, arranged on a two-dimensional translational lattice, allow the magnetic field to "penetrate". Each flux tube is surrounded by superconducting ring currents and encloses one magnetic flux quantum

found experimentally [10.19]. With neutron scattering it has been possible to measure the microscopic magnetic field in the Shubnikov phase of the type II superconductor niobium. The field distribution (Fig. 10.26) clearly shows the presence of a hexagonal array of flux tubes.

Our treatment of the Shubnikov phase has so far assumed the free mobility of the flux tubes. In reality, however, this only holds as a limiting case. Defects in the lattice, such as dislocations, grain boundaries, precipitates, etc., have the effect of limiting the mobility. As a result of the many possible interactions with such defects there exist preferred sites which spatially "pin" the flux vortices. As a consequence, one observes distortions of the magnetization curves (Fig. 10.23) and hysteresis effects which depend upon the pre-treatment of the material (annealing etc.).

On the other hand, the pinning of flux vortices by crystallographic defects can offer technical advantages. If a type II superconductor is carrying current in the Shubnikov phase, then a Lorentz force acts on the flux tubes in a direction perpendicular to both the current density and the magnetic field. The migration of flux tubes due to this effect causes losses, i.e., the transformation of electrical energy into heat. This energy can only be extracted from the current if an electrical voltage appears in the sample; thus an electrical resistance is created. This effect is reduced by the pinning of the flux tubes. Unlike a type I superconductor, a type II superconductor does not show a simple relation of the type (10.93) between critical current and critical magnetic field on the surface of the conductor. The strongly inhomogeneous structure of the Shubnikov phase in such a conductor forces the magnetic field and current carrying regions into the interior of the material. The critical current and critical magnetic field depend in a complicated manner on a variety of material parameters, such as coherence length, mean free path, degree of crystallographic perfection, etc.

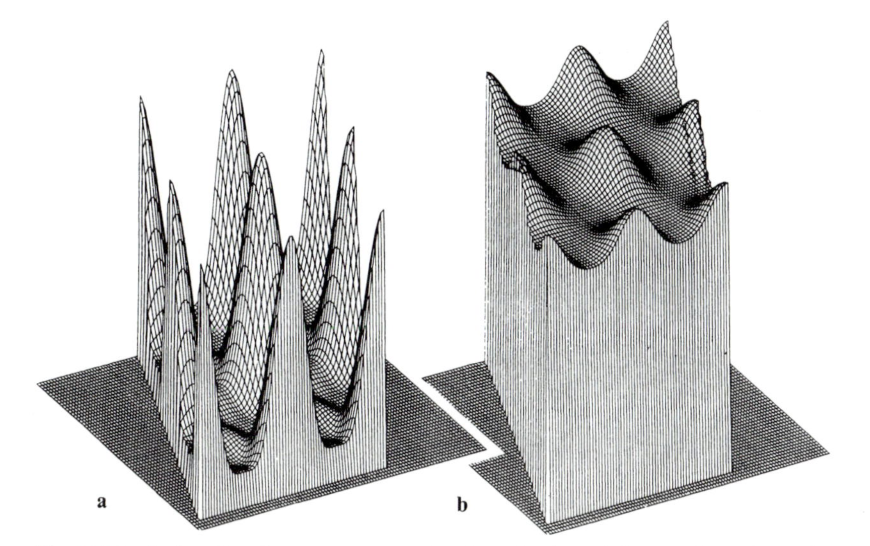

Fig. 10.26 a, b. Results of a neutron scattering investigation of the two-dimensional flux tube lattice of the type II superconductor niobium in the Shubnikov phase (4.2 K). A detailed evaluation of the diffraction pattern yields the distribution of the microscopic magnetic field around a flux vortex as shown. Macroscopic field strength $B=0.056$ T (**a**) and $B=0.220$ T (**b**). In (**a**) the nearest neighbor distance in the hexagonal lattice amounts to 206 nm, the maximum field at the center of the flux tube is 0.227 T. The corresponding values for (**b**) are 104 nm and 0.255 T (the upper critical field B_{c2} is 0.314 T). (After [10.19])

Finally it should be emphasized that important technological applications of superconductors are based chiefly on the possibility of creating very high magnetic fields via the loss-free superconducting currents. Once the magnetic field is created, then in principle no electrical power is required to maintain it. A condition for the functioning of such a high power superconducting magnet is of course the existence of materials with sufficiently high critical magnetic field $B_{c2}(T)$. This is only possible in the case of type II conductors. At present, materials such as Nb-Ti alloys are used to produce the solenoids. At $T \simeq 0$ they have a critical field of about 130 kG. Even higher fields can be created with Nb_3Sn, whose maximum critical field is above 200 kG. In operation, the solenoid must of course remain below the critical temperature T_c; in general it is held at the temperature of liquid helium (4.2 K).

10.11 "High-Temperature" Superconductors

Large-scale applications of superconductors, particularly their use for the transport of high electrical power over large distances, have so far been hindered by the necessity to cool the material below the critical temperature T_c to achieve the superconducting state. Because T_c is generally very small this usually means cooling to 4.2 K, the temperature of liquid helium. Exciting technological applications could be imagined if cooling were necessary only to a temperature of about 70 K, the temperature of liquid nitrogen. Ever since investigations of superconductivity began, a major aim has been to find superconductors with the highest possible transition temperature T_c, ideally close to room temperature. Based on the general postulates and conclusions of BCS theory there are two main possibilities:

i) A particularly strong electron-phonon interaction V_0, possibly with an associated tendency to lattice instability, could lead to a high transition temperature through a relation of the type (10.79). A very high density of electronic states $Z(E_F^0)$ at the Fermi level would have a particularly favorable effect.
ii) A novel many-electron interaction, not necessarily due to phonons, could make possible a “condensation” of the Fermi sea into Cooper pairs. If this were to happen through the interaction of (quasi) particles (e.g. the electrons themselves) with considerably lower mass than the lattice particles (phonons), one could also expect a much higher critical temperature T_c due to the much higher ω_D (10.79).

A decisive breakthrough in superconductivity research occurred in 1986, when Bednorz and Müller [10.20] discovered that metallic, oxygen-deficient copper oxide compounds of the Ba-La-Cu-O system showed a transition temperature of about 30 K. Figure 10.27 shows the original measurements of the resistance of the polycrystalline oxide $Ba_{0.75}La_{4.25}Cu_5O_{5(3-y)}$ ($y>0$, exact

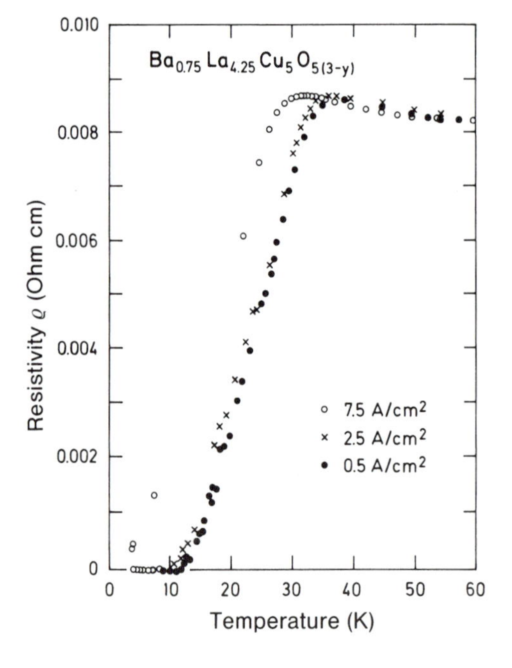

Fig. 10.27. Temperature dependence of the resistivity of a polycrystalline ceramic oxide of the type $Ba_{0.75}La_{4.25}Cu_5O_{5(3-y)}$, measured at different current densities. (After [10.20])

value unknown), which at room temperature is quasi metallic. Depending on the current density, a significant drop in the resistivity was found to begin between 20 K and 30 K. The fall in the resistance is not sharp as in the case of "normal" superconductors (Fig. 10.1). It is possible to interpret these results in terms of a mixture of superconducting and normally conducting phases. With this work, Bednorz and Müller set the scene for an explosion of research interest in this new class of materials and may have discovered a new mechanism for superconductivity (Nobel Prize 1987).

Shortly after this Chu and coworkers succeeded in reaching transition temperatures of 90 K in ceramics of the $Ba_{1-x}Y_xCuO_{3-y}$ system [10.21]. Following the development of improved preparation methods for single crystals and layers, and better methods of analysis, the alloy $YBa_2Cu_3O_{7-x}$, sometimes known as YBCO, emerged as the most interesting material. The Y atom can also be replaced by other rare-earth atoms such as Eu or Gd. This defines a whole class of materials, which, on account of their stoichiometry, are called 123 materials, with critical temperatures of up to 95 K. Figures 10.28 and 10.29 show typical results for the temperature-dependent electrical resistivity and the magnetic behavior as measured by numerous groups throughout the world. For an oxygen deficiency of x=0–0.1, the specific electrical resistance falls at about 90 K to immeasurably small values whereas, at higher temperatures, it shows metallic behavior (Fig. 10.28). With increasing oxygen deficiency x, the transition temperature shifts to lower values, whilst in the temperature range of normal conductivity the resistance increases further. Figure 10.29 shows results for the temperature-dependent magnetization of a YBCO film grown by ion sputtering. The measurements were made using a vibration magnetometer in which the sample vibrates in an external magnetic field B_{ext} produced by coils. The vibrational frequency, which is a function of the sample's magnetization, M, is then measured. The results were ob-

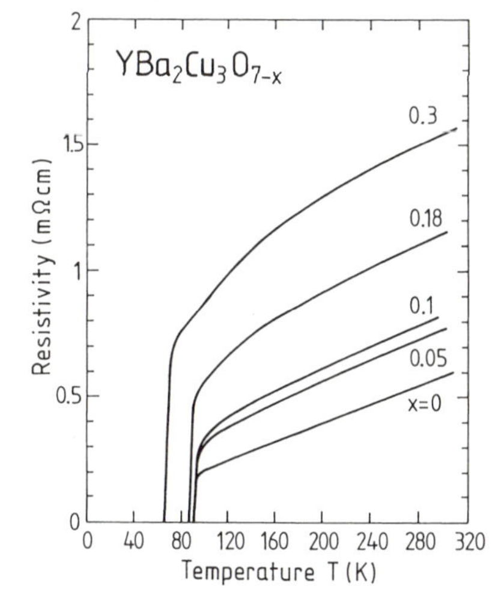

Fig. 10.28. Resistivity of sintered $YBa_2Cu_3O_{7-x}$ samples of various oxygen deficiencies x=0–0.3. The oxygen content was fixed by tempering the samples in a stream of oxygen at various temperatures and then quenching in liquid nitrogen. (After [10.22])

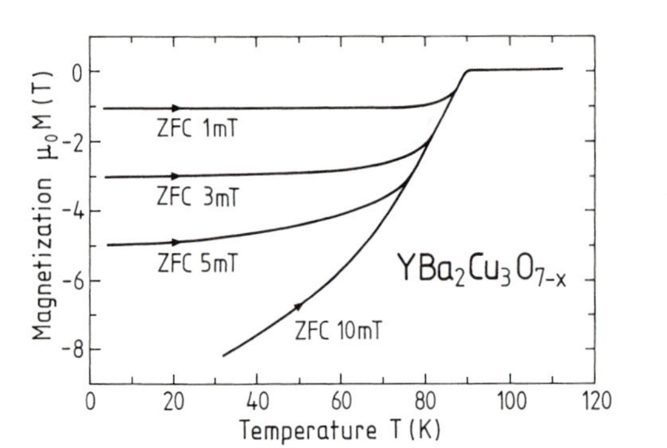

Fig. 10.29. Magnetization as a function of temperature for a YBCO film produced by sputtering. The sample was cooled in zero magnetic field (ZFC). The magnetic field then applied ranged from 1 to 10 mT ($B \parallel$ c-axis). (After [10.23])

tained by cooling the sample with no external field (ZFC: zero field cooled). For ideal diamagnetism and on account of the relation $B = \mu_0 M + B_{ext}$, the magnetization $\mu_0 M$ below the transition temperature should reach the same magnitude as the external field. However, as can be seen in Fig. 10.29, at low temperatures $T \ll T_c$, i.e., below 10 K, the values of $\mu_0 M$ reach something like a thousand times the value of the external field B_{ext} (also indicated on the figure). This is due to the demagnetization factor, which, for the case of a thin film, is markedly different to that of a long rod. Furthermore, in the ideal case one would expect an abrupt transition for $\mu_0 M$ at the transition temperature T_c. The "dragging out" of the transition to $T < T_c$ is partly attributable to inhomogeneities in the sample. Other factors include the pinning of flux lines at crystal defects and a thermal activation of the motion of flux lines. Results such as those depicted in Fig. 10.29 are typical for ceramic high-temperature superconductors of the YBCO type.

A further class of superconductors that display even higher transition temperatures are the Bi- and Tl-containing materials discovered in 1988. Their typical stoichiometry is $Bi_2Sr_2Ca_2Cu_3O_{10}$ [10.24] or $Tl_2Ba_2Ca_2Cu_3O_{10}$ [10.25]. Table 10.2 gives a compilation of the main classes of ceramic superconductors with transition temperatures above 90 K [10.26]. It has become common to give the Bi- and Tl-containing materials names such as 2212 or 2223, according to their stoichiometry.

Although the crystal structure of the YBCO ceramics differs in detail from that of the Bi and Tl materials, both classes display a characteristic layer structure that plays an essential role in the mechanism of the superconductivity (Fig. 10.30). This layer structure can be clearly seen in high resolution electron micrographs of the materials. Figure 10.31 is an image of a $Bi_2Sr_2CaCu_2O_{8+x}$ layer taken with high resolution transmission electron microscopy (TEM). According to Fig. 10.30a, the YBCO materials are axial crystals with alternating CuO_2 planes [Cu(2), O(2)] and oxygen atoms in both pyramid-type and rectangular planar co-ordination along the c-axis. These rectangular planar structure elements form oxygen chains along the b-axis of the crystal. It is the oxygen vacancies in this chain of oxygen atoms that are decisive for the superconductivity. The Bi and Tl ceramics

Table 10.2. Important families of high temperature superconductors. Corresponding to their stoichiometries they have become known by abbreviated names such as Tl–2212. (After [10.26])

Formula	Abbreviation	Highest T_c reached so far [K]
$REBa_2Cu_3O_7$	RE BCO or 123	92 (YBCO)
(RE = Rare Earths = Y, Eu, Gd, ...)		
$Bi_2Sr_2Ca_{n-1}Cu_nO_{2n+4}$ (+Pb doping)	BSCCO or Bi-22 $(n-1)n$	90 (Bi 2212) 122 (Bi-2223) 90 (Bi-2234)
$Tl_2Ba_2Ca_{n-1}Cu_nO_{2n+4}$	TBCCO or Tl-22 $(n-1)n$	110 (Tl-2212) 127 (Tl-2223) 119 (Tl-2234)
$TlBa_2Ca_{n-1}Cu_nO_{2n+3}$ (A=Sr, Ba)	Tl-12 $(n-1)n$	90 (Tl-1212) 122 (Tl-1223) 122 (Tl-1234) 110 (Tl-1245)
$HgBa_2Ca_{n-1}Cu_nO_{2n+2}$	Hg-12 $(n-1)n$	96 (Hg-1201) 128 (Hg-1212) 135 (Hg-1223)

display a similar layer structure along the c-axis, but with a significantly larger unit cell (Fig. 10.30b). Here one can clearly identify two regions of the unit cell: In the one region (above) the copper atoms are centered and in the other (below) they lie at the corners of the Cu–O planes. The two parts of the cell are separated by two Bi–O (or Tl–O) planes. The various Bi and Tl ceramics (Table 10.2) differ from one another in the number of "inserted" CuO_2 planes. This in turn has a strong influence on the transition temperature.

As might be expected from their strongly anisotropic crystal structures, the ceramic superconductors also display highly anisotropic electronic properties, in particular the superconductivity. Temperature-dependent resistance curves for YBCO layers (Fig. 10.32) show large differences in the specific resistivities ϱ_c and ϱ_{ab} measured along, and perpendicular to the c-axis, respectively. The resistivity ϱ_c has a broad minimum at about 150 K, the markedness of which is probably connected to the degree of oxygen depletion.

The ceramic high-temperature superconductors also have very interesting properties in relation to their critical magnetic fields. All the materials known to date are type II superconductors (Sect. 10.10) which display a Shubnikov phase between a lower critical field B_{c1} (or H_{c1}) and an upper

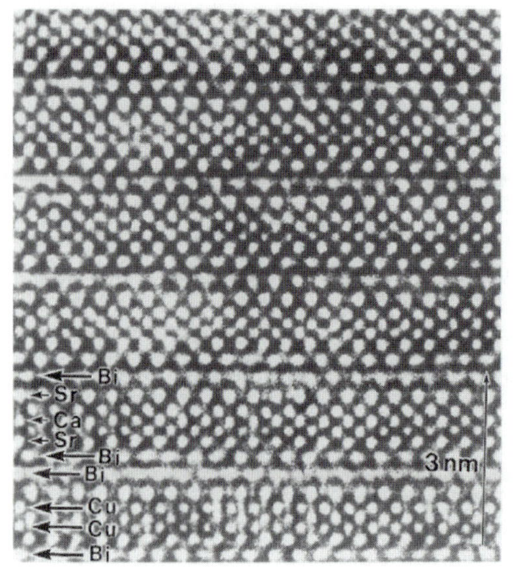

Fig. 10.30. Crystal structure of YBCO and Bi and Tl superconducting ceramics. (**a**) Unit cell of a $YBa_2Cu_3O_7$ crystal. The rectangular planar and pyramid-like coordination of the O atoms along the c-axis is indicated. The numbers in brackets denote the special sites of O and Cu atoms in the CuO_2 planes. (**b**) Unit cell of $Bi_2Sr_2Ca_2Cu_3O_{10}$ or $Tl_2Ba_2Ca_2Cu_3O_{10}$ ceramics (2223 phases)

critical field B_{c2} (or H_{c2}). Although the lower critical field B_{c1} lies below 10 mT, the upper critical field B_{c2} of $YBa_2Cu_3O_{7-x}$ perpendicular to the c-axis exceeds that of all other known superconductors with an estimated value of 340 T. Here again one can observe the high anisotropy of the ceramic superconductors: For external magnetic fields parallel to the c-axis, the B_{c2}

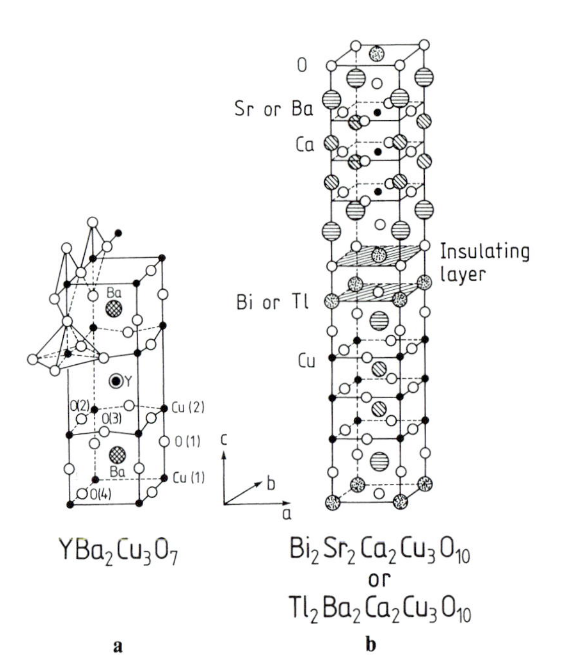

Fig. 10.31. High-resolution transmission electron microscope (TEM) picture of a $Bi_2Sr_2CaCu_2O_{8+x}$ film. For transmission perpendicular to the c-axis of the crystal the layer structure of the superconductor is visible. Light and dark points correspond to the positions of rows of atoms. (After [10.27])

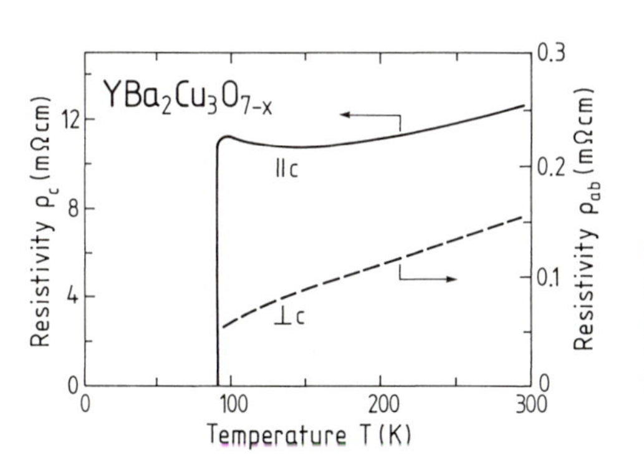

Fig. 10.32. Resistivity of YBCO films as a function of temperature measured parallel to the c-axis (ρ_c, *left-hand scale*) and in the ab-plane (ρ_{ab}, *right-hand scale*). (After [10.28])

values are about a factor of 5 to 7 times lower than those found for a field orientation perpendicular to the c-axis. The critical magnetic fields allow one to estimate a coherence length ξ_c of about 3–5 Å in the c-direction, whereas, in the basal plane, a value of 20–30 Å is found for ξ_{ab}. For YBCO layers grown with good crystallinity and well oriented on suitable substrate material, critical current densities of 10^6–10^7 A/cm^2 have been measured at zero magnetic field and at a temperature of 77 K.

There is as yet no generally accepted theoretical description of the mechanism of superconductivity in these ceramic materials. Nonetheless, a great many results have been gathered and there is good reason to hope that these will soon enable a consistent theoretical picture to be constructed. It is known from the observation of flux quantization (Sect. 10.9) that Cooper pairs are responsible for the superconductivity, as in the case of BCS superconductors. For most high-temperature superconductors these Cooper pairs consist not of two electrons but of two holes. As in semiconductors too (Sects. 9.2, 12.2), a missing electron in a chemical bond or an empty state in a nearly full electronic band acts as a positive charge carrier, or hole, in the conduction process. One speaks of hole conductivity or p-conductivity. The positive charge of the Cooper pairs, i.e., the p-type superconductivity is deduced from the sign of the Hall constant (Panel XIV). Corresponding to the high critical temperatures T_c, onc finds band-gap energies Δ in the range 20–30 meV. These are established using various experimental methods including tunneling spectroscopy (Panel IX). The characteristic ratio $\Delta(0)/kT_c$ of (10.67) thus reaches values between 3 and 4 which are relatively high compared to the BCS value of 1.764.

Another important feature that will have to be explained by any complete theory of high-temperature superconductivity is the dependence of the transition temperature T_c on the oxygen deficiency x in the stoichiometry $YBa_2Cu_3O_{7-x}$, as seen in Fig. 10.28. With increasing loss of oxygen x the transition temperature changes in a step-like fashion from 90 K to about 30 K (Fig. 10.33). Above $x \simeq 0.7$ the YBCO ceramics lose their superconducting properties and display insulating antiferromagnetic behavior. These observations, together with a series of experimental investigations [10.29,

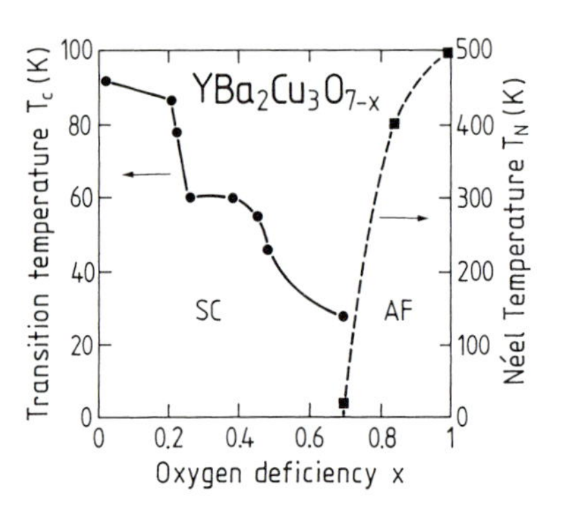

Fig. 10.33. Experimentally determined phase diagram of the ceramic $YBa_2Cu_3O_{7-x}$, i.e., the dependence of the critical temperature T_c on the oxygen deficiency x. In the region marked SC the material is superconducting, whereas AF denotes an antiferromagnetic phase. (After [10.29])

10.30], support the following conclusions: Oxygen atoms with their high electronegativity act as electron acceptors in the YBCO structure. The metal ions Y and Ba each donate two electrons to the bonds to the CuO_2 planes. The oxygen atoms "swallow" these electrons. At low oxygen depletion, i.e., small x, plenty of oxygen is available to accept the electrons. Thus more holes are provided to form the hole Cooper pairs in the CuO_2 planes. This idea leads to a general concept which also can be applied to the Tl and Bi superconductors: CuO_2 layers are responsible for a quasi-two-dimensional charge transport by means of holes which form Cooper pairs. Between the CuO_2 layers are layers containing alkali or rare-earth metals and oxygen or halogens. These act as charge-carrier reservoirs. These layers can take up electrons from the CuO_2 layers thus creating Cooper pairs of holes in the so-called conduction planes.

More recently, materials have been discovered which display high-temperature superconductivity with electron Cooper pairs, i.e., conventional n-type superconductivity. These are the compounds Nd_2CuO_4 and $Nd_{2-x}Ce_xCuO_4$ [10.31]. Although there is still no generally valid theoretical description of the superconductivity in the copper oxide materials, very interesting technical applications are already being put through their paces. An example is the use of "high temperature" SQUIDS (Panel X) at measuring temperatures of 77 K for detecting and recording the electrical currents in the human heart and brain, thus opening up new diagnostic possibilities [10.32].

In 1991 a further novelty arrived in the field of superconductivity: The fullerenes, themselves only discovered in 1985 (Sect. 1.2), also display superconductivity under certain conditions, with transition temperatures above 15 K [10.33]. A single fullerene molecule consists of 60 carbon atoms (C_{60}) which form a closed ball of diameter 7.1 Å between carbon nuclei (Fig. 1.5). The carbon atoms form a slightly buckled structure of 20 hexagons and 12 pentagons, similar to the structures found in graphite, benzene, and other organic molecules. As already described in Sect. 1.2, the carbon electrons occupy sp^2 hybrid states and p_z states. In the molecule the p_z states form occupied π orbitals and unoccupied π^* orbitals which

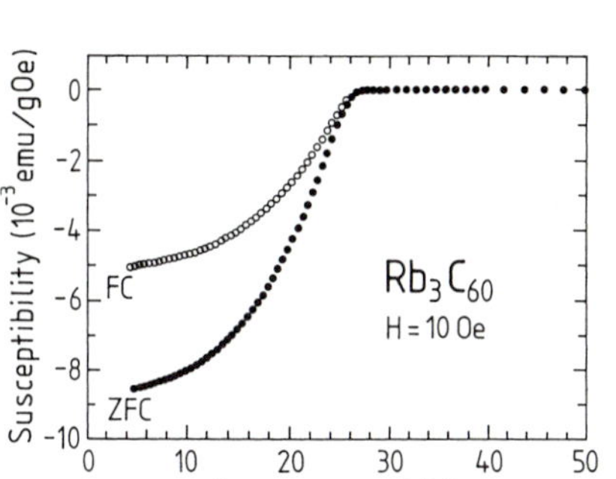

Fig. 10.34. Temperature dependence of the magnetic susceptibility of Rb_3C_{60}. The measurement was carried out in an external magnetic field of 10 Oe. ZFC: Zero-field cooled. FC: field cooled. (After [10.34])

are delocalized over the entire surface of the ball. At room temperature and below C_{60} molecules crystallize with a face-centered cubic structure of lattice constant 14.16 Å. In the crystallization the strongest interaction is between the outer valence electrons of the C_{60} molecules, i.e., between the π and π^* orbitals. This interaction gives rise to a relatively narrow π band which is fully occupied and an empty π^* band. The two bands are separated in energy by about 1.5 eV making the C_{60} crystal a semiconductor.

If a crystalline film of C_{60} is exposed to potassium vapor then the layer becomes metallic. Potassium atoms are incorporated into the gaps between the C_{60} spheres. A maximum conductivity of about 500 S cm^{-1} is obtained for a stoichiometry of K_3C_{60}, whereas a "full doping" of K_6C_{60} causes the conductivity to vanish once more. The optimum metallic properties of K_3C_{60} are attributed to the fact that the originally empty π^* states of the C_{60} are three-fold degenerate, enabling them to accept a maximum of 6 electrons per molecule. The incorporation of three K atoms per C_{60} atom means that the three valence electrons of the potassium atoms produce a half-full π^* band. This is the prerequisite for metallic conductivity. The metallic material K_3C_{60} also turns out to be superconducting with a transition temperature of $T_c = 19.3$ K. Even higher critical temperatures have been found for other similar fullerenes, such as: Rb_3C_{60} ($T_c \simeq 28$ K), $RbCs_2C_{60}$ ($T_c \simeq 33$ K), and $Rb_{2.7}Tl_{2.2}C_{60}$ ($T_c \simeq 48$ K). Because these materials react strongly with atmospheric oxygen, the studies that have so far been carried out have involved mainly magnetic measurements. Figure 10.34 shows the measured temperature dependence of the magnetic susceptibility of Rb_3C_{60} [10.34]. As in Fig. 10.29, the sample was cooled in zero magnetic field (ZFC). One observes, similar to the case of ceramic high-temperature superconductors, a gradual onset of diamagnetic behavior below T_c.

Problems

10.1 Superconductivity is a thermodynamic state described at constant pressure by the two state variables temperature T and magnetic field strength H or B.

a) Starting with the differential form of the Gibb's free energy

$$dG = -SdT - MdB$$

(S entropy, M magnetization) derive the following relation

$$\frac{dH_c}{dT} = \frac{S_n - S_s}{\mu_0(M_s - M_n)}$$

for the temperature dependence of the critical magnetic field $H_c(T)$. Use the continuity of $G(T)$ across the phase boundary between normal (n) and superconducting (s) states (μ_0 magnetic permeability).

b) Using the fact that the superconducting state displays perfect diamagnetism ($B=0$) while the normal state has negligible diamagnetism ($M \cong 0$), show that the entropy changes across the phase boundary by an amount

$$S_n - S_s = -\frac{V}{\mu_0} H_c \frac{dH_c}{dT}$$

and derive an expression for the latent heat of the transition from the normal to the superconducting state in a magnetic field.

10.2 Discuss the high frequency behavior of a superconductor by using the complex conductivity $\tilde{\sigma}(\omega) = \sigma_1 - i\sigma_2$ (Sect. 11.1) to describe time-dependent transport phenomena.

a) Show within the framework of the London equations that a superconductor has a purely imaginary conductivity at temperatures $T \ll T_c$ and finite frequencies. What is the physical reason?

   What consequences are expected for circuits based on superconducting devices?

b) Derive the frequency dependence of the imaginary part of the conductivity σ_2 and give a numerical estimate of σ_2 for a superconductor with a concentration of superconducting electrons $n_s = 10^{20}\,\mathrm{cm}^{-3}$ at a frequency of 10 GHz.

10.3 Discuss the temperature dependence of the thermal conductivity for a material changing from the normal state to the superconducting state. Consider both the "free" electron gas condensing into the BCS ground state and the phonon system.

10.4 Consider a simple BCS ground state of only two Cooper pairs (Sect. 10.4):

$$|\phi_{\rm BCS}\rangle \cong (u_k|0\rangle_k + v_k|1\rangle_k)(u_{k'}|0\rangle_{k'} + v_{k'}|1\rangle_{k'}) \,.$$

Calculate the energy gain $\langle\phi_{\rm BCS}|\mathcal{H}|\phi_{\rm BCS}\rangle$ by means of the Hamiltonian $\mathcal{H}$ in which scattering processes of the pairs $\boldsymbol{k}\rightleftarrows\boldsymbol{k}'$ are described by spin matrices. Compare the result with the general expression (10.41) for many Cooper pairs.

10.5 There is a close analogy between BCS theory and ferromagnetism:

a) Show that the BCS ground state energy $W_{\rm BCS}$ (10.42) can be derived from a BCS Hamiltonian as the expectation value $\langle\phi_{\rm BCS}|\mathcal{H}_{\rm BCS}|\phi_{\rm BCS}\rangle$, where

$$\mathcal{H}_{\rm BCS} = 2\sum_k \xi_k \sigma_k^{(3)} - \frac{V_0}{4L^3}\sum_{kk'}(\sigma_k^{(1)}\sigma_k^{(1)} + \sigma_{k'}^{(2)}\sigma_{k'}^{(2)})$$

and

$$\sigma_k^{(1)} = \begin{pmatrix} 0 & 1 \\ 1 & 0 \end{pmatrix}_k, \quad \sigma_k^{(2)} = \begin{pmatrix} 0 & -\mathrm{i} \\ \mathrm{i} & 0 \end{pmatrix}_k,$$

$$\sigma_k^{(3)} = \begin{pmatrix} 1 & 0 \\ 0 & -1 \end{pmatrix}_k$$

are the Pauli spin matrices in the x, y, z (or 1, 2, 3) directions.

b) The Hamiltonian $\mathcal{H}_{\rm BCS}$ resembles formally the Heisenberg model Hamiltonian (8.57) for interacting spins in an external magnetic field. In analogy to the theory of ferromagnetism (Sect. 8.6) construct the linearized Hamiltonian

$$\mathcal{H}_{\rm BCS} = -\sum_k \boldsymbol{\sigma}_k \cdot \boldsymbol{H}_k \,,$$

where the model field $\boldsymbol{H}_k$ consists of a component parallel to the z direction (3) and a generalized mean field component (average over neighboring spins) in the x, y plane (2, 3). Show that the formalism of BCS theory is obtained when the mean field component is identified with the gap energy Δ of the superconductor, and $2\theta_k$ in (10.48, 49) corresponds to the angle between the effective field $\boldsymbol{H}_k$ and the direction z (3).

c) Using the mean field approximation (Sect. 8.6) for the temperature dependence of the magnetization, derive a self-consistency relation for determining the temperature dependence of the superconductor gap $\Delta(T)$.

10.6 Calculate the reduction in energy $W^0_{\rm BCS} - W^0_{\rm n}$ (10.53) due to the transition to the superconducting state. $W^0_{\rm n}$ is the energy of the normally conducting state.

Hint: Use the form (10.52a) for the energy W^0_{BCS} of the BCS ground state and rewrite all summations as integrals ($L^{-3}\sum \to \int d\boldsymbol{k}/4\pi^3$).

For determining the limits of integration note that

$$u_k^2 = 0 \quad \text{for} \quad E < E_{\mathrm{F}}^0 - \hbar\omega_{\mathrm{D}} \ ,$$

$$v_k^2 = 0 \quad \text{for} \quad E > E_{\mathrm{F}}^0 + \hbar\omega_{\mathrm{D}} \ .$$

One also has $\Delta \ll \omega_{\mathrm{D}} \ll E_{\mathrm{F}}^0$, and the density of states $Z(E)$ close to the Fermi edge can be taken to be a constant. For the gap energy one should use the following representation which stems from (10.46, 10.49, 10.60):

$$\Delta = \frac{V_0}{L^3} Z(E_{\mathrm{F}}^0) \int\limits_{-\hbar\omega_{\mathrm{D}}}^{\hbar\omega_{\mathrm{D}}} d\varepsilon \frac{\Delta}{\sqrt{\varepsilon^2 + \Delta^2}} \ .$$

Panel IX
One-Electron Tunneling in Superconductor Junctions

Tunnel experiments play an important role in superconductor research. Firstly, they are an important means of determining the gap Δ (or 2Δ) in the excitation spectrum of the superconductor, and secondly, inelastic contributions to the tunnel current provide information about the characteristic excitations in the insulating barrier itself. Tunneling is a general quantum mechanical phenomenon: An atomic particle, e.g. an electron (mass m) penetrates a potential barrier (height $\hat{V}_0$, width d, Fig. IX.1), although classically its kinetic energy is not sufficient to surmount the barrier ($E<\hat{V}_0$). Wave mechanically, however, there is a finite probability of finding the particle in the region c behind the barrier when it approaches from region a (Fig. IX.1). A quantitative description is obtained by solving the Schrödinger equations

$$\frac{d^2\psi}{dx^2}+\frac{2m}{\hbar^2}E\psi=0 \qquad \text{(IX.1 a)}$$

in regions a and c, and

$$\frac{d^2\psi}{dx^2}+\frac{2m}{\hbar^2}(E-\hat{V}_0)\psi=0 \qquad \text{(IX.1 b)}$$

in region b. A wave approaches in region a and is partly reflected from the barrier, similarly in region b. One obtains the following solutions to the Schrödinger equation (IX.1) in the different regions:

$$\psi_a=A_1\mathrm{e}^{\mathrm{i}kx}+B_1\mathrm{e}^{-\mathrm{i}kx}$$

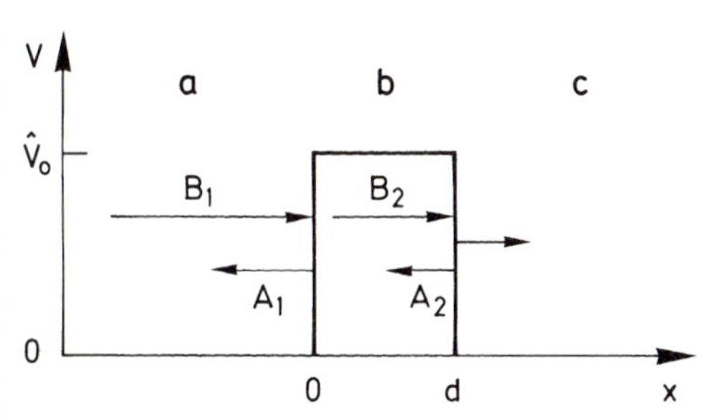

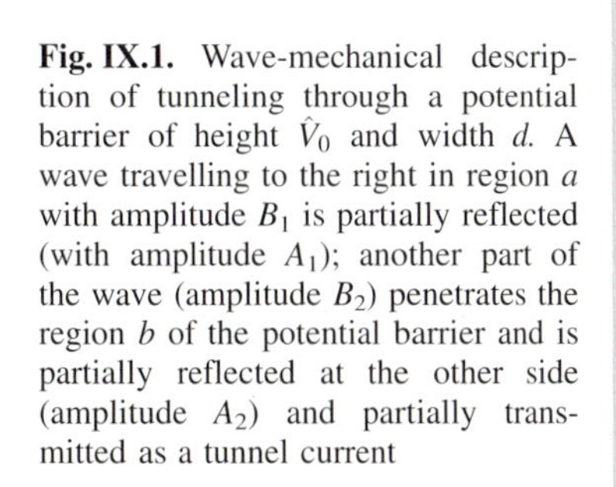
Fig. IX.1. Wave-mechanical description of tunneling through a potential barrier of height $\hat{V}_0$ and width d. A wave travelling to the right in region a with amplitude B_1 is partially reflected (with amplitude A_1); another part of the wave (amplitude B_2) penetrates the region b of the potential barrier and is partially reflected at the other side (amplitude A_2) and partially transmitted as a tunnel current

with

$$k = \frac{1}{\hbar}\sqrt{2mE}\ , \tag{IX.2 a}$$

$$\psi_b = A_2 e^{ik'x} + B_2 e^{-ik'x}$$

with

$$k' = \frac{1}{\hbar}\sqrt{2m(E - \hat{V}_0)}\ , \tag{IX.2 b}$$

$$\psi_c = e^{-ikx}\ . \tag{IX.2 c}$$

The wave amplitudes are normalized to the region c, since this is the interesting region. At the interfaces of the potential barrier $x=0$ and $x=d$, the wavefunctions (IX.2) and their first derivatives must be continuous in order to conserve total current. This leads to the conditions

$$A_1 = \frac{1}{2}\left(1 + \frac{k'}{k}\right)A_2 + \frac{1}{2}\left(1 - \frac{k'}{k}\right)B_2\ , \tag{IX.3 a}$$

$$B_1 = \frac{1}{2}\left(1 - \frac{k'}{k}\right)A_2 + \frac{1}{2}\left(1 + \frac{k'}{k}\right)B_2 \tag{IX.3 b}$$

for $x=0$, and analogously, for $x=d$, we have

$$A_2 = \frac{1}{2}\left(1 - \frac{k}{k'}\right)e^{-i(k+k')d}\ , \tag{IX.4 a}$$

$$B_2 = \frac{1}{2}\left(1 + \frac{k}{k'}\right)e^{-i(k-k')d}\ . \tag{IX.4 b}$$

Since we are interested in tunneling from a to c, the most important quantity is the ratio of the wave amplitudes in a and c. By inserting A_2 and B_2 from (IX.4) into (IX.3), the amplitude of the incoming wave in a (relative to that in c) is

$$B_1 = \frac{1}{4}\left(1 - \frac{k'}{k}\right)\left(1 - \frac{k}{k'}\right)e^{-i(k+k')d} + \frac{1}{4}\left(1 + \frac{k'}{k}\right)\left(1 + \frac{k}{k'}\right)e^{-i(k-k')d}\ . \tag{IX.5}$$

For the case $E<\hat{V}_0$ relevant to a tunnel experiment, we insert in (IX.2b) $k'=i\kappa$, so that

$$k' = i\kappa = \frac{1}{\hbar}\sqrt{2m(\hat{V}_0 - E)}\ . \tag{IX.6}$$

The probability of finding the incoming particle in region a, relative to its probability of being in c, is thus

$$B_1 B_1^* = \frac{1}{2} - \frac{1}{8}\left(\frac{k}{\kappa} - \frac{\kappa}{k}\right)^2 + \frac{1}{8}\left(\frac{k}{\kappa} + \frac{\kappa}{k}\right)^2 \cosh 2\kappa d \; . \tag{IX.7}$$

This probability increases approximately exponentially (for large κd) with the probability of finding the particle in region c. Conversely it follows that for a given intensity of the incoming wave $B_1 B_1^*$, the intensity of the wave penetrating the barrier decreases approximately exponentially with the thickness d and height $\hat{V}_0$ of the barrier

$$\psi_c \psi_c^* \sim \exp\left[-\frac{2}{\hbar} d \sqrt{2m(\hat{V}_0 - E)}\right] \; . \tag{IX.8}$$

Electrons are thus able to tunnel through an insulating layer between two conductors provided the layer has a sufficiently small thickness d. The tunnel current through such a layer depends exponentially on the energetic height $\hat{V}_0$ of the barrier and the kinetic energy E of the particles, i.e., on the voltage applied across the tunnel junction (IX.8). In real tunnel experiments, one uses two conductors (e.g. metals) separated by an insulating layer (typically a metal oxide film with a thickness of 10–100 Å). A voltage U applied between the two conductors causes a tunnel current $I(U)$ which depends exponentially on U over a wide voltage range. Tunnel experiments can be conveniently discussed using the potential well model (Fig. 6.13) for metals (normally conducting state), or for the case of superconductors, using the electronic density of states of the excitation spectrum (Fig. 10.12). If a voltage U is applied between two normally conducting metals separated by an insulating barrier of thickness d, then the Fermi levels are shifted by eU relative to one another. Occupied electronic states in the negatively biased metal match empty states at the same energy in the other metal, and a current can flow through the barrier (Fig. IX.2a). This charge transport is largely "elastic" since the electrons experience no change in their energy on crossing the barrier. For small U the exponential voltage dependence $I(U)$ is almost linear (Fig. IX.2b). If the negatively biased metal is in the superconducting state (Fig. IX.2d), there is a gap Δ in the excitation spectrum between the BCS many-body ground state (occupied by Cooper pairs) and the continuum of one-particle states. At $U=0$, only fully occupied states match and elastic tunneling is not possible (Fig. IX.2c), because the breaking up of a Cooper pair is associated with a change in energy. For small voltages $U<\Delta/e$, current flow is thus impossible. The situation changes when the bias voltage reaches the value Δ/e. The Fermi level of the normally conducting metal now lies Δ below the BCS ground state (Fig. IX.2d). The breaking up of a Cooper pair and the simultaneous flow of current become possible. One electron of a Cooper pair is excited into the continuum of one-electron states in the supercon-

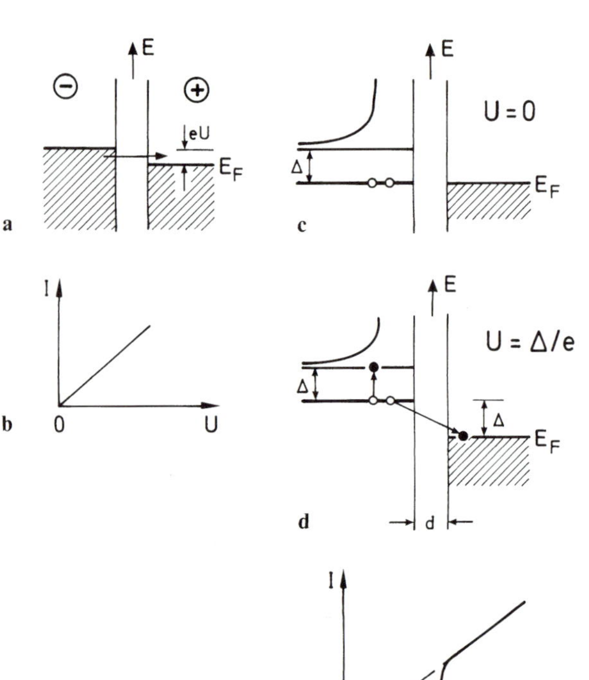

Fig. IX.2 a–e. Schematic representation of the tunneling of a single electron through an insulating layer (barrier) of thickness *d*. The occupied conduction band of the normally conducting metal is shown shaded up to the Fermi energy E_F. Superconductors are characterized by their BCS ground state (a single line on the energy axis E) and by the density of one-particle states ($E>E_F+\Delta$). (**a**) Tunneling of single electrons between two normally conducting metals through an insulating layer across which an external voltage U is applied. (**b**) Schematic I–U characteristic for one-electron tunneling between two normal conductors. (**c**) Tunnel contact between a superconductor (*left*) and a normal conductor (*right*) in thermal equilibrium ($U=0$). A Cooper pair in the BCS ground state is indicated schematically. (**d**) Splitting up of a Cooper pair and elastic tunneling of a single electron from the superconductor into the normal conductor on application of an external voltage $U=\Delta/e$. (**e**) Schematic I–U characteristic for the elastic tunneling of single electrons between a superconductor and a normal conductor

ductor, while the other gains the corresponding energy by tunneling into an empty state of the normally conducting metal. At this threshold voltage, the current jumps and then increases almost linearly with U (Fig. IX.2e). As U rises beyond the value of Δ/e, ever more Cooper pairs are "dissociated" and can contribute to the tunnel current.

A measurement of the characteristic $I(U)$ as in Fig. IX.2e directly yields the gap energy Δ of the excitation spectrum of the superconductor. "Elastic" tunneling between two superconductors separated by an insulating film can be treated in a similar way (Fig. IX.3). For small voltages U (e.g. $U=0$ in Fig. IX.3) either empty or full states match in energy ($T=0$); no

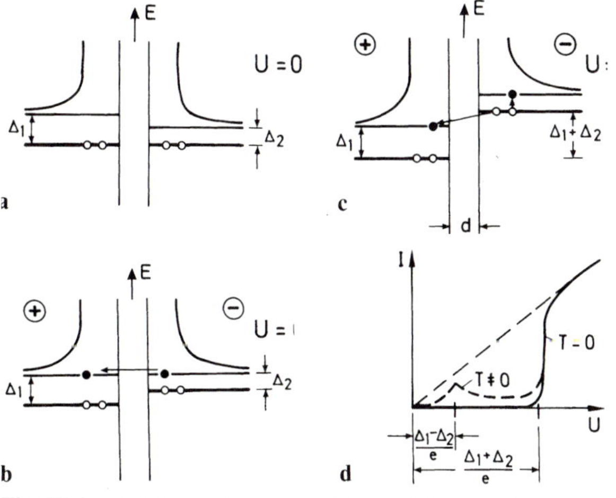

Fig. IX.3 a–d. Schematic representation of the tunneling process of single electrons between two different superconductors ($\Delta_1 > \Delta_2$) separated by a tunnel barrier of thickness d. The superconductors are characterised by their BCS ground state (single line on the energy axis) and by the one particle density of states ($E > E_F + \Delta$). (**a**) The situation at thermal equilibrium ($U=0$). (**b**) An external voltage $U=(\Delta_1-\Delta_2)/e$ is applied, so that at finite temperature ($0<T<T_c$) single electrons can tunnel. Cooper pairs can not yet break up. (**c**) An external voltage $U=(\Delta_1-\Delta_2)/e$ is applied, so that in the right-hand superconductor Cooper pairs can break up; one electron tunnels to the left and the other is simultaneous excited above the gap Δ_2. (**d**) Schematic I–U characteristic for elastic tunneling of single electrons between the two superconductors. At vanishingly low temperature ($T \approx 0$) only processes of type **c** exist with an onset at $(\Delta_1+\Delta_2)/e$, while at finite temperature ($0<T<T_c$) additional processes of type **b** are possible (*dashed line*)

tunnel current can flow. When the bias voltage reaches the value $U=(\Delta_1-\Delta_2)/e$ at $T \simeq 0$, then the Cooper pairs in one of the two superconductors can break up and occupy one-electron states in both superconductors by overcoming the tunnel barrier (Fig. IX.3 c); the tunnel current sets in suddenly (Fig. IX.3 d). If the tunnel junction is at a finite temperature below the critical temperatures T_c of both superconductors, then the one-electron states are partly occupied and above a voltage of $U=(\Delta_1-\Delta_2)/e$ an "elastic" tunnel current of normally conducting electrons can flow (Fig. IX.3 b). The tunnel characteristic $I(U)$ shows a maximum at $U=(\Delta_1-\Delta_2)/e$ (Fig. IX.3 d) because here the singularities in the two one-electron densities of states are at the same energy. With increasing bias voltage U these two points shift away from one another and the "elastic" tunnel current decreases again until the threshold $(\Delta_1-\Delta_2)/e$ is reached. By measuring the characteristics of the tunneling between two superconductors at $T \neq 0$, one can determine the gap energies Δ_1 and Δ_2 as in Fig. IX.3 d. Simple arguments show that the tunnel characteristics (Fig. IX.3 d, 2 e) are symmetric about $U=0$.

Because of the importance of such tunnel experiments for the determination of the gap energy Δ, we will say a few words about how they are carried out in practice. Tunnel junctions are generally produced by the successive evaporation of two metal films in the shape of a cross. The first film is evaporated (pressure $< 10^{-4}$ Pa) through a mask to form a strip on an insulating substrate (e.g. quartz). Then the tunnel barrier is formed by surface oxidation of the film. Oxygen can be admitted to the evaporation chamber (10^3–10^4 Pa), or alternatively the oxidation is carried out by a glow discharge in an oxygen atmosphere of 10–100 Pa. The second metal strip is then evaporated through a different mask, this time perpendicular to the first metal strip. Typical resistances of such tunnel arrangements lie in the range 100–1000 Ω. The actual tunnel barrier between the two perpendicular metal strips has an area of about 0.1–0.2 mm^2. After preparation of this "sandwich" structure in an evaporation chamber, the "tunnel package" and its contacts are mounted in a cryostat like that in Fig. IV.1. The cur-

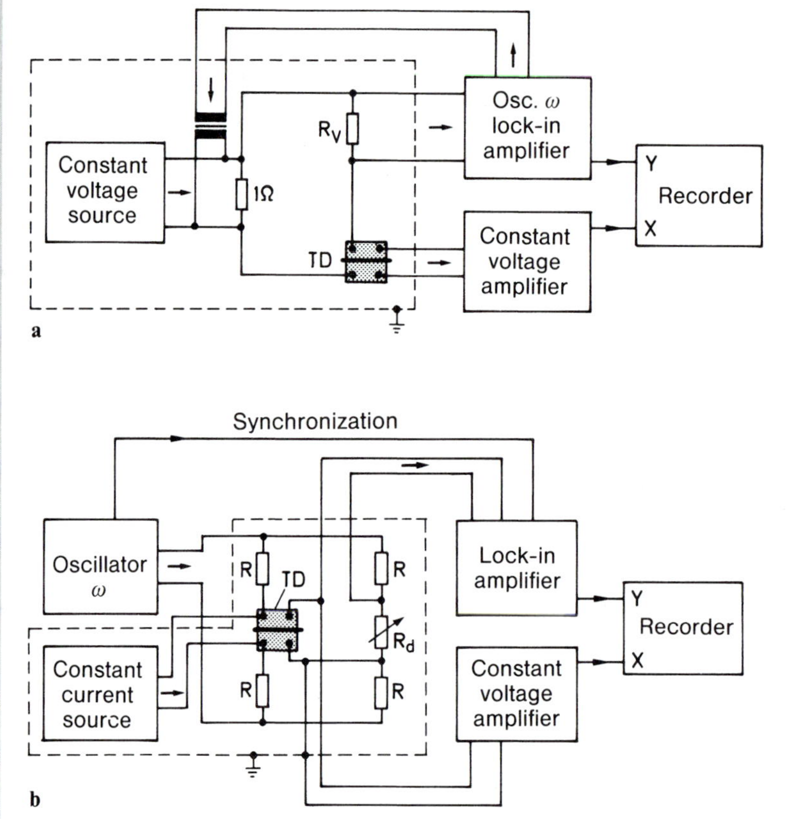

Fig. IX.4 a, b. Block diagrams of the circuits for measuring tunnel characteristics. (**a**) Circuit for determining dU/dI curves (gap curves) as a function of tunnel voltage U at small energies eU. The tunnel current I is proportional to the voltage drop across R_V. The tunnel diode is marked TD. (**b**) Bridge circuit for recording dU/dI and d^2U/dI^2 curves as a function of the tunnel voltage for energies $eU > \Delta$. This arrangement is particularly suited for measuring small inelastic signals on a background of strongly elastic tunnel current. The resistance R_d is for balancing the bridge and TD is the tunnel diode. (After [IX.1])

rent-voltage characteristic $I(U)$ is measured at low temperature ($T<4$ K). For studying the novel oxide superconductors (Sect. 10.11) temperatures in the region of 20–70 K are of course sufficient. To determine the precise thresholds of $I(U)$, which are characteristic for the gap energies, it is best to measure the first derivative dI/dU or dU/dI. For this, a circuit such as that in Fig. IX.4a can be used. A bias voltage U is applied to the tunnel diode (TD) from a dc source, and this voltage is registered on the x-axis of the recorder. A small ac voltage $v \cos \omega t$ is superimposed on the dc voltage U, and a "lock-in" amplifier measures the component of the current at frequency ω flowing through the tunnel contact and sends this to the y-axis of the recorder. The total current, comprising dc and ac components, can be written

$$I(U + v\cos\omega t) = I(U) + \frac{dI}{dU} v\cos\omega t + \frac{1}{2}\frac{d^2I}{dU^2} v^2(1+\cos 2\omega t) + \dots , \tag{IX.9}$$

and thus the signal detected at frequency ω is proportional to the derivative dI/dU. Such experiments employ frequencies in the kHz region and voltages of 20–50 μV (which determine the resolution).

Figure IX.5 shows, as a typical example, the tunnel characteristic dI/dU of a $Hg/Al_2O_3/Al$ tunnel diode, taken at 1.21 K [XI.1]. It is clearly symmetric about $U=0$. In this tunnel diode, the rapidly condensed Al is present in microcrystalline form, and its transition temperature T_c lies between 2.5 and 1.8 K, depending on the degree of crystallinity (for crystalline bulk Al, $T_c=1.18$ K). The Hg was quenched on the oxidized Al, i.e. on the Al_2O_3 insulating layer. According to its degree of crystal perfection its transition temperature lies between 4 and 4.5 K. Thus in Fig. IX.5, both components

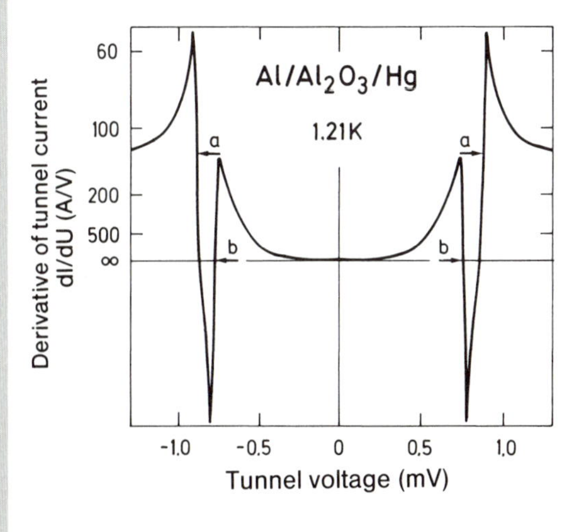

Fig. IX.5. Typical derivative dI/dU of a characteristic curve for elastic one-electron tunneling between superconductors (gap curve), measured with the arrangement shown in Fig. IX.4a. The measurement was performed on a $Hg/Al_2O_3/Al$ tunnel diode at a temperature of 1.21 K after tempering at 150 K. The arrows labelled a and b indicate the intervals $2(\Delta_{Hg}+\Delta_{Al})/e$ and $2(\Delta_{Hg}-\Delta_{Al})/e$ on the voltage axis. (After [IX.1])

of the tunnel diode are superconducting. One sees qualitatively that the curve for $U>0$ represents the first derivative of an $I(U)$ curve of the type shown in Fig. IX.3 d ($T\neq 0$). The separations a and b marked in Fig. IX.5 immediately provide the information $2(\Delta_{Hg}+\Delta_{Al})/e$ and $2(\Delta_{Hg}-\Delta_{Al})/e$ about the energy gaps of the two superconductors.

Besides the elastic tunneling described up until now, which is important for the determination of gap energies, there are also higher order inelastic processes which can influence the tunnel current in the voltage region $U>\Delta/e$ (Fig. IX.6). In the insulating barrier of a tunnel junction there are

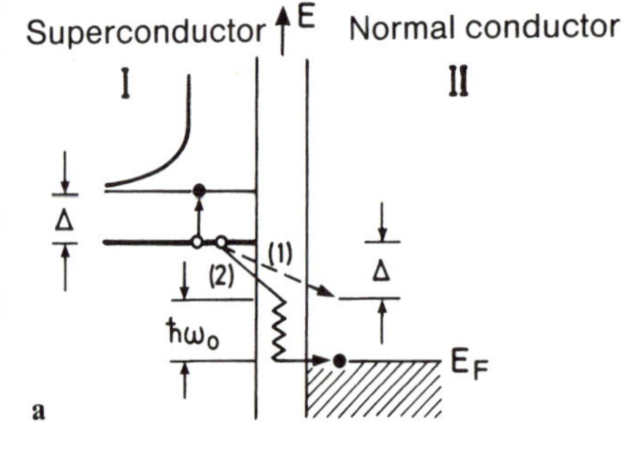

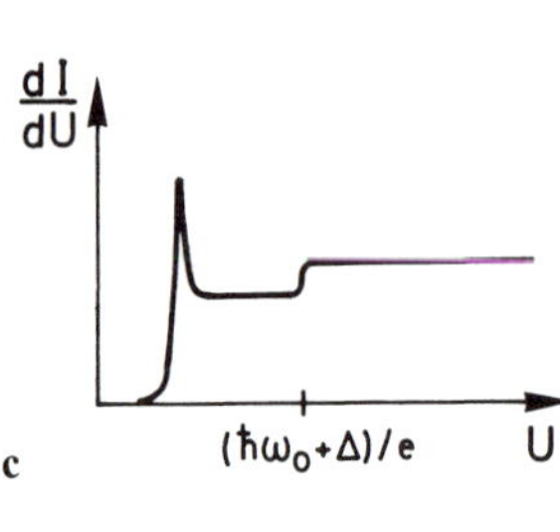

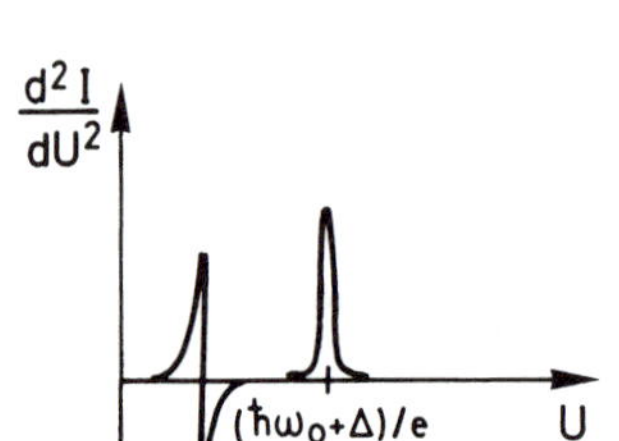

Fig. IX.6a–d. Schematic representation of an inelastic tunnel process between a superconductor I and a normal conductor II. (**a**) Representation in terms of energy bands. The superconductor is characterized by its BCS ground state (*single fat line*) and by the density of one-particle states ($E>E_F+\Delta$), and the occupied conduction band of the normal conductor is shown shaded. In addition to the elastic process (*1*), inelastic processes (*2*) also become possible when the tunnel voltage U exceeds the value $\Delta+\hbar\omega_0$. Here $\hbar\omega_0$ is the quantum energy of a single characteristic excitation in the barrier (molecular vibration, phonon, etc.). (**b**) Schematic I–U characteristic for one-electron tunneling between a superconductor and a normal conductor. At the tunnel voltage Δ/e, the Cooper pairs break up and elastic tunneling processes begin. At $(\Delta+\hbar\omega_0)/e$, inelastic processes begin. (**c**) Schematic representation of the first derivative dI/dU of the I–U characteristic in (**b**). (**d**) Schematic representation of the second derivative d^2I/dU^2 of the I–U characteristic in (**b**)

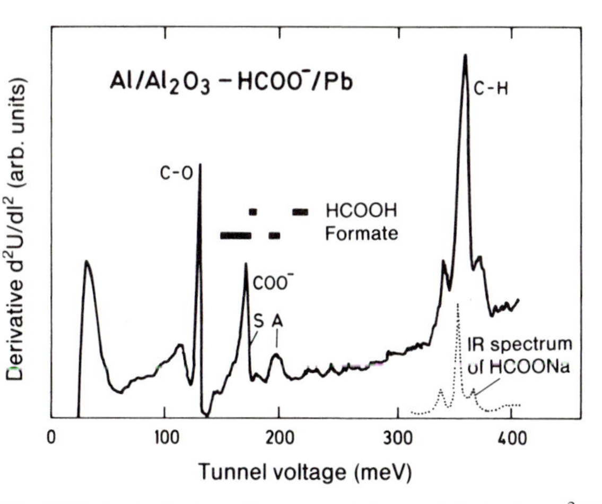

Fig. IX.7. Inelastic tunneling current (second derivative d^2U/dI^2) measured at a temperature of 2 K on an Al/oxide/Pb tunnel contact, in which the Al_2O_3 layer was "doped" with formic acid (HCOOH) from the gas phase before deposition of the Pb. Comparison of the infrared vibrational bands shows that on absorption of HCOOH a formate ion $HCOO^-$ is formed, whose vibrations explain the tunnel spectrum. The symmetric (S) and asymmetric (A) stretching vibrations are characteristic of the COO^- group. (After [IX.2])

many ways in which the tunneling electron can create excitations. In addition to lattice vibrations, the electron can also excite vibrations of embedded atoms or impurity molecules. If $\hbar\omega_0$ is the vibrational excitation energy of such an embedded molecule, then besides the elastic tunneling path (1) in Fig. IX.6a, there is also an inelastic path (2) available to an electron passing from the superconductor (I) to the normal conductor (II) after the breaking up of a Cooper pair. This inelastic tunneling (2) via the excitation of a molecular vibration $\hbar\omega_0$ becomes possible when the applied voltage U exceeds the value $(\hbar\omega_0 + \Delta)/e$. The tunnel characteristic $I(U)$ shows a very slight increase in gradient above this voltage (Fig. IX.6b). In the first derivative dI/dU a step appears (Fig. IX.6c), which, however, can only be reliably measured as a peak in the second derivative (Fig. IX.6d). Experimental investigations of such inelastic processes are therefore carried out by recording of the second derivative d^2I/dU^2 or d^2U/dI^2 of the tunnel characteristic above the gap threshold Δ/e. In the relevant voltage range the elastic background only changes slowly compared with the inelastic contributions. The measurement sensitivity can thus be increased by placing the tunnel diode in the middle arm of a Wheatstone bridge, which suppresses the applied voltage (Fig. IX.4b). The second derivative d^2U/dI^2 is measured, according to (IX.9), by superimposing a small alternating voltage (frequency ω) and detecting the signal produced at 2ω with a lock-in amplifier (Fig. IX.4b). Figure IX.7 shows the inelastic tunneling spectrum of an $Al/Al_2O_3/Pb$ tunnel diode, in which formic acid (HCOOH) was adsorbed onto the insulating Al_2O_3 layer prior to Pb deposition [IX.2]. In the measured d^2U/dI^2 curve one sees sharp peaks; they are due to the inelastic scattering of tunneling electrons by excitations in or near the Al_2O_3 barrier. In addition to the phonon contribution at about 35 meV, one observes vibrational bands that can be attributed to dissociatively adsorbed formic acid. A comparison with the known infrared bands of undissociated formic acid (HCOOH), and with those of sodium formate (HCOONa), shows that the formic acid molecules dissociate into formate ions upon adsorption onto the Al_2O_3. Experiments of this kind have been used to study adsorption processes and excitations in thin, insulating films. The tunnel barrier consists thereby of the insulating material whose adsorption properties are to be studied. It should be emphasized that, for this particular application, the superconducting nature of one or both tunnel electrodes is only important in that it yields a singularity in the density of states in the vicinity of the gap ($E \gtrsim \Delta$). This in turn provides a high resolution (~ 2 meV)in the vibrational spectra measured for adsorbed or incorporated molecules.

References

IX.1 U. Roll: Diplomarbeit, Tunnelexperimente an supraleitenden Hg/Al_2O_3/Al-Übergängen, RWTH Aachen (1976)

IX.2 J. Klein, A. Léger, M. Belin, D. Defourneau, M.J.L. Sangster: Phys. Rev. B **7**, 2336 (1973)

Panel X
Cooper-Pair Tunneling – The Josephson Effect

In Panel IX we discussed superconductor tunnel experiments in which single electrons tunnel through the barrier. In 1962 *Josephson* showed theoretically that Cooper pairs can also tunnel [X.1]. In this so-called Josephson effect, the characteristic high coherence of the Cooper pairs in the BCS ground state becomes particularly evident. The experiments are carried out on very thin tunnel barriers ($d<30$ Å), which couple the two superconductors to one another. The coupling must be sufficiently weak that there is only a very low probability of finding a Cooper pair in the barrier region. Experimentally, this is achieved, for example, in a tunnel "sandwich" arrangement, where a thin oxide barrier (10–20 Å) or a thin normally conducting layer separates two superconducting metal films (Fig. X.1a). Micro-bridges (Fig. X.1b), i.e., very thin connections between two superconducting materials, are also used. Another simple realization of such a micro-bridge consists of a point contact, in which the tip of one superconductor is pressed against the surface of another. In the basic circuit for observing the Josephson effect (Fig. X.2) an external voltage U_{ext} is applied across an external resistance R to the "tunnel contact" S1/S2. One can measure both the current I through the contact and the voltage drop U across the tunnel barrier. Because of the strong influence of magnetic fields on the highly coherent Cooper pair states (Sects. 10.8, 10.9), magnetic

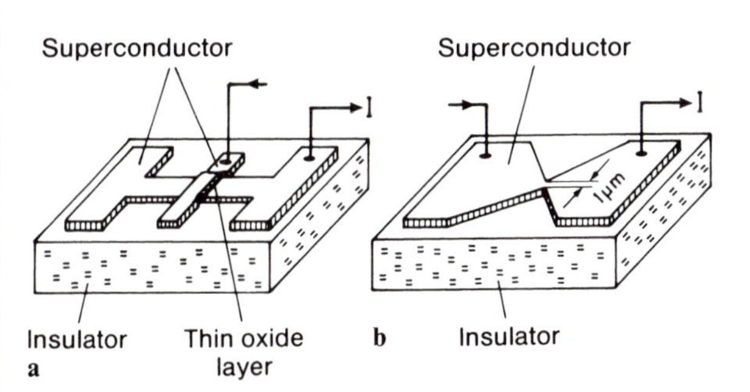

Fig. X.1 a, b. Two possible arrangements that yield a weakly coupled contact between two superconductors. (**a**) Tunnel contact in which two superconducting strips perpendicular to one another are separated by a thin oxide barrier (10–20 Å) or by a thin normally conducting metallic layer (100–1000 Å). (**b**) Microbridge in which the superconducting layers are separated by a thin bridge of about 1 μm, which can be formed, for example, by lithographic and etching processes

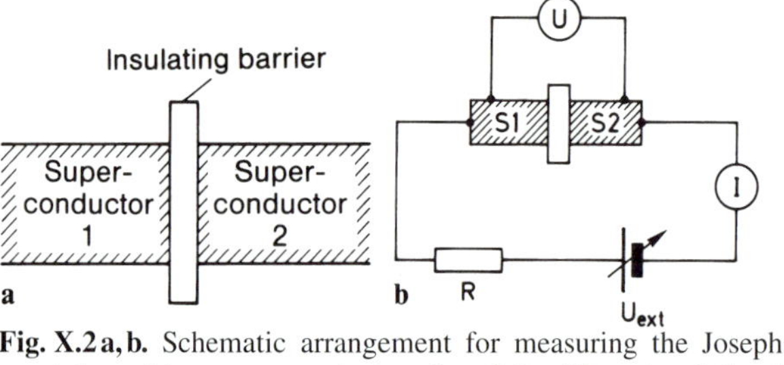

Fig. X.2 a, b. Schematic arrangement for measuring the Josephson tunnel effect at a contact consisting of two superconductors 1 and 2 with an insulating barrier (**a**). The circuit (**b**) enables one to measure the tunnel current I and the voltage drop U across a tunnel diode (tunnel diode voltage)

fields in the vicinity of the tunnel barrier must be screened. The Earth's magnetic field alone is sufficient to completely obscure the Josephson effect. In the arrangement in Fig. X.2, only the single particle tunnel effects discussed in Panel IX (Fig. IX.3) would be observed if there were no magnetic screening.

For a theoretical description of the effects observed in a Josephson junction we assume the simple model shown in Fig. X.2 a. Two superconductors 1 and 2 of the same material are connected via a thin tunnel barrier. The temperature is so low that we only need to consider the electrons in their superconducting (BCS) ground state. The Cooper pair states in the two superconductors are described by the many-body wavefunctions Φ_1 and Φ_2. If $\mathscr{H}_1$ and $\mathscr{H}_2$ are the Hamiltonians (energies) of the respective isolated superconductors, then for the system connected by a barrier we have

$$i\hbar \frac{\partial \Phi_1}{\partial t} = \mathscr{H}_1 \Phi_1 + T\Phi_2 \,, \tag{X.1 a}$$

$$i\hbar \frac{\partial \Phi_2}{\partial t} = \mathscr{H}_2 \Phi_2 + T\Phi_1 \,. \tag{X.1 b}$$

Here T is the characteristic coupling constant for the tunnel junction. The terms $T\Phi_2$ and $T\Phi_1$ give the energy contributions due to the fact that the many-body wave-function Φ_2 or Φ_1 of one superconductor does not completely vanish in the other (tunneling of the BCS state or Cooper pairs). For a completely impenetrable barrier ($T=0$), equation (X.1) gives two uncoupled, time-dependent, many-body Schrödinger equations for the isolated superconductors.

For simplicity we consider two identical superconductors in (X.1). In the spirit of a perturbation theory treatment, we assume that Φ_1 is an approximate solution with respect to $\mathscr{H}_1$. We therefore replace $\mathscr{H}_1$ by the energy of the superconducting state. A voltage U applied across the diode causes the energies in the two superconductors to shift by qU with respect to one another ($q=-2e$). If the zero of the energy scale is taken midway between the energies of two superconductors, then (X.1) reduces to

$$\mathrm{i}\hbar\dot{\Phi}_1 = \frac{qU}{2}\Phi_1 + T\Phi_2 \ , \tag{X.2 a}$$

$$\mathrm{i}\hbar\dot{\Phi}_2 = \frac{-qU}{2}\Phi_2 + T\Phi_1 \ . \tag{X.2 b}$$

Since $|\Phi|^2 = n_s/2 = n_c$ is the density of Cooper pairs (n_s, is the density of single electrons), the wave functions Φ_1 and Φ_2 (see Sect. 10.8) can be represented as

$$\Phi_1 = \sqrt{n_{c1}}\mathrm{e}^{\mathrm{i}\varphi_1} \ , \tag{X.3 a}$$

$$\Phi_2 = \sqrt{n_{c2}}\mathrm{e}^{\mathrm{i}\varphi_2} \ , \tag{X.3 b}$$

where φ_1 and φ_2 are the phases of the BCS wave functions in the two superconductors. Inserting (X.3) in (X.2 a) and separating the real and imaginary parts yields

$$\frac{\hbar}{2}\dot{n}_{c1}\sin\varphi_1 - \left(\hbar\dot{\varphi}_1 + \frac{qU}{2}\right)n_{c1}\cos\varphi_1 = T\sqrt{n_{c1}n_{c2}}\cos\varphi_2 \ , \tag{X.4 a}$$

$$\frac{\hbar}{2}\dot{n}_{c1}\cos\varphi_1 - \left(\hbar\dot{\varphi}_1 + \frac{qU}{2}\right)n_{c1}\sin\varphi_1 = T\sqrt{n_{c1}n_{c2}}\sin\varphi_2 \ . \tag{X.4 b}$$

Similar equations follow when (X.3) is inserted in (X.2 b). Multiplying (X.4 a and b) by $\cos\varphi_1$ and $\sin\varphi_1$, or by $\sin\varphi_1$ and $\cos\varphi_1$, and then subtracting the equations from one another or adding them yields the following equations, together with the analogous equations from (X.2 b),

$$\dot{n}_{c1} = \frac{2}{\hbar}T\sqrt{n_{c1}n_{c2}}\sin(\varphi_2 - \varphi_1) \ , \tag{X.5 a}$$

$$\dot{n}_{c2} = -\frac{2}{\hbar}T\sqrt{n_{c1}n_{c2}}\sin(\varphi_2 - \varphi_1) \ , \tag{X.5 b}$$

and

$$\dot{\varphi}_1 = \frac{1}{\hbar}T\sqrt{\frac{n_{c2}}{n_{c1}}}\cos(\varphi_2 - \varphi_1) - \frac{qU}{2\hbar} \ , \tag{X.6 a}$$

$$\dot{\varphi}_2 = \frac{1}{\hbar}T\sqrt{\frac{n_{c1}}{n_{c2}}}\cos(\varphi_2 - \varphi_1) + \frac{qU}{2\hbar} \ . \tag{X.6 b}$$

For two identical, symmetrically disposed superconductors as in Fig. X.2, with $n_{c1} = n_{c2} = n_c$, we obtain

$$\dot{n}_{c1} = (2T/\hbar) n_c \sin(\varphi_2 - \varphi_1) = -\dot{n}_{c2} \,, \tag{X.7 a}$$

$$\hbar(\dot{\varphi}_2 - \dot{\varphi}_1) = -qU \,. \tag{X.7 b}$$

These equations have a non-zero solution for the current flux even for a vanishing voltage ($U=0$). Experimentally this means that, immediately after closing the contact in Fig. X.2b, a tunnel current of Cooper pairs flows. Its direction depends on $\sin(\varphi_2 - \varphi_1)$ (X.7a), i.e., on the phase difference of the BCS many-body states in the two superconductors. There is no voltage drop U at the diode, although an external voltage U_{ext} is applied in the external circuit. The current measured in the circuit is the tunnel current I_{CP}. Due to the current, a charge asymmetry results which would eventually stop the current if it were not collected and replaced at the two terminals of the external battery. The tunnel current therefore remains constant ($n_{c1}=n_{c2}=n_c$). Plotted against the tunnel diode voltage U (Fig. X.2b), this dc Josephson current I_{CP} initially has only one value at $U=0$ (Fig. X.3). As the external voltage U_{ext} increases, the current I_{CP} also increases up to a maximum value I_{CP}^{max}. Below I_{CP}^{max} the current is regulated by current supplied from the external circuit. If this maximum Josephson current I_{CP}^{max} is exceeded by further increasing U_{ext}, then the state becomes unstable and a voltage $U \neq 0$ appears on the tunnel diode. The current I then has a value $I(R)$, determined by U_{ext}, U, the external resistance R and the tunnel characteristics for one-particle tunneling between two superconductors (Fig. IX.3). This current I no longer originates from the tunneling of Cooper pairs, but instead from the single electron tunneling associated with the breaking up of Cooper pairs. In the case of a micro-bridge, an increase above I_{CP}^{max} does not lead to the usual tunnel characteristics, but to an I–U curve displaying quasi-ohmic behavior.

From (X.7a), the maximum current I_{CP}^{max} is stable so long as the phase difference $(\varphi_2 - \varphi_1)$ does not vary with time. From (X.7b), we see that this is true provided there is no voltage drop U across the diode. If the current

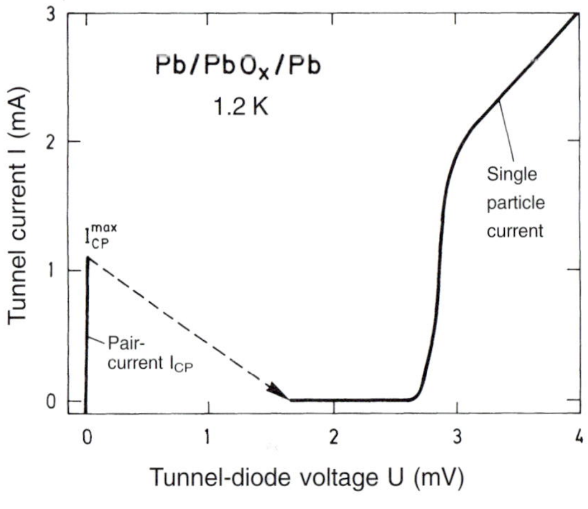

Fig. X.3. Current-voltage characteristic of a Pb/PbO$_x$/Pb Josephson tunnel diode, measured with a setup as shown in Fig. X.2. At vanishingly small diode voltage U, the Cooper pair tunnel current I_{CP} rises with increasing external voltage U_{ext} to a maximum value I_{CP}^{max}, and then changes to the one-particle characteristic with a finite voltage drop across the diode. (After [X.2])

I_{CP}^{max} collapses, i.e., a voltage U is created, the phase difference between the two states of the superconductors begins to grow with time t. Integration of (X.7 b) gives (q=–2 e)

$$\varphi_2 - \varphi_1 = \frac{1}{\hbar} 2eUt + \Delta\varphi_{init} \; . \tag{X.8}$$

Inserting this in (X.7 a) gives an alternating current $I_{CP}^{\approx}$ in the tunnel diode in addition to the direct current resulting from one electron tunneling

$$I_{CP}^{\approx} \sim \dot{n}_{c1} = \frac{1}{\hbar} 2Tn_c \sin(\omega_{CP}t + \Delta\varphi_{init}) \; , \tag{X.9 a}$$

with

$$\omega_{CP} = \frac{1}{\hbar} 2eU \; . \tag{X.9 b}$$

For a typical voltage drop of about 1 mV across the tunnel diode, (X.9 b) gives a circular frequency ω_{CP} for this alternating Josephson current of about 3×10^{12} s^{-1}, i.e., oscillations in the infrared region of the spectrum.

The Josephson tunneling of Cooper pairs in the presence of magnetic fields enables one to experimentally demonstrate the high coherence of the superconducting state, in a similar way to the interference of coherent light. In the experiment, a highly coherent supercurrent is divided so that it flows along two paths (Fig. X.4), both containing Josephson tunnel diodes (a and b). Connecting the two current paths afterwards leads to closure of the supercurrent circuit which is interrupted only by the tunnel barriers a and b. According to (X.8), the total wavefunction of the superconducting state in region II experiences a phase shift at the barriers compared to that in region I. When the magnetic field penetrating the superconducting loop

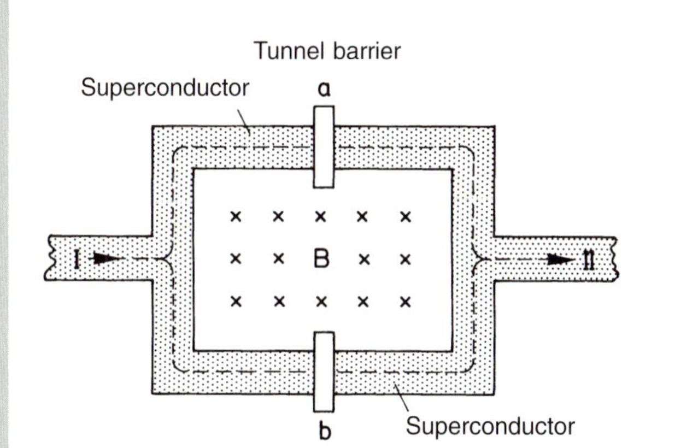

Fig. X.4. Schematic illustration of two parallel Cooper pair tunnel barriers a and b which are weakly coupled to one another by the superconductors I and II. The phase difference between the total Cooper pair wavefunctions in I and II, i.e. around the ring, can be controlled by the magnetic flux (flux density B) which penetrates the ring. The oscillations due to interference between the two supercurrents flowing from I to II can be used to measure weak magnetic fields (Superconducting Quantum Interferometer Device, SQUID)

(Fig. X.4) is zero, the phase shifts are denoted as δ_a and δ_b. From (X.7 a, X.9 a) the currents across the tunnel barriers a and b can then be written as

$$I_a = I_0 \sin\delta_a \,, \tag{X.10 a}$$

$$I_b = I_0 \sin\delta_b \,. \tag{X.10 b}$$

In a magnetic field, however, the field $\boldsymbol{B}$ (or its vector potential $\boldsymbol{A}$) also affects the phase difference in the two superconductors. This can easily be seen from the equation for the supercurrent (10.110) by performing a line integration along a path between two points X and Y deep inside the superconductor, i.e., where $\boldsymbol{j}_s=0$. In this case the phase difference due to a magnetic vector potential $\boldsymbol{A}$ is

$$\Delta\varphi|_X^Y = \int_X^Y \nabla\varphi \cdot d\boldsymbol{l} \equiv \frac{2e}{\hbar}\int_X^Y \boldsymbol{A} \cdot d\boldsymbol{l} \,. \tag{X.11}$$

From this the total phase shifts of the superconducting state function along the two paths a or b from region I to region II are

$$\Delta\varphi|_{\mathrm{I}}^{\mathrm{II}} = \delta_a + \frac{2e}{\hbar}\int_a \boldsymbol{A} \cdot d\boldsymbol{l} \,, \tag{X.12 a}$$

$$\Delta\varphi|_{\mathrm{I}}^{\mathrm{II}} = \delta_b - \frac{2e}{\hbar}\int_b \boldsymbol{A} \cdot d\boldsymbol{l} \,. \tag{X.12 b}$$

Since the wavefunction must have a unique value at every point, the two phase shifts must be identical and subtraction yields

$$\delta_b - \delta_a = \frac{2e}{\hbar}\oint \boldsymbol{A} \cdot d\boldsymbol{l} = \frac{2e}{\hbar}\int \boldsymbol{B} \cdot d\boldsymbol{S} \,. \tag{X.13}$$

The two line integrals in (X.12) are in opposite directions and thus together they give an integral over a closed path and finally the magnetic flux $\int \boldsymbol{B} \cdot d\boldsymbol{S}$ through the ring. The total phase difference around the ring can therefore be controlled by varying the magnetic flux. The situation is similar to that of two coherent light beams which are made to interfere (over paths a and b); their relative phase is controlled by varying the path difference. The phenomenon of supercurrent interference becomes evident if we consider the total current I due to the sum of the currents I_a and I_b in the presence of a magnetic field. On the basis of (X.13) we introduce a constant but arbitrary phase shift δ_0, which depends on the nature of the tunnel barriers and the applied voltage. We obtain

$$\delta_a = \delta_0 - \frac{e}{\hbar}\int \boldsymbol{B} \cdot d\boldsymbol{S} \,, \tag{X.14 a}$$

$$\delta_b = \delta_0 + \frac{e}{\hbar}\int \boldsymbol{B} \cdot d\boldsymbol{S} \tag{X.14 b}$$

For the total current we then find

$$I = I_0\left[\sin\left(\delta_0 + \frac{e}{\hbar}\int \boldsymbol{B} \cdot d\boldsymbol{S}\right) + \sin\left(\delta_0 - \frac{e}{\hbar}\int \boldsymbol{B} \cdot d\boldsymbol{S}\right)\right],$$

$$I = 2I_0 \sin\delta_0 \cos\left(\frac{e}{\hbar}\int \boldsymbol{B} \cdot d\boldsymbol{S}\right). \tag{X.15}$$

The supercurrent through two parallel Josephson tunnel barriers therefore varies as the cosine of the magnetic flux through the superconducting loop (X.15). Maxima in the current occur when

$$\int \boldsymbol{B} \cdot d\boldsymbol{S} = N\frac{\pi\hbar}{e} = N\frac{h}{2e}\ , \quad N = 1, 2, 3 \ldots, \tag{X.16}$$

i.e., whenever a further magnetic flux quantum (10.114) is enclosed in the ring. Figure X.5 shows the experimentally measured current through a pair of Josephson contacts as a function of the magnetic field between the two contacts. The oscillations each correspond to a flux quantum and result from the cosine interference term in (X.15).

There are a number of obvious applications of the Josephson effects discussed here. The two stable states of a Josephson junction ($I_{CP} \neq 0$ at $U=0$ and one-electron tunneling at $U \neq 0$, Fig. X.3) allow the construction of binary switching devices in microelectronics, e.g., for computer data storage. These devices are extremely fast, but require cooling. The interference between the Cooper pair tunneling currents of two parallel Josephson contacts (Fig. X.4) is exploited to produce extremely sensitive magnetometers, so-called SQUIDs (Superconducting Quantum Interferometer De-

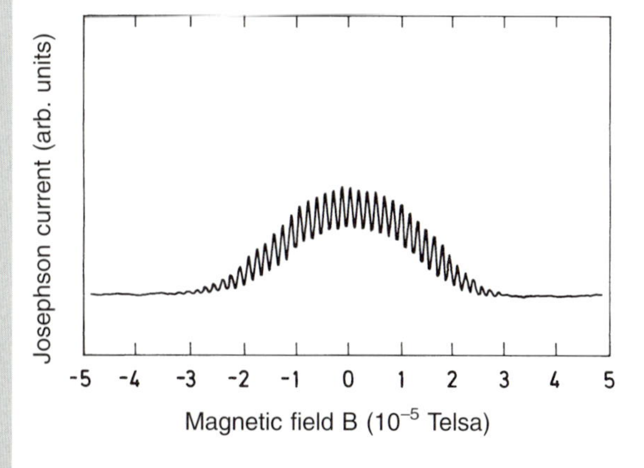

Fig. X.5. The Cooper pair current through a circuit of two parallel Josephson tunnel barriers plotted against the magnetic flux B penetrating the ring (circuit as in Fig. X.4). (After [X.3])

vices). With these devices one can even measure the extremely small magnetic fields due to currents in the human brain.

References

X.1 B.D. Josephson: Phys. Lett. **1**, 251 (1962)
X.2 D.N. Langenberg, D.J. Scalapino, B.N. Taylor: IEEE Proc. **54**, 560 (1966)
X.3 R.C. Jaklevic, J. Lambe, J.E. Mercereau, A.H. Silver: Phys. Rev. **140**, A 1628 (1965)

11 Dielectric Properties of Materials

The interaction of electromagnetic radiation with a solid may be described either microscopically or macroscopically according to requirements. In a microscopic picture one would speak, for instance, of the absorption of a photon via the creation of a phonon or an electron-hole pair. The Maxwellian theory, on the other hand, is a macroscopic approach and the solid is described by material constants. The relation between the two approaches will be developed in this chapter in a discussion limited to linear phenomena. It should be pointed out, however, that the area of nonlinear optics has become increasingly important in the last decade, but a discussion of nonlinear materials is beyond the scope of this introductory text.

11.1 The Dielectric Function

An essential aspect of the interaction of an electric field with a solid is the fact that the field induces a flow of "quasi-free" charge carriers. The simplest case of such a process, an ohmic current in a metal, was treated in Chap. 9. This description leads phenomenologically to a material-specific conductivity σ. Another kind of interaction of the solid with electromagnetic field consists of the finite spatial displacement of local charges (e.g., the more strongly bound valence electrons in the field of positive ion cores) to form local dipole moments. These two processes are accounted for in the classical macroscopic Maxwell equations by the two terms that appear as the origin of curl $\boldsymbol{H}$:

$$\mathrm{curl}\,\mathscr{E} = -\dot{\boldsymbol{B}}\,, \tag{11.1 a}$$

$$\mathrm{curl}\,\boldsymbol{H} = \boldsymbol{j} + \dot{\boldsymbol{D}}\,. \tag{11.1 b}$$

Within the range of validity of Ohm's law, the current density carried by electrons or holes in partially filled bands is related to the electric field by

$$\boldsymbol{j} = \sigma\mathscr{E}\,. \tag{11.2}$$

In general, all of the fields appearing in (11.1) are time-dependent quantities. An equivalent, and frequently more useful, representation is obtained by Fourier transforming the electric field strength $\mathscr{E}$ and the dielectric displacement $\boldsymbol{D}$

$$\mathscr{E}(t) = \int_{-\infty}^{\infty} \mathscr{E}(\omega) e^{-i\omega t} d\omega \,, \tag{11.3}$$

$$\boldsymbol{D}(t) = \int_{-\infty}^{\infty} \boldsymbol{D}(\omega) e^{-i\omega t} d\omega \,. \tag{11.4}$$

Because $\mathscr{E}(t)$ and $\boldsymbol{D}(t)$ are real functions, we have

$$\mathscr{E}(\omega) = \mathscr{E}^*(-\omega) \quad \text{and} \quad \boldsymbol{D}(\omega) = \boldsymbol{D}^*(-\omega) \,. \tag{11.5}$$

The Fourier coefficients of $\boldsymbol{D}$ and $\mathscr{E}$ are related by a frequency-dependent dielectric function $\varepsilon_0\varepsilon(\omega)$. Like σ, the quantity $\varepsilon(\omega)$ is generally a second rank tensor, which, however, becomes a scalar for isotropic media and cubic crystals, i.e.

$$\boldsymbol{D}(\omega) = \varepsilon_0\varepsilon(\omega)\mathscr{E}(\omega) \,. \tag{11.6}$$

Just as we denoted properties of the solid connected with σ as conductivity phenomena, so we denote properties associated with ε as *dielectric properties*. The present chapter concentrates on the microscopic processes that determine the spectral variation of $\varepsilon(\omega)$. In oscillating fields, one property can be expressed in terms of the other one, because for harmonic fields (11.3, 11.4) the second Maxwell equation (11.1 b) can be written as

$$\operatorname{curl} \boldsymbol{H}(\omega) = \sigma\mathscr{E}(\omega) - i\omega\varepsilon_0\varepsilon(\omega)\mathscr{E}(\omega) \,. \tag{11.7}$$

We can define a generalized frequency-dependent conductivity that also takes into account the dielectric effects:

$$\tilde{\sigma} = \sigma - i\omega\varepsilon_0\varepsilon \,. \tag{11.8}$$

Equation (11.7) can also be expressed

$$\operatorname{curl} \boldsymbol{H}(\omega) = -i\varepsilon_0\tilde{\varepsilon}(\omega)\omega\mathscr{E}(\omega) = -i\omega\boldsymbol{D}(\omega) \,, \tag{11.9}$$

where the generalized dielectric constant [cf. (11.9) and (11.7)] is

$$\tilde{\varepsilon}(\omega) = \varepsilon(\omega) + i\sigma/\varepsilon_0\omega \,. \tag{11.10}$$

Conductivity phenomena are accounted for by the σ term. For example, in Sect. 11.9 this is done explicitly for the free electron gas in a metal. The fact that the dielectric description can be replaced by (11.8) or (11.10) has its physical origin in the fact that the distinction between free and bound charges becomes poorly defined in oscillating fields ($\omega \neq 0$). In both cases we are concerned with periodic displacements of charge. In a constant field ($\omega = 0$) the behavior of free and bound charge carriers is naturally different, and the fundamental difference between σ and ε then becomes evident.

If one allows $\mathcal{E}$ and $\boldsymbol{D}$ to have a spatial as well as a time dependence, then a decomposition in terms of plane waves replaces the Fourier decompositions (11.3) and (11.4). The dielectric function ε then depends on ω and on the wavevector $\boldsymbol{k}$. In this chapter, we restrict the spatial dependence of $\mathcal{E}$ and $\boldsymbol{D}$ to a slow variation between neighboring unit cells. This approximation remains valid for interactions with light far into the ultraviolet, but is not appropriate for X-ray radiation, whose wavelength is comparable to the dimensions of a unit cell. Thus we shall be concerned with cases in which only long wavelengths appear in the Fourier decomposition, i.e. waves for which $k \ll G$. We may then neglect the dependence of the dielectric function on k. We also disregard nonlocal effects in which the displacement $\boldsymbol{D}$ at a position $\boldsymbol{r}$ may depend on the electric field at a position $\boldsymbol{r}'$. The equation (11.6) is written in S.I. units with ε_0 as the dielectric permittivity of vacuum. $\varepsilon(\omega)$ is then a complex, dimensionless function, which is defined in the whole interval $-\infty < \omega < \infty$. From (11.5, 11.6) one also has

$$\varepsilon^*(-\omega) = \varepsilon(\omega) \ . \tag{11.11}$$

Instead of the dielectric displacement $\boldsymbol{D}$, one can also consider the polarization $\boldsymbol{P} = \boldsymbol{D} - \varepsilon_0 \mathcal{E}$ and introduce a dielectric susceptibility:

$$\chi(\omega) = \varepsilon(\omega) - 1 \ . \tag{11.12}$$

The complex function $\chi(\omega)$ is analytic and has its poles in the negative imaginary half-plane (Problem 11.1). If one chooses the integration path shown in Fig. 11.1, the integral

$$\oint \frac{\chi(\omega')}{\omega' - \omega} d\omega' = 0 \ . \tag{11.13}$$

vanishes.

For very large ω, $\chi(\omega)$ approaches zero because the polarization can no longer follow the field. If the semicircle is extended in the complex plane, the only contributions that remain are from the real axis and from the integral around the pole, and one obtains

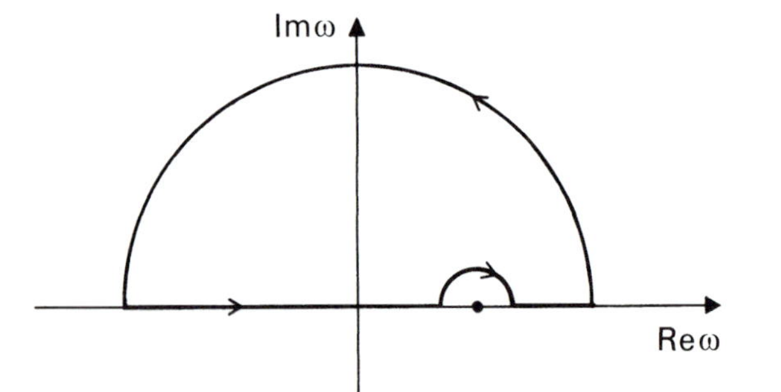

Fig. 11.1. Integration paths in the complex ω plane

$$-\mathrm{i}\pi\chi(\omega) + \mathscr{P}\int \frac{\chi(\omega')}{\omega'-\omega}d\omega' = 0 . \tag{11.14}$$

Here $\mathscr{P}$ denotes the principal value of the integral.

If $\varepsilon(\omega)$ is separated into real and imaginary parts,

$$\varepsilon(\omega) = \varepsilon_1(\omega) + \mathrm{i}\,\varepsilon_2(\omega) , \tag{11.15}$$

then (11.14) yields the equations:

$$\varepsilon_1(\omega) - 1 = \frac{1}{\pi}\mathscr{P}\int_{-\infty}^{\infty} \frac{\varepsilon_2(\omega')}{\omega'-\omega}d\omega' , \tag{11.16}$$

$$\varepsilon_2(\omega) - \frac{1}{\pi}\mathscr{P}\int_{-\infty}^{\infty} \frac{\varepsilon_1(\omega')-1}{\omega'-\omega}d\omega' . \tag{11.17}$$

These are the so-called *Kramers-Kronig* relations which relate the real and imaginary parts of the dielectric function. They can be used, for instance, to calculate one part of the dielectric function from an exact measurement of the other over a wide spectral region. The residues of the integrals for very high or very low frequencies can usually be estimated from approximate solutions for $\varepsilon(\omega)$.

11.2 Absorption of Electromagnetic Radiation

In this section we shall examine the absorption of electromagnetic waves as they pass through a dielectric layer. This will enable us to relate the dielectric properties to the results of a typical absorption experiment. Electromagnetic waves in a dielectric medium are described by an electric field:

$$\mathscr{E} = \mathscr{E}_0\,\mathrm{e}^{-\mathrm{i}\,\omega(t-\tilde{n}x/c)} \tag{11.18}$$

with a complex refractive index

$$\tilde{n}(\omega) = n + \mathrm{i}\kappa = \sqrt{\varepsilon(\omega)} , \tag{11.19}$$

$$n^2 - \kappa^2 = \varepsilon_1 , \tag{11.20}$$

$$2n\kappa = \varepsilon_2 . \tag{11.21}$$

The sign of the exponents in the Fourier decomposition (11.3) and in the decomposition of $\varepsilon(\omega)$ and $\tilde{n}(\omega)$ are chosen such that ε_2 and κ have positive signs, i.e., the wave amplitude decreases in the $+x$ direction. If we were to use a Fourier decomposition with positive exponents, then the refractive index would have to be written $\tilde{n} = n - \mathrm{i}\kappa$ with $\kappa > 0$. Both representations are in common use.

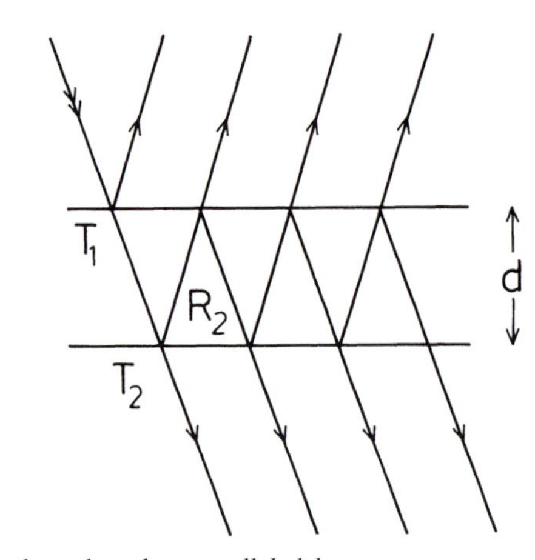

Fig. 11.2. Rays through a plane parallel slab

On passing through a plane, parallel slab (Fig. 11.2), the radiation is partly reflected and partly transmitted at each interface. For the special case of normal incidence, the transmission and reflection coefficients for the amplitude of $\mathscr{E}$ are:

$$T_1 = \frac{2}{\tilde{n}+1} \ , \quad T_2 = \frac{2\tilde{n}}{\tilde{n}+1} \ , \quad R_2 = \frac{\tilde{n}-1}{\tilde{n}+1} \ . \tag{11.22}$$

The total amplitude of the transmitted beam is

$$\begin{aligned} \mathscr{E} &= \mathscr{E}_0 \, T_1 T_2 \mathrm{e}^{\mathrm{i}(\tilde{n}\omega/c)d}(1 + R_2^2 \mathrm{e}^{2\mathrm{i}(\tilde{n}\omega/c)d} + \dots) \ , \\ \mathscr{E} &= \mathscr{E}_0 \, T_1 T_2 \frac{\mathrm{e}^{\mathrm{i}(\tilde{n}\omega/c)d}}{1 - R_2^2 \mathrm{e}^{2\mathrm{i}(\tilde{n}\omega/c)d}} \ . \end{aligned} \tag{11.23}$$

We discuss two limiting cases. In the first, n is assumed to be close to one and we are thus dealing with an optically thin medium:

$$\tilde{n} - 1 + \Delta \ , \quad |\Delta| \ll 1 \ . \tag{11.24}$$

In the linear approximation we then have $T_1 T_2 \sim 1$ and $R_2^2 \sim 0$, and thus

$$\mathscr{E} = \mathscr{E}_0 \, \mathrm{e}^{\mathrm{i}(\tilde{n}\omega/c)d} \tag{11.25}$$

and the transmitted intensity is

$$\mathscr{E}\mathscr{E}^* \propto I = I_0 \mathrm{e}^{-(2\kappa\omega/c)d} \ . \tag{11.26}$$

Since $|\Delta \ll 1$, it follows that $2\kappa \sim \varepsilon_2$, whence

$$I = I_0 \mathrm{e}^{-(\varepsilon_2 \omega/c)d} \ . \tag{11.27}$$

For optically dense media, and in the case of transmission through a sufficiently thin, film (11.23) is approximated by

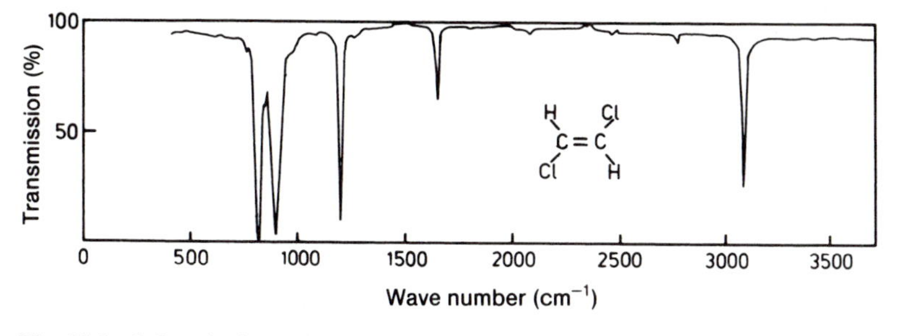

Fig. 11.3. Infrared absorption spectrum of 1,2-trans-dichloroethene [11.1]. For absorption which is not too strong, the curves correspond to the spectral form of $\omega\varepsilon_2(\omega)$ at the normal mode resonances of the molecule

$$\left|\frac{\tilde{n}\omega}{c}d\right| \ll 1 . \tag{11.28}$$

After inserting (11.28) in (11.23) and performing some algebra, we obtain, in a linear approximation:

$$I = I_0\left(1 - \frac{\varepsilon_2\omega}{c}d\ldots\right) . \tag{11.29}$$

The quantity

$$K(\omega) = \frac{\omega\varepsilon_2(\omega)}{c} \tag{11.30}$$

is also called the *absorption coefficient.* It determines the spectrum of absorbed intensity, and contains the imaginary part of $\varepsilon(\omega)$ as the most important frequency-dependent factor. Figure 11.3 shows a typical absorption spectrum of dichloroethene [11.1]. The absorption maxima correspond to maxima in $\varepsilon_2(\omega)$ at the frequencies of the normal mode vibrations of the dichloroethene molecule. In the following sections, the spectral properties of the dielectric function will be related to the atomic properties of the system.

The equations (11.27) and (11.29) represent a particularly simple relation between experiment and the dielectric properties of a system. In principle, the spectral behavior of the reflectivity of a solid is also determined by $\varepsilon(\omega)$. The mathematical relations are however more complicated since both ε_1 and ε_2 enter the reflectivity. A calculation of ε_1 and ε_2 from the reflectivity is therefore not possible, unless one knows the reflectivity over the entire spectral range, in which case the Kramers-Kronig relations (11.16, 11.17) can be applied.

11.3 The Dielectric Function for a Harmonic Oscillator

We will now investigate the dielectric properties of a harmonic oscillator for which a displacement u from the equilibrium position produces a dipole moment:

$$p = e^* u \,. \tag{11.31}$$

The quantity e^* is called the effective ionic charge; it is not necessarily equal to the charge responsible for the static dipole of the oscillator. Such a dynamic dipole moment arises, for example, when the positive and negative sub-lattices of an ionic crystal (e.g., CsCl, see Fig. 1.6) are shifted relative to one another, as is the case for the displacements associated with transverse and longitudinal optical phonons at $q=0$. Other examples are the normal mode vibrations of molecules possessing a dipole moment. The equation of motion for an oscillator in an external field $\mathscr{E}$ is then

$$\ddot{u} + \gamma \dot{u} = -\omega_0^2 u + \frac{e^*}{\mu} \mathscr{E} \,. \tag{11.32}$$

Furthermore, the polarization P is given by

$$P = \frac{N}{V} e^* u + \varepsilon_0 \frac{N}{V} \alpha \mathscr{E} \,, \tag{11.33}$$

where μ is the reduced mass, N/V the number density, α the electronic polarizability, and γ a damping constant describing the finite lifetime of the normal mode vibration. Strictly speaking, the external field $\mathscr{E}$ should be replaced by the local field $\mathscr{E}_{\mathrm{loc}}$. The local field may be different because the dipoles created by the external field also contribute to the field and amplify it (Sect. 11.7). This correction is not negligible for dense media, i.e. solids and liquids. However the basic spectral form of $\varepsilon(\omega)$ is not affected, and so we shall disregard the field amplification for now. Transforming (11.32, 11.33) to the Fourier representation, we obtain

$$u(\omega)(\omega_0^2 - \omega^2 - \mathrm{i}\,\gamma\,\omega) = \frac{e^*}{\mu} \mathscr{E}(\omega) \,, \tag{11.34}$$

$$P(\omega) = \frac{N}{V} e^* u(\omega) + \varepsilon_0 \frac{N}{V} \alpha \mathscr{E}(\omega) \,. \tag{11.35}$$

We have assumed here that the electronic polarizability is frequency independent in the vicinity of the eigenfrequencies of the lattice. From (11.34, 11.35), we obtain for $\varepsilon(\omega)$

$$\varepsilon(\omega) = 1 + \frac{N}{V}\alpha + \frac{\dfrac{N}{V}\dfrac{e^{*2}}{\varepsilon_0 \mu}}{\omega_0^2 - \omega^2 - \mathrm{i}\,\gamma\,\omega} \,. \tag{11.36}$$

After introducing the static dielectric constant $\varepsilon_{st}=\varepsilon(\omega=0)$ and the high frequency limit ε_∞, we obtain $\varepsilon(\omega)$ in the form

$$\varepsilon(\omega) = \varepsilon_\infty + \frac{\omega_0^2(\varepsilon_{st} - \varepsilon_\infty)}{\omega_0^2 - \omega^2 - \mathrm{i}\,\gamma\,\omega} \,. \tag{11.37}$$

The damping constant γ is positive. If it were negative, the damping forces in (11.32) would have the same sign as the direction of movement, which would lead to an increase in amplitude with time. The sign of γ therefore ensures that $\varepsilon(\omega)$ does not have a pole in the positive imaginary half-plane, a fact that we have already used in the derivation of the Kramers-Kronig relations (see also Problem 11.1).

It should again be emphasized that all statements about the signs of quantities depend on the sign chosen in the Fourier representation (11.3).

For the following discussion, we separate $\varepsilon(\omega)$ into real and imaginary parts, $\varepsilon(\omega)=\varepsilon_1(\omega)+\mathrm{i}\,\varepsilon_2(\omega)$, with

$$\varepsilon_1(\omega) = \varepsilon_\infty + \frac{(\varepsilon_{st} - \varepsilon_\infty)\omega_0^2(\omega_0^2 - \omega^2)}{(\omega_0^2 - \omega^2)^2 + \gamma^2\omega^2} \,, \tag{11.38}$$

$$\varepsilon_2(\omega) = \frac{(\varepsilon_{st} - \varepsilon_\infty)\omega_0^2\gamma\omega}{(\omega_0^2 - \omega^2)^2 + \gamma^2\omega^2} \,. \tag{11.39}$$

The behavior of ε_1 and ε_2 is shown in Fig. 11.4; ε_2 has the form of a damped resonance curve with γ the full width at half maximum. For the important case of weak damping, $\gamma \ll \omega_0$, a convenient approximate formula can be given. We apply the mathematical identity

$$\lim_{\gamma\to 0} \frac{1}{z - \mathrm{i}\gamma} = \mathscr{P}\,\frac{1}{z} + \mathrm{i}\,\pi\,\delta(z) \,. \tag{11.40}$$

to the two poles $\omega=\omega_0$ and $\omega=-\omega_0$ in (11.37) and obtain

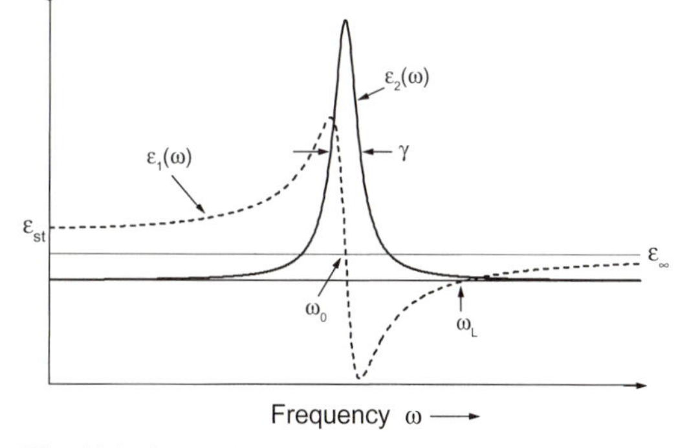

Fig. 11.4. General form of $\varepsilon_1(\omega)$ and $\varepsilon_2(\omega)$ for a dipole oscillator. For finite damping, the zero crossings of $\varepsilon_1(\omega)$ give only approximate values of the frequencies ω_0 and ω_L

$$\varepsilon(\omega) = \varepsilon_\infty + \frac{\omega_0^2(\varepsilon_{st} - \varepsilon_\infty)}{\omega_0^2 - \omega^2} + i\frac{\pi}{2}\omega_0(\varepsilon_{st} - \varepsilon_\infty)[\delta(\omega - \omega_0) - \delta(\omega + \omega_0)] \, . \tag{11.41}$$

One can convince oneself that this special form of $\varepsilon(\omega)$ also obeys the Kramers-Kronig relations if $\varepsilon_\infty = 1$ such that the function $\varepsilon(\omega)$–1 vanishes for $\omega \to \infty$. With this approximation, the integral absorption over the absorption band can readily be calculated (Fig. 11.3). For optically thin media ($\mathscr{E}_{loc} = \mathscr{E}$) one has

$$\int \ln \frac{I_0}{I(\omega)} d\omega = d \int K(\omega) d\omega = \pi d \frac{\omega_0^2}{c} (\varepsilon_{st} - \varepsilon_\infty) \, . \tag{11.42}$$

From (11.36), this is equivalent to

$$\int_{-\infty}^{+\infty} K(\omega) d\omega = \frac{N}{V} \frac{\pi}{\varepsilon_0 c} \frac{e^{*2}}{\mu} \, . \tag{11.43}$$

For spectroscopic applications, it must be remembered that the dipoles of the oscillators are, in general, not oriented parallel to the $\mathscr{E}$ field. Averaging over a random orientation of dipoles yields a prefactor of 1/3. After integrating over positive frequencies only, one finally obtains

$$\int_0^{\infty} K(\omega) d\omega = \frac{1}{6} \frac{N}{V} \frac{\pi}{\varepsilon_0 c} \frac{f^2 e^2}{\mu} \tag{11.44}$$

where $f^2 e^2 = e^{*2}$. The factor f is the ratio of the oscillator charge to the elementary charge e.

In spectroscopy, the wave number ($1/\lambda$) is often used instead of the frequency. Equation (11.43) then becomes

$$\int K\left(\frac{1}{\lambda}\right) d\frac{1}{\lambda} = \frac{1}{12\varepsilon_0 c^2} \frac{N}{V} \frac{f^2 e^2}{\mu} \, . \tag{11.45}$$

This is the quantitative formulation of Beer's Law. The integrated absorption is proportional to the number density of oscillators.

11.4 Longitudinal and Transverse Normal Modes

In this section we discuss longitudinal and transverse normal modes of the polarization in a dielectric with a resonance at ω_0, but without damping. In place of the polarization $\boldsymbol{P}$ one may also consider the displacement s of the system of harmonic oscillators. If we consider waves travelling in the positive x-direction, then

$$P_x = P_{x0} e^{-i(\omega t - qx)} \tag{11.46}$$

represents a longitudinal wave, and

$$P_y = P_{y0} e^{-i(\omega t - qx)} \tag{11.47}$$

a transverse wave. The longitudinal wave satisfies the conditions

$$\operatorname{curl} \boldsymbol{P}_\mathrm{L} = \boldsymbol{0} \quad \operatorname{div} \boldsymbol{P}_\mathrm{L} \neq 0 , \tag{11.48}$$

while for a transverse wave

$$\operatorname{curl} \boldsymbol{P}_\mathrm{T} \neq \boldsymbol{0} \quad \operatorname{div} \boldsymbol{P}_\mathrm{T} = 0 . \tag{11.49}$$

Equations (11.48) and (11.49) can also be regarded as a generalized definition of longitudinal and transverse waves. We first examine the longitudinal waves in more detail.

In a dielectric without charge carriers or other sources of space charge ϱ, the divergence of the dielectric displacement must vanish:

$$\operatorname{div} \boldsymbol{D} = \varrho = 0 = \varepsilon_0 \varepsilon(\omega) \operatorname{div} \mathscr{E} = \varepsilon(\omega) \frac{\operatorname{div} \boldsymbol{P}}{\varepsilon(\omega) - 1} . \tag{11.50}$$

According to (11.48) a longitudinal wave can then exist only at an eigenfrequency determined by the condition

$$\varepsilon(\omega_\mathrm{L}) = 0 . \tag{11.51}$$

Furthermore, we have

$$\mathscr{E}_\mathrm{L} = -\frac{1}{\varepsilon_0} \boldsymbol{P}_\mathrm{L} . \tag{11.52}$$

Hence, the field and polarization vectors are exactly 180° out of phase. Such longitudinal waves cannot interact with transverse light waves but they can interact with electrons (Sect. 11.12).

The solution for *transverse* waves can be found only in the context of the complete set of Maxwell equations, since these waves couple to electromagnetic radiation ($\operatorname{curl} \mathscr{E} \neq \boldsymbol{0}$)

$$\operatorname{curl} \mathscr{E} = -\mu_0 \dot{\boldsymbol{H}} , \quad (\mu \approx 1) , \tag{11.53}$$

$$\operatorname{curl} \boldsymbol{H} = \dot{\boldsymbol{D}} . \tag{11.54}$$

With a plane wave ansatz of the type (11.47), and after replacing $\mathscr{E}$ and $\boldsymbol{D}$ by $\boldsymbol{P}$, we obtain, using the dielectric constants,

$$q P_{y0} - \omega \varepsilon_0 \mu_0 [\varepsilon(\omega) - 1] H_{z0} = 0 , \tag{11.55}$$

$$-\omega \frac{\varepsilon(\omega)}{\varepsilon(\omega) - 1} P_{y0} + q H_{z0} = 0 . \tag{11.56}$$

This system of equations has solutions for frequencies ω at which the determinant vanishes. Thus we find

$$\omega^2 = \frac{1}{\varepsilon(\omega)} c^2 q^2 \ . \tag{11.57}$$

These solutions represent coupled electromagnetic and mechanical waves, and are called *polaritons*. The behavior of the polariton dispersion relation (11.57) depends on the form of $\varepsilon(\omega)$. We discuss here the dispersion relation for the case of an ionic structure, i.e., a system of harmonic oscillators, with an eigenfrequency ω_0. We insert into (11.57) the expression (11.37) for $\varepsilon(\omega)$ without damping. One obtains the solutions sketched in Fig. 11.5, all of which have transverse character. For large q, the lower branch asymptotically approaches the value ω_0, i.e., the eigenfrequency of the harmonic oscillators corresponds to the frequency of the transverse waves for large q and is therefore denoted ω_T in the following.

Because of the large value of the velocity of light, the value ω_T is already a good approximation even for q values that are very small (10^{-4} G) compared with a reciprocal lattice vector $\boldsymbol{G}$ (Fig. 11.5). If the velocity of light were infinite, then ω_T would be the only solution. Thus the dispersion arises because of the retarding effect of the finite signal velocity of electromagnetic waves. The solution $\omega_T = \omega_0$ is therefore also called the "unretarded" solution. Between ω_T and ω_L, (11.57) has no solutions, but a second branch exists for $\omega \geq \omega_L$.

The phonon-polaritons shown in Fig. 11.5 are eigenmodes (normal modes) of the dielectric, and can be observed by means of Raman scattering (Panel III). To correctly interpret an absorption experiment, one must consider Maxwell's equations inside and outside the solid, and of course the boundary conditions. As a result of the boundary conditions, other

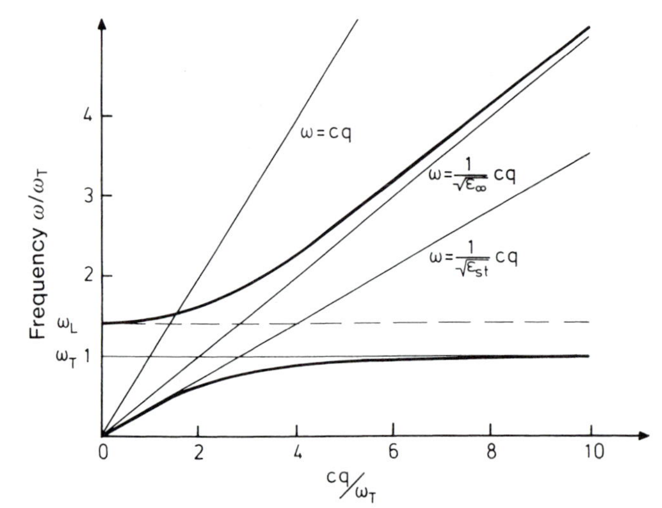

Fig. 11.5. Dispersion curves for a phonon-polariton. The range of q values shown here is small compared to a reciprocal lattice vector. The lattice dispersion can therefore be neglected

kinds of polaritons may appear (see also Sect. 11.5 and Panel XIII) that are different to those discussed here. The problem of absorption in a thin film has already been solved in Sect. 11.2, but without an explicit calculation of the normal modes. It was shown there that absorption in a thin film is mainly determined by the behavior of $\varepsilon_2(\omega)$. The maximum of the absorption then lies at the frequency ω_T of the transverse optical phonons.

We have already seen that, for q values which are not too small, the frequency of the transverse waves is equal to the resonance frequency ω_0, while the frequency of the longitudinal waves is determined by the zero of $\varepsilon_1(\omega)$. Using (11.37) ($\gamma \ll \omega_0$), we obtain the important relation

$$\frac{\omega_L^2}{\omega_T^2} = \frac{\varepsilon_{st}}{\varepsilon_\infty} , \tag{11.58}$$

(Lyddane-Sachs-Teller relation), which gives the frequency separation of the longitudinal and transverse waves. The physical origin of the displacement of the longitudinal wave frequency to higher values is the strengthening of the effective force constant by the field accompanying the longitudinal wave (11.52). For a crystal that possesses optical phonon branches (i.e., more than one atom per elementary cell), but is not infrared active because of the absence of a dynamic effective ionic charge, the frequencies ω_L and ω_T must be degenerate at $q=0$. An example is the case of silicon (Fig. 4.4).

11.5 Surface Waves on a Dielectric

In the last section, we became acquainted with the wave-like normal modes of an infinitely extended lattice. As a model for a finite solid, we consider a dielectric half-space whose boundary surface is the x,y plane. In addition to bulk waves, there now exist normal modes which have wave character with respect to the x,y coordinates, but whose amplitude decays exponentially normal to the surface. For simplicity, we consider the unretarded solutions which is formally obtained by letting $c \rightarrow \infty$. Then it is always true that

$$\operatorname{curl} \mathscr{E} = \mathbf{0} . \tag{11.59}$$

Since we are no longer interested in obtaining the longitudinal bulk waves, we also set, in contrast to (11.48),

$$\operatorname{div} \mathscr{E} = 0 , \quad \text{for} \quad z \neq 0 . \tag{11.60}$$

The two conditions (11.59) and (11.60) are fulfilled if $\mathscr{E}$ can be written as the gradient of a potential φ that satisfies the Laplace equation

$$\Delta \varphi = 0 , \quad \text{for} \quad z \neq 0 . \tag{11.61}$$

The Laplace equation is satisfied both outside and inside the solid by surface waves of the form

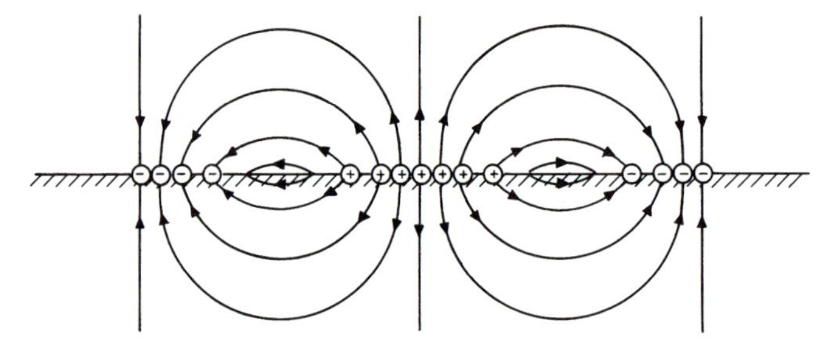

Fig. 11.6. Schematic representation of the electric field lines associated with a surface wave on the planar surface of a dielectric medium

$$\varphi = \varphi_0 e^{-q|z|} e^{i(qx-\omega t)} . \tag{11.62}$$

The associated field lines are sketched in Fig. 11.6. Inside the dielectric ($z<0$) the field is associated with a polarization. To be a valid solution (11.62) must also obey the boundary condition that the normal component of the dielectric displacement be continuous:

$$D_z = -\varepsilon_0 \varepsilon(\omega) \left.\frac{\partial \varphi}{\partial z}\right|_{z \leqq 0} = -\varepsilon_0 \left.\frac{\partial \varphi}{\partial z}\right|_{z \geqq 0} . \tag{11.63}$$

and thus we require

$$\varepsilon(\omega) = -1 . \tag{11.64}$$

The frequency of the surface wave is therefore lower than that of the longitudinal bulk wave ω_L (Fig. 11.4). For the harmonic oscillator and for ionic lattices one obtains

$$\omega_S = \omega_T \left(\frac{\varepsilon_{st} + 1}{\varepsilon_\infty + 1} \right)^{1/2} . \tag{11.65}$$

Surface waves can also be calculated for geometries other than a half-space (small spheres, ellipses, thin films). Their frequencies always lie between ω_T and ω_L. For particles whose dimensions are small compared to the wavelength, surface waves can be observed with light absorption (Fig. 11.7). The surface wave calculated here for a plane is a non-radiative solution at the interface, and thus does not couple to the light field. A more exact discussion taking into account retardation shows that the surface waves exist only for q values to the right of the light line. Surface waves can be observed by the method of "frustrated" total reflection (Panel XIII). Charged particles (electrons) also interact with the surface waves of the dielectric (Sect. 11.12).

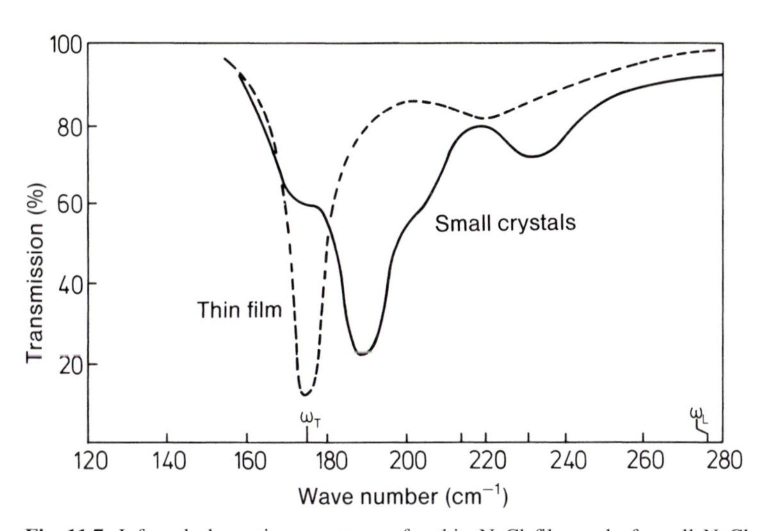

Fig. 11.7. Infrared absorption spectrum of a thin NaCl film and of small NaCl cubes of side 10 μm, after Martin [11.3]. The thin film absorbs at $\omega_T = 175\ \mathrm{cm}^{-1}$ [cf. (11.30, 11.41)]. The additional absorption at $220\ \mathrm{cm}^{-1}$ is due to two-phonon processes. The absorption maximum for small particles is clearly shifted to higher frequencies

11.6 Reflectivity of a Dielectric Half-Space

While the imaginary part of $\varepsilon(\omega)$ determines the absorption for propagation through a sufficiently thin film, the reflectivity contains both real and imaginary parts. For normal incidence, the fraction of the intensity reflected is [cf. (11.22)]

$$R = \frac{(n-1)^2 + \kappa^2}{(n+1)^2 + \kappa^2} \ . \tag{11.66}$$

We now consider the frequency dependence of the reflection coefficient for an ionic lattice, and, for simplicity, we first treat the case of low damping. Then $\varepsilon(\omega)$ is real for $\omega < \omega_T$ and as the frequency approaches ω_T, the reflectivity rises to the limiting value $R=1$ (Fig. 11.8). Between ω_T and ω_L, ε_1 is real but negative. For this reason, the real part of the complex refractive index n is zero [cf. (11.9)], and the reflectivity remains equal to 1. Above ω_L, $\varepsilon_1 > 0$, and the reflectivity falls until it reaches the value zero at $\varepsilon_1 = +1$. At higher frequencies, it rises again to the limiting value $(\varepsilon_\infty^{1/2}-1)^2/(\varepsilon_\infty^{1/2}+1)^2$. Thus, in the region where $\varepsilon_1 < 0$, the wave cannot penetrate the dielectric from the exterior.

This result is less strictly fulfilled for finite damping, but the reflectivity remains high between ω_T and ω_L. The measured curve for InAs in Fig. 11.8 shows this behavior. After multiple reflection from surfaces only the frequency range between ω_T and ω_L remains. This particular frequency band is thus called *reststrahlen*, the German term for "residual radiation".

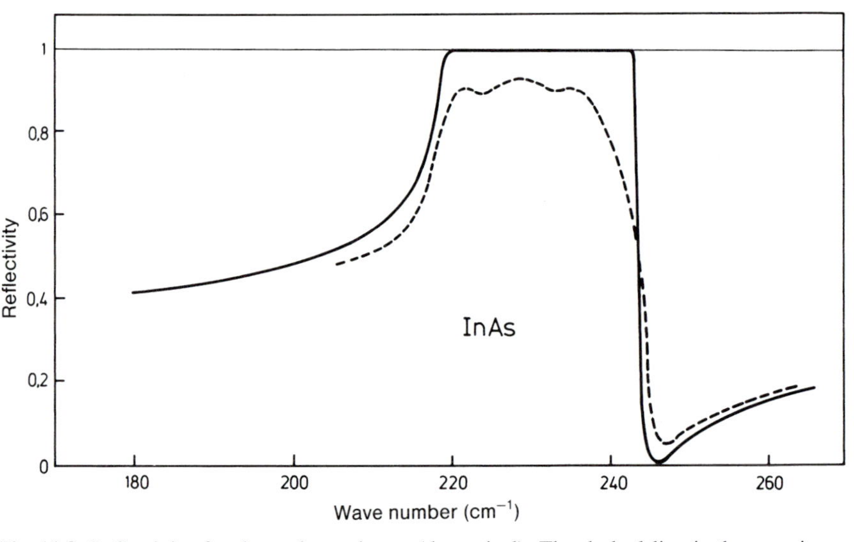

Fig. 11.8. Reflectivity for the undamped case (theoretical). The dashed line is the experimentally measured curve for a real ionic crystal (InAs). (After [11.4])

11.7 The Local Field

Only in the case of a single dipole is the effective field at the position of the dipole equal to the externally applied field. For an ensemble of many dipole oscillators, the neighboring dipoles also contribute to the local field. This contribution can be neglected for very dilute systems, but not for solids. If we choose the external field along the z axis, then the *local effective field* is

$$\mathscr{E}_{\mathrm{loc}} = \mathscr{E} + \frac{1}{4\pi\,\varepsilon_0}\sum_i p_i \frac{3z_i^2 - r_i^2}{r_i^5} \; . \tag{11.67}$$

The sum extends over all neighboring dipoles. If p_i is the dipole moment per unit cell in an ionic crystal, then i sums over all neighboring cells. The sum in (11.67) is not as complicated as it might appear. For a spherical shape of the solid and homogeneous polarization (origin at the center of the sphere), one has

$$\sum_i p_i \frac{x_i^2}{r_i^5} = p\sum_i \frac{x_i^2}{r_i^5} = p\sum_i \frac{y_i^2}{r_i^5} = p\sum_i \frac{z_i^2}{r_i^5} = \frac{1}{3}p\sum_i \frac{r_i^2}{r_i^5} \; . \tag{11.68}$$

The sum in (11.67) thus vanishes in this case and the local field at the center is equal to the external field. Unfortunately, this simple case does not correspond to typical experimental conditions; for the absorption of electromagnetic waves, the assumption of homogeneous polarization is only fulfilled for spheres with a diameter small compared to the wavelength of light. However, the sphere model is still useful for describing transverse and longitudinal waves. We can imagine a sphere cut out of the solid, within which the polarization is approximately constant, but which nonetheless contains many elementary cells. The only condition now is that the wavelength is large compared with an elementary cell, which is generally true for interactions with light. The contribution to the local field at the center of the sphere due to the spherical region is zero, as we saw above. For regions outside the sphere, the distances to the center are so large that one can consider the dipole distribution to be continuous, and thus use a macroscopic polarization ***P***. The field due to this polarization is described in terms of polarization charges on the surface of the sphere, whose charge density is

$$\varrho_{\mathrm{P}} = -P_n \,, \tag{11.69}$$

i.e., equal to the normal component of ***P***. The charge contained in a circular element (Fig. 11.9) at polar angle θ is then

$$dq = -P\cos\theta \, 2\pi a \sin\theta \, a \, d\theta \tag{11.70}$$

and its contribution to the field at the center of the sphere is

$$d\mathscr{E} = -\frac{1}{4\pi\varepsilon_0}\frac{dq}{a^2}\cos\theta \,. \tag{11.71}$$

The total additional field $\mathscr{E}_{\mathrm{P}}$ is therefore

$$\mathscr{E}_{\mathrm{P}} = \frac{P}{2\varepsilon_0}\int\limits_0^\pi \cos^2\theta \, \sin\theta \, d\theta = \frac{1}{3\varepsilon_0}P \tag{11.72}$$

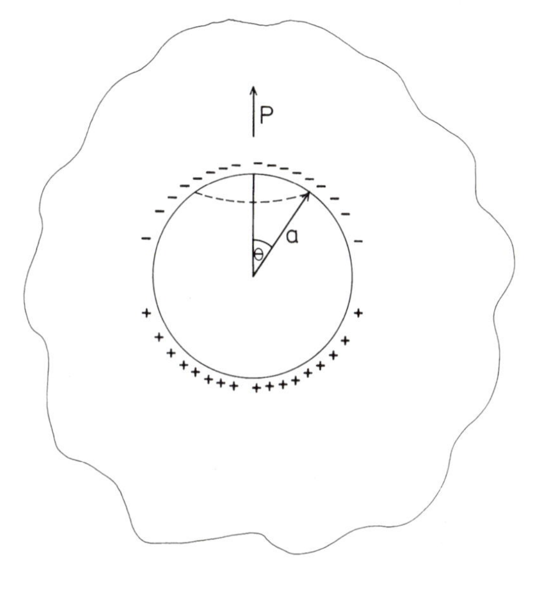

Fig. 11.9. Polarization and the local field (see text)

and hence the local field becomes

$$\mathcal{E}_{\mathrm{loc}} = \mathcal{E} + \frac{1}{3\,\varepsilon_0} P \,. \tag{11.73}$$

This local field is to be inserted in the equations of motion for an ionic crystal, (11.32) and (11.33). In place of (11.34) and (11.35), one obtains, after some rearrangement,

$$u(\omega)\left(\omega_0^2 - \frac{1}{\mu}\frac{\frac{1}{3\,\varepsilon_0}\frac{N}{V}e^{*2}}{1-\frac{1}{3}\frac{N}{V}\alpha} - \omega^2 - \mathrm{i}\,\gamma\,\omega\right) = \frac{e^*}{\mu}\mathcal{E}(\omega)\frac{1}{1-\frac{N}{V}\frac{\alpha}{3}} \,, \tag{11.74}$$

$$P(\omega) = \frac{\frac{N}{V}e^*}{1-\frac{1}{3}\frac{N}{V}\alpha}u(\omega) + \frac{\varepsilon_0\frac{N\,\alpha}{V}}{1-\frac{1}{3}\frac{N}{V}\alpha}\mathcal{E}(\omega) \,. \tag{11.75}$$

The dielectric function $\varepsilon(\omega)$ which these equations yield has the same form as in (11.37), although the relationship of the macroscopic quantities ω_{T}, $\varepsilon_{\mathrm{st}}$ and ε_∞ to the microscopic properties of the system ω_0, e^* is different.

Let us consider the high frequency limit $(\omega \gg \omega_{\mathrm{T}})$ as an example. Clearly, since $u(\omega) \to 0$, one has

$$\varepsilon_\infty = 1 + \frac{\frac{N}{V}\alpha}{1-\frac{1}{3}\frac{N}{V}\alpha} \quad \text{or} \tag{11.76}$$

$$\frac{\varepsilon_\infty - 1}{\varepsilon_\infty + 2} = \frac{1}{3}\frac{N}{V}\alpha \,. \tag{11.77}$$

This is the Clausius-Mossotti equation, which relates the dielectric constant ε_∞ to the electronic polarizability.

11.8 The Polarization Catastrophe and Ferroelectrics

One effect of the local field is to reduce the frequency of the transverse normal mode of the lattice – see (11.74) –

$$\omega_{\mathrm{T}}^2 = \omega_0^2 - \frac{1}{\mu}\frac{\frac{1}{3\,\varepsilon_0}\frac{N}{V}e^{*2}}{1-\frac{1}{3}\frac{N}{V}\alpha} \,. \tag{11.78}$$

For sufficiently large effective charge e^*, high electronic polarizability α, and/or relatively weak coupling to nearest neighbors (small ω_0), the transverse frequency can even become zero. The local field associated with the movement of an ion in the lattice is then so strong that the repulsive forces due to nearest neighbor interactions are smaller than the forces due to this field. The resultant force displaces the ion from its usual position until it reaches a new equilibrium position. The forces occurring in this situation are not included in our harmonic approximation. The new state of the solid is characterized by its decomposition into regions with permanent polarization. This effect is called the polarization catastrophe, and in analogy to ferromagnetism, the new state is called the *ferroelectric* state [although it has nothing to do with ferrum (Latin; iron)]. Typical ferroelectric materials are the perovskites (e.g. $BaTiO_3$, $SrTiO_3$) and hydrogen phosphates or arsenates (e.g. KH_2PO_4). The ferroelectric state can be regarded as a "frozen in" transverse phonon. A prerequisite for this freezing is that the thermal motion of the lattice is small. With increasing temperature, the thermal fluctuations become ever larger. The effective field at the ion is thereby reduced until, at a critical temperature T_c, the right side of (11.78) becomes positive. As the temperature increases beyond T_c, the transverse frequency increases according to

$$\omega_T^2 \propto T - T_c \,. \qquad (11.79)$$

From the Lyddane-Sachs-Teller relation (11.58), the static dielectric constant is then approximately

$$\varepsilon_{st}^{-1} \propto T - T_c \,. \qquad (11.80)$$

In practice, however, the dielectric constant remains finite on approaching T_c. A schematic representation of these relationships is shown in Fig. 11.10.

On account of their large dielectric constants above the Curie point, ferroelectric materials, in particular ferroelectric ceramics, are of great technological importance as dielectrics in capacitors. Below the Curie point ferroelectric ceramics have a large piezoelectric constant (Problem 5.3). Ceramics with suitable Curie points are utilized as piezoelectric actuators.

Ferroelectric nanostructures serve as non-volatile memories in advanced computers.

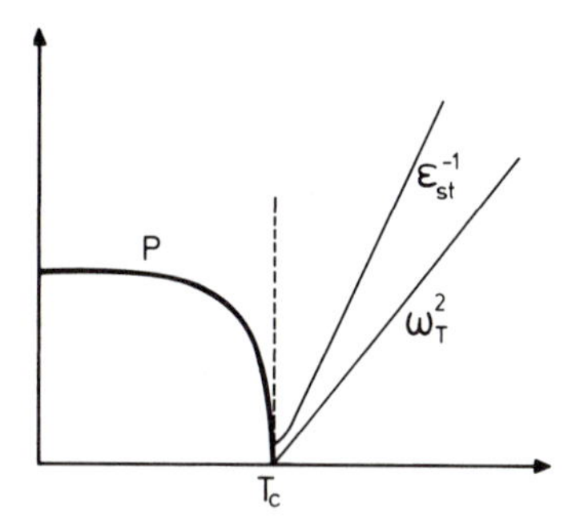

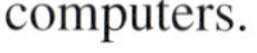

Fig. 11.10. Temperature dependence of the permanent polarization P, the reciprocal dielectric constant ε_{st}^{-1}, and the square of the transverse frequency ω_T^2 for a ferroelectric material (schematic)

11.9 The Free-Electron Gas

The dielectric behavior of metals, and also of semiconductors with high electron concentrations, is largely determined by collective excitations of the free carriers. As in the case of an ionic lattice, the dielectric function can be derived from an equation of motion. If u is the homogeneous displacement of the electron gas relative to the ion cores, then the equation of motion is

$$n m \ddot{u} + \gamma \dot{u} = -n e \mathcal{E} \ , \tag{11.81}$$

where n is the electron concentration and m is the electron mass. In contrast to (11.32), equation (11.81) does not contain any mechanical restoring force. The damping constant γ is now related to the conductivity σ, since, for a stationary flux ($\ddot{u} = 0$), we have

$$j = -e n \dot{u} = \frac{n^2 e^2}{\gamma} \mathcal{E} \quad \text{and thus} \quad \sigma = \frac{j}{\mathcal{E}} = \frac{n^2 e^2}{\gamma} \ . \tag{11.82}$$

Thus (11.81) becomes

$$n m \ddot{u} + \frac{n^2 e^2}{\sigma} \dot{u} = -n e \mathcal{E} \ . \tag{11.83}$$

The Fourier-transformed equation then reads

$$\left(-n m \omega^2 - \mathrm{i} \frac{n^2 e^2}{\sigma(\omega)} \omega \right) u(\omega) = -n e \mathcal{E}(\omega) \ . \tag{11.84}$$

The conductivity must also be treated here as a function of frequency. If we replace the displacement $u(\omega)$ by the polarization

$$P(\omega) = -e n u(\omega) \ , \tag{11.85}$$

then the dielectric function becomes

$$\varepsilon(\omega) = 1 - \frac{\omega_\mathrm{p}^2}{\omega^2 + \mathrm{i} \dfrac{\omega_\mathrm{p}^2 \varepsilon_0}{\sigma(\omega)} \omega} \tag{11.86}$$

with

$$\omega_\mathrm{p}^2 = \frac{n e^2}{m \varepsilon_0} \ . \tag{11.87}$$

For weak damping ω_p is just the frequency at which $\varepsilon(\omega) = 0$. It denotes the frequency of a longitudinal oscillation of the free electron gas. These oscillations are called *plasmons*. There is no transverse eigenfrequency because of the lack of a restoring force, i.e., ω_T is zero. Typical values for the plasmon energy lie between 3 and 20 eV. They can be calculated from

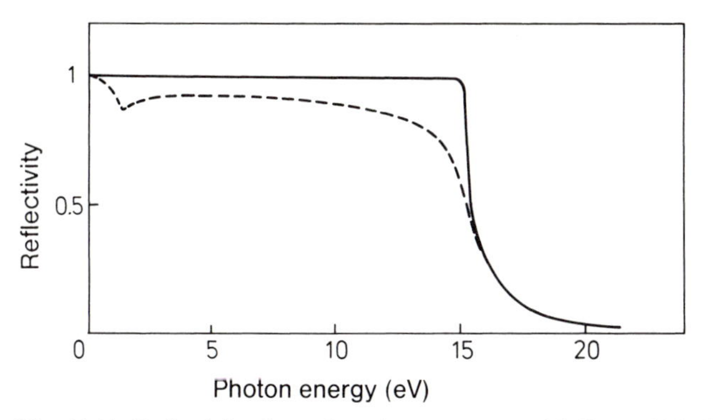

Fig. 11.11. Reflectivity for a free electron gas model ($\hbar\omega_p = 15.2$ eV, $\sigma = 3.6 \times 10^5\ \Omega^{-1}\ \mathrm{cm}^{-1}$) and for aluminium (– – –)

the electron density (with an appropriate value for the effective mass of the electrons, Sect. 9.1), but such calculations have little value. In very many metals, the longitudinal oscillations are so strongly damped by interband transitions (Sect. 11.10), that the plasmons are no longer well-defined excitations. In cases where they can be observed, the zero-crossing of $\varepsilon_1(\omega)$ is strongly shifted from the free-electron-gas value by interband transitions. Nevertheless the functional form of $\varepsilon(\omega)$ derived from the free-electron-gas model provides a qualitative understanding of one of the important universal properties of metals, namely, their high reflectivity. In Fig. 11.11 the reflectivity is plotted for $\hbar\omega_p = 15.2$ eV and $\sigma = \sigma(\omega = 0) = 3.6 \times 10^5\ (\Omega\ \mathrm{cm})^{-1}$. These values apply for aluminium, the metal which most closely approximates to a free electron gas system. The measured reflectivity is plotted for comparison. The dip at 1.5 eV is due to an interband transition. This is also the origin of the strong damping, which limits the maximum reflectivity to 90% for visible and higher frequency light.

An interesting application of the free-electron-gas model is to semiconductors (see also Panel III). In semiconductors, the concentrations of free electrons can be varied over a wide range by doping (Chap. 12). One can thereby vary the plasma edge of the reflectivity. In Fig. 11.12 the reflectivity of a thin layer of In_2O_3 doped with Sn is plotted. Such a layer transmits visible light almost completely, but is an effective barrier to the propagation of infrared radiation. Such coatings are applied in Na vapor lamps and as heat reflecting windows [11.5]. The reflectivity is well-characterized here within the free-electron-gas model provided that the 1 in (11.86) is replaced by the actual ε_∞ in the visible region. The reason is that (without free carriers) In_2O_3 shows no absorption in the visible. The interband transitions begin only above ~ 2.8 eV.

It should also be mentioned that semiconductors, besides the plasma oscillations of free electrons, also display plasma oscillations of the valence electrons. As is the case for many metals, the excitation energies lie between 10 and 20 eV.

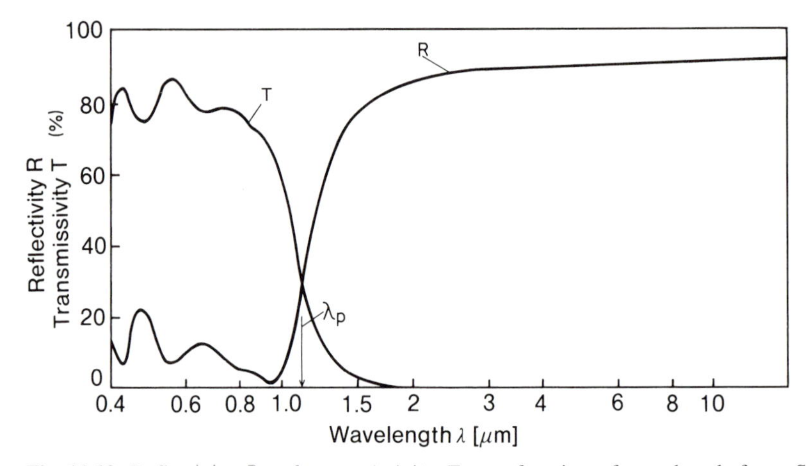

Fig. 11.12. Reflectivity R and transmissivity T as a function of wavelength for a Sn-doped In_2O_3 layer. The thickness of the layer is 0.3 μm and the electron concentration is $1.3\times10^{21}\,cm^{-3}$. λ_p marks the plasma wavelength. (After [11.5])

11.10 Interband Transitions

In Sect. 11.9 we treated the collective excitations of the free electron gas within the framework of classical mechanics. Even for the simple free electron gas model, this is only possible provided one is not interested in the so-called spatial dispersion, i.e., the $\boldsymbol{k}$-dependence of $\varepsilon(\omega,\boldsymbol{k})$. Otherwise the problem of screening must be considered explicitly and single electron excitations within a partially filled band may occur. One also speaks of *intraband transitions*, which determine $\varepsilon(\omega,\boldsymbol{k})$ of the free electron gas.

For sufficiently high ω, transitions can take place between states in different bands. In this case one speaks of *interband transitions*. Corresponding to the discrete nature of the allowed and forbidden energy bands, one expects an energetically discrete structure of these excitation processes and of the resulting $\varepsilon(\omega)$ curve. Within the tight-binding approximation (Sect. 7.3), interband transitions correspond merely to excitations between occupied and unoccupied levels of the atoms, with the discrete energy levels of the single atom being broadened into bands by the overlap of the wavefunctions. An adequate description of the resulting $\varepsilon(\omega)$ function is obtained by summing over all possible individual excitations. For example, a single excitation could be the transition from an occupied $1s$ to an unoccupied $2p$ energy level of a Na atom. The modification due to the crystal structure manifests itself in the broadening and shifting of the $1s$ and $2p$ levels.

The connection between the macroscopic dielectric function $\varepsilon(\omega)$ and the electron excitation spectrum will now be illustrated with the help of a quantum mechanical calculation. As in the remainder of this chapter, we will assume here that the wavelength of the exciting light is large compared to atomic dimensions (lattice constant). The spatial dependence of the electromagnetic field can then be neglected in the Schrödinger equation.

Unfortunately, it is still not possible to treat absorption directly using perturbation theory. This is because the "perturbation", i.e., the electromagnetic wave, loses energy. With the exception of the actual absorption however, one can deduce the properties of the quantum mechanical system in the field by means of perturbation theory.

For this we assume an electromagnetic field of the form

$$\mathscr{E}_x = \mathscr{E}_{0x}\, \boldsymbol{e}_x \cos\omega t \; . \tag{11.88}$$

This perturbation yields a time-dependent Schrödinger equation for the one-electron problem:

$$\left(-\frac{\hbar^2}{2m}\Delta + V(\boldsymbol{r}) - \frac{1}{2}e\,\mathscr{E}_{0x}x(\mathrm{e}^{\mathrm{i}\omega t} + \mathrm{e}^{-\mathrm{i}\omega t})\right)\psi(\boldsymbol{r},t) = \mathrm{i}\hbar\dot{\psi}(\boldsymbol{r},t) \; . \tag{11.89}$$

We let $\psi_{0\mathrm{i}}(\boldsymbol{r},t)=\exp\left[(-\mathrm{i}/\hbar)E_i t\right]\varphi_i(\boldsymbol{r})$ be the solutions in the absence of a perturbation. The solution we are seeking can then be expanded in the form

$$\psi(\boldsymbol{r},t) = \sum_i a_i(t)\mathrm{e}^{-(\mathrm{i}/\hbar)E_i t}\varphi_i(\boldsymbol{r}) \tag{11.90}$$

with time-dependent coefficients $a_i(t)$. The time derivative of $\psi(\boldsymbol{r},t)$ can now be taken directly and, together with the ansatz (11.90), we substitute this into the Schrödinger equation (11.89). Using an arbitrary state $\mathrm{e}^{(\mathrm{i}/\hbar)E_j t}\varphi_j^*(\boldsymbol{r})$ we then form a matrix element with the entire Schrödinger equation. Making use of the orthonormality of the solutions of the unperturbed problem, we thereby obtain

$$\mathrm{i}\hbar\dot{a}_j(t) = -\tfrac{1}{2}e\,\mathscr{E}_{0x}\sum_i\langle j|x|i\rangle a_i(t)(\mathrm{e}^{\mathrm{i}\omega t} + \mathrm{e}^{-\mathrm{i}\omega t})\mathrm{e}^{\mathrm{i}(E_j-E_i)t/\hbar} \; , \tag{11.91}$$

with

$$\langle j|x|i\rangle = \int \varphi_j^*(\boldsymbol{r})x\varphi_i(\boldsymbol{r})\, d\boldsymbol{r} \; .$$

We assume that at time $t=0$ the system is in a particular state i, so that $a_i(t=0)=1$ and all other coefficients are zero. We then have

$$\begin{aligned} a_j(t) &= \frac{\mathrm{i}}{2\hbar}e\,\mathscr{E}_{0x}\langle j|x|i\rangle \int_0^t (\mathrm{e}^{\mathrm{i}\omega t} + \mathrm{e}^{-\mathrm{i}\omega t})\mathrm{e}^{\mathrm{i}(E_j-E_i)t/\hbar}dt \\ &= \frac{e}{2\hbar}\mathscr{E}_{0x}\langle j|x|i\rangle\left(\frac{\mathrm{e}^{\mathrm{i}(\omega_{ji}-\omega)t}-1}{\omega_{ji}-\omega} + \frac{\mathrm{e}^{\mathrm{i}(\omega_{ji}+\omega)t}-1}{\omega_{ji}+\omega}\right) \; , \end{aligned} \tag{11.92}$$

where we have introduced $\omega_{ji}=(E_j-E_i)/\hbar$. With this expression for the time-dependent expansion coefficients we can calculate the expectation value of the dipole moment in the perturbed state. We neglect quadratic terms in the coefficients $a_j(t)$.

$$
\begin{aligned}
\langle e x\rangle &= \int d\boldsymbol{r}\, \psi^*(\boldsymbol{r},t) e\, x\, \psi(\boldsymbol{r},t) \\
&= e x_{ii} + e \sum_j a_j(t) x_{ij} \mathrm{e}^{-\mathrm{i}\omega_{ji} t} + a_j^*(t) x_{ij}^* \mathrm{e}^{\mathrm{i}\omega_{ji} t} \\
&= e x_{ii} + (e \mathscr{E}_{0x}/\hbar) \sum_j |x_{ij}|^2 \frac{2\omega_{ji}}{\omega_{ji}^2 - \omega_0^2} (\cos\omega t - \cos\omega_{ji} t)\ ,
\end{aligned}
$$

$$\text{with} \quad x_{ij} = \int \varphi_i^*(\boldsymbol{r}) x \varphi_j(\boldsymbol{r})\, d\boldsymbol{r}\ . \tag{11.93}$$

The first term in this equation describes a field-independent contribution to the dipole moment, which vanishes for systems with inversion symmetry. The second term is linear in the field and contains a component that oscillates with the same frequency as the field producing the excitation. This term thus describes the polarizability of the medium. There is a further linear component that oscillates at an "atomic" frequency. It cannot be detected in a macroscopic measurement as it averages over atomic frequencies. Equation (11.93) therefore yields the real part of the polarizability, or alternatively $\varepsilon_1(\omega)-1$, for all frequencies except at the poles. Here the perturbation theory breaks down even for vanishingly small fields; see (11.92). It is admissible, however, to now make use of the analytical properties of the complex dielectric constants. One can determine the corresponding imaginary parts, which describe absorption, by means of the Kramers-Kronig relation (11.17). If we take for example (11.41), which in terms of the poles has the same structure as (11.93), we can find $\varepsilon_2(\omega)$ directly:

$$\varepsilon_2(\omega) = \frac{\pi e^2}{\varepsilon_0 V} \sum_{ij} |x_{ij}|^2 \{\delta(\hbar\omega - (E_j - E_i)) - \delta(\hbar\omega + (E_j - E_i))\}\ . \tag{11.94}$$

In this, the summation over i is the sum over all initial states of the system in volume V, and the summation over j runs over all final states. For an ensemble of oscillators in states with the quantum number n, one can use (11.94) to derive the n-independent value of $\varepsilon_2(\omega)$ in agreement with the classical value (11.41). One simply needs to insert the value of the matrix element of a harmonic oscillator, $|x_{n,n+1}|^2 = (n+1)\hbar/2m\omega_0$ or $|x_{n,n-1}|^2 = n\hbar/2m\omega_0$; see Problem 11.7.

We note at this point that the matrix element of the momentum operator $p_{ij} = \mathrm{i} m\omega x_{ij}$ is often used in place of the matrix element of the spatial coordinate x_{ij}. Indeed, this would have appeared directly in our derivation had we used the vector potential $\dot{\boldsymbol{A}} = -\mathscr{E}$ instead of the field $\mathscr{E}$, whereby we would also have had to replace the operator $p_x (= (\hbar/\mathrm{i})\partial/\partial x$ in the real-space representation) in the Schrödinger equation by $p_x + A_x$.

The expression (11.94) for $\varepsilon_2(\omega)$ applies to any quantum mechanical system. For the particular case of a periodic solid, we can further evaluate the matrix element of the momentum operator in (11.94). The states can then be described by Bloch waves characterized by the band index and the $\boldsymbol{k}$-vector (cf. Sect. 7.1). For the indexing of the bands we retain the indices i,j. In this notation the real space representations of Bloch states $|i,\boldsymbol{k}_i\rangle$ and $|j,\boldsymbol{k}_j\rangle$ are

$$\begin{aligned}\langle \boldsymbol{r}|i,\boldsymbol{k}_i\rangle &= \frac{1}{\sqrt{V}}u_{\boldsymbol{k}_i}(\boldsymbol{r})\mathrm{e}^{\mathrm{i}\boldsymbol{k}_i\cdot\boldsymbol{r}} ,\\ \langle \boldsymbol{r}|j,\boldsymbol{k}_j\rangle &= \frac{1}{\sqrt{V}}u_{\boldsymbol{k}_j}(\boldsymbol{r})\mathrm{e}^{\mathrm{i}\boldsymbol{k}_j\cdot\boldsymbol{r}} .\end{aligned} \tag{11.95}$$

In calculating the matrix element as an integral over all space, we can, as in Sect. 3.6, split the integral up into an integral over a unit cell and a summation over all cells

$$\langle i,\boldsymbol{k}_i|\boldsymbol{p}|j,\boldsymbol{k}_j\rangle = \langle i,\boldsymbol{k}_i|\boldsymbol{p}|j,\boldsymbol{k}_j\rangle_{\mathrm{cell}}\frac{1}{N}\sum_n \mathrm{e}^{\mathrm{i}(\boldsymbol{k}_j-\boldsymbol{k}_i)\cdot\boldsymbol{r}_n} . \tag{11.96}$$

As was discussed in Chap. 3, the sum has non-zero values only for $\boldsymbol{k}_j=\boldsymbol{k}_i$. This is a special form of wave number or quasi-momentum conservation, which we have already met a number of times, and which is a consequence of the lattice periodicity. It is valid when the wavenumber of the light k_{L} can be regarded as small; this holds for light well into the ultraviolet (cf. Sect. 4.5). Optical transitions, when represented in $\boldsymbol{k}$ space, are therefore "vertical transitions" (Fig. 11.13).

With quasi-momentum conservation, we can now evaluate $\varepsilon_2(\omega)$ further. The sum over all states is now carried out as a sum over all $\boldsymbol{k}$ and all bands

$$\varepsilon_2(\omega) = \frac{\pi}{\varepsilon_0}\frac{e^2}{m^2\omega^2}\frac{1}{V}\sum_{ijk}|\langle i,\boldsymbol{k}|\boldsymbol{p}|j,\boldsymbol{k}\rangle|^2\delta(E_j - E_i - \hbar\omega) . \tag{11.97}$$

Replacing the sum over $\boldsymbol{k}$ by an integral, we have

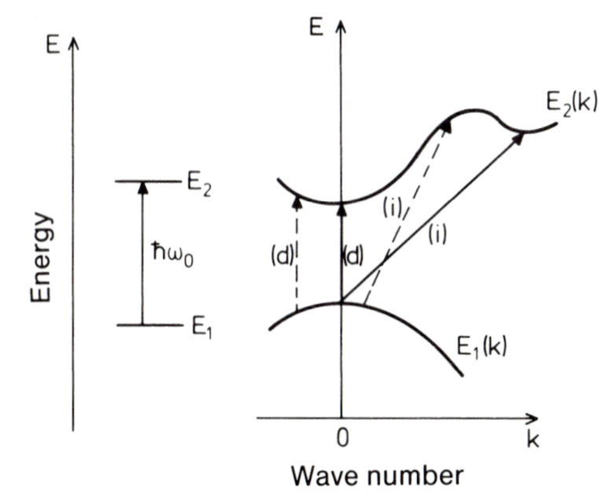

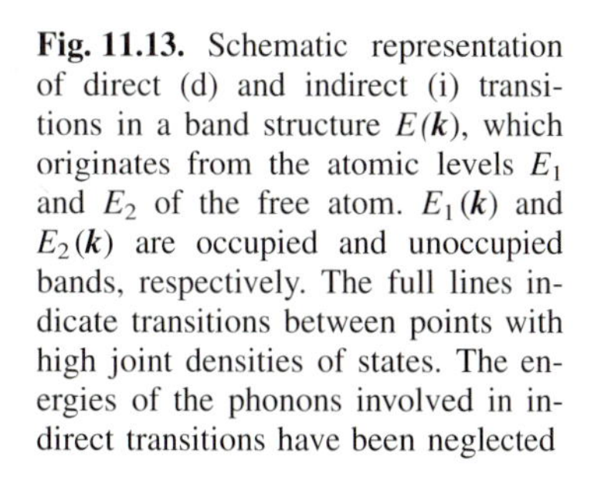

Fig. 11.13. Schematic representation of direct (d) and indirect (i) transitions in a band structure $E(\boldsymbol{k})$, which originates from the atomic levels E_1 and E_2 of the free atom. $E_1(\boldsymbol{k})$ and $E_2(\boldsymbol{k})$ are occupied and unoccupied bands, respectively. The full lines indicate transitions between points with high joint densities of states. The energies of the phonons involved in indirect transitions have been neglected

$$\frac{1}{V}\sum_{k} \Rightarrow \frac{1}{(2\pi)^3}\int d\boldsymbol{k} \ , \tag{11.98}$$

$$\varepsilon_2(\omega) = \frac{\pi}{\varepsilon_0}\frac{e^2}{m^2\omega^2}\frac{1}{(2\pi)^3}\sum_{ij}\int |\langle i,\boldsymbol{k}|\boldsymbol{p}|j,\boldsymbol{k}\rangle|^2 \delta(E_j(\boldsymbol{k})-E_i(\boldsymbol{k})-\hbar\omega)\, d\boldsymbol{k} \ . \tag{11.99}$$

The δ-function converts the volume integral in $\boldsymbol{k}$ space to an area integral over a surface of constant energy difference $E_j(\boldsymbol{k})$–$E_i(\boldsymbol{k})=\hbar\omega$ [cf. (7.41)]

$$\varepsilon_2(\omega) = \frac{\pi}{\varepsilon_0}\frac{e^2}{m^2\omega^2}\frac{1}{(2\pi)^3}\sum_{ij}\int_{\hbar\omega=E_j-E_i} |\langle i,\boldsymbol{k}|\boldsymbol{p}|j,\boldsymbol{k}\rangle|^2 \times \frac{df_\omega}{|\mathrm{grad}_{\boldsymbol{k}}[E_j(\boldsymbol{k})-E_i(\boldsymbol{k})]|} \ . \tag{11.100}$$

The imaginary part of the dielectric function – insofar as it is determined by interband transitions – may be decomposed into a matrix element and a *joint density of states*

$$Z_{ij}(\hbar\omega) = \frac{1}{(2\pi)^3}\int_{\hbar\omega=E_j-E_i} \frac{df_\omega}{|\mathrm{grad}_{\boldsymbol{k}}[E_j(\boldsymbol{k})-E_i(\boldsymbol{k})]|} \ . \tag{11.101}$$

This decomposition is meaningful if, as is frequently assumed, $\langle i,\boldsymbol{k}|\boldsymbol{p}|j,\boldsymbol{k}\rangle$ shows no significant $\boldsymbol{k}$ dependence. The joint density of states is high for energies at which two energy surfaces with this separation lie parallel to one another at a particular $\boldsymbol{k}$ value. At such points one has so-called critical points or van Hove singularities, as was shown in Sect. 5.1 for phonon dispersion branches, and in Sect. 7.5 for electronic bands. These critical points cause the prominent structures in the $\varepsilon_2(\omega)$ and thus in the optical absorption spectrum. As an example, Fig. 11.14 shows the experimentally determined $\varepsilon_2(\omega)$ spectrum for Ge.

As in the case of electronic transitions in free atoms, we differentiate between *allowed* and *forbidden* transitions. A transition is allowed when the matrix element (11.96) does not vanish. If we assume that in a semiconductor the minimum of the conduction band is at the same position in $\boldsymbol{k}$-space as the maximum of the valence band, then one has for the energies of the conduction band E_c and the valence band E_v the expansions

$$E_c = E_g + \frac{\hbar^2}{2m_c^*}k^2 \ , \quad E_v = \frac{\hbar^2}{2m_v^*}k^2 \ , \tag{11.102}$$

Hence one inserts

$$E_j - E_i = E_c - E_v = E_g + \frac{\hbar^2}{2}(m_c^{*-1} - m_v^{*-1})k^2 \tag{11.103}$$

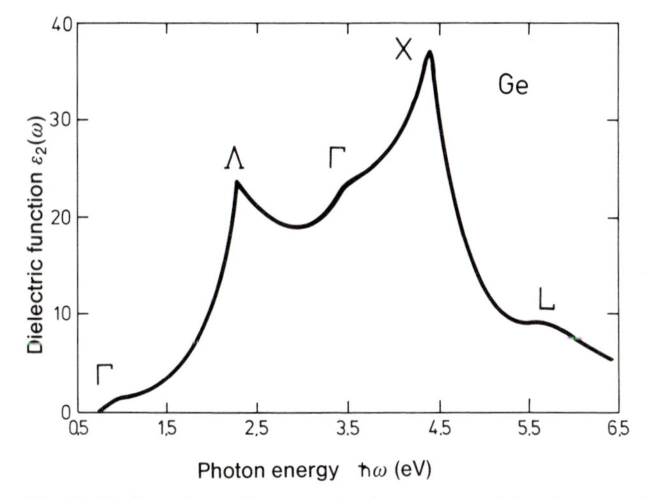

Fig. 11.14. Experimentally determined spectrum of the dielectric function $\varepsilon_2(\omega)$ for germanium. Γ,X and L denote critical points in the Brillouin zone, to which the measured high densities of states can be assigned. A large contribution originates from transitions along the Λ direction (corresponding to $\langle 111 \rangle$) (see also Fig. 7.13). (After [11.6])

in the joint density of states Z_{ij} $(\hbar\omega)$ (11.101) so that $Z_{ij}(\hbar\omega)$ becomes proportional to $(\hbar\omega - E_g)^{1/2}$, analogously to the calculation of the density of states of a free electron gas (Sect. 6.1). This term determines the form of $\varepsilon_2(\omega)$ in the vicinity of the band gap E_g for the case of an allowed transition.

If the matrix element at the band edge vanishes (forbidden transition), then in this vicinity the matrix element can be expanded as a series in $(\hbar\omega - E_g)$; truncation after the linear term of the series leads, therefore, for forbidden transitions, to a characteristic dependence $\varepsilon_2(\omega) \sim (\hbar\omega - E_g)^{3/2}$, because of the square-root dependence of the joint density of states. As Fig. 11.15 shows, the $\boldsymbol{k}$ dependence of the matrix element must, in general, also be considered, so that the discussion based on the joint density of states only provides a qualitative starting point.

Until now we have omitted the electron-phonon interaction from the discussion. Within an improved approximation, the phonon system of the crystal can no longer be treated as decoupled from the optical transition. The matrix element (11.96) then contains, instead of the purely electronic wavefunction $\langle E_i, \boldsymbol{k}_i|$, the total wavefunction $\langle E_i, \boldsymbol{k}_i, \omega_{\boldsymbol{q}}, \boldsymbol{q}|$ of the coupled electron-phonon system. As a perturbation one now must consider both the interaction with the light field $\boldsymbol{A}(\omega)$, and the interaction with the phonon system. In the particle picture, one is dealing with a three-particle interaction involving photons, electrons and phonons. Transitions of this kind provide a much smaller contribution to $\varepsilon_2(\omega)$ than the previously considered "two-particle interaction". Here, too, we require conservation of wave number $\boldsymbol{k}$ and of energy:

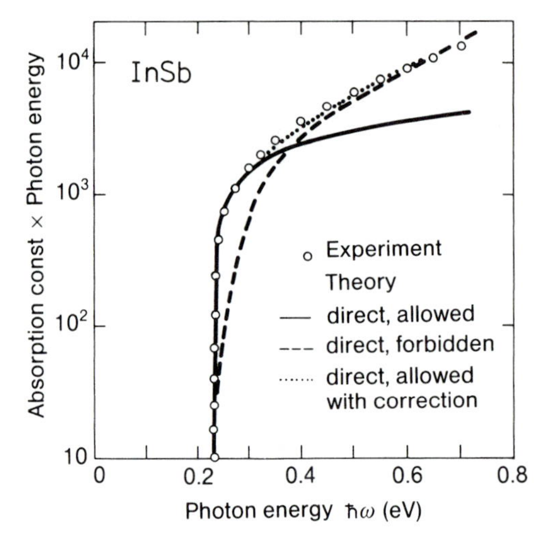

Fig. 11.15. Product of the photon energy $\hbar\omega$ and the absorption constant plotted against photon energy for InSb. The experimentally determined points (after [11.7]) are compared with calculations for a direct allowed transition and a forbidden transition. The best description is obtained for allowed transitions in which the matrix element is corrected (for $\boldsymbol{k}$ dependence), and the non-parabolic form of the conduction band is considered. (After [11.8])

$$E_j - E_i = \hbar\omega \pm \hbar\omega_{\boldsymbol{q}} \ , \tag{11.104}$$

$$\boldsymbol{k}_j - \boldsymbol{k}_i = \boldsymbol{k}_{\mathrm{L}} \pm \boldsymbol{q} \ . \tag{11.105}$$

Energy conservation (11.104) entails merely a minor modification compared with the case of direct transitions, because for phonons $\hbar\omega_{\boldsymbol{q}}$ is typically two orders of magnitude smaller than the electronic transition energies (E_j–E_i) at critical points. However, because phonons can contribute $\boldsymbol{q}$ vectors from the entire Brillouin zone, indirect transitions are possible between arbitrary initial and final states within the Brillouin zone. In Fig. 11.13 such indirect transitions (marked (i)) are indicated without considering the negligibly small phonon energy $\hbar\omega_{\boldsymbol{q}}$. Such transitions can contribute significantly to the $\varepsilon_2(\omega)$ spectrum only if they take place between critical points of the valence and conduction bands, and provided, of course, that they do not overlap with direct transitions. For Ge, the onset of interband absorption is due to a so-called "indirect band gap" because the minimum of the conduction band is at L (Fig. 7.13) and the maximum of the valence band is at Γ. However, for Ge the first direct transition $\Gamma_{25'} \rightarrow \Gamma_{2'}$ lies only slightly higher in energy.

The example of silver in Figs. 11.16, 11.17 shows that interband transitions also contribute substantially to the optical spectra of metals. As in the case of copper, the energetically rather narrow d-bands of silver lie well below the Fermi level. Transitions from these d-bands to regions of high density of states above the Fermi level are responsible for the structures in Fig. 11.16. These are superimposed on a free electron gas part $\varepsilon_1^{(\mathrm{f})}$ (Fig. 11.17). The characteristic color of the noble metals (Ag, Cu, Au) has its origin in these transitions. We note further that the point at which $\varepsilon_1(\omega)$ passes through zero, and therefore the "plasma frequency" is strongly shifted due to the interband transitions superposed on the spectrum.

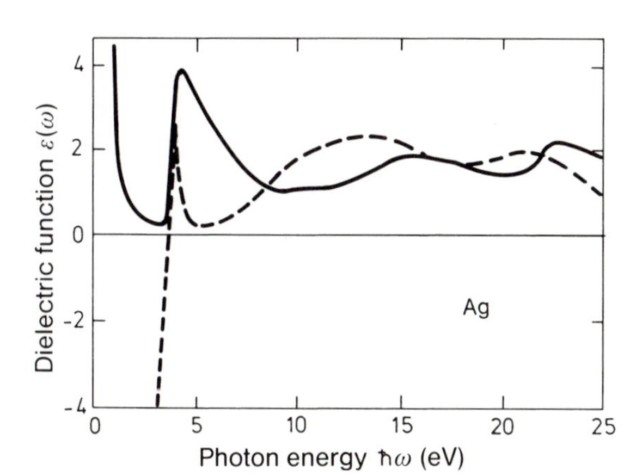

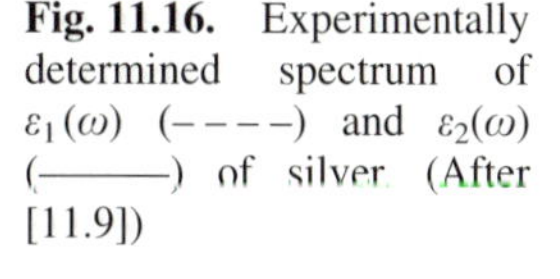

Fig. 11.16. Experimentally determined spectrum of $\varepsilon_1(\omega)$ (– – – –) and $\varepsilon_2(\omega)$ (———) of silver. (After [11.9])

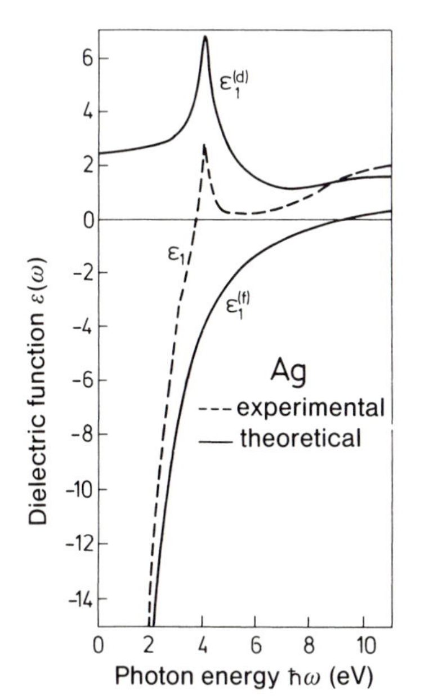

Fig. 11.17. Mathematical decomposition of the experimentally determined $\varepsilon_1(\omega)$ spectrum of Ag from Fig. 11.16 into a part $\varepsilon_1^{(f)}$ due to the free electron gas, and a part $\varepsilon_1^{(d)}$, due to interband transitions from d states

11.11 Excitons

For semiconductors at low temperatures, a spectrum of $\varepsilon_2(\omega)$ like that shown in Fig. 11.14 is rarely found at the onset of optical absorption in the vicinity of the band gap energy E_g. One frequently observes a sharply structured onset of optical absorption, as shown in Fig. 11.18 for GaAs. These structures are due to the so-called excitons. Excitons are bound states between an electron excited into the conduction band and the hole that remains in the valence band, with the Coulomb interaction being responsible for the binding energy.

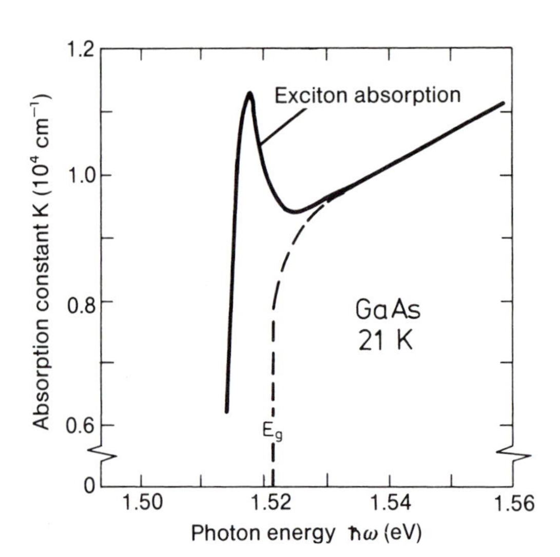

Fig. 11.18. Absorption constant of GaAs measured at 21 K in the vicinity of the band gap energy E_g. The dashed curve would be the result in the absence of exciton excitations. (After [11.10])

The simplest mathematical description of such an excitonic state is given by the so-called hydrogenic model. In this model, the possible energy states $E_{n,\boldsymbol{K}}$ of the bound electron-hole pair are essentially the Bohr eigenstates:

$$E_{n,\boldsymbol{K}} = E_g - \frac{\mu^* e^4}{32\pi^2\hbar^2\varepsilon^2\varepsilon_0^2}\frac{1}{n^2} + \frac{\hbar^2 K^2}{2(m_c^* + m_v^*)} \,. \tag{11.106}$$

The gap energy E_g appears as an additive term because the zero of energy is usually taken to be the top of the valence band edge. The electron mass is replaced by the reduced effective mass μ^* derived from the effective masses of electrons (m_c^*) and holes (m_v^*). Furthermore, the dielectric constant ε_0 of the vacuum is modified by the dielectric constant ε of the semiconductor material, to take into account screening by the surrounding dielectric. Because an electron-hole pair can move almost freely in the crystal, the kinetic energy of the center of mass must also be included in the energy of the two particle system – third term in (11.106) – where $\boldsymbol{K}$ is the wave vector corresponding to the center of mass motion of the two particles. One notes that (11.106) is concerned with two-particle energy levels, which cannot be directly drawn in a band scheme in which only single particle energies are plotted. However, via the excitation of an electron it is possible to reach energy levels $E_{n,\boldsymbol{K}}$ which lie just below the band edge, i.e. just falling short of the ionization threshold of the exciton (separation of electron and hole). Since for semiconductors ε is of the order 10, the exciton spectrum for $\boldsymbol{K}=\boldsymbol{0}$ is a strongly compressed hydrogen atom spectrum, whose binding energies lie below 0.1 eV. For the same reason, the spatial extent of the combined electron-hole wavefunction of the exciton is much larger than that of the hydrogen atom (about 10 Bohr radii). This, in hindsight, justifies the application of the simple hydrogenic model with a macroscopic dielectric constant. For germanium the excitonic binding energy E_{ex} for the lowest direct transition at Γ and for $\boldsymbol{K}=0$ is calculated as

$$E_{ex} = -\frac{\mu^* e^4}{32\pi^2\hbar^2\varepsilon^2\varepsilon_0^2} = -0.0017\,\mathrm{eV} \tag{11.107}$$

and the corresponding Bohr radius is 47×10^{-8} cm. The experimental value for the binding energy is –0.0025 eV. In an optical absorption experiment, one sees, of course, only exciton states with $\boldsymbol{K}\simeq0$, (11.107), because light quanta can only transfer a negligibly small momentum, as discussed above. Since the energetic separation of the exciton state from the band edge E_g is rather small, measurements to show the excitonic character of optical absorption must be carried out at low temperature. The weakly bound excitons discussed here are also called *Mott-Wannier excitons.* For the sake of completeness it should also be mentioned that so-called *Frenkel excitons* are observed in molecular, noble-gas and ionic crystals. Their binding energy is of the order of 1 eV, and, in contrast to the Mott-Wannier excitons, the wavefunction of the electron-hole pair is then localized on an atom or molecule.

11.12 Dielectric Energy Losses of Electrons

Besides optical methods, the inelastic scattering of electrons had also become established as a tool for investigating dielectric properties. On passing through a material, electrons suffer characteristic energy losses. These may be caused by collisions with, and resonance excitations of individual atoms; this component however is small compared to the so-called dielectric energy losses. In a semiclassical model, dielectric losses are due to power dissipation by the electric field which accompanies the moving electron. When viewed from a fixed point in the solid, the field of a moving electron is time dependent and therefore contains a broad frequency spectrum in a Fourier analysis. The individual Fourier components experience dissipation in the dielectric medium. The energy loss per unit volume, with $\mathscr{E}$ and D from (11.3) and (11.4) is

$$\mathrm{Re}\left\{\int_{-\infty}^{\infty} \mathscr{E}\dot{D}dt\right\} = \mathrm{Re}\left\{2\pi\mathrm{i}\int_{-\infty}^{\infty} \omega\,\mathscr{E}(\omega)\,D^*(\omega)\,d\omega\right\}. \tag{11.108}$$

This energy must be supplied by the kinetic energy of the electrons. The electron acts as a source of dielectric displacement, whose Fourier components can be determined from the path of the electron. We do not perform the calculation in detail here; see Ref. [11.11]. The essential part of a rather lengthy calculation becomes apparent though, if we replace $\mathscr{E}(\omega)$ in (11.108) by $D(\omega)/\varepsilon_0\varepsilon(\omega)$

$$\mathrm{Re}\left\{\int_{-\infty}^{\infty}\mathcal{E}\dot{D}dt\right\} = -2\pi\int_{-\infty}^{\infty}\omega\,\mathrm{Im}\left\{\frac{1}{\varepsilon(\omega)}\right\}\frac{|D(\omega)|^2}{\varepsilon_0}d\omega\,. \tag{11.109}$$

In this classical picture, there are as yet no discrete energy losses. We can, however, assign a discrete energy loss $\hbar\omega$ to each frequency ω. The spectral function

$$I(\omega) \propto -\omega\,\mathrm{Im}\left\{\frac{1}{\varepsilon(\omega)}\right\}, \quad \omega > 0 \tag{11.110}$$

then describes the distribution of energy losses. For the case of a free electron gas, this loss function has a very simple and intuitively clear interpretation. After inserting the dielectric constant of the free electron gas from (11.86), one obtains

$$I(\omega) \propto \omega\,\mathrm{Im}\left\{\frac{\omega^2 + \mathrm{i}\dfrac{\omega_\mathrm{p}^2\varepsilon_0}{\sigma}\omega}{\omega_\mathrm{p}^2 - \omega^2 - \mathrm{i}\dfrac{\omega_\mathrm{p}^2\varepsilon_0}{\sigma}\omega}\right\}. \tag{11.111}$$

For the case of weak damping we make use of the δ-function representation as in (11.41), and obtain

$$I(\omega) \propto \frac{\pi}{2}(\delta(\omega - \omega_\mathrm{p}) - \delta(\omega + \omega_\mathrm{p}))\,, \tag{11.112}$$

i.e., a pole at the frequency of the longitudinal plasma wave ("plasmon"). We have already seen that the description as a free electron gas is particularly appropriate for aluminium. As a result, the plasmon losses in Al are especially prominent (Fig. 11.19). However, many other materials also display a clear maximum in the loss function, which is likewise called a "plasmon loss" in deference to the free electron gas model.

It can easily be seen that the maximum interaction between the electron and the plasma wave occurs when the velocity of the electron matches the

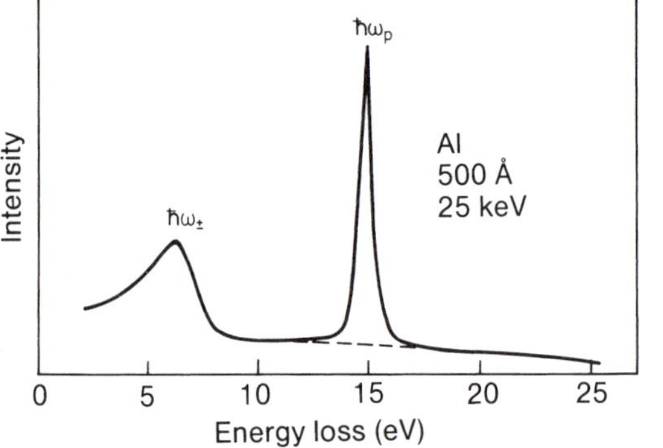

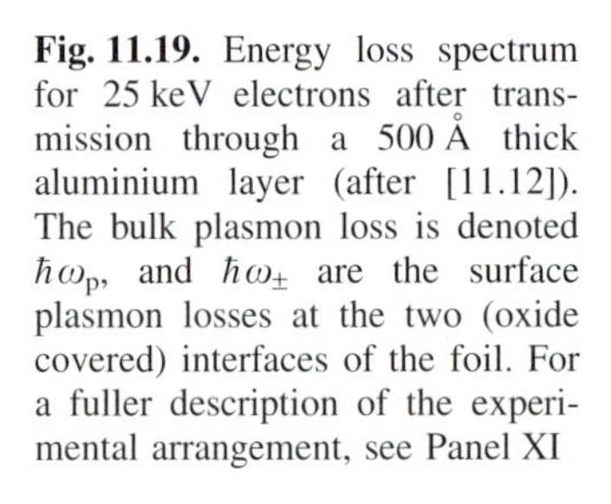
Fig. 11.19. Energy loss spectrum for 25 keV electrons after transmission through a 500 Å thick aluminium layer (after [11.12]). The bulk plasmon loss is denoted $\hbar\omega_\mathrm{p}$, and $\hbar\omega_\pm$ are the surface plasmon losses at the two (oxide covered) interfaces of the foil. For a fuller description of the experimental arrangement, see Panel XI

phase velocity of the longitudinal wave. Like a surfer, the electron then continuously loses (or gains) energy depending on the relative phase. A quantitative consideration of the resonance condition

$$\omega/k = v_{\mathrm{el}} \tag{11.113}$$

leads to the result that for $\hbar\omega = 15$ eV and an electron primary energy of 20 keV, $k \sim 2.7\times10^{-2}$ Å^{-1}, which is small compared to the dimensions of the Brillouin zone. This is also the justification for neglecting the k dependence of $\varepsilon(\omega)$. Through a measurement of the energy losses, it is thus possible to determine "optical" data for solids. This method of energy loss spectroscopy has the advantage that a wide spectral region can be probed (cf. Panel XI).

Electric fields in solids are created not only by electrons inside the solids but also by electrons traveling outside the solid in the vicinity of a surface. The screening factor for the field is then, however, not $1/\varepsilon(\omega)$, but $1/(\varepsilon(\omega)+1)$. Correspondingly, the loss function is

$$I(\omega) \propto -\omega\,\mathrm{Im}\left\{\frac{1}{\varepsilon(\omega)+1}\right\}, \tag{11.114}$$

and the associated losses are called surface losses. For weak damping, the position of the pole in the loss function now lies at $\varepsilon_1(\omega) = -1$. This was exactly the condition for the existence of a surface wave of the dielectric medium. Electrons outside the medium can therefore excite the surface waves (phonons and plasmons) represented in Fig. 11.6. Such losses can also be observed experimentally. For example, in Fig. 11.19 surface plasmons of a thin oxide-covered film are shown. Figure 11.20 illustrates the loss spectrum due to the excitation of surface phonons.

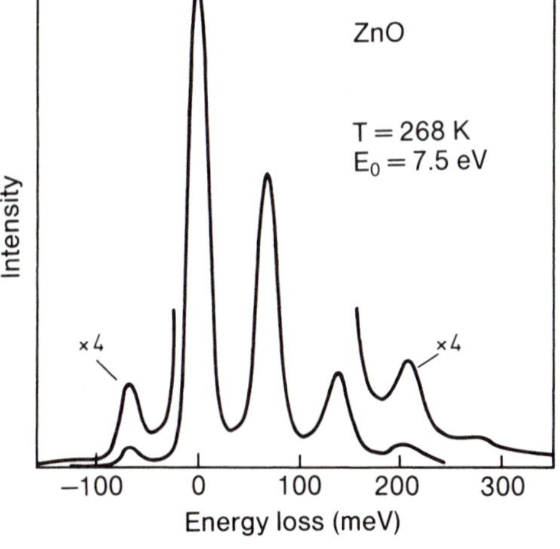

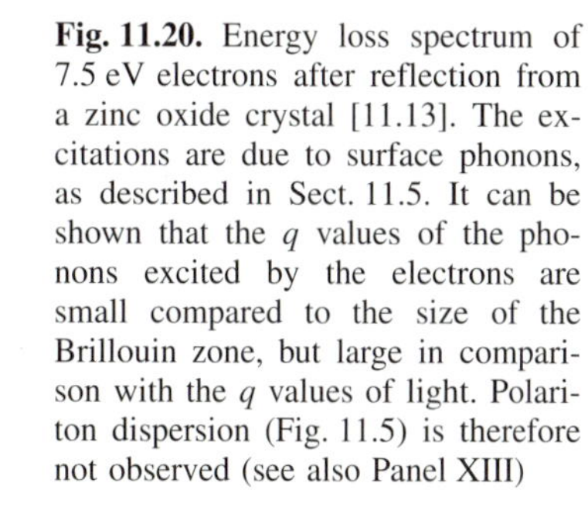

Fig. 11.20. Energy loss spectrum of 7.5 eV electrons after reflection from a zinc oxide crystal [11.13]. The excitations are due to surface phonons, as described in Sect. 11.5. It can be shown that the q values of the phonons excited by the electrons are small compared to the size of the Brillouin zone, but large in comparison with the q values of light. Polariton dispersion (Fig. 11.5) is therefore not observed (see also Panel XIII)

The electron can also gain energy if the temperature is high enough for the first excited state of the phonons to be occupied with sufficient probability. The wave vector of the surface phonons excited by electrons is much larger than that of those excited by light (Panel XIII), although it is still small compared to a reciprocal lattice vector. This results here from the fact that optimum excitation occurs when the phase velocity of the surface wave ω/q is equal to the component of the electron velocity parallel to the surface $v_{\parallel}$. For 5 eV electrons and a 45° angle of incidence, one calculates, for example, $q_{\parallel} \cong 1\times 10^{-2}$ Å^{-1}.

Problems

11.1 The principle of causality demands that a dielectric polarization of a medium at a particular time t can only be caused by electric fields $\mathscr{E}(t')$ with $t' \leq t$. Show that, as a consequence of this causality principle, the susceptibility $\chi(\omega)$ has its poles exclusively in the negative imaginary half-space. In order to arrive at this conclusion it is necessary to choose the Fourier-representation

$$\boldsymbol{P}(t) = \int_{-\infty}^{\infty} \boldsymbol{P}(\omega) e^{-i\omega t} d\omega$$

$$\mathscr{E}(t) = \int_{-\infty}^{\infty} \mathscr{E}(\omega) e^{-i\omega t} d\omega .$$

(If one were to use a Fourier representation with a positive sign in the exponent, the poles of $\chi(\omega)$ would be in the positive imaginary half-space.) *Hint:* By appropriate choice of a complex integration path one can demonstrate that $\boldsymbol{P}(t) \equiv 0$ for $t' \geq t$, provided that $\chi(\omega)$ has no pole in the positive half-space and that $\chi(\omega) \to 0$ as $\omega \to \infty$.

11.2 Show that (11.38) and (11.39) fulfil the Kramers-Kronig relation. *Hint:* Integration is performed by separation of the fractions:

$$\frac{1}{(\omega-\omega_1)(\omega-\omega_2)} = \frac{A}{\omega-\omega_1} + \frac{B}{\omega-\omega_2} .$$

11.3 Suppose one has an absorption represented by a sum of δ-functions

$$\varepsilon_2(\omega) = \frac{\pi}{2}\sum_i f_i(\delta(\omega-\omega_1) - \delta(\omega+\omega_i))$$

What is the real part of $\varepsilon(\omega)$?

11.4 Calculate the normal incidence reflectivity of a free electron gas for the cases of weak and strong (frequency-independent) damping. Repeat the calculation with the assumption of an additional absorption line in the regime $\omega \ll \omega_p$, i.e., below the plasma frequency.

11.5 Derive the dispersion relation $\omega(q_{\parallel})$ of the eigenmodes of a dielectric slab with vacuum on either side by invoking the ansatz for the electrostatic potential

$$\varphi(x,z,t) = e^{i(q_{\parallel}x-\omega t)}\left(A\,e^{-q_{\parallel}|z-d/2|} + B\,e^{q_{\parallel}|z+d/2|}\right),$$

where $q_{\parallel}$ is the wave vector parallel to the surface of the slab and d is the thickness. The appropriate boundary conditions lead to a secular equation. The dispersion relation follows from the condition that the determinant vanishes. Explain why the plasma frequency goes to zero when $q_{\parallel}d \ll 1$ for a polarization parallel to the slab. Discuss also the other limits.

11.6 The real part of the polarizability of an electron system may be written as

$$\alpha(\omega) = \frac{e^2}{m\,\varepsilon_0 V}\sum_{i,j}\frac{f_{ij}}{\omega_{ij}^2 - \omega^2},$$

where ω_{ij} are the absorption frequencies $\omega_{ij}=(E_i-E_j)/\hbar$ and f_{ij} are the oscillator strengths (11.93). In the limit of $\omega \to \infty$ all electrons may be considered as free electrons. Show that $\sum_j f_{ij}=1$ by comparison to $\alpha(\omega)$ of a free electron gas (f-sum rule, a general proof follows from the commutator $[x,p]=i\hbar$).

11.7 Calculate $\varepsilon_2(\omega)$ for an ensemble of harmonic oscillators of the density N/V using the general equation (11.94)

$$\varepsilon_2(\omega) = \frac{\pi e^2}{\varepsilon_0 V}\sum_{ij}|x_{ij}|^2 \times \{\delta(\hbar\omega - (E_j - E_i)) - \delta(\hbar\omega + (E_j - E_i))\}\ !$$

Use the operator representation

$$\boldsymbol{x} = \left(\frac{\hbar}{2m\omega_0}\right)^{1/2}(\boldsymbol{a}^+ + \boldsymbol{a})$$

with $\boldsymbol{a}^+$ and $\boldsymbol{a}$ the creation and annihilation operators which act as follows:

$$\boldsymbol{a}^{+}|n\rangle = \sqrt{n+1}|n+1\rangle\,,$$
$$\boldsymbol{a}|n\rangle = \sqrt{n}|n-1\rangle\,.$$

Show that the result is independent of the initial state of the harmonic oscillator. Compare with the classical expression (11.39).

11.8 Discuss $\varepsilon_2(\omega)$ for a two-level system with occupation probabilities P_0 and P_1 for the ground state and excited state respectively. What is your result for $P_1 > P_0$? What is the meaning of a negative $\varepsilon_2(\omega)$? Why does one need at least three levels for a laser and what are the conditions imposed on the spontaneous transitions between the three levels if one wants laser action?

11.9 Consider the dielectric function

$$\varepsilon_1(\omega) = 1 + \frac{e^2}{\varepsilon_0 V}\sum_{ij}\frac{f_{ij}}{\omega_{ij}^2 - \omega^2}$$

and try to estimate the static dielectric constant for an insulator with an energy gap E_g by assuming that the total oscillator strength is accumulated at $\hbar\omega = E_g$. Apart from a factor of the order of one, the result is equivalent to the result of the so-called Penn model with a spherical "Brillouin zone". Calculate ε_1 for Si ($E_g = 3.4$ eV) and Ge ($E_g = 0.85$ eV). Why is the calculated result too high? Why does one have to insert the energy of the direct gap, not that of the indirect gap?

11.10 By complex integration show that

$$\phi(\boldsymbol{r}) = \frac{e}{r}\mathrm{e}^{-k_{\mathrm{TF}} r}$$
$$= \frac{e}{2\pi^2}\int \frac{1}{q^2 + k_{\mathrm{TF}}^2}\mathrm{e}^{-\mathrm{i}\boldsymbol{q}\cdot\boldsymbol{r}} d\boldsymbol{q}\ !$$

This is the Fourier representation in q-space of the Thomas-Fermi screened potential (Sect. 6.5). Hence $\phi(q) = e/[2\pi^2(q^2 + k_{\mathrm{TF}}^2)]$. One may introduce a dielectric constant in q-space via

$$\phi(q) \equiv \frac{e}{2\pi^2}\frac{1}{q^2 + k_{\mathrm{TF}}^2} = \frac{1}{\varepsilon(q)}\phi_{\mathrm{ext}}(q)\,,$$

where ϕ_{ext} is the potential of the unscreened electric charge e in q-space ($k_{\mathrm{TF}} = 0$!). Derive $\varepsilon(q)$.

11.11 The plasma frequency of a homogeneous distribution of positive ions of mass M is

$$\Omega_{\mathrm{p}}^2 = \frac{ne^2}{M\varepsilon_0}$$

with n the density. Since the restoring force is provided by the electric field due to the charge density fluctuations in the plasma wave, and since the field is screened by $1/\varepsilon(q)$, one may derive a q-dependent frequency when the field between the ions is screened by a free electron gas. Derive $\Omega(q)$ using the result of Problem 11.10 for $\varepsilon(q)$. Calculate the sound velocity for a longitudinal acoustic wave in this model (Bohm-Staver relation). Compare with the experimental results for Na, K, and Al.

11.12 Calculate the bulk and surface loss functions for electrons, $-\mathrm{Im}\{1/\varepsilon\}$ and $-\mathrm{Im}\{1/(\varepsilon+1)\}$, respectively, for a free electron gas and an ensemble of harmonic oscillators with a vibrational frequency ω_0. Interpret the result in the context of Sect. 11.5.

11.13 As a simple model for an infrared active crystal consider a diatomic linear chain (two different masses with alternating charges $\pm e$ at equal distances, and with equal force constants). Electrostatic interactions between the ions are not taken into account. Calculate and discuss the vibrational amplitude of the atoms under the action of an external electric field $\mathscr{E} = \mathscr{E}_0 \exp(-\mathrm{i}\omega t)$. Why is it legitimate to restrict the calculation to phonon excitations near the center of the Brillouin zone ($q \simeq 0$) when ω is a frequency in the infrared spectral range?

a) Calculate the frequency-dependent polarization $P(\omega)$ and the dielectric function $\varepsilon(\omega)$, without and with damping.
b) Derive the Lyddane-Sachs-Teller relation (11.58).

Panel XI
Spectroscopy with Photons and Electrons

By far the greatest amount of information about solids has been obtained by spectroscopic methods, among which spectroscopy with photons is particularly important. The intensity and spectral range of different light sources are compared in Fig. XI.1, which illustrates the available photon flux per eV of bandwidth. In such a comparison, however, the widely varying properties of the sources must also be considered. For the black-body radiation curve, it has been assumed that angular resolution relative to the angle of incidence of the radiation is not necessary, or that the sample is

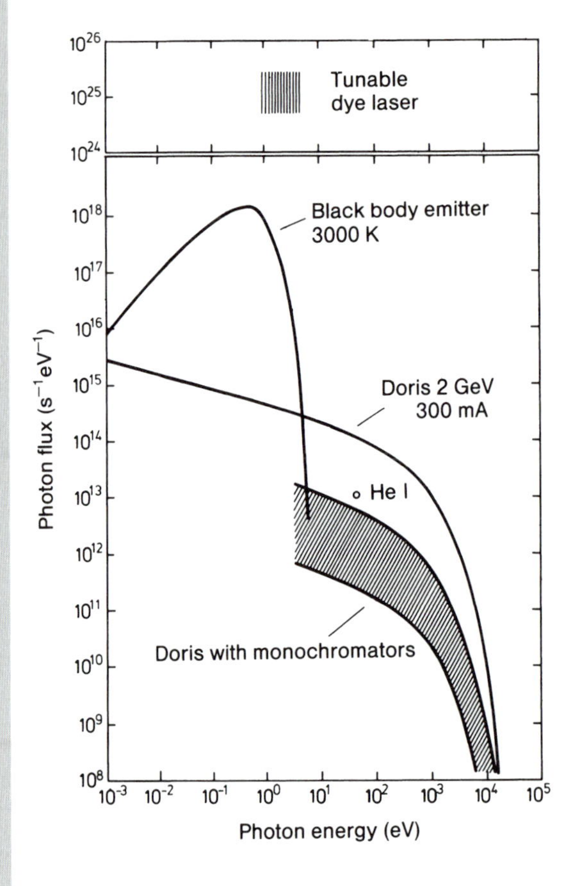

Fig. XI.1. Typical photon flux (per eV) at the sample for various light sources. The sample surface area is taken to be 1 cm^2. For a legitimate comparison, the very different characteristics of the various kinds of radiation must also be taken into account. The figure serves therefore only as a qualitative guide. Synchrotron radiation and laser radiation are highly collimated. For experiments where high angular resolution is required, they are thus particularly advantageous. For normal absorption measurements, which require neither angular resolution nor high energy resolution, a black-body radiator can also be useful

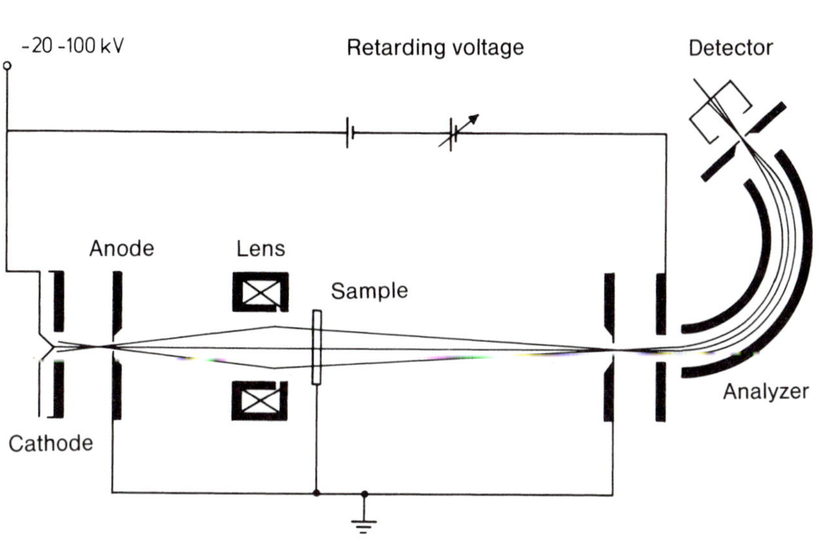

Fig. XI.2. Electron spectrometer for transmission spectroscopy (after [XI.4])

sufficiently large. Synchrotron radiation is strongly concentrated in a particular direction. Fig. XI.1 gives the photon flux of the storage ring "Doris" (near Hamburg, FRG), which has a usable angular aperture of about 1 mrad×1 mrad. For smaller machines, the usable angular aperture is larger. In this case, synchrotron radiation can also be useful in the far infrared. Monochromators in the UV region have a strongly varying and generally not very good efficiency, depending on design and spectral region. This is indicated by the shaded curve. Some experiments can also be run with fixed frequency sources, gas discharge lamps producing characteristic X-ray emission. As an example, the photon flux of the He I line of a gas discharge lamp is plotted. This source is frequently used for photoemission experiments (cf. Panel V). The light yield is then comparable to that of a synchrotron source. Extremely intense dye lasers are unfortunately only available in the region 1–3.6 eV. When comparing dye lasers with other light sources, it must be remembered that their relative (tuned) linewidth is only 10^{-6} eV. If this energy resolution is not used, then the advantage of a high photon flux is largely wasted (data from [XI. 1,2]).

As described in Sect. 11.12, one can also use electron energy loss spectroscopy to obtain information about absorption behavior. Such an experiment is shown schematically in Fig. XI.2. The energy resolution is limited here by the thermal width of the cathode emission ($\sim$0.5 eV), but systems with monochromators and a total energy resolution of 2 meV have also been built [XI.3]. In order for the radiation to penetrate, the sample must be in the form of a thin foil, and the electron energy must be sufficiently high (20–100 keV). Figure XI.3 compares the frequency dependence of $\varepsilon(\omega)$ derived from a Kramers-Kronig analysis of the energy loss function (Sect. 11.1) with the results of optical reflectivity measurements. Electron

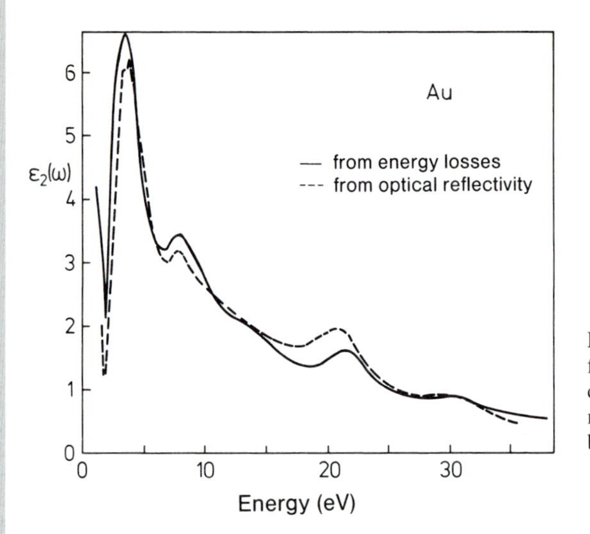

Fig. XI.3. Comparison of $\varepsilon_2(\omega)$ for gold from electron spectroscopy (——) and from measurements of the optical reflectivity by Daniels et al. [(XI. 5])

energy loss spetroscopy in transmission is frequently used in electron microscopes for a local material analysis. In reflection geometry with electron energies between 1–100 eV the technique is employed to probe vibrational and electronic excitations at surfaces [XI.6].

References

XI.1 C. Kunz: In *Photoemission in Solids II*, ed. by L. Ley, M. Cardona. Topics Appl. Phys., Vol. 27 (Springer, Berlin, Heidelberg 1979) p. 299 ff.

XI.2 F.P. Schäfer (Ed.): *Dye Lasers*, 3rd edn. Topics Appl. Phys., Vol. 1 (Springer, Berlin, Heidelberg 1990)

XI.3 H. Boersch, J. Geiger, W. Stickel: Z. Phys. 212, 130 (1968)

XI.4 M. Raether: *Springer Tracts Mod. Phys.* **38**, 84 (Springer, Berlin, Heidelberg 1965)

XI.5 J. Daniels, C. v. Festenberg, M. Raether, K. Zeppenfeld: *Springer Tracts Mod. Phys.* **54**, 77 (Springer, Berlin, Heidelberg 1970)

XI.6 H. Ibach, D.L. Mills: *Electron Energy Loss Spectroscopy and Surface Vibrations* (Academic Press, New York 1982)

Panel XII
Infrared Spectroscopy

For spectroscopy in the infrared (IR) spectral region, a special experimental technique has been developed. Because glass and quartz are not transparent at long wavelengths ($\lambda > 4\ \mu m$), mirrors are preferred as optical elements. Up to the mid-IR ($10\ \mu m < \lambda < 100\ \mu m$), alkali halide prisms (NaF, KCl, KI etc.) can be used as dispersive elements in monochromators. Otherwise metal-coated reflection gratings are used. Radiation sources up to the mid-IR include rods of silicon carbide ("Globar") and zirconium or yttrium oxide ("Nernst rods"); they are heated electrically to yellow heat. In the far IR ($\lambda > 200\ \mu m$), where quartz becomes transparent again, high pressure mercury lamps are useful. Depending on the details of the experimental requirements, synchrotron radiation can also be advantageous [XII.1].

Detectors include thermocouples, low-temperature semiconductor photoresistors, and pneumatic detectors ("Golay cells": based on the thermal expansion of a gas volume). The intensity maximum in the thermal background radiation at room temperature lies at about $\lambda = 10\ \mu m$, and this background would interfere with measurements. To suppress this stray light in IR spectrometers (Fig. XII.1), the radiation is periodically interrupted ("chopped") to produce a modulated electrical signal at the detector.

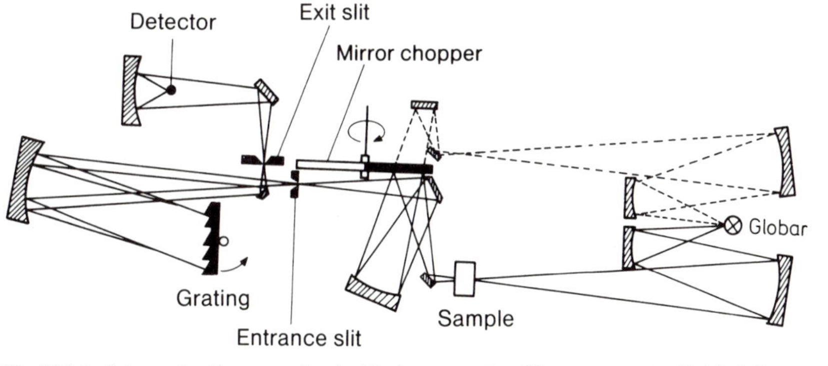

Fig. XII.1. Schematic diagram of a double-beam grating IR spectrometer. Behind the rotating sector mirror (chopper), the reference beam (– – –) traverses an identical path to the measurement beam (——), the latter passing through the sample whose absorption is to be measured. For reflection measurements, a mirror must be used to reflect the measurement beam from the sample. To avoid absorption by molecular vibrations of atmospheric water, the whole spectrometer must be flushed with dry air or even evacuated

The signal is measured with a narrow-band and/or phase-sensitive amplifier ("lock-in" amplifier) tuned to the modulation frequency. Two-beam spectrometers are often used, in which a reference beam is compared with the beam transmitted through the sample. In Fig. XII.1, the two rays are transmitted to the detector by a sector mirror which alternately transmits the reference beam and the measurement beam. The latter undergoes an additional reflection which phase-shifts it with respect to the reference beam. The electrical signals corresponding to the intensities can then be electronically separated and compared.

In conventional spectroscopy, one measures the intensity distribution $I(\omega)$. Since IR sources have relatively low intensity, the alternative approach of Fourier Transform Spectroscopy has found wide application. Its advantage is that, during the measurement time, the whole spectrum is recorded, and not just a region $\Delta\omega$ (limited by the resolution). The Michelson system shown in Fig. XII.2 is frequently used to decompose the light beam into its Fourier components. In this arrangement one mirror of the Michelson interferometer can be moved whilst being kept precisely parallel. The transmission $T(\omega)$ through the sample is measured for a wavelength-independent intensity I_0 (white light source), via

$$dI = T(\omega) I_0 d\omega \ . \tag{XII.1}$$

As a function of the mirror displacement x, the electric fields of the two interfering monochromatic beams are

$$\mathscr{E}_1 = \sqrt{I_0 T(\omega)} \sin \omega t \ ; \quad \mathscr{E}_2 = \sqrt{I_0 T(\omega)} \sin(\omega t + 2\omega x/c)$$

The time-averaged intensity measured by the detector thus has an x dependence

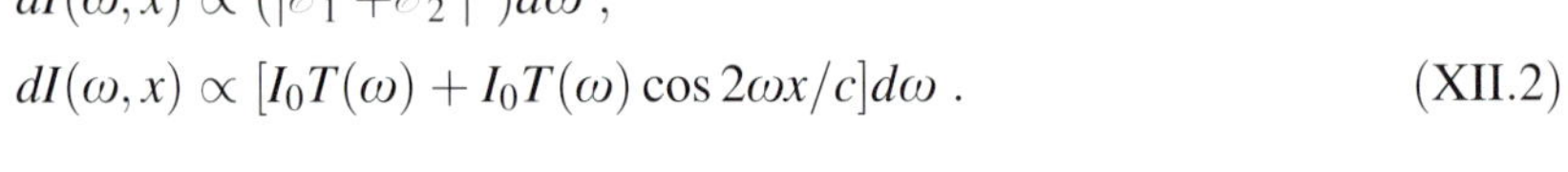

$$dI(\omega, x) \propto (|\mathscr{E}_1 + \mathscr{E}_2|^2) d\omega \ ,$$

$$dI(\omega, x) \propto [I_0 T(\omega) + I_0 T(\omega) \cos 2\omega x/c] d\omega \ . \tag{XII.2}$$

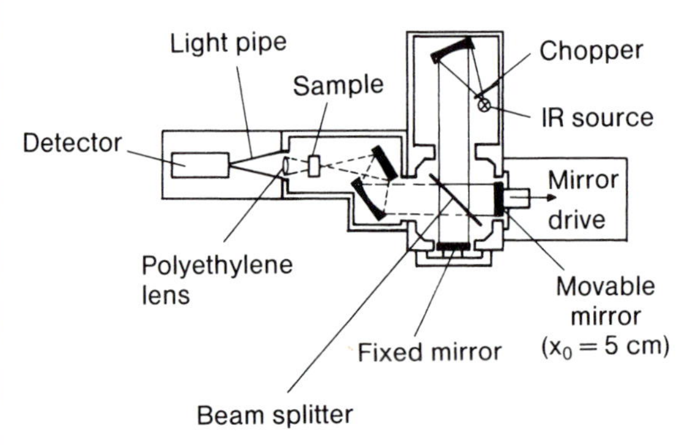

Fig. XII.2. Schematic illustration of a Michelson Fourier spectrometer

Because all light frequencies are processed simultaneously, when illuminated with a white light spectrum I_0, the detector measures the following intensity as a function of mirror position x:

$$I(x) = \int dI = I_0 \int_0^\infty T(\omega) d\omega + I_0 \int_0^\infty T(\omega) \cos \frac{2\omega x}{c} d\omega \,. \qquad \text{(XII.3)}$$

The first term gives a constant $I(\infty)$, and the second represents the Fourier transform of the transmission $T(\omega)$. The procedure therefore consists of measuring the intensity $I(x)$ as a function of the position x of a mirror (up to a maximum of x_0). The Fourier transform of $(I(x)-I(\infty))$ is performed by computer and gives the transmission spectrum $T(\omega)$ directly. The Fourier transform is only exact for infinitely large movements x. The accuracy, i.e., the resolution $\Delta\omega$, of this spectroscopy is determined by the maximum mirror movement x_0 according to $\Delta\omega/c \cong x_0^{-1}$.

Reference

XII.1 C.J. Hirschmugl, G.P. Williams, F.M. Hoffmann, Y.J. Chabal: Phys. Rev. Lett. **65**, 480 (1990)

Panel XIII
The Frustrated Total Reflection Method

In the absorption of electromagnetic waves by a (crystalline) solid, the wave number is conserved, i.e., a light wave of frequency ω can only be absorbed by solid state excitations which have a wave vector $q = \omega/c$. The surface waves discussed in Sect. 11.5, however, exist only for $q > \omega/c$. They are an example of excitations that cannot be detected by a normal absorption experiment. But it is possible to detect them using the technique of "frustrated total reflection". Figure XIII.1 shows the experimental arrangement employed by Marschall and Fischer [XIII.1]. Light is totally reflected from the inner surface of a prism (in this case, made of Si). The wave vector parallel to the surface is now $q_{\parallel} = (\omega/c)\, n \sin\alpha$ (with $n = 3.42$). For a very small air gap d, the exponentially decaying field outside the prism can be used to excite surface waves in a GaP crystal. This excitation appears as a dip in the intensity reflected in the prism (Fig. XIII.2). According to the

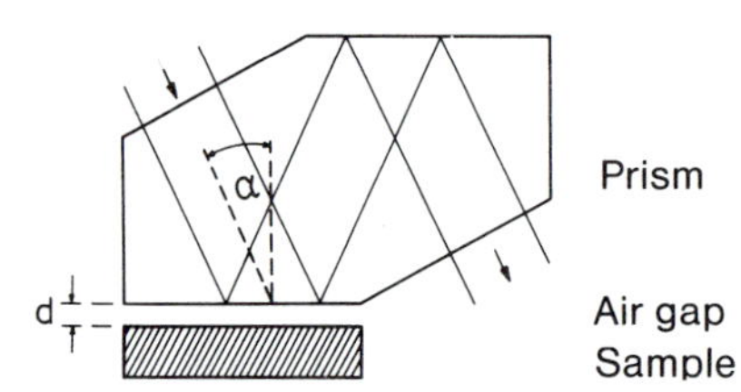

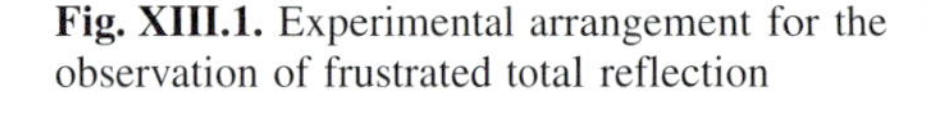

Fig. XIII.1. Experimental arrangement for the observation of frustrated total reflection

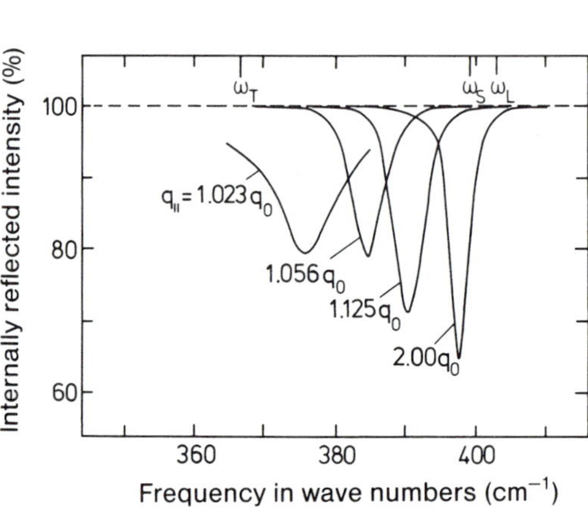

Fig. XIII.2. Reflected intensity as a function of frequency for a GaP sample, using the arrangement shown in Fig. XIII.1. The variable parameter is the angle of incidence α. It determines – together with the frequency – the parallel component of the wave vector $q_{\parallel}$. The minima are due to coupling with surface polaritons. For optimal observation, the strength of the coupling must be matched to the value of $q_{\parallel}$ by varying the air gap d

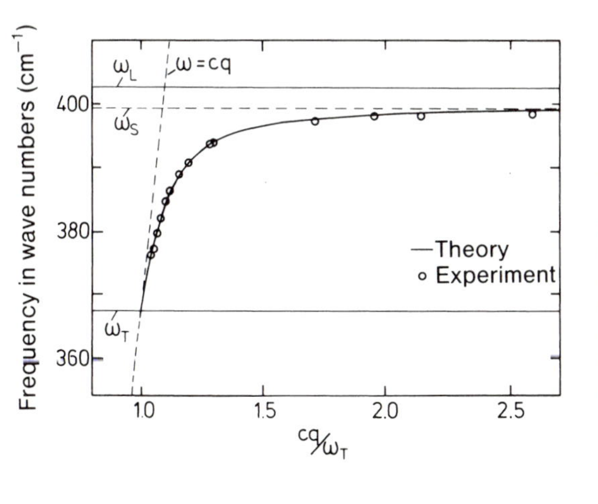

Fig. XIII.3. Dispersion of surface polaritons in GaP. ω_L and ω_T are the frequencies of the longitudinal and transverse optical phonons respectively, ω_S is the frequency of the surface phonon

chosen angle α, that is according to $q_{\|}$, the minima appear at different frequencies. In this way, the complete dispersion curve of the surface polariton can be determined (Fig. XIII.3). The theoretical form of the dispersion curve can be found by a full consideration of Maxwell's equations (retarded solutions), as discussed in Sect. 11.5.

Reference

XIII.1 N. Marschall, B. Fischer: Phys. Rev. Lett. **28**, 811 (1972)

12 Semiconductors

In Sect. 9.2 we learned that only a partly filled electronic band can contribute to electric current. Completely filled bands and completely empty bands do not contribute to electrical conductivity and a material which has only completely full and completely empty bands is therefore an insulator. If the distance between the upper edge of the highest filled band (valence band) and the lower edge of the lowest empty band (conduction band) is not too large (e.g. ~ 1 eV), then the finite width of the region over which the Fermi distribution changes rapidly has observable consequences at moderate and high temperatures: A small fraction of the states in the vicinity of the upper edge of the valence band is unoccupied and the corresponding electrons are found in the conduction band. Both these "thermally excited" electrons and the holes that they leave in the valence band can carry electric current. In this case one speaks of a semiconductor. In Fig. 12.1 the differences between a metal, a semiconductor and an insulator are summarized schematically.

A special property of semiconductor materials that is not found in metals is that their electrical conductivity can be altered by many orders of magnitude by adding small quantities of other substances. These additives also determine whether the conductivity is of electron or hole character.

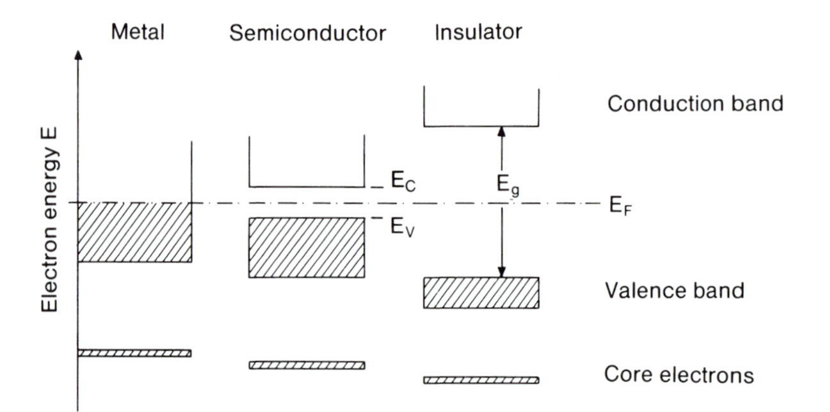

Fig. 12.1. Energy level diagrams for a metal, a semiconductor and an insulator. Metals have a partly occupied band (*shaded*) even at $T=0$ K. For semiconductors and insulators the Fermi level lies between the occupied valence band and the unoccupied conduction band

The entire field of solid state electronics relies on this particular property. Because of their importance, we shall dedicate a whole chapter to the subject of semiconductors.

12.1 Data for a Number of Important Semiconductors

In Sect. 7.3 we discussed the origin of the band structure of the typical elemental semiconductors diamond (C), Si, and Ge. Due to a mixing of the s- and p-wavefunctions, tetrahedral bonding orbitals (sp^3) are formed, which, for a bonding distance near equilibrium, lead to a splitting into bonding and antibonding orbitals. The bonding orbitals constitute the valence band and the antibonding orbitals the conduction band (Fig. 7.9). Assigning all four s- and p-electrons to the lowest available states leads to a completely filled valence band and a completely empty conduction band. The result is an insulator such as diamond, or, for smaller energy gaps between the valence and conduction band, a semiconductor like silicon or germanium.

From Fig. 7.9 one can deduce an important physical property of the energy gap between the valence and conduction bands: the size of the gap must be temperature dependent. With increasing temperature the lattice parameter increases due to thermal expansion. The splitting between the bonding and antibonding states therefore decreases and the band gap becomes smaller (Table 12.1). A more exact treatment of this effect must also consider the influence of lattice vibrations on E_g. The overall behavior of the band gap with temperature is a linear dependence at room temperature and a quadratic dependence at very low temperature.

Although Fig. 7.9 represents a qualitatively similar picture for the important semiconductors Si and Ge, in the $E(\boldsymbol{k})$ representation, i.e. in reciprocal space, a quite different picture of the electronic bands emerges due to the different atomic properties of the bands ($3s$, $3p$ or $4s$, $4p$ wavefunctions). These differences can be seen in Fig. 12.2. The curves come from calculations that have been fitted to experimental quantities such as band gap, position of the critical points and effective masses (band curvature).

From this $E(\boldsymbol{k})$ representation along the high symmetry directions in $\boldsymbol{k}$-space, it follows that both semiconductors are so-called indirect-gap semi-

Table 12.1. Energy gaps (width of forbidden bands) between the valence and conduction bands of germanium and silicon

	E_g (T=0 K) [eV]	E_g (T=300 K) [eV]
Si	1.17	1.12
Ge	0.75	0.67

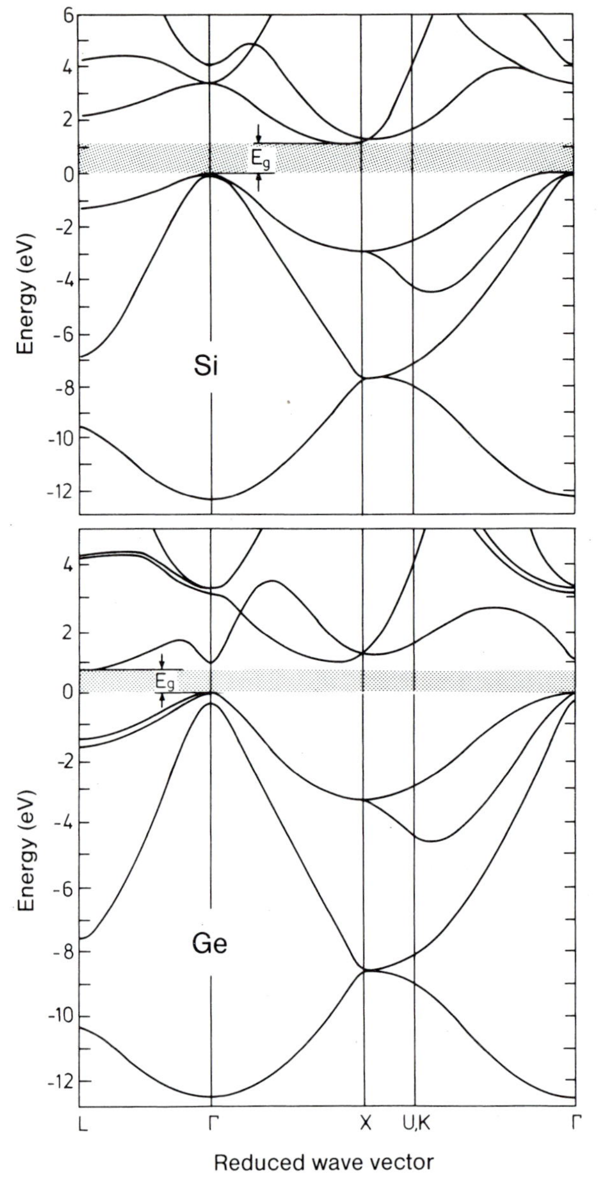

Fig. 12.2. Calculated bandstructures of silicon and germanium. For germanium the spin-orbit splitting is also taken into account. (After [12.1]). Both semiconductors are so-called indirect semiconductors, i.e. the maximum of the valence band and the minimum of the conduction band are at different positions in the Brillouin zone. The minimum of the conduction band of silicon lies along the ΓX=[100] direction and that of germanium along the ΓL=[111] direction. Note that the form of the Ge bands is very similar to that of Fig. 7.13, although the calculations were performed differently

conductors: the minimum distance between the conduction and valence bands (band gap E_g) is between states with different $\boldsymbol{k}$-vectors (Γ, or $\boldsymbol{k}=[000]$ at the valence band maximum of both materials, and $\boldsymbol{k}$ along [111] for Ge, or $\boldsymbol{k}$ along [100] for Si at the conduction band minimum). The conduction band electrons of lowest energy thus have $\boldsymbol{k}$-vectors along the [100] direction in Si, and along the [111] direction in Ge. These regions of $\boldsymbol{k}$-space containing the conduction electrons of Si and Ge are shown in Fig. XV.2.

If $E(\boldsymbol{k})$ is expressed in the parabolic approximation, i.e. retaining terms up to the order k^2, then the surfaces of constant energy are ellipsoids around the [111] or [100] direction. In the principal axis representation (the principal axes are [100] for Si and [111] for Ge), the energy surfaces of the conduction electrons are thus

$$E(\boldsymbol{k}) = \hbar^2 \left(\frac{k_x^2 + k_y^2}{2m_t^*} + \frac{k_z^2}{2m_l^*} \right) = \text{const} . \tag{12.1}$$

Here m_t^* and m_l^* are the "transverse" and "longitudinal" effective masses respectively. The zero of the energy scale is taken at the conduction band minimum.

Measurements by cyclotron resonance of the effective mass of the electrons at these points relative to the mass m of free electrons give the values in Table 12.2.

A detailed study (see Panel XV, "Cyclotron Resonance") of the properties of holes in Si and Ge shows that the structure of the valence band maximum in the vicinity of $\Gamma(\boldsymbol{k}=0)$ is more complicated than Fig. 12.2 suggests: besides the two valence bands with different curvatures which can be seen in Fig. 12.2, there exists another valence band at Γ which is split off slightly from the other two by $\Delta=0.29$ eV for Ge and $\Delta=0.044$ eV for Si. This splitting of the bands stems from spin-orbit interaction, which was not considered in the calculations for Si shown in Fig. 12.2. The qualitative behavior of the bands near Γ is shown in Fig. 12.3 for Si and Ge. In the parabolic approximation one can identify three different effective masses for the holes at Γ which contribute to charge transport. One speaks of heavy and light holes with masses m_{hh}^* or m_{lh}^*, corresponding to the different band curvatures. The holes of the split-off band are called split-off holes and their mass is denoted by m_{soh}^*.

Table 12.2. The transverse (m_t^*) and longitudinal (m_l^*) effective masses relative to the mass m of the free electron for silicon and germanium

	m_t^*/m	m_l^*/m
Si	0.19	0.92
Ge	0.082	1.57

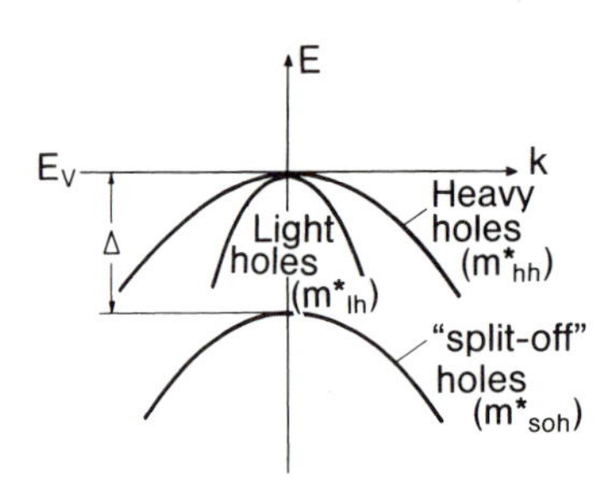

Fig. 12.3. Qualitative bandstructure for Si or Ge in the vicinity of the top of the valence band, including the effect of spin-orbit interaction. Δ is the spin-orbit splitting

Since the formation of sp^3 hybrids is obviously important in the chemical bonding of Si and Ge, one might well expect semiconducting properties in other materials with tetrahedral crystal structure, i.e. with sp^3 hybridization. Based on this consideration, one can correctly identify another important class, the *III–V semiconductors*, which are compound semiconductors comprising elements from the third and fifth groups of the periodic table. Typical examples are InSb, InAs, InP, GaP, GaAs, GaSb, and AlSb. In these compound crystals the bonding is mixed ionic and covalent (cf. Chap. 1). The mixed bonding can be imagined as a superposition of two extreme cases, the *ionic*, in which the electron transfer from Ga to As gives an ionic structure Ga^+As^-, and the *covalent*, in which electron displacement from As to Ga leaves both Ga and As with four electrons in the outer shell, and thereby allows the formation of sp^3 hybrids, just as in Si and Ge. This latter covalent structure is evidently dominant since the crystal would not otherwise be tetrahedrally bonded with the ZnS structure.

In contrast to the elemental semiconductors, the most important representatives of the III–V semiconductors possess a so-called *direct* band gap, i.e., the valence band maximum and conduction band minimum both lie at Γ (Fig. 12.4). There are again three distinct valence bands with a qualitatively similar form at Γ to those of the tetrahedral elemental semiconductors (Fig. 12.3). Important data for a few III–V semiconductors with direct band gaps are summarized in Table 12.3.

For the sake of completeness, we should mention that GaP and AlSb have indirect band gaps similar to Si and Ge (2.32 eV and 1.65 eV, respectively, at $T=0$ K).

Similar arguments to those presented for the case of III–V compounds lead to an understanding of the so-called II–VI semiconductors, such as ZnO (3.2 eV), ZnS (3.6 eV), CdS (2.42 eV), CdSe (1.74 eV) and CdTe (1.45 eV), where the value in brackets is the direct band gap E_g at 300 K. In these compounds there is also a mixed ionic-covalent bonding, but now with a larger ionic component than for the III–V semiconductors. The crystal structure is either that of the III–V semiconductors (ZnS) or that of wurtzite (Sect. 2.5). In both cases the local structure is tetrahedral, which can again be attributed to the sp^3 hybridization of the bonding partners.

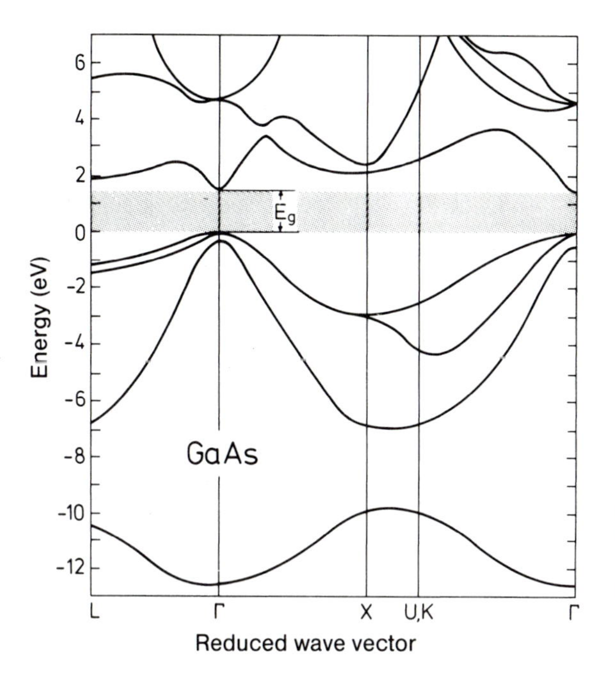

Fig. 12.4. Typical bandstructure of a III–V semiconductor, in this case GaAs. (After [12.1])

Table 12.3. Band gap E_g, effective mass m^* and spin-orbit splitting Δ for a few III–V semiconductors: m is the mass of the free electron, m_n^* the effective mass of the electrons, m_{lh}^* the effective mass of the light holes, m_{hh}^* that of the heavy holes, and m_{soh}^* that of the split-off holes

	$E_g(0\,\mathrm{K})$ [eV]	$E_g(300\,\mathrm{K})$ [eV]	m_n^*/m	m_{lh}^*/m	m_{hh}^*/m	m_{soh}^*/m	Δ [eV]
GaAs	1.52	1.43	0.07	0.08	0.5	0.15	0.34
GaSb	0.81	0.7	0.047	0.05	0.3	0.14	0.8
InSb	0.24	0.18	0.015	0.02	0.4	0.11	0.8
InAs	0.43	0.35	0.026	0.025	0.4	0.14	0.4
InP	1.42	1.35	0.073	0.12	0.6	0.12	0.11

12.2 Charge Carrier Density in Intrinsic Semiconductors

According to the definition of mobility in Sect. 9.5, the electrical conductivity σ of a semiconductor in which electrons and holes contribute to the flow of current, can be written as

$$\sigma = |e|(n\mu_n + p\mu_p)\ . \tag{12.2}$$

Here μ_n and μ_p are the mobilities of the electrons and holes, respectively, and n and p are the corresponding volume concentrations of the charge carriers. The form of (12.2) implies the neglect of any energy dependence

($\boldsymbol{k}$-dependence) of the quantities μ_n and μ_p in a first approximation. This is because it is generally sufficient to consider only charge carriers in the parabolic part of the bands, where the effective mass approximation is valid (i.e. m_n^* and m_p^* are constant). Because of the opposite signs of both the drift velocity and the electrical charge e for holes and electrons, both types of charge carrier contribute with the *same* sign to σ.

In contrast to metallic conductivity, the conductivity of semiconductors is strongly temperature dependent. This is because the band gap E_g, across which "free" charge carriers must be thermally excited, causes (via the Fermi distribution – see below) a strong dependence of the charge carrier concentrations n and p on the temperature.

Semiconductors are called *intrinsic* when "free" electrons and holes can be created only by electronic excitations from the valence band to the conduction band. (In Sect. 12.3 we will also consider excitations from defects and impurities.) As in any solid, the occupation of the energy levels in semiconductors must obey Fermi statistics $f(E,T)$ (Sect. 6.3), i.e.

$$n = \int_{E_c}^{\infty} D_C(E) f(E,T)\, dE\,, \tag{12.3 a}$$

and for holes

$$p = \int_{-\infty}^{E_V} D_V(E)[1 - f(E,T)]\, dE\,. \tag{12.3 b}$$

The ranges of integration should really extend only to the upper and lower band edges, respectively. However, because the Fermi function $f(E,T)$ decreases sufficiently rapidly, the ranges can be extended to infinity. The functions $D_C(E)$ and $D_V(E)$ are the densities of states in the conduction and valence bands, respectively. In the parabolic approximation (m^*=const) one has [cf. (6.11)]:

$$D_C(E) = \frac{(2m_n^*)^{3/2}}{2\pi^2\hbar^3}\sqrt{E - E_C}\,, \quad (E > E_C)\,; \tag{12.4 a}$$

$$D_V(E) = \frac{(2m_p^*)^{3/2}}{2\pi^2\hbar^3}\sqrt{E_V - E}\,, \quad (E < E_V)\,. \tag{12.4 b}$$

The density in the region of the forbidden band $E_V < E < E_C$ is of course zero. In an intrinsic semiconductor all "free" electrons in the conduction band originate from states in the valence band, and thus the concentration of holes p must be equal to the concentration of "free" electrons n. The situation is sketched in Fig. 12.5. If the effective masses m_n^* and m_p^* and therefore also the densities of states D_C and D_V are equal, the Fermi level E_F must lie in the middle of the forbidden band. If D_C and D_V are different

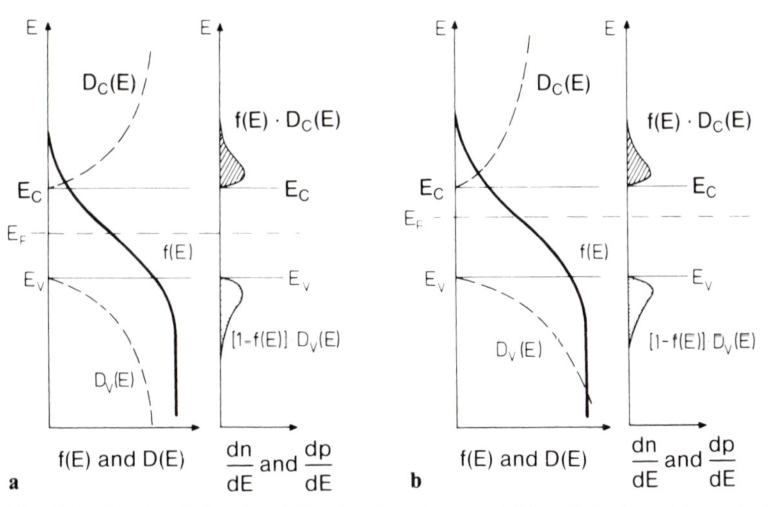

Fig. 12.5. (**a**) Fermi function $f(E)$, density of states $D(E)$ and electron (n) and hole (p) concentrations in the conduction and valence bands for the case of equal densities of states in the conduction and valence bands (schematic); (**b**) the same figure for the case of differing densities of states in the conduction and valence bands. The number of holes must again be equal to the number of electrons, and thus the Fermi level no longer lies in the middle of the gap between conduction and valence bands; its position then becomes temperature dependent

from one another, then E_F shifts slightly towards one of the band edges, such that the occupation integrals (12.3) remain equal.

Because the "width" of the Fermi function ($\sim 2kT$) is, at normal temperatures, small compared to the gap width (~ 1 eV), the Fermi function $f(E,T)$ can be approximated by Boltzmann occupation statistics within the bands ($E > E_C$ and $E < E_V$), i.e. for the conduction band:

$$\frac{1}{\exp[(E-E_F)/kT]+1} \sim \exp\left(-\frac{E-E_F}{kT}\right) \ll 1 \quad \text{for} \quad E-E_F \gg 2kT\,. \tag{12.5}$$

From (12.3a) and (12.4a), the electron concentration n in the conduction band follows as

$$n = \frac{(2m_n^*)^{3/2}}{2\pi^2\hbar^3} e^{E_F/kT} \int_{E_C}^{\infty} \sqrt{E-E_C} \cdot e^{-E/kT} dE\,. \tag{12.6}$$

Substituting $X_C = (E-E_C)/kT$, we obtain the formula

$$n = \frac{(2m_n^*)^{3/2}}{2\pi^2\hbar^3} (kT)^{3/2} \exp\left(-\frac{E_C-E_F}{kT}\right) \int_0^{\infty} X_C^{1/2} e^{-X_C} dX_C\,. \tag{12.7}$$

With an analogous calculation for the valence band, one finally obtains the expressions

$$n = 2\left(\frac{2\pi m_n^* k T}{h^2}\right)^{3/2} \exp\left(-\frac{E_C - E_F}{k T}\right) = N_{\text{eff}}^C \exp\left(-\frac{E_C - E_F}{k T}\right) , \tag{12.8 a}$$

$$p = 2\left(\frac{2\pi m_p^* k T}{h^2}\right)^{3/2} \exp\left(\frac{E_V - E_F}{k T}\right) = N_{\text{eff}}^V \exp\left(\frac{E_V - E_F}{k T}\right) . \tag{12.8 b}$$

The prefactors N_{eff}^C and N_{eff}^V are the well-known partition functions for translational motion in three dimensions. One sees that small concentrations of free charge carriers in semiconductors can be approximately described by Boltzmann statistics (the approximation to Fermi statistics for $E - E_F \gg 2 k T$). If one compares Fig. 12.5 with the potential well model of Chap. 6, the conduction band can be formally regarded as a potential well, in which the Fermi level E_F lies well below ($k T$) the bottom of the potential well.

The use of the so-called "effective densities of states" N_{eff}^C and N_{eff}^V in (12.8) allows a further formal interpretation in which the whole conduction (valence) band can be characterized by a single energy level E_C (E_V) (i.e., the band edge) with the density of states N_{eff}^C (N_{eff}^V) (temperature dependent!). The occupation densities n and p of these bands are determined by the Boltzmann factor (with the energy in each case measured from E_F). This approximation, which is often valid for semiconductors, is called the *approximation* of non-degeneracy. High densities of charge carriers can be created in semiconductors by means of high densities of impurities (Sect. 12.3). The approximation of non-degeneracy is then no longer valid and one speaks of *degenerate semiconductors.*

From (12.8 a, b) the following generally valid relationship can be derived ($E_g = E_C - E_V$):

$$n p = N_{\text{eff}}^C N_{\text{eff}}^V \mathrm{e}^{-E_g/k T} = 4\left(\frac{k T}{2\pi \hbar^2}\right)^3 (m_n^* m_p^*)^{3/2} \mathrm{e}^{-E_g/k T} . \tag{12.9}$$

This equation implies that for a particular semiconductor, which is completely characterized by its absolute band gap E_g and the effective masses m_n^* and m_p^* in the conduction and valence bands, the electron and hole concentrations behave as a function of temperature according to the *law of mass action.*

If we further assume an intrinsic semiconductor ($n = p$), then the intrinsic charge carrier concentration n_i varies with temperature as follows

$$n_i = p_i = \sqrt{N_{\text{eff}}^C N_{\text{eff}}^V} \mathrm{e}^{-E_g/2k T} = 2\left(\frac{k T}{2\pi \hbar^2}\right)^{3/2} (m_n^* m_p^*)^{3/4} \mathrm{e}^{-E_g/2k T} . \tag{12.10}$$

Values of n_i and E_g for the important materials Ge, Si and GaAs are summarized in Table 12.4.

Table 12.4. Band gap E_g and the intrinsic carrier concentration n_i for germanium, silicon and gallium arsenide at 300 K

	E_g [eV]	n_i [cm^{-3}]
Ge	0.67	2.4×10^{13}
Si	1.1	1.5×10^{10}
GaAs	1.43	5×10^{7}

According to (12.8a,b), the Fermi level at a particular temperature adopts the position necessary to yield charge neutrality, i.e.

$$n = p = N_{\text{eff}}^{\text{C}} e^{-E_{\text{C}}/kT} e^{E_{\text{F}}/kT} = N_{\text{eff}}^{\text{V}} e^{E_{\text{V}}/kT} e^{-E_{\text{F}}/kT} , \tag{12.11}$$

$$e^{2E_{\text{F}}/kT} = \frac{N_{\text{eff}}^{\text{V}}}{N_{\text{eff}}^{\text{C}}} e^{(E_{\text{V}}+E_{\text{C}})/kT} , \tag{12.12}$$

$$E_{\text{F}} = \frac{E_{\text{C}}+E_{\text{V}}}{2} + \frac{kT}{2}\ln(N_{\text{eff}}^{\text{V}}/N_{\text{eff}}^{\text{C}}) = \frac{E_{\text{C}}+E_{\text{V}}}{2} + \frac{3}{4}kT\ln(m_p^*/m_n^*) . \tag{12.13}$$

If the effective densities of states and effective masses (i.e. the band curvature of the conduction and valence bands) are equal, then the Fermi level of an intrinsic semiconductor lies exactly in the middle of the forbidden band, and this is true for all temperatures. If, however, the effective densities of states in the conduction and valence bands are different, then the Fermi function lies asymmetrically with respect to the band edges E_{C} and E_{V} (Fig. 12.5) and the Fermi level shows a weak temperature dependence according to (12.13).

12.3 Doping of Semiconductors

The intrinsic carrier concentration n_i of 1.5×10^{10} cm^{-3} (at 300 K) of Si is not nearly large enough to yield the current densities necessary for practical semiconductor devices. Concentrations that are orders of magnitude higher than n_i can be created by doping, i.e. by the addition of electrically active impurities to the semiconductor. Most semiconductors cannot be grown as single crystals with sufficient purity that one can observe intrinsic conductivity at room temperature. Unintentional doping, even in the purest commercially available GaAs single crystals, leads to carrier densities of about 10^{16} cm^{-3} (at 300 K), compared with a corresponding intrinsic concentration of $n_i = 5\times10^7$ cm^{-3}.

Electrically active impurities in a semiconductor raise the concentration of either the “free” electrons or the “free” holes by donating electrons to the conduction band or by accepting them from the valence band. These

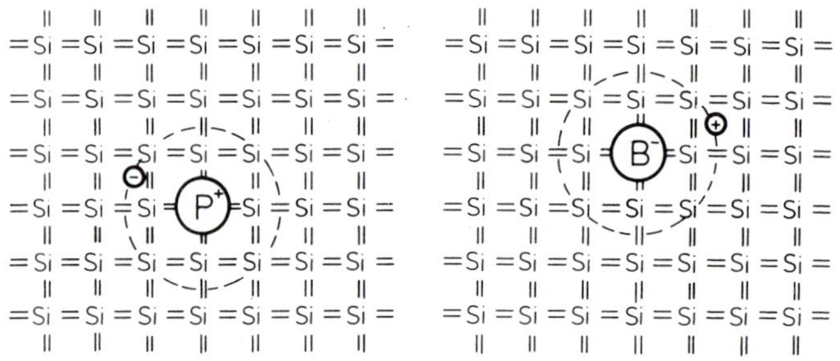

a n-doped silicon **b** p-doped silicon

Fig. 12.6 a, b. Schematic representation of the effect of a donor (**a**) and an acceptor (**b**) in a silicon lattice. The valence-five phosphorus atom is incorporated in the lattice at the site of a silicon atom. The fifth valence electron of the phosphorus atom is not required for bonding and is thus only weakly bound. The binding energy can be estimated by treating the system as a hydrogen atom embedded in a dielectric medium. The case of an acceptor (**b**) can be described similarly: the valence-three boron accepts an electron from the silicon lattice. The hole that is thereby created in the valence band orbits around the negatively charged impurity. The lattice constant and the radius of the defect center are not drawn to scale. In reality the first Bohr radius of the "impurity orbit" is about ten times as large as the lattice parameter

impurities are called *donors* and *acceptors*, respectively. A donor in a Si lattice is created, for example, when a valence-four Si atom is replaced by a valence-five atom such as P, As or Sb. The electronic structure of the outermost shell of these impurities is s^2p^3 instead of the $3s^23p^2$ Si configuration. To adopt the tetrahedral bonding structure of an sp^3 hybrid in the lattice, only the s^2p^2 electrons of the valence-five atom are necessary; the excess electron in the p shell has no place in the sp^3 hybrid bond. This electron can be imagined as weakly bound to the positively charged and tetrahedrally bonded donor core, which has replaced a Si atom in the lattice (Fig. 12.6 a).

A valence-five donor impurity in the Si lattice can be pictured, to a good approximation, as a positively charged monovalent core, to which one electron is bound. The electron can become dissociated and can then move "freely" through the lattice; i.e., on ionization, this electron will be excited from the impurity into the conduction band. The donor impurity can be described as a hydrogen-like center, in which the Coulomb attraction between the core and the valence electron is screened by the presence of Si electrons in the vicinity.

To estimate the excitation and ionization energy of the extra phosphorus electron (Fig. 12.6 a), one can approximate the screening effect of the surrounding Si by inserting the dielectric constant of Si ($\varepsilon_{\rm Si} = 11.7$) into the expression for the energy levels of the hydrogen atom. The energy levels for the Rydberg series of the hydrogen atom are

$$E_n^{\rm H} = \frac{m_{\rm e}\, e^4}{2(4\pi\varepsilon_0\hbar)^2}\,\frac{1}{n^2} \qquad (12.14)$$

and for the $n = 1$ level, the ionization energy is 13.6 eV. For the P donor, the mass m_e of the free electron must be replaced by the effective mass $m_n^* = 0.3\, m_e$ of a silicon conduction electron, and the dielectric constant ε_0 of vacuum by $\varepsilon_0 \varepsilon_{Si}$. A value for the ionization energy E_d of the donor of ~ 30 meV results. The energy level E_D of the donor electron in the bound state should thus lie about 30 meV below the conduction band edge E_C. A still smaller value is obtained for germanium. Here, $\varepsilon_{Ge} = 15.8$ and $m_n^* \sim 0.12\, m_e$. An estimate of $(E_C - E_D) \simeq 6$ meV is obtained. The situation is depicted in a bandstructure scheme in Fig. 12.7, where only the ground state of the donor is drawn. Between this ground state and the conduction band edge are a series of excited states [$n > 1$ in (12.14)], whose spacing decreases with increasing energy, and which finally join the continuum of the conduction band. The situation is very similar to that of the H atom, where the conduction band continuum corresponds to the unbound states above the vacuum level. The energy of the excited states can be determined, for example, from optical spectra. Figure 12.8 shows an absorption spectrum of the Sb donor in Ge. The bands below 9.6 meV correspond to excitations from the ground state to higher, excited states. The spectrum is actually more complicated than would be expected from the simple hydrogenic model; this is because the crystal field lifts the partial degeneracy of the hydrogen-like states. Above a photon energy of 9.6 meV, the electrons are excited to the continuum of the conduction band.

As can be seen from the experimental example in Fig. 12.8, the simple description of the donor by a hydrogenic model allows one to estimate the order of magnitude of the ionization energy E_d. Within this model all donor impurities, such as P, As and Sb, should give the same ionization energy E_d when present in the same semiconductor host. The experimental values of E_d in Table 12.5 show, however, that the values vary somewhat from donor to donor.

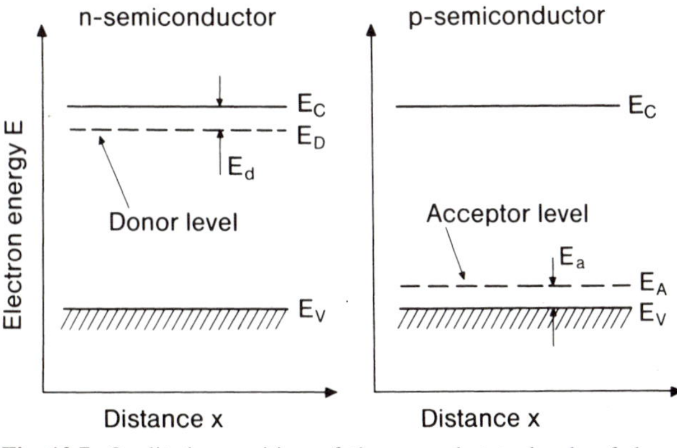

Fig. 12.7. Qualitative position of the ground state levels of donors and acceptors relative to the minimum of the conduction band E_C and the maximum of the valence band E_V. The quantities E_d and E_a are the ionization energies of the donor and acceptor, respectively

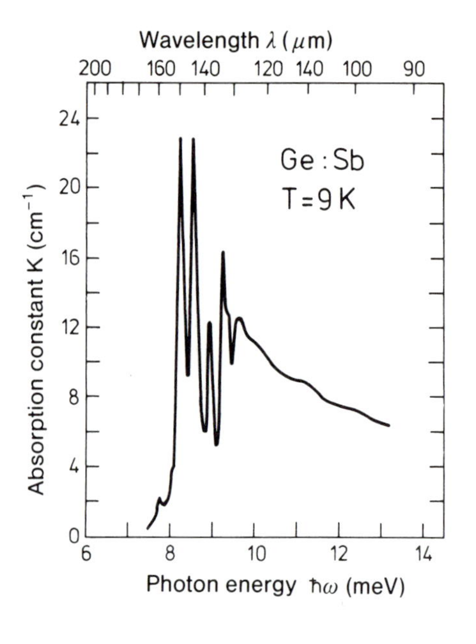

Fig. 12.8. Optical absorption spectrum of a Sb donor in germanium, measured at T=9 K. (After [12.2])

Table 12.5. Ionization energies E_d for a few donor species in silicon and germanium

	P [meV]	As [meV]	Sb [meV]
Si	45	54	43
Ge	13	14	10

It is not surprising to find that the crude description of screening in terms of the dielectric constant of the semiconductor is inadequate to describe the finer details resulting from atomic effects.

That the description by a macroscopic dielectric constant nonetheless yields good values for E_d, is because the screening leads to a wavefunction which is "smeared out" over many lattice constants. Inserting the dielectric constant of the semiconductor ε_s in the formula for the Bohr radius

$$r = \varepsilon_0 \varepsilon_s \frac{h^2}{\pi m_n^* e^2} \qquad (12.15)$$

expands this radius by a factor of ε_s (~ 12 for Si) compared to the (hydrogen) Bohr-radius.

The bound valence electron of the donor impurity is therefore "smeared" over about 10^3 lattice sites.

If a valence-three impurity atom (B, Al, Ga, In) is present in the lattice of a valence-four elemental semiconductor (Si, Ge), the sp^3 hybrid responsible for the tetrahedral bonding can easily accept an electron from the valence

Table 12.6. Ionization energies E_a for a few acceptor species in silicon and germanium

	B [meV]	Al [meV]	Ga [meV]	In [meV]
Si	45	67	74	153
Ge	11	11	11	12

band and leave behind a hole (Fig. 12.6b). Such impurities are called *acceptors*. An acceptor is neutral at very low temperature. It becomes ionized when an electron obtains sufficient energy to be lifted from the valence band to the so-called acceptor level. Acceptors therefore have the charge character "from neutral to negative", while donors are ionized from neutral to positive. The hole that exists in the vicinity of an ionized acceptor is in the screened Coulomb field of the fixed, negative impurity, and the energy needed to separate the hole from the ion and create a "free" hole in the valence band can be estimated using the hydrogenic model as in the case of donors. The relationships derived for donors and acceptors are fundamentally similar except for the sign of the charge. Table 12.6 shows that the ionization energies E_a for acceptors are in fact quite close to those of donors. Semiconductors doped with donors and acceptors are called *n*- and *p*-type materials, respectively.

The lowest impurity concentrations that can presently be achieved in semiconductor single crystals are of the order of $10^{12}\,\mathrm{cm}^{-3}$. Ge with an intrinsic charge carrier concentration n_i of $2.4\times10^{13}\,\mathrm{cm}^{-3}$ (at 300 K) is therefore obtainable as intrinsic material at room temperature, whereas Si ($n_i = 1.5\times10^{10}\,\mathrm{cm}^{-3}$ at 300 K) does not show intrinsic conductivity at room temperature. In addition to the electrically active impurities discussed here, a semiconductor may of course contain many impurities and defects that cannot be ionized as easily and thus do not affect the electrical conductivity.

12.4 Carrier Densities in Doped Semiconductors

In a doped semiconductor, an electron in the conduction band can originate either from the valence band or from the ionization of a donor; likewise, a hole in the valence band may correspond either to the electron in the conduction band or to a negatively charged (ionized) acceptor. For a non-degenerate semiconductor, the occupation of the conduction and valence bands must nonetheless be governed by the Boltzmann approximation (12.8a,b). Therefore the so-called "law of mass action" must also be valid for the doped semiconductor (12.9)

$$np = N_{\mathrm{eff}}^{\mathrm{C}} N_{\mathrm{eff}}^{\mathrm{V}} \mathrm{e}^{-E_g/kT} ,$$

in which the position of the Fermi level E_F no longer appears. In comparison with an intrinsic semiconductor, the position of E_F is now governed by a rather more complicated "neutrality condition", which also takes account of the charge on the impurities: the terms used in the following discussion are represented schematically in Fig. 12.9. The density of all available donors N_D and acceptors N_A is composed of the density of the neutral donors and acceptors, N_D^0 and N_A^0, and the density of ionized donors N_D^+ (which are then positively charged) and ionized acceptors N_A^- (negatively charged). In a homogeneous semiconductor, the negative charge density $n+N_A^-$ must be compensated by an equally large positive charge density $p+N_D^+$ (see Fig. 12.9). Hence, the following neutrality condition governs the position of the Fermi level E_F in a homogeneously doped semiconductor:

$$n + N_A^- = p + N_D^+ \ , \tag{12.16}$$

in which

$$N_D = N_D^0 + N_D^+ \ , \tag{12.17a}$$

$$N_A = N_A^0 + N_A^- \ . \tag{12.17b}$$

For typical impurity concentrations (10^{13}–10^{17} cm^{-3}), in which the individual donors and acceptors do not influence one another, the occupation of the donors by electrons (n_D) and of acceptors by holes (p_A) is given to a good approximation by

$$n_D = N_D^0 = N_D[1 + \exp(E_D - E_F)/k T]^{-1} \ , \tag{12.18a}$$

$$p_A = N_A^0 = N_A[1 + \exp(E_F - E_A)/k T]^{-1} \ . \tag{12.18b}$$

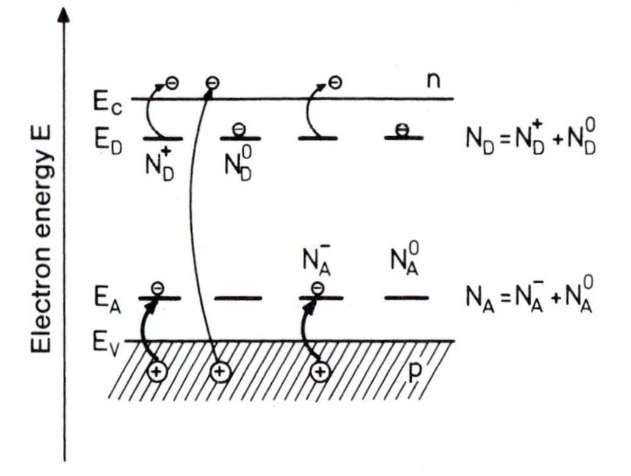

Fig. 12.9. Explanation of the notation commonly used for carrier and impurity concentrations in n- and p-type semiconductors: n and p are the concentrations of "free" electrons and holes. The total concentrations N_D and N_A of donors and acceptors consist of the density of neutral, N_D^0 or N_A^0, and ionized, N_D^+ or N_A^-, donors and acceptors, respectively. Electrons in the conduction band (density n) and holes in the valence band (density p) originate either from interband excitations or from impurities

We neglect here a modification of the Fermi function (multiplication of the exponential term by $\frac{1}{2}$), which takes into account the possibility of the capture of a single electron only, however, with two distinguishable spin states.

The general case, in which both donors and acceptors are considered simultaneously, can only be treated numerically. We therefore restrict the present treatment to the case of a pure *n-type semiconductor*, in which only donors are available. Equations (12.8 a, 12.17 a, 12.18 a) can then be used to calculate the concentration of charge carriers. For convenience these are repeated below:

$$n = N_{\mathrm{eff}}^{\mathrm{C}} \mathrm{e}^{-(E_{\mathrm{C}}-E_{\mathrm{F}})/kT} , \tag{12.8 a}$$

$$N_{\mathrm{D}} = N_{\mathrm{D}}^{0} + N_{\mathrm{D}}^{+} , \tag{12.17 a}$$

$$N_{\mathrm{D}}^{0} = N_{\mathrm{D}}[1 + \exp(E_{\mathrm{D}} - E_{\mathrm{F}})/kT]^{-1} . \tag{12.18 a}$$

"Free" electrons in the conduction band can only originate from donors or from the valence band, i.e.,

$$n = N_{\mathrm{D}}^{+} + p . \tag{12.19}$$

As a further simplification we assume that the main contribution to the conductivity stems from ionized donors, i.e. that $N_{\mathrm{D}}^{+} \gg n_{\mathrm{i}}$ ($np = n_{\mathrm{i}}^{2}$). For Si for example ($n_{\mathrm{i}} = 1.5 \times 10^{10}\,\mathrm{cm}^{-3}$ at 300 K) this is readily fulfilled even at low doping levels. For this simple case, (12.19) is replaced by

$$n \approx N_{\mathrm{D}}^{+} = N_{\mathrm{D}} - N_{\mathrm{D}}^{0} , \tag{12.20}$$

i.e. with (12.18 a) one has

$$n \approx N_{\mathrm{D}} \left(1 - \frac{1}{1 + \exp[(E_{\mathrm{D}} - E_{\mathrm{F}})/kT]} \right) . \tag{12.21}$$

With the help of (12.8 a), E_{F} can be expressed via

$$(n/N_{\mathrm{eff}}^{\mathrm{C}}) \mathrm{e}^{E_{\mathrm{C}}/kT} = \mathrm{e}^{E_{\mathrm{F}}/kT} , \tag{12.22}$$

to yield

$$n \approx \frac{N_{\mathrm{D}}}{1 + \mathrm{e}^{E_{\mathrm{d}}/kT} n/N_{\mathrm{eff}}^{\mathrm{C}}} , \tag{12.23}$$

where $E_{\mathrm{d}} = E_{\mathrm{C}} - E_{\mathrm{D}}$ is the energetic distance of the donor level from the conduction band edge. Expressing (12.23) as a quadratic equation for n we have

$$n + \frac{n^2}{N_{\mathrm{eff}}^{\mathrm{C}}} \mathrm{e}^{E_{\mathrm{d}}/kT} \approx N_{\mathrm{D}} . \tag{12.24}$$

The physically meaningful solution is

$$n \approx 2N_{\rm D}\left(1+\sqrt{1+4\frac{N_{\rm D}}{N_{\rm eff}^{\rm C}}\,{\rm e}^{E_{\rm d}/\not k T}}\right)^{-1} . \tag{12.25}$$

This expression for the conduction electron concentration in n-type semiconductors contains the following limiting cases:

I) If the temperature T is so low that

$$4(N_{\rm D}/N_{\rm eff}^{\rm L}){\rm e}^{E_{\rm d}/\not k T} \gg 1 \tag{12.26}$$

then one has

$$n \approx \sqrt{N_{\rm D}N_{\rm eff}^{\rm C}}\,{\rm e}^{-E_{\rm d}/2\not k T} . \tag{12.27}$$

In this region, where a sufficiently large number of donors still retain their valence electrons, i.e. are not ionized, one speaks of the *freeze-out range* of carriers. The similarity between (12.27) and (12.10) should be noted: instead of the valence band quantities $N_{\rm eff}^{\rm V}$ and $E_{\rm V}$, one now has the corresponding donor quantities $N_{\rm D}$ and $E_{\rm d}$ (i.e. $E_{\rm d}$ instead of $E_{\rm g}$). In this low-temperature regime, the electron concentration depends exponentially on the temperature T as in an intrinsic semiconductor; here, however, it is the much smaller donor ionization energy $E_{\rm d}$ that appears in place of $E_{\rm g}$. The special case treated here of a pure n-type doping is of course rarely realized in practice. Trace quantities of acceptors are almost always present. As a consequence, the Fermi level lies below $E_{\rm D}$, and an activation energy of $E_{\rm d}$ is therefore found in most cases.

II) At temperatures T, for which

$$4(N_{\rm D}/N_{\rm eff}^{\rm C}){\rm e}^{E_{\rm d}/\not k T} \ll 1 , \tag{12.28}$$

Eq. (12.25) becomes

$$n \approx N_{\rm D} = {\rm const} , \tag{12.29}$$

i.e. the concentration of donor electrons in the conduction band has reached the maximum possible value, equal to the concentration of donors; all donors are ionized, and one speaks of the *saturation range*. In a first approximation, one may still neglect electrons excited from the valence band.

III) At yet higher temperatures the concentration of electrons excited from the valence band across the gap $E_{\rm g}$ increases, and eventually outweighs the electron density due to donors. In this region the n-type material behaves as an intrinsic semiconductor and one speaks of the *intrinsic region* of the carrier concentration. The various temperature and carrier concentration regimes are portrayed in Fig. 12.10 together with the corresponding position of the Fermi level $E_{\rm F}$.

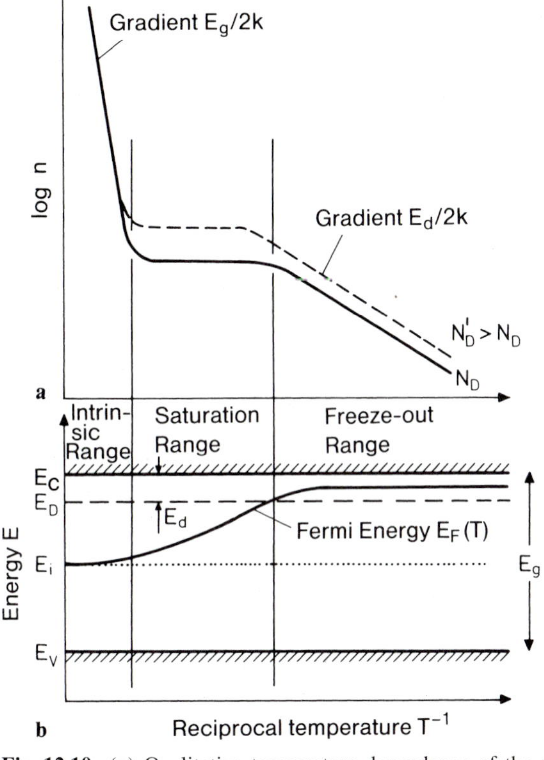

Fig. 12.10. (**a**) Qualitative temperature dependence of the concentration n of electrons in the conduction band of an n-type semiconductor for two different donor concentrations $N'_D > N_D$. The width of the forbidden band is E_g and E_d is the ionization energy of the donors; (**b**) qualitative temperature dependence of the Fermi energy $E_F(T)$ in the same semiconductor. E_C and E_V are the lower edge of the conduction band and the upper edge of the valence band, respectively, E_D is the position of the donor levels and E_i is the Fermi level of an intrinsic semiconductor

The position of the Fermi level E_F as a function of temperature can be discussed analogously to the case of an intrinsic semiconductor, but it will not be explicitly worked out here. In the freeze-out range, one may replace the valence band edge by the donor level. In the region of very high temperature, i.e. in the intrinsic region, the laws pertinent to an intrinsic semiconductor apply.

For n-doped Si with a phosphorus concentration of $3\times10^{14}\,\mathrm{cm}^{-3}$, the saturation range stretches between 45 and 500 K, i.e., at room temperature all donors are ionized. Figure 12.11 shows experimental results for the electron concentration $n(T)$, determined by Hall effect measurements, of n-doped Ge with impurity concentrations between 10^{13} and $10^{18}\,\mathrm{cm}^{-3}$. The relationships sketched qualitatively in Fig. 12.10 are clearly recognizable.

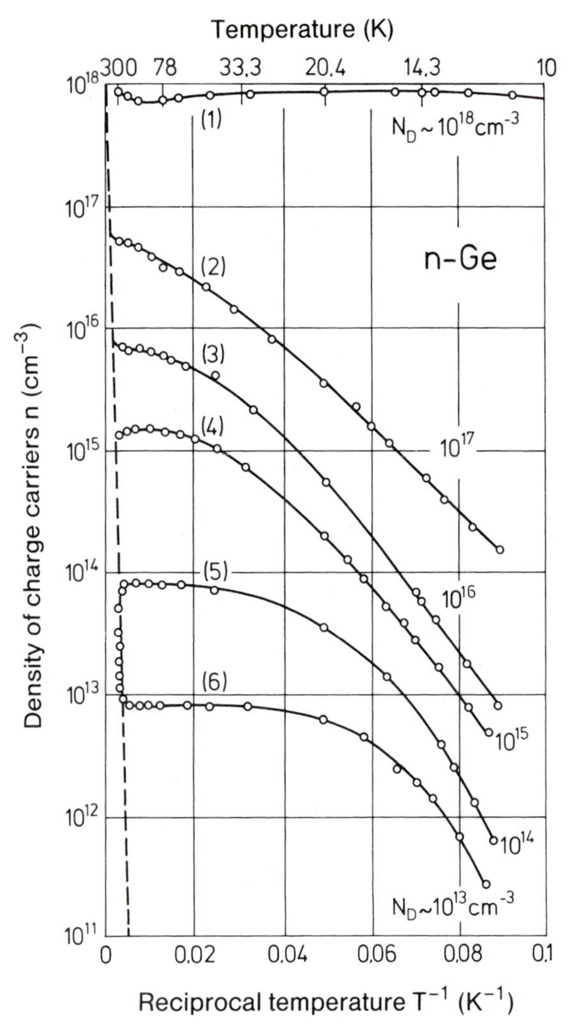

Fig. 12.11. The concentration n of free electrons in n-type germanium, measured using the Hall effect (Panel XIV). For the samples *(1)* to *(6)*, the donor concentration N_D varies between 10^{18} and 10^{13} cm^{-3}. The temperature dependence of the electron concentration in the intrinsic region is shown by the dashed line. (After [12.3])

12.5 Conductivity of Semiconductors

As was discussed more fully in Sect. 9.5, a calculation of the electrical conductivity σ as a function of temperature T requires a more detailed treatment of the mobility. For a semiconductor, one must consider both the electrons in the lower conduction band region (concentration: n, mobility: μ_n) and the holes at the upper valence band edge (p, μ_p). The current density $\boldsymbol{j}$ in an isotropic semiconductor (non-tensorial conductivity σ) is therefore

$$\boldsymbol{j} = e(n\,\mu_n + p\,\mu_p)\ . \tag{12.30}$$

In contrast to a metal, where only electrons at the Fermi edge need be considered, i.e. $\mu = \mu(E_F)$ (Sect. 9.5), the mobilities μ_n and μ_p in a semiconductor are average values for the electron and hole occupied states in the lower conduction band and upper valence band, respectively. In non-degenerate semiconductors, Fermi statistics can be approximated by Boltzmann statistics in this region. Within the framework of this approximation, the averaging, which will not be carried out in more detail, yields the following expression

$$\mu_n = \frac{1}{m_n^*}\frac{\langle \tau(\boldsymbol{k})v^2(\boldsymbol{k})\rangle}{\langle v^2(\boldsymbol{k})\rangle}e\ . \tag{12.31}$$

Here $v(\boldsymbol{k})$ is the velocity of an electron in an electric field $\mathcal{E}$ at the point $\boldsymbol{k}$ in reciprocal space, and $\tau(\boldsymbol{k})$ is its relaxation time (a more exact definition is given in Sect. 9.4). For holes, the quantities in (12.31) are taken at corresponding points in the valence band, and the effective mass m_n^* of the electrons is replaced by that of holes.

Instead of deriving a rigorous solution of the Boltzmann equation and an exact further treatment of (12.31), we restrict ourselves in the present case, as we did for metals (Sect. 9.5), to a more qualitative discussion of the scattering processes undergone by electrons and holes in semiconductors. In this respect, electrons and holes behave in a qualitatively similar way. After considerable simplification, (12.31) yields a proportionality between the mobility and the relaxation time ($\mu \propto \tau$). For metals, this proportionality holds exactly, see (9.58 b). Because τ is also proportional to the average time between collisions, it follows that

$$\frac{1}{\tau} \propto \langle v\rangle \Sigma\ , \tag{12.32}$$

where Σ represents the scattering cross section for electrons and holes at a scattering center. In contrast to the case of metals (Sect. 9.5), $\langle v\rangle$ is to be considered as a thermal average (according to Boltzmann statistics) over all electron or hole velocities in the lower conduction band or upper valence band, respectively. Equation (12.32) is essentially the scattering probability for electrons and holes. Because of the validity of Boltzmann statistics in semiconductors, we have

$$\langle v\rangle \propto \sqrt{T}\ . \tag{12.33}$$

If we now consider *scattering from acoustic phonons*, we can estimate the cross section Σ_{ph}, as in the case of metals (Sect. 9.5), from the square of the average vibrational amplitude $\langle s^2(\boldsymbol{q})\rangle$ of a phonon $(\boldsymbol{q}, \omega_{\boldsymbol{q}})$, i.e. for temperatures $T \gg \Theta$ (Θ is the Debye temperature), we have (Sect. 9.5)

$$M\,\omega_{\boldsymbol{q}}^2\langle s^2(\boldsymbol{q})\rangle = \not{k}T\ , \tag{12.34 a}$$

$$\Sigma_{\mathrm{ph}} \sim T \,. \tag{12.34b}$$

Using (12.32, 12.33), one then arrives at the following estimate

$$\mu_{\mathrm{ph}} \sim T^{-3/2} \,. \tag{12.35}$$

In addition to the usual scattering from phonons considered here, the scattering in piezoelectric semiconductors (e.g. III–V and II–VI compounds) may contain substantial contributions from phonons that are associated with a polarization (piezoelectric scattering). Charge carriers can also be scattered from optical phonons of higher energy. In this case the description becomes extremely complicated, since the relaxation time approximation (Sect. 9.4) loses its validity.

A further important source of scattering in semiconductors is *scattering from charged defects* (ionized donors or acceptors). A carrier moving past a point-like charged defect experiences a Coulomb interaction, and the scattering cross section Σ_{def} for this "Rutherford scattering" is

$$\Sigma_{\mathrm{def}} \propto \langle v \rangle^{-4} \,, \tag{12.36}$$

where the thermal average $\langle v \rangle \propto \sqrt{T}$ is assumed for the velocity. Equation (12.36) follows from a classical or quantum mechanical treatment of the scattering process (see introductory textbooks on mechanics or quantum mechanics). Because the total scattering probability must also be proportional to the concentration N_{def} of impurities, it follows from (12.32, 12.33, 12.36) that

$$\frac{1}{\tau_{\mathrm{def}}} \propto N_{\mathrm{def}}/T^{3/2} \,, \tag{12.37}$$

and for the mobility, provided the scattering is due only to charged defects, one thus has

$$\mu_{\mathrm{def}} \propto T^{3/2} \,. \tag{12.38}$$

The reciprocal of the total mobility for scattering from defects and phonons is given by the sum of the reciprocal mobilities in (12.35) and (12.38), i.e. the qualitative behavior shown in Fig. 12.12 results. Figure 12.13 shows the experimentally measured dependence $\mu(T)$ for the electron mobility in *n-Ge*, as determined from Hall effect and conductivity measurements; for the purest crystals ($N_{\mathrm{D}} \simeq 10^{13}\,\mathrm{cm}^{-3}$), $\mu(T)$ approaches the theoretically expected dependence, (12.35), for pure phonon scattering. With increasing donor concentration N_{D}, the additional contribution from impurities (12.38) becomes evident (see also Fig. 12.12).

Figure 12.14 shows that the characteristic trend of the mobility in Fig. 12.12 also manifests itself in the temperature dependence of the conductivity in the region of carrier saturation, where $n(T)$ is approximately constant (Figs. 12.10, 12.11). In this region $\sigma(T)$ shows a maximum, while in the regions of intrinsic conductivity (high T) and in the freeze-out range

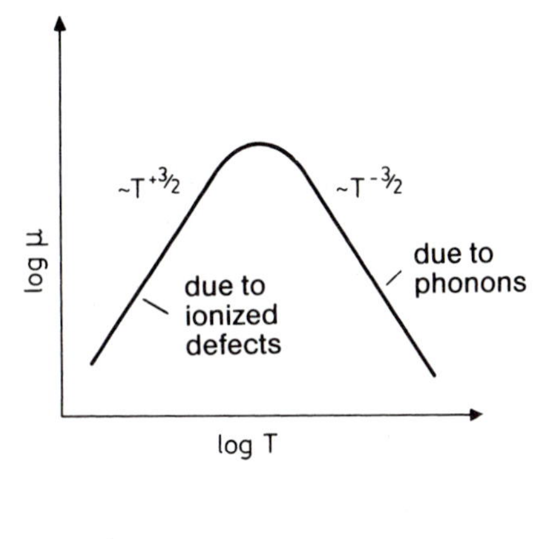

Fig. 12.12. Schematic temperature dependence of the mobility μ for a semiconductor in which scattering from phonons and charged impurities occurs

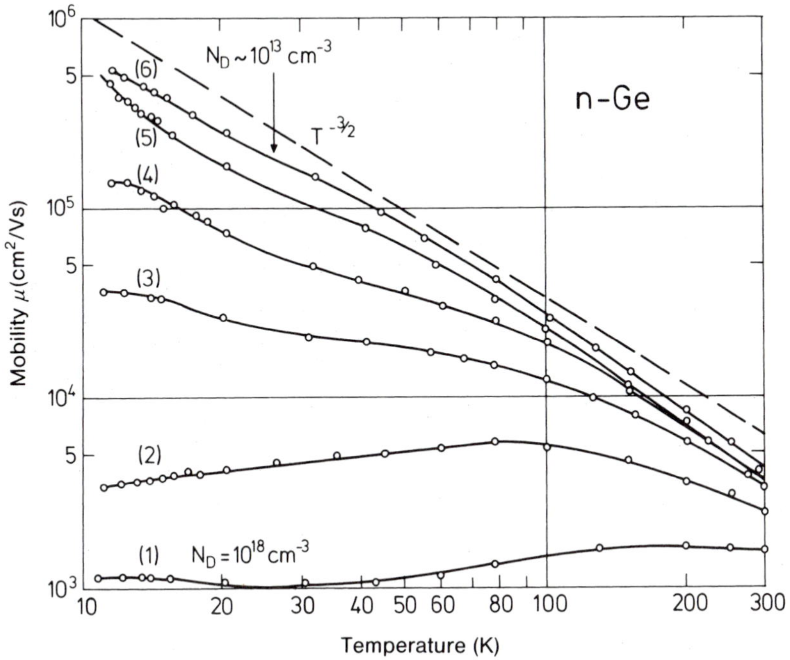

Fig. 12.13. Experimentally determined temperature dependence of the mobility μ of free electrons. For the samples *(1)* to *(6)*, the donor concentration N_D varies between 10^{18} and 10^{13} cm^{-3}. The samples are the same as those used for the measurements in Fig. 12.11. (After [12.3])

(low T), the exponential dependence of the carrier concentration $n(T)$ (Fig. 12.11) dominates the weak temperature dependence of the mobility. The experimental results of Figs. 12.11, 12.13 and 12.14 were all obtained from the same Ge samples [numbered from (1) to (6)], so that the conductivity σ can be calculated directly from $n(T)$ and $\mu(T)$.

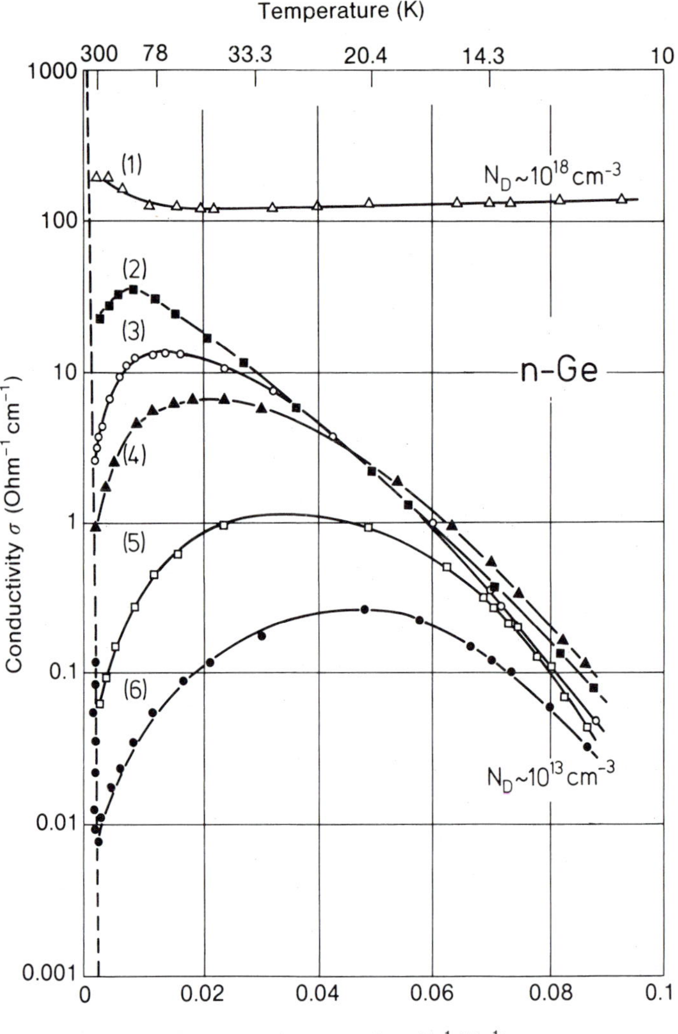

Fig. 12.14. Experimentally measured conductivity σ of n-type germanium as a function of temperature. For the samples (*1*) to (*6*), which were also used for the measurements in Figs. 12.11 and 12.13, the donor concentration N_D varies between 10^{18} and 10^{13} cm^{-3}. (After [12.3])

So far the discussion of the conductivity of semiconductors was concerned merely with the ohmic behavior at relatively low electric fields $\mathscr{E}$ where the carrier drift velocity v_D is proportional to the electric field and the mobility $\mu = v_D/\mathscr{E}$ is a constant. In modern semiconductor devices with dimensions in the sub-micrometer range electric fields are frequently in excess of 10^5 V/cm and Ohm's law is no longer valid since the average drift velocity v_D is no longer proportional to the field strength. According to experiment and theoretical calculations, the proportionality $v_D \propto \mathscr{E}$ remains valid up to fields of about 2×10^3 V/cm for the important semiconductors Si, Ge, GaAs, etc. (Fig. 12.15). For higher fields v_D eventually saturates at a velocity $v_S \approx 10^7$ cm/s for the indirect semiconductor Si, and similarly for Ge. The energy that is continuously transferred to the carriers by the electric field is lost essentially via phonon scattering processes and is thus converted to heat. Scattering due to optical phonons is particularly efficient in this energy range. The carriers are accelerated in the external electric field along the energy band profiles $E(\boldsymbol{k})$, until they reach the energy (referred to E_F) of optical phonons with high density of states (about 60 meV in Si, 36 meV in GaAs). These phonons are then efficiently excited and any further energy gain of the carriers is immediately lost to phonons; the drift velocity saturates as a result. The optical phonons excited in this process are thermalized to low-energy acoustic phonons.

A peculiar behavior is observed in direct semiconductors such as GaAs, InP, GaN. These semiconductors display a negative differential conductivity $\sigma = \partial j/\partial\mathscr{E} = en\partial v_D/\partial\mathscr{E} < 0$ for higher electric fields (Fig. 12.15). In the low field regime ($<10^3$ V/cm) electrons have a high mobility because of the low effective mass of $0.068\,m_0$ (for GaAs) in the Γ-minimum. As the

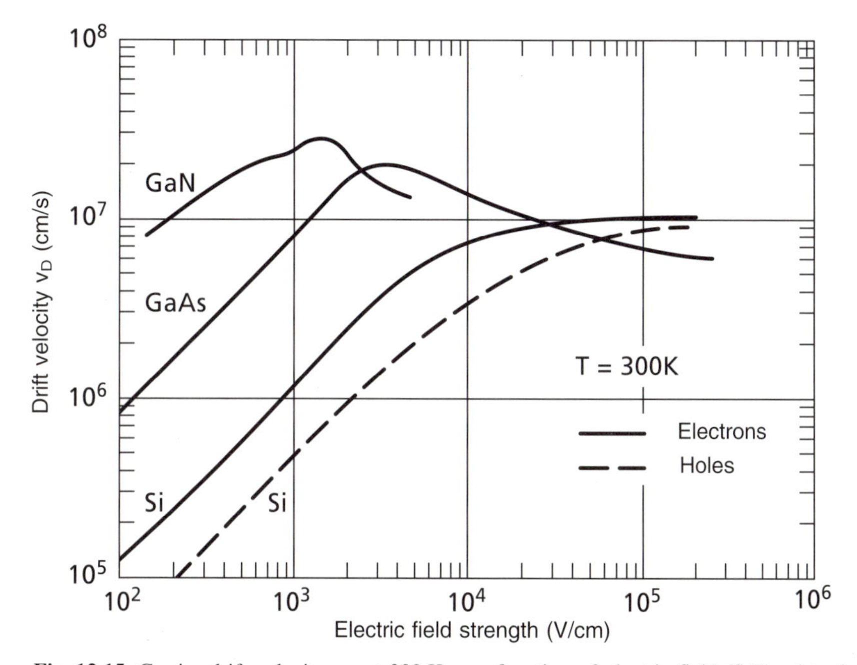

Fig. 12.15. Carrier drift velocity v_D at 300 K as a function of electric field $\mathscr{E}$. The data for Si and GaAs are from a compilation (Sze [12.4]) of experimental results on highly pure, crystalline samples. The curve for GaN is calculated by means of Monte Carlo simulations (after Gelmont et al. [12.5])

electrons are accelerated to higher kinetic energies effective phonon scattering into the side minima at L and X sets in. Electrons in these minima possess a higher effective mass. Their mobility is therefore lower, and the average drift velocity of the whole ensemble of conduction electrons is thus thereby reduced. As a consequence, the drift velocity v_D passes through a maximum with increasing field strength and eventually saturates at a lower value. The saturation value is then similar to that of Si. With even further increasing field strength, above 3×10^5 V/cm, the electrons in the side valleys of the band structure with high effective mass are accelerated and the drift velocity v_D would again increase proportional to the field strength albeit with much lower mobility than in the low-field regime ($<10^3$ V/cm). In the range of extremely high fields, above some 10^5 V/cm this effect is, however, superimposed by the onset of the avalanche breakthrough. The accelerated electrons gain so much energy that they excite more and more electrons from the valence into the conduction band. The conductivity of the semiconductor increases abruptly, avalanche-like, by the multiplication of the number of free carriers. This effect is also used in modern devices.

It is remarkable that the indirect semiconductor Si has a similar saturation velocity at field strengths above 3×10^4 V/cm as the indirect semiconductor GaAs (Fig. 12.15). Hence, the mobility advantage of GaAs with respect to Si at low fields is lost in small devices in which rather high fields occur. A further remarkable aspect of Fig. 12.15 is the high drift velocity of the wide band gap semiconductor GaN. Because of its additional high thermal stability GaN is well suited for applications in the area of fast "high-temperature" devices. Furthermore, the large band gap energies (>3 eV) of the group III-nitrides in general (GaN, AlN, $Al_xGa_{1-x}N$) enable high breakthrough voltages (up to 100 V) and this class of material is employed in fast high-power devices.

12.6 The *p–n* Junction and the Metal/Semiconductor Schottky Contact

Modern solid state physics is closely associated with the development of semiconductor devices, i.e. solid state electronics. The operation of almost all semiconductor devices relies on phenomena that are due to inhomogeneities in semiconductors. Inhomogeneous concentrations of donor and acceptor impurities cause particularly interesting conductivity phenomena, which enable the construction of semiconductor devices.

The most important building blocks in semiconductor devices are the *p–n* junction and the metal/semiconductor contact. In a *p–n* junction, we have a semiconductor crystal (usually Si), which is *p*-type on one side, and *n*-type on the other (Fig. 12.16). In the ideal case (which cannot, of course, be practically realized), the transition from one zone to the other would take the form of a step function.

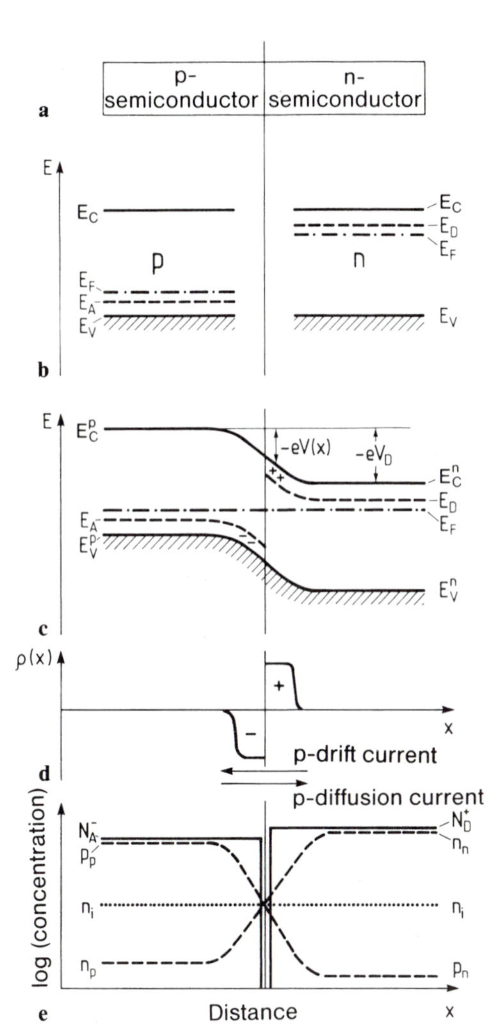

Fig. 12.16a–e. Schematic representation of a p–n junction in thermal equilibrium: (**a**) a semiconductor crystal doped on one side with acceptors (N_A) and on the other side with donors (N_D); (**b**) band scheme for the n and p sides for the imaginary case of total decoupling of the two sides. E_A and E_D indicate the ground states of the acceptors and donors; E_F is the Fermi level; (**c**) band scheme of the p–n junction when the two sides are in thermal equilibrium with one another. The transition from the p to the n doping is assumed to be abrupt. The position of conduction and valence band edges are denoted E_C^p and E_V^p deep in the p region, and E_C^n and E_V^n deep in the n region. V_D is the diffusion voltage. In the region of the p–n junction, a so-called macropotential $V(x)$ is induced; (**d**) the fixed space charge $\varrho(x)$ in the region of the p–n junction due to the ionized impurities; (**e**) qualitative behavior of the concentrations of acceptors N_A, donors N_D, holes p and free electrons n. The intrinsic carrier concentration is n_i and p_p and p_n denote the hole concentrations deep in the p and n regions, respectively (and similarly for n_p and n_n). Considered here is the frequently occurring case in which almost all of the donors and acceptors in the interior of the crystal are ionized

In practice, doping inhomogeneities can be created using a variety of techniques. For example, acceptor impurities can be diffused into one region while donors are diffused into another. The dopant atoms can also be implanted as ions: the ions of the required dopant elements are "fired" into the semiconducting material with high kinetic energy (using high electric fields).

In the following we will consider the conduction behavior of a p–n junction, consisting for example of a Si crystal, which is doped in the left half with acceptors (e.g. B, Al, Ga) and in the right half with donors (e.g. P, As, Sb). We begin by considering this inhomogeneously doped semiconductor in thermal equilibrium, i.e. without an externally applied voltage (Fig. 12.16).

The p–n Junction in Thermal Equilibrium

Let us first of all imagine that the p- and n-type halves of the crystal are isolated from one another (Fig. 12.16b), such that the Fermi levels in the two regions lie at different points on the common energy scale. In reality, however, we are concerned with one and the same crystal which simply has an abrupt doping junction. The Fermi level, i.e. the electrochemical potential, must therefore be common to both crystal halves at thermal equilibrium. In the transition zone between the n and p regions there must therefore be a so-called band bending, as shown in Fig. 12.16c. Within the present semiclassical description, the situation in the transition layer is described by a position-dependent *macropotential* $V(x)$, which reflects the bending of the band structure. This description is possible because the potential $V(x)$ changes only slightly over a lattice parameter. According to the Poisson equation, the macropotential $V(x)$ corresponds to a space charge $\varrho(x)$

$$\frac{\partial^2 V(x)}{\partial x^2} = -\frac{\varrho(x)}{\varepsilon\,\varepsilon_0} \,. \tag{12.39}$$

For the limiting case $T \simeq 0$, where, deep inside the crystal, E_F lies either near the acceptor level (p-type region) or near the donor level (n-type region), one can understand the origin of the space charge from the following qualitative argument: The bending of the bands in the transition region has the effect that acceptors in the p region are pushed below the Fermi level, i.e. are occupied by electrons, whereas in the n region donors are lifted above the Fermi level, i.e. are unoccupied and thus positively charged. The space charge that results from (12.39) thus consists of ionized acceptors and donors that are fixed in space and give rise to a charged double layer across the jump in the doping profile. In the case of carrier freeze-out shown in Fig. 12.16 ($T \simeq 300$ K, E_F between the impurity levels and the middle of the forbidden band), the space charge stems from the fact that the charge of the fixed donors and acceptors in the vicinity of the doping jump is no longer compensated by the mobile electrons and holes of the conduction and valence bands. One therefore speaks of a space-charge region where $\varrho(x) \neq 0$. The concentrations of charge carriers and impurities associated with this space-charge region are represented in Fig. 12.16e. Well outside the space-charge zone, the donors (N_D^+ if charged) or acceptors (N_A^-) are compensated by equally large electron (n_n) or hole (p_p) concentrations. The subscripts n,p indicate whether the electrons and holes are situated in the n or p regions. These carriers correspond to the type of doping in the respective regions and they are denoted *majority carriers*. Because the electrons and holes are "freely" mobile, electrons diffuse into the p region and holes into the n region. There they are called *minority carriers* and their concentrations are denoted by n_p (electrons) and p_n (holes). In thermal equilibrium, the law of mass action ($n_i^2 = np$) must be fulfilled at each point.

For the concentration of majority carriers (electrons in the n region, n_n, or holes in the p region, p_p), it follows from the arguments in Sect. 12.3 that

$$n_n = N_{\mathrm{eff}}^{\mathrm{C}} \exp\left(-\frac{E_{\mathrm{C}}^n - E_{\mathrm{F}}}{k T}\right) , \tag{12.40a}$$

$$p_p = N_{\mathrm{eff}}^{\mathrm{V}} \exp\left(-\frac{E_{\mathrm{F}} - E_{\mathrm{V}}^p}{k T}\right) . \tag{12.40b}$$

Furthermore we have

$$n_i^2 = n_n p_n = N_{\mathrm{eff}}^{\mathrm{V}} N_{\mathrm{eff}}^{\mathrm{C}} \exp\left(-\frac{E_{\mathrm{C}}^n - E_{\mathrm{V}}^n}{k T}\right) . \tag{12.41}$$

The *diffusion voltage* V_{D} – the difference between the maximum and minimum of the macropotential $V(x)$ (Fig. 12.16c) – which is built up in thermal equilibrium, is thus related to the carrier density by

$$eV_{\mathrm{D}} = -(E_{\mathrm{V}}^n - E_{\mathrm{V}}^p) = k T \ln \frac{p_p n_n}{n_i^2} . \tag{12.42}$$

At low temperature in the carrier freeze-out regime it is evident that $|eV_{\mathrm{D}}| \sim E_{\mathrm{g}}$ (Fig. 12.16c). Here, E_{V}^n and E_{V}^p are the valence band edges in the n and p regions.

The state of a semiconductor as represented in Fig. 12.16b–e must be understood as a steady state, because the concentration profiles of "free" carriers as in Fig. 12.16e imply diffusion current (electrons diffusing from right to left and holes from left to right). On the other hand, a space charge as in Fig. 12.16d is associated with an electric field $\mathscr{E}(x)$ and therefore with drift currents of electrons and holes. We represent the corresponding (charge) current densities as:

$$j^{\mathrm{diff}} = j_n^{\mathrm{diff}} + j_p^{\mathrm{diff}} = e\left(D_n \frac{\partial n}{\partial x} - D_p \frac{\partial p}{\partial x}\right) , \tag{12.43}$$

$$j^{\mathrm{drift}} = j_n^{\mathrm{drift}} + j_p^{\mathrm{drift}} = e(n \mu_n + p\mu_p) \mathscr{E}_x . \tag{12.44}$$

Here D_n and D_p denote the diffusion constants for electrons and holes, respectively. In (dynamic) thermal equilibrium, the currents exactly compensate one another. In the p and n regions electron-hole pairs are continually created due to the finite temperature, and subsequently recombine. The total current density obeys

$$j^{\mathrm{diff}} + j^{\mathrm{drift}} = 0 \tag{12.45}$$

and thus the separate contributions of the electrons and holes must vanish individually, i.e. for electrons it follows from (12.43–12.45) that

$$D_n \frac{\partial n}{\partial x} = n\mu_n \frac{\partial V(x)}{\partial x} , \tag{12.46}$$

where $\mathscr{E}_x = -\partial V/\partial x$ is used.

If, instead of considering, as in (12.40), the space-charge concentration far outside the space-charge layer where E_C^n and E_V^p are constant, we consider it in the space-charge zone itself, then the conduction band edge is of course described by $[E_C^p - eV(x)]$ and the concentration of the electrons is position dependent with

$$n(x) = N_{\text{eff}}^{C} \exp\left(-\frac{E_C^p - eV(x) - E_F}{k T}\right) , \tag{12.47}$$

from which it follows that

$$\frac{\partial n}{\partial x} = n \frac{e}{k T} \frac{\partial V}{\partial x} \tag{12.48}$$

or, by substituting into (12.46),

$$D_n = \frac{k T}{e} \mu_n . \tag{12.49}$$

This so-called *Einstein relation* between carrier diffusion constant and mobility is valid whenever diffusion currents and drift currents are carried by one and the same type of carrier. A relation analogous to (12.49) also applies of course for holes, since the total hole current must vanish too.

A rigorous treatment of the *p–n* junction is not simple because, in the Poisson equation (12.39), the exact form of the space-charge density $\varrho(x)$ depends on the interplay of diffusion and drift currents [which in turn depends on $V(x)$]. For the "abrupt" *p–n* boundary treated here, the following approximate solutions can be given; they are known as the *Schottky model of the space-charge zone*. If we imagine that the zero of the x-axis in Fig. 12.16 is at the junction between the n and the p regions, where the donor (N_D) and the acceptor (N_A) concentrations abruptly meet, then for the space charge we have the general relations

$$\varrho(x > 0) = e(N_D^+ - n + p) \quad \text{in the } n\text{-region} , \tag{12.50 a}$$

$$\varrho(x < 0) = -e(N_A^- + n - p) \quad \text{in the } p\text{-region} . \tag{12.50 b}$$

The position-dependent concentrations $n(x)$ and $p(x)$ of "free" charge carriers adjust themselves of course according to the respective distances of the conduction and valence band edges from the Fermi level (Fig. 12.16c). Although this distance changes slowly and monotonically over the entire space-charge zone, the Fermi function and therefore the occupation in an energy region $\sim 2 k T\,(300\,\text{K}) \simeq 0.05\,\text{eV}$, which is small compared to the band gap, changes from approximately zero to its maximum value. If one neglects the so-called transition zone of the Fermi function, then the concentration N_D^+ of charged donors not compensated by free electrons, and the corresponding concentration of charged acceptors N_A^-, can be approximated by a step function (Fig. 12.17a), i.e. the space-charge density becomes

$$\varrho(x) = \begin{cases} 0 & \text{for} \quad x < -d_p \\ -eN_\mathrm{A} & \text{for} \quad -d_p < x < 0 \\ eN_\mathrm{D} & \text{for} \quad 0 < x < d_n \\ 0 & \text{for} \quad x > d_n \,. \end{cases} \tag{12.51}$$

With this piecewise constant space-charge density, the Poisson equation, for example for the n region ($0<x<d_n$)

$$\frac{d^2V(x)}{dx^2} \simeq -\frac{eN_\mathrm{D}}{\varepsilon\,\varepsilon_0} \tag{12.52}$$

can be readily integrated. The calculation for the p region is of course completely analogous. For the electric field $\mathscr{E}_x(x)$ and the potential $V(x)$ in the n region of the space charge zone, one obtains (Fig. 12.17b,c)

$$\mathscr{E}_x(x) = -\frac{e}{\varepsilon\,\varepsilon_0} N_\mathrm{D}(d_n - x) \tag{12.53}$$

and

$$V(x) = V_n(\infty) - \frac{eN_\mathrm{D}}{2\varepsilon\,\varepsilon_0}(d_n - x)^2 \,. \tag{12.54}$$

Outside the "Schottky space-charge zone", the potentials are $V_n(\infty)$ in the n region and $V_p(-\infty)$ in the p region (Fig. 12.17c). One notes the inverted form of the band edges $E_\mathrm{V}(x)$ and $E_\mathrm{C}(x)$ (Fig. 12.16c) compared with the potential energy [$E(x)=-eV(x)$]. Within the Schottky model, the lengths d_n and d_p give the spatial extent of the space-charge zone in the n and p regions respectively.

From charge neutrality it follows that

$$N_\mathrm{D} d_n = N_\mathrm{A} d_p \,, \tag{12.55}$$

and the continuity of $V(x)$ at $x=0$ demands

$$\frac{e}{2\,\varepsilon\,\varepsilon_0}(N_\mathrm{D} d_n^2 + N_\mathrm{A} d_p^2) = V_n(\infty) - V_p(-\infty) = V_\mathrm{D} \,. \tag{12.56}$$

If the impurity concentrations are known, one can thus calculate the spatial extent of the space-charge layer from the diffusion voltage V_D and the difference between the positions of the band edges in the p and n regions.

From (12.55) and (12.56) it follows that

$$d_n = \left(\frac{2\,\varepsilon\,\varepsilon_0 V_\mathrm{D}}{e}\,\frac{N_\mathrm{A}/N_\mathrm{D}}{N_\mathrm{A}+N_\mathrm{D}}\right)^{1/2} , \tag{12.57a}$$

$$d_p = \left(\frac{2\,\varepsilon\,\varepsilon_0 V_\mathrm{D}}{e}\,\frac{N_\mathrm{D}/N_\mathrm{A}}{N_\mathrm{A}+N_\mathrm{D}}\right)^{1/2} . \tag{12.57b}$$

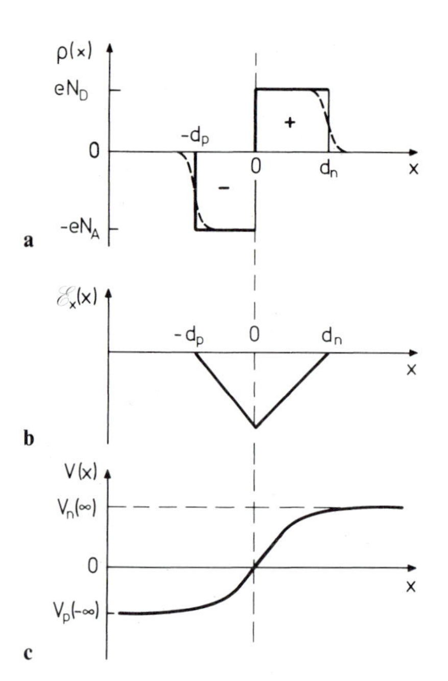

Fig. 12.17 a–c. The Schottky model for the space-charge zone of a p–n junction (at $x=0$). (**a**) Spatial variation of the space-charge density $\varrho(x)$ produced by the ionized acceptors (N_A) and donors (N_D). The real form of the curve (*dashed*) is approximated by the rectangular (*full line*) form; (**b**) behavior of the electric field strength $\mathcal{E}_x(x)$; (**c**) the potential $V(x)$ in the region of the p–n junction

Since diffusion potentials are typically of the order of E_g, that is about 1 eV, impurity concentrations of 10^{14}–10^{18} cm^{-3} result in a space-charge zone of size d_n or d_p between 10^4 and 10^2 Å. The electric field strength is therefore between 10^4 and 10^6 V/cm in such space charge zones.

The Biased p–n Junction – Rectification

If a time-independent external electrical voltage U is applied to a p–n junction, thermal equilibrium is destroyed, and the situation in the p–n junction can be described as a stationary state in the vicinity of thermal equilibrium. Because of the depletion of free carriers (depletion zone), the space-charge zone between $-d_p$ and d_n has a considerably higher electrical resistance than the region outside the p–n junction. As a result, the potential drop across the space-charge zone accounts for nearly all of the externally applied voltage U. Thus the band scheme of Fig. 12.16 and the form of the potential in Fig. 12.17c do not alter except in the region of the space-charge zone; elsewhere, $E_C(x)$, $E_V(x)$ and $V(x)$ are constant and therefore remain horizontal. The potential drop across the space-charge zone, instead of being equal to the diffusion voltage V_D (at equilibrium: $U=0$ V), now has the value

$$V_n(\infty) - V_p(-\infty) = V_D - U \, . \tag{12.58}$$

Here U is taken to be positive when the potential on the p side is raised relative to the n side. The externally applied voltage now influences the size of the space-charge zone, since the quantity V_D in (12.57) is now replaced by $V_D - U$. One thus has

$$d_n(U) = d_n(U = 0)(1 - U/V_D)^{1/2} , \quad (12.59\,a)$$

$$d_p(U) = d_p(U = 0)(1 - U/V_D)^{1/2} . \quad (12.59\,b)$$

In thermal equilibrium, the drift and diffusion currents of electrons are equal and opposite, and the same is true for the hole currents. If an external voltage U is applied, then equilibrium is destroyed. Let us consider, for example, the balance in the electron currents: we are concerned on the one hand with the drift currents of the minority carriers coming from the p region (where electrons are the minority carriers), which are drawn across into the n region by the diffusion voltage V_D. Because these minority carriers are continually generated in the p region by thermal excitation, this current is called the *generation current*, I_n^{gen}. For a sufficiently thin space-charge zone and a sufficiently small recombination rate in this region, each electron coming from the p region that finds its way into the field of the space-charge zone will be drawn from this into the n region. This effect is largely independent of the value of the diffusion voltage and therefore also of the external voltage.

The diffusion current of electrons from the n region, where the electrons are majority carriers, into the p region (called the *recombination current* I_n^{rec}) behaves differently. In this direction the electrons are moving against the potential threshold of the diffusion voltage. The fraction of electrons which can overcome the potential threshold is determined by the Boltzmann factor $\exp[-e(V_D - U)/k T]$, and is therefore strongly dependent on the externally applied voltage U. The following relations describe the electron currents I_n through a p–n junction with an externally applied voltage U:

$$I_n^{rec}(U = 0) \approx I_n^{gen}(U \neq 0) , \quad (12.60)$$

$$I_n^{rec} \propto e^{-e(V_D - U)/k T} , \quad (12.61)$$

which together yield

$$I_n^{rec} = I_n^{gen} e^{eU/k T} , \quad (12.62)$$

and therefore a total electron current of

$$I_n = I_n^{rec} - I_n^{gen} = I_n^{gen}(e^{eU/k T} - 1) . \quad (12.63)$$

The same analysis applies to the hole currents I_p, so that for the characteristic of a p–n junction one obtains

$$I(U) = (I_n^{gen} + I_p^{gen})\left(\exp\frac{eU}{k T} - 1\right) . \quad (12.64)$$

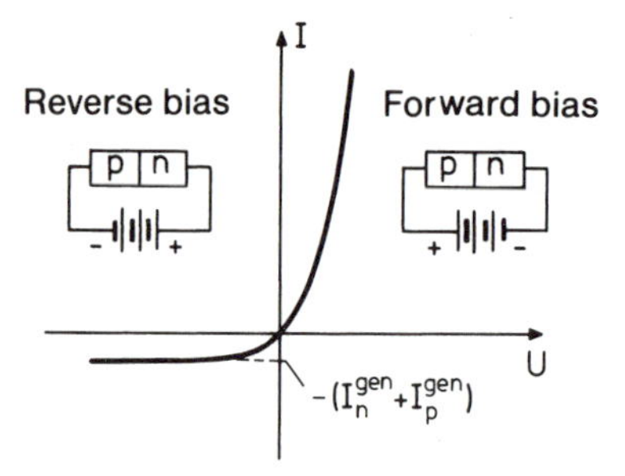

Fig. 12.18. Schematic representation of the current voltage (I–U) characteristic of a p–n junction, together with the corresponding circuit. The maximum current in the reverse direction is given by the sum of the generation currents for electrons and holes

The extremely asymmetric form of this typical rectifying characteristic for the two polarities can be seen in Fig. 12.18.

To derive a quantitative value of the saturation current $-(I_n^{\text{gen}}+I_p^{\text{gen}})$ in the reverse bias direction, a rather more exact treatment of the stationary state in the presence of a dc voltage U is necessary. It follows from the treatment above that the disturbance of thermal equilibrium is due largely to a change in the diffusion currents, while the influence of the external voltage U on the drift currents can, to a good approximation, be neglected. In the framework of the so-called *diffusion current approximation*, it is sufficient to consider only the diffusion currents under the influence of the voltage U.

If we now consider a p–n junction biased in the forward direction (Fig. 12.19), the carrier concentration rises in the space-charge region, as can be seen from Fig. 12.19b. In this situation, the law of mass action $np=n_i^2$ is no longer fulfilled. The Fermi levels well outside the space-charge zone are different by exactly the applied voltage U, corresponding to an energy $-eU$ (Fig. 12.19a). In the space-charge zone, which is no longer in thermal equilibrium, a true Fermi level can no longer be defined. If, as is assumed here, the stationary state is close to thermal equilibrium, then it remains possible to describe the situation approximately with Boltzmann statistics. However, instead of a single Fermi level, two so-called quasi Fermi levels, for electrons (dotted line in Fig. 12.19a) and for holes (dashed line in Fig. 12.19a), must be introduced. Further details of this approach are beyond the scope of this treatment.

Using the approximation that recombination of electrons with holes can be neglected in the space-charge zone, it is sufficient to consider the change in the diffusion current density at the edges of the space-charge zone, i.e. at $x=-d_p$ and $x=d_n$. Here the calculation is particularly easy. We restrict ourselves to the treatment of hole diffusion current densities j_p^{diff}, because the calculation for electrons is analogous.

For the diffusion current at $x=d_n$, it follows from (12.43) that

$$j_p^{\text{diff}}(x=d_n) = -e\,D_p \left.\frac{\partial p}{\partial x}\right|_{x=d_n} . \tag{12.65}$$

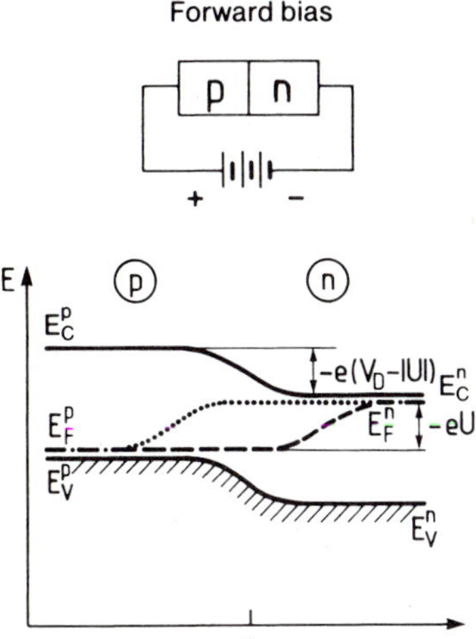

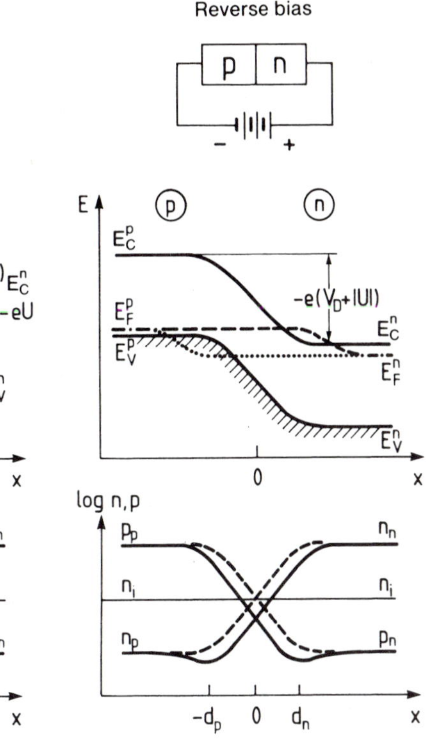

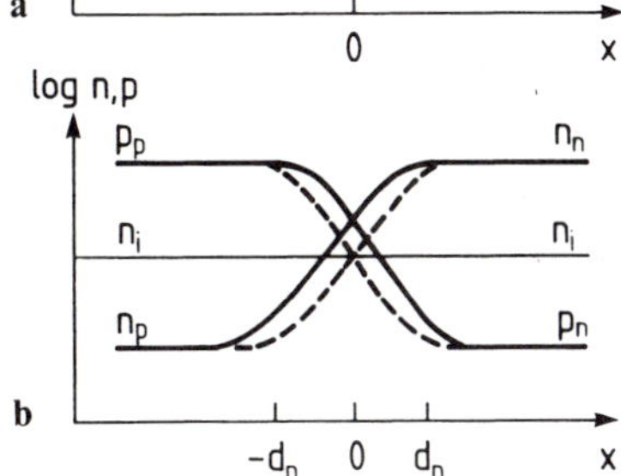

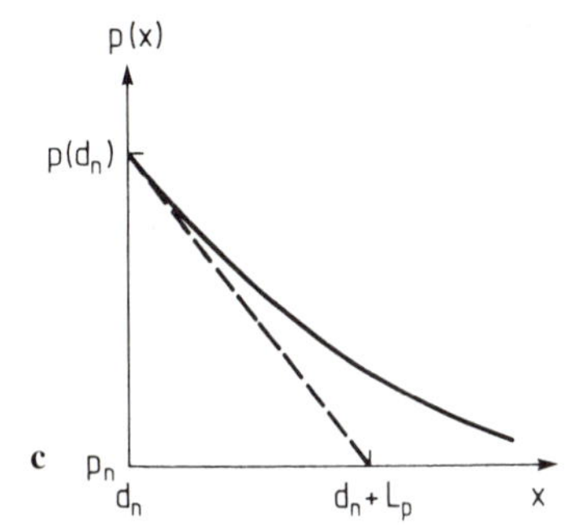

Fig. 12.19 a–c. Forward- and reverse-biased p–n junctions (non-equilibrium state). (**a**) Band scheme in the presence of an external voltage $+U$ or $-U$. The Fermi levels E_F^p and E_F^n in the p and n regions are shifted with respect to one another by eU. In the region of the p–n junction, the equilibrium Fermi level (–·–) splits into so-called quasi Fermi levels for electrons (···) and for holes (– – –); (**b**) spatial variation of the concentration of holes p and electrons n in a biased p–n junction (*full line*) and without bias at thermal equilibrium (– – –). The lengths $-d_p$ and d_n give the range of the space charge zone in thermal equilibrium, i.e. without bias voltage. The carrier concentrations deep in the p and n regions are denoted p_p, n_p and p_n, n_n, respectively; (**c**) spatial variation of the hole concentration $p(x)$ in a forward biased p–n junction in the region outside the thermal equilibrium space-charge zone ($x > d_n$). (Enlarged from Fig. 12.19 b for forward bias)

As will be shown in the following, diffusion theory yields a simple relationship between the concentration gradient $\partial p/\partial x$ and the increase in the hole concentration at $x = d_n$. The resulting hole concentration $p(x = d_n)$ (Fig. 12.19b) at $x = d_n$, due to the applied bias U, follows from Boltzmann statistics as

$$p(x = d_p) = p_p \exp\left(-e\frac{V_D - U}{k T}\right) , \tag{12.66}$$

$$p(x = d_n) = p_n \exp(e\,U/k T) . \tag{12.67}$$

Here d_n and d_p are the widths of the space-charge zone in thermal equilibrium (U=0 V). The increased hole concentration leads to increased recombination in the n region, which results in a larger electron current. Far into the n region, the current is thus carried by electrons and far into the p region by holes.

The continuity condition implies that the hole concentration in a volume element can only change if holes flow in or out, recombine or are thermally created. The greater the amount by which the hole concentration exceeds the equilibrium concentration p_n, the higher is the recombination rate. With an average lifetime τ_p for a hole, the continuity relation for the non-equilibrium carriers is

$$\frac{\partial p}{\partial t} = -\frac{1}{e}\,\mathrm{div}\,\boldsymbol{j}_p^{\,\mathrm{diff}} - \frac{p - p_n}{\tau_p} . \tag{12.68}$$

In the stationary case considered here, $\partial p/\partial t$ must be equal to zero and from (12.65, 12.68) we have

$$\frac{\partial p}{\partial t} = D_p\frac{\partial^2 p}{\partial x^2} - \frac{p - p_n}{\tau_p} = 0 , \tag{12.69 a}$$

and thus

$$\frac{\partial^2 p}{\partial x^2} = \frac{1}{D_p\tau_p}(p - p_n) . \tag{12.69 b}$$

Solution of this differential equation yields the diffusion profile

$$p(x) \sim \exp(-x/\sqrt{D_p\tau_p}) . \tag{12.70}$$

Here $L_p = \sqrt{D_p\tau_p}$ is the diffusion length for holes. This is the distance over which the hole concentration decreases from a certain value, e.g. $p(x = d_n)$ in Fig. 12.19c, by a factor 1/e. From (12.70) and referring to Fig. 12.19c, we obtain

$$-\left.\frac{\partial p}{\partial x}\right|_{x=d_n} = \frac{p(x = d_n) - p_n}{L_p} , \tag{12.71}$$

where, in the case of an externally applied voltage, we have from (12.67)

$$p(x = d_n) - p_n = p_n[\exp(e\,U/k\,T) - 1] \,. \tag{12.72}$$

From (12.65, 12.71, 12.72), we finally arrive at the hole diffusion current density due to the externally applied voltage U:

$$j_p^{\text{diff}}\bigg|_{x=d_n} = \frac{e\,D_p}{L_p} p_n[\exp(e\,U/k\,T) - 1] \,. \tag{12.73}$$

Analogous calculations can be carried out for the diffusion current carried by electrons, and the total current density flowing through the p–n junction is simply the sum of these two components. As shown above, it is not necessary to consider the drift currents. Their components from the p and n regions are, within the present approximation, unchanged from their thermal equilibrium values and serve merely to compensate the equilibrium part of the diffusion current, i.e.

$$j(U) = \left(\frac{e\,D_p}{L_p} p_n + \frac{e\,D_n}{L_n} n_p\right)\left(\exp\frac{e\,U}{k\,T} - 1\right) \,. \tag{12.74}$$

We have thus expressed the generation currents appearing in (12.64) in terms of the diffusion constants and diffusion lengths of electrons and holes, and the minority carrier concentrations p_n and n_p. Figure 12.20 shows an experimentally measured current-voltage characteristic $I(U)$ of a p–n junction. The form of the curve sketched qualitatively in Fig. 12.18 for a rectifier is clearly recognizable.

When an external voltage U is applied to a p–n junction, the spatial extent of the space-charge zone d_n from (12.59), and thus also the charge stored in the space-charge zone, are altered

$$Q_{\text{sc}} \simeq e\,N_{\text{D}} d_n(U) A \,. \tag{12.75}$$

This relation is valid within the framework of the above "Schottky approximation" for the p–n junction, where A is the cross-sectional area of the junction (perpendicular to the current flux) and N_{D} is the concentration of donors as above.

A p–n junction thus has a voltage-dependent "space-charge capacitance" C_{sc}:

$$C_{\text{sc}} = \left|\frac{dQ_{\text{sc}}}{dU}\right| = e\,N_{\text{D}} A \left|\frac{d}{dU} d_n(U)\right| \,. \tag{12.76}$$

From (12.59 a) and (12.57 a) it follows that

$$\left|\frac{d}{dU} d_n(U)\right| = \frac{1}{2V_{\text{D}}}\left(\frac{2\,\varepsilon\,\varepsilon_0 V_{\text{D}} N_{\text{A}}/N_{\text{D}}}{e(N_{\text{A}} + N_{\text{D}})} \frac{1}{1 - U/V_{\text{D}}}\right)^{1/2} , \tag{12.77}$$

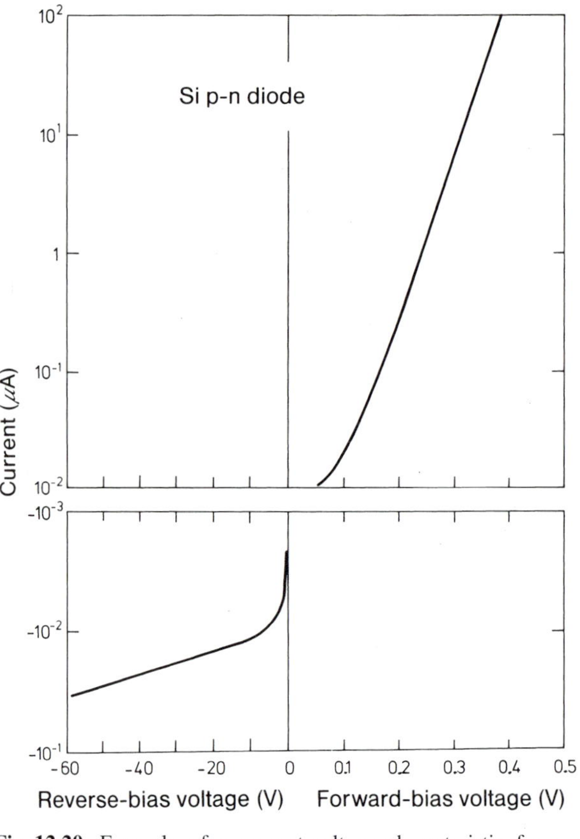

Fig. 12.20. Example of a current-voltage characteristic for a silicon *p–n* junction. Reverse voltages and currents are given as negative values. (From the advanced lab. course of the II. Physics Institute of the RWTH Aachen)

and so the space-charge capacitance (12.76) can be written

$$C_{sc} = \frac{A}{2}\left(\frac{N_A N_D}{N_A + N_D}\,\frac{2\,e\,\varepsilon\,\varepsilon_0}{(V_D - U)}\right)^{1/2} . \tag{12.78}$$

The relationship (12.78) implies that

$$C_{sc}^2 \sim \frac{1}{V_D - U} , \tag{12.79}$$

which explains why a measurement of space-charge capacitance as a function of external voltage is often used to determine the impurity concentrations.

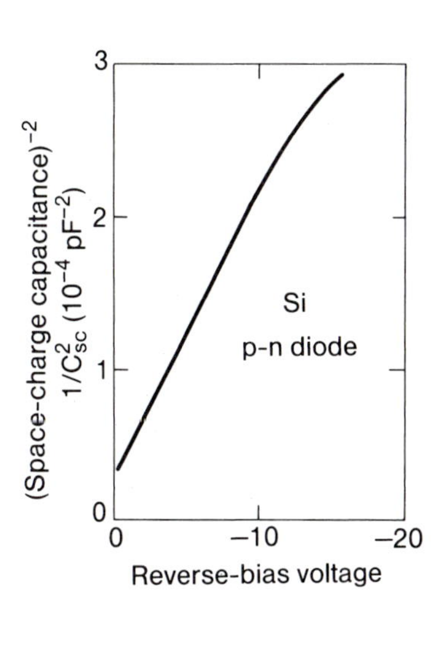

Fig. 12.21. Experimentally determined relationship between the space-charge capacitance and the reverse voltage (indicated by negative values) for the Si *p–n* diode discussed in Fig. 12.19. (From the advanced lab. course of the II. Physics Institute of the RWTH Aachen)

Figure 12.21 shows experimental results for space-charge capacitance as a function of external voltage.

The Metal/Semiconductor Schottky Contact

Because of its rectifying function – similar to the *p–n* junction – the metal/semiconductor contact was used as a device already at the beginning of the 20th century with the advent of microelectronics. When a metal is evaporated onto a clean semiconductor surface under good vacuum conditions, mostly an electronic band scheme as in Fig. 12.22 is established in the case of *n*-doping. The reason is found in the existence of electronic interface states that are formed at the metal/semiconductor interface. Their spatial extension is limited to a few atomic layers around the interface and their energetic distribution is fixed with respect to the conduction and valence band edges of the semiconductor. These interface states, sometimes called MIGS (metal-induced gap states) originate from the Bloch waves in the metal. Bloch waves are extended states in the metal that can not abruptly end at the interface, they must tail into the semiconductor, even in the energy range of the forbidden band, where no electronic bulk states exist on the semiconductor side. In the energy range of the forbidden band these tails of the Bloch states (MIGS) must therefore be mathematically represented by a superposition (Fourier series) of bulk-wave functions of the conduction and valence band, respectively. The energetic band of interface states (MIGS) has thus more conduction-band character (close to the conduction-band edge) or more valence band character (close to the valence-band edge) depending on the percentage of the respective bulk states that constitute the MIGS.

Conduction band states are negatively charged when occupied by electrons and neutral in the unoccupied state (acceptor-like); valence band states, however have a donor-like charging character, i.e. positive when empty and neutral in the occupied state. Thus, a neutrality level exists

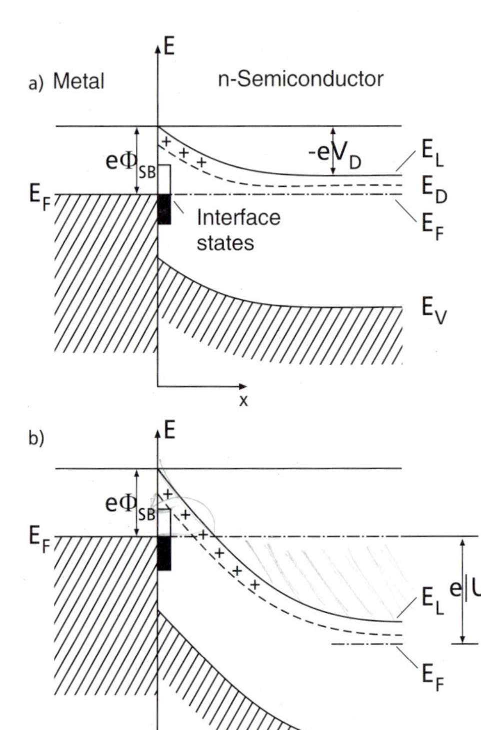

Fig. 12.22. Electronic band scheme of a metal/semiconductor (n-doped) junction; pinning of the Fermi-level E_F in interface states near the neutrality level causes the formation of a Schottky-barrier $e\phi_{SB}$ and a depletion space charge layer within the semiconductor. V_D is the "built-in" diffusion voltage. (**a**) In thermal equilibrium, (**b**) under external bias U

within the band of interface states, which separates the more acceptor-like states in the upper part from the more donor-like states in the lower part of the band. When the Fermi energy E_F crosses the band of interface states just at the neutrality level, the interface states as a whole are neutral. Slight deviations of E_F from the neutrality level cause interface charge within the MIGS: negative if E_F lies higher in energy than the neutrality level, positive for E_F below the neutrality level. The band bending within the semiconductor in Fig. 12.22 is thus determined by a charge balance between negative charge Q_{is} in the interface states and the positive space charge Q_{SC} originating from the ionized bulk donor states in the depletion layer.

For atom area densities at the interface of some 10^{14} cm^{-2} electronic interface state densities can reach a similar order of magnitude. Typically, these states are distributed over an energy range of 0.1 to 1 eV within the forbidden band. Interface densities of states per energy can thus easily amount to 10^{15} (eV)$^{-1}$ cm^{-2}. Such a high density of interface states will practically fix (pin) the Fermi level very close to the neutrality level (Fermi-level pinning). Larger deviations of E_F from the neutrality level would cause too high a charge density within the interface that can not be compensated by the space charge within the semiconductor. Changes of the space charge Q_{SC} by variation of doping or by means of external electric fields thus can shift the Fermi level within the band of interface states with respect to the neutrality level only by amounts of 10^{-2} to 10^{-3} eV. The Schottky barrier (Fig. 12.22) is therefore a quantity that is characteristic for a particular metal-semiconductor junction. Within certain limits it is, of course, slightly affected by atomistic details of the interface structure and morphology.

From a comparison of Fig. 12.22a with Fig. 12.16b it is evident that a Schottky barrier at a metal-semiconductor junction can be described in a simplifying fashion as one half of a *p–n* junction, where the *p*-type semiconductor is replaced by the metal. Obviously, one can interchange the situation; the *n*-side of the *p–n* junction can be replaced by a metal and a Schottky barrier in the *p*-doped semiconductor with hole depletion results. In a *p*-doped semiconductor E_F is close to the valence band edge E_V deep in the bulk (Fig. 12.22). The band curvature in the space charge region of the Schottky contact on *p*-doped material is thus inverse to that of an *n*-type semiconductor. Under the condition of ideal pinning of E_F Schottky barriers for electrons and holes in *n*- and *p*-type material of the same semiconductor sum up to the band gap energy.

As in a *p–n* junction the depletion space-charge region in a Schottky contact exhibits a resistance. An external bias produces a voltage drop essentially across the space-charge zone (Fig. 12.22b). Apart from this spatial range, i.e. within the metal and deep in the semiconductor the voltage drop is negligible and thermal equilibrium is nearly achieved. There, Fermi energies E_F can be defined, the energetic difference of which corresponds to an externally applied voltage. Within the Schottky barrier space-charge region quasi-Fermi levels can be defined similarly as for *p–n* junctions.

The mathematical description of the space-charge region below a metal-semiconductor junction is analogous to a *p–n* junction. The thickness of the Schottky contact space charge region in thermal equilibrium, e.g., is obtained from (12.57a) for the *p–n* junction with the assumption $N_D \ll N_A$ (much higher metallic conductivity as compared with the *p*-type semiconductor) as

$$d = \left(\frac{2\,\varepsilon\,\varepsilon_0 V_D}{e\,N_D}\right)^{1/2} . \tag{12.80}$$

Similarly, the capacity of a metal-semiconductor junction as a function of external bias can be obtained from (12.78) as

$$C = \frac{A}{2}\left(\frac{2\,e\,\varepsilon\,\varepsilon_0 N_D}{V_D - U}\right)^{1/2} , \tag{12.81}$$

where A is the area of the contact.

The rectifying properties of a *p–n* junction are analogously found on a metal-semiconductor Schottky contact, as is evident from Fig. 12.22. Electron transport from the metal into the semiconductor requires that the carriers overcome the Schottky barrier $e\phi_{SB}$, while in the inverse direction electrons can penetrate the metal from the semiconductor side without any barrier for an external bias $|U| > \phi_{SB}$.

12.7 Semiconductor Heterostructures and Superlattices

Using modern epitaxial methods, such as molecular beam epitaxy (MBE, Panel XVII) or metal organic chemical vapor deposition (MOCVD, Panel XVII), it is today possible to deposit two different semiconductors on one another in a crystalline form. These semiconductors will generally have different electronic properties, and in particular different band gaps. Such layer structures play a particularly important role in devices made from III–V semiconductors such as GaAs, InP, etc. It is also significant that using such epitaxial methods, ternary and quaternary alloys of the type $Al_xGa_{1-x}As$ or $Ga_xIn_{1-x}As_yP_{1-y}$ can be deposited, whose band gaps lie between those of the corresponding binary compounds. By controlled variation of the composition x in $Al_xGa_{1-x}As$, the electronic bandstructure can be continuously adjusted between that of GaAs and that of AlAs. At the composition $x=0.45$, the alloy changes from a direct-gap semiconductor (like GaAs) to an indirect-gap semiconductor (like AlAs) (Fig. 11.13). Figure 12.23 plots the band gap energies at 300 K for the most important binary and elemental semiconductors against the lattice constant. This plot is of particular interest because, as one might expect, it is possible to produce

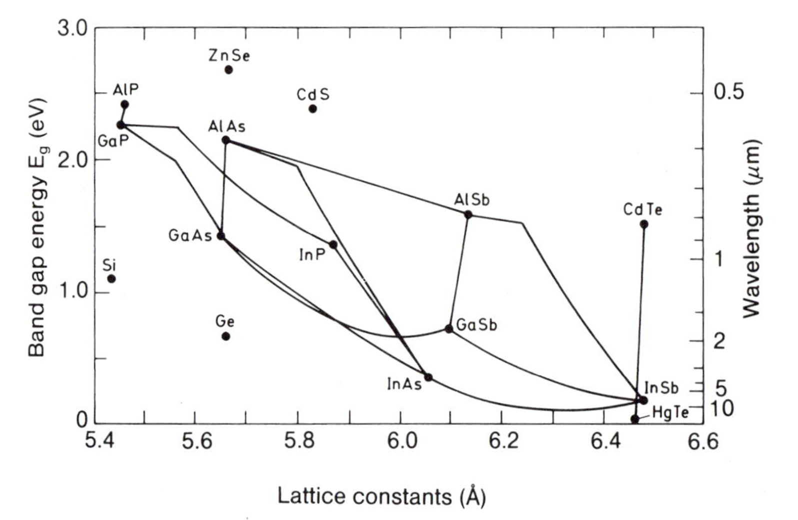

Fig. 12.23. Band gap E_g of some important elemental and binary compound semiconductors plotted against the lattice parameter at 300 K. The right-hand scale gives the light wavelength λ corresponding to the band gap energy. The connecting lines give the energy gaps of the ternary compounds composed of various ratios of the corresponding binary materials

particularly good and defect-free epitaxial layers from semiconductors whose lattice constants match. From Fig. 12.23 it can be seen that both band gap and lattice parameter change when one chooses a composition for epitaxy which lies between two binary semiconductors. It can easily be seen that the epitaxy of Ge on GaAs, and vice versa, leads to particularly good lattice matching, and that the alloy system AlGaAs allows a variation of the band gap between 1.4 and 2.2 eV, whereby it is expected that the extremely good lattice matching of the two components GaAs and AlAs should lead to excellent crystalline quality in growing one semiconductor on another. CdTe and HgTe are also an excellent pair of semiconductors which can be grown on one another largely free from defects. The structure consisting of layers of two different semiconductors grown epitaxially on one another is called a semiconductor heterostructure. The width of the transition regions between the two semiconductors can be as little as one atomic layer if a suitable epitaxial method (MBE, MOCVD) is used. In such a heterostructure the band gap changes over distances of atomic dimensions. What does the electronic bandstructure of such a semiconductor heterostructure look like (Fig. 12.24)? Two important questions need to be answered:

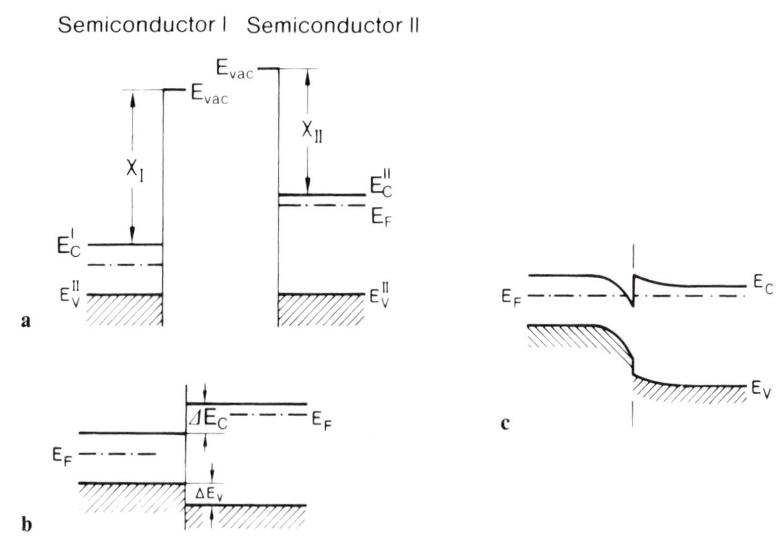

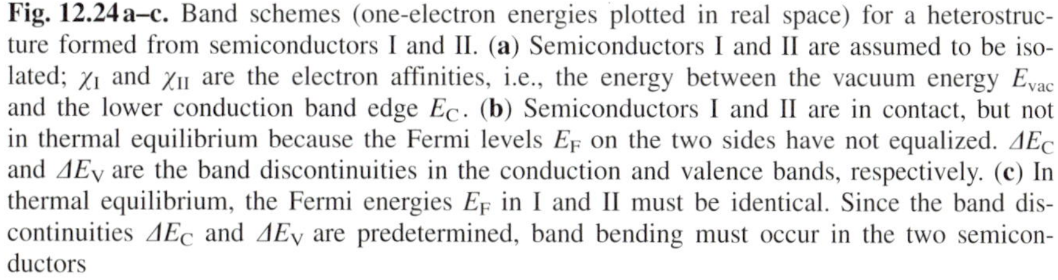

Fig. 12.24a–c. Band schemes (one-electron energies plotted in real space) for a heterostructure formed from semiconductors I and II. (**a**) Semiconductors I and II are assumed to be isolated; χ_I and χ_{II} are the electron affinities, i.e., the energy between the vacuum energy E_{vac} and the lower conduction band edge E_C. (**b**) Semiconductors I and II are in contact, but not in thermal equilibrium because the Fermi levels E_F on the two sides have not equalized. ΔE_C and ΔE_V are the band discontinuities in the conduction and valence bands, respectively. (**c**) In thermal equilibrium, the Fermi energies E_F in I and II must be identical. Since the band discontinuities ΔE_C and ΔE_V are predetermined, band bending must occur in the two semiconductors

a) How are the valence band edges E_V and conduction band edges E_C to be "lined up"? This question addresses the so-called band discontinuity or band offset ΔE_V (Fig. 12.24 b).
b) What band bending occurs in the two semiconductors I and II to the left and right of the junction (Fig. 12.24 c)?

In nearly all cases it is possible to treat these two questions separately and independently of one another, because the relevant phenomena involve different energy and length scales. Matching of the two bandstructures takes place within an atomic distance. Interatomic forces and energies are decisive for this process and the electric fields are of the order of the atomic fields ($\gtrsim 10^8$ V/cm). The band bending, on the other hand, takes place over hundreds of Ångstroms, so that in thermal equilibrium the Fermi level has the same value on both sides of the semiconductor interface, and is determined deep inside each semiconductor by the doping level (Fig. 12.24 c). As was the case for the *p–n* junction, the decisive condition is that no net current flows in the heterostructure at thermal equilibrium, see (12.45). The band bending in semiconductors I and II is, as in the *p–n* junction, associated with space charge, and the space-charge field strengths are of the order of 10^5 V/cm.

The most important material-related parameters of a semiconductor heterostructure are therefore the valence and conduction band discontinuities, ΔE_V and ΔE_C. The classical, but nowadays revised, assumption was that in the ideal case the two bandstructures of the semiconductors line up such that the vacuum energy levels E_{vac} match one another. This gave a conduction band discontinuity ΔE_C equal to the difference in electron affinities $\Delta\chi$ (Fig. 12.24)

$$\Delta E_C = \chi_I - \chi_{II} = \Delta\chi \,. \tag{12.82}$$

The electron affinity χ, which is the difference between the vacuum energy and the bottom of the conduction band, is a theoretically well-defined quantity in the bulk of a material. It is determined, however, with the aid of surface experiments, in which the bulk value of χ can be strongly altered by surface dipoles (e.g. due to the different atomic coordination at the surface). For the interface between two semiconductors χ cannot be related in a simple way to the bulk quantities. Band offsets are meanwhile well explained by models in which the electronic bands in ideal, abrupt semiconductor heterostructures are lined up so that no atomic dipoles are created, e.g. due to electronic interface states (similar to MIGS, p. 428) or charge transfer in the chemical bonds at the interface [12.7]. A detailed theoretical treatment of these models requires a microscopic description of the electronic properties of the few atomic layers at the semiconductor junction. This is beyond the scope of the present book. We therefore treat the band discontinuities ΔE_C and ΔE_V as phenomenological quantities, which may be determined experimentally. A few well-established values are listed in Table 12.7 [12.7]. From a knowledge of the valence band discontinuity

Table 12.7. Compilation of a few experimentally determined valence band discontinuities ΔE_V. The semiconductor named first is the substrate, on which the second semiconductor is deposited, generally as a thin strained layer. (After [12.7])

Hetero-structure	Valence band discontinuity ΔE_V [eV]	Hetero-structure	Valence band discontinuity ΔE_V [eV]	Hetero-structure	Valence band discontinuity ΔE_V [eV]
Si-Ge	0.28	InAs-Ge	0.33	CdTe-α-Sn	1.1
AlAs-Ge	0.86	InAs-Si	0.15	ZnSe-Ge	1.40
AlAs-GaAs	0.34	InP-Ge	0.64	ZnSe-Si	1.25
AlSb-GaSb	0.4	InP-Si	0.57	ZnSe-GaAs	1.03
GaAs-Ge	0.49	InSb-Ge	0.0	ZnTe-Ge	0.95
GaAs-Si	0.05	InSb-Si	0.0	ZnTe-Si	0.85
GaAs-InAs	0.17	CdS-Ge	1.75	GaSe-Ge	0.83
GaP-Ge	0.80	CdS-Si	1.55	GaSe-Si	0.74
GaP-Si	0.80	CdSe-Ge	1.30	CuBr-GaAs	0.85
GaSb-Ge	0.20	CdSe-Si	1.20	CuBr-Ge	0.7
GaSb-Si	0.05	CdTe-Ge	0.85		

ΔE_V and the band gap energies of the two semiconductors, the conduction band discontinuity ΔE_C can of course easily be determined.

The calculation of the space-charge zones is performed in complete analogy to the calculations for a simple *p–n* junction (Sect. 12.6), except that the positive and negative space-charges now occur in two different semiconductor materials I and II with different dielectric constants ε^{I} and ε^{II}.

The corresponding quantities for a *p–n* heterojunction are depicted in Fig. 12.25. The simplest description is again within the framework of the Schottky model, in which the space charges $-eN_A^{\text{I}}$ and eN_D^{II} are assumed to be constant over their respective space-charge zones d^{I} and d^{II}. The Poisson equation (12.39, 12.52) can once again be integrated piecewise in semiconductors I and II. It is important to remember that the diffusion voltage V_D is now partitioned between two different semiconductors, i.e. divided into V_D^{I} and V_D^{II}

$$V_D = V_D^{\text{I}} + V_D^{\text{II}} . \qquad (12.83)$$

The same is true for an external voltage applied across the *p–n* heterojunction:

$$U = U^{\text{I}} + U^{\text{II}} . \qquad (12.84)$$

To match the two solutions of the Poisson equation at the interface $(x = 0)$, the continuity of the dielectric displacement must be considered

$$\varepsilon_0 \varepsilon^{\text{I}} \mathscr{E}^{\text{I}}(x = 0) = \varepsilon_0 \varepsilon^{\text{II}} \mathscr{E}^{\text{II}}(x = 0) . \qquad (12.85)$$

In analogy to (12.57) the thicknesses of the space-charge zones in semiconductors I and II are given by the formulae

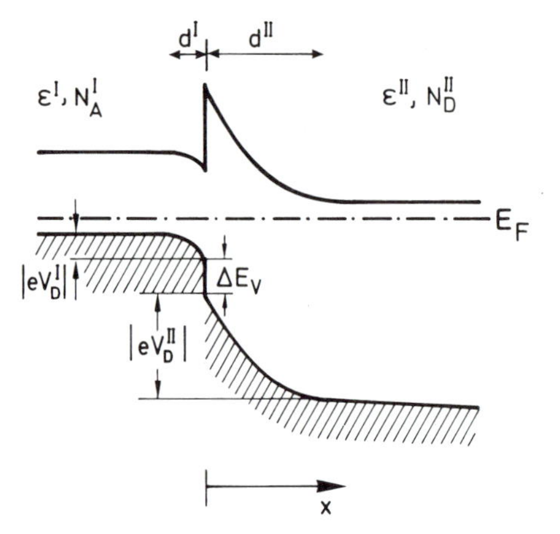

Fig. 12.25. Band scheme of a semiconductor heterojunction; semiconductor I with dielectric function ε^{I} is p-doped with an acceptor concentration $N_{\mathrm{A}}^{\mathrm{I}}$, semiconductor II with a dielectric function $\varepsilon^{\mathrm{II}}$ is n-doped with a donor concentration $N_{\mathrm{D}}^{\mathrm{II}}$. The space-charge zones have thicknesses d^{I} and d^{II}; the diffusion voltages associated with the space-charge zones are $V_{\mathrm{D}}^{\mathrm{I}}$ and $V_{\mathrm{D}}^{\mathrm{II}}$. ΔE_{V} is the valence band discontinuity. Such a heterojunction consisting of a p-type semiconductor with a small band gap and an n-type semiconductor with a large band gap is also called a p–N heterojunction (small p for the small band gap, large N for the large band gap)

$$d^{\mathrm{I}} = \left(\frac{2N_{\mathrm{D}}^{\mathrm{II}}\,\varepsilon_0\varepsilon^{\mathrm{I}}\varepsilon^{\mathrm{II}}(V_{\mathrm{D}}-U)}{eN_{\mathrm{A}}^{\mathrm{I}}(\varepsilon^{\mathrm{I}}N_{\mathrm{A}}^{\mathrm{I}}+\varepsilon^{\mathrm{II}}N_{\mathrm{D}}^{\mathrm{II}})}\right)^{1/2}, \tag{12.86a}$$

$$d^{\mathrm{II}} = \left(\frac{2N_{\mathrm{A}}^{\mathrm{I}}\,\varepsilon_0\varepsilon^{\mathrm{I}}\varepsilon^{\mathrm{II}}(V_{\mathrm{D}}-U)}{eN_{\mathrm{D}}^{\mathrm{II}}(\varepsilon^{\mathrm{I}}N_{\mathrm{A}}^{\mathrm{I}}+\varepsilon^{\mathrm{II}}N_{\mathrm{D}}^{\mathrm{II}})}\right)^{1/2}. \tag{12.86b}$$

Appearing here, in addition to the terms in (12.57), is the externally applied voltage U. The ratio of the voltage drops in the two semiconductors is

$$\frac{V_{\mathrm{D}}^{\mathrm{I}}-U^{\mathrm{I}}}{V_{\mathrm{D}}^{\mathrm{II}}-U^{\mathrm{II}}} = \frac{N_{\mathrm{D}}^{\mathrm{II}}\varepsilon^{\mathrm{II}}}{N_{\mathrm{A}}^{\mathrm{I}}\varepsilon^{\mathrm{I}}}. \tag{12.87}$$

For the case of a simple p–n junction ($\varepsilon^{\mathrm{I}}=\varepsilon^{\mathrm{II}}$) in thermal equilibrium (U=0), equation (12.86) reduces of course to (12.57).

Of particular interest are heterojunctions between two different semiconductors with the same doping, so-called isotypic heterojunctions (e.g. n–N in Fig. 12.24c). In this case, because of the continuity conditions for the Fermi level, an accumulation space-charge zone for electrons is created on the side of the semiconductor with a smaller forbidden gap, which leads to an extremely large increase in local electron concentration. This is true even when this side of the heterostructure is only very weakly doped, i.e. almost intrinsic (Fig. 12.26a). The high concentration of free electrons in this space-charge zone (semiconductor II) is compensated by a depletion space-charge zone in semiconductor I. This has a strong positive space-charge as a result of the high concentration of ionized donors. These donors have given up their valence electrons to the energetically more favorable potential well in the accumulation zone of semiconductor II. In this way the high density of free electrons is spatially separated from the ion-

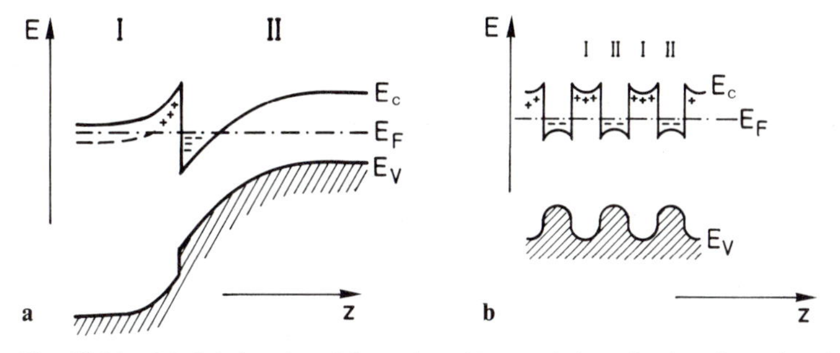

Fig. 12.26 a. Modulation-doped heterojunction consisting of a heavily n-doped semiconductor I with a large band gap and a weakly (or nearly intrinsic) n-doped semiconductor II with a small band gap (N–n heterojunction); (**b**) band scheme of a modulation-doped composition superlattice; the layers of the semiconductor I are in each case highly n-doped, while the layers of type II are weakly doped or nearly intrinsic

ized impurities from which they originate. Impurity scattering, which is an important source of electrical resistance at low temperature, is therefore strongly reduced for this free electron gas. In a homogeneously doped semiconductor, an increase of the carrier concentration requires a simultaneous increase of the doping level, thus leading to increased impurity scattering, which in turn reduces the conductivity. This necessary correlation of higher impurity scattering with increasing carrier concentration does not occur for heterostructures such as that shown in Fig. 12.26. This type of "one-sided" doping in a heterostructure is called "modulation doping".

Electron mobilities of a modulation-doped $Al_xGa_{1-x}As/GaAs$ heterostructure are shown in Fig. 12.27. If the electron mobility were governed by scattering from ionized impurities (ID) and phonons (P) in the same way as in homogeneously doped GaAs, then there would be a characteristic limit resulting from these processes, decreasing at higher and lower temperature. A maximum in the mobility of about 4×10^3 cm^2/Vs would be obtained for a donor concentration of N_D of 10^{17} cm^{-3} at about 150 K. The experimentally determined mobilities of modulation-doped structures (shaped region) show, however, no reduction in the mobility at low temperature. The variation of the mobility in the shaded region is due to differences in the perfection of the heterojunction achieved by different authors. At temperatures below 10 K, extremely high mobilities of up to 2×10^6 cm^2/Vs are achieved commonly. Best state-of-heart values are around 10^7 cm^2/Vs. The increase in mobility can be further enhanced if an undoped $Al_xGa_{1-x}As$ layer with a thickness of about 100 Å is placed between the highly n-doped AlGaAs and the weakly doped GaAs during epitaxy. In this way, scattering processes at impurities in the immediate vicinity of the semiconductor junction are eliminated.

For n-doping concentrations in AlGaAs of about 10^{18} cm^{-3}, typical thicknesses of the electron enrichment layer are in the region 50–100 Å. The free electrons are confined here in a narrow triangular potential well in

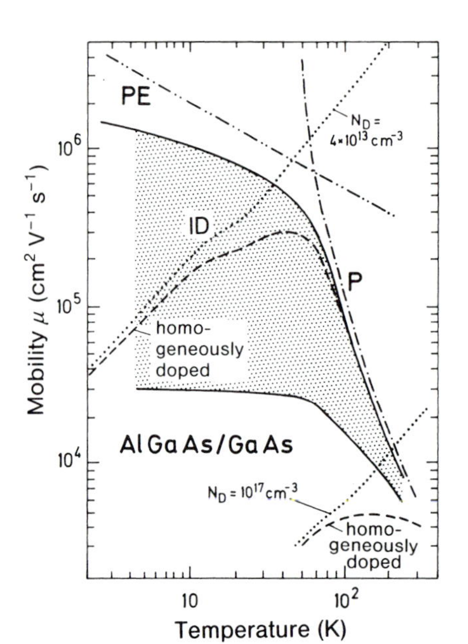

Fig. 12.27. Temperature dependence of the electronic mobility in a quasi 2D electron gas in modulation-doped AlGaAs/GaAs structures; the shaded region represents a great many experimental data. For comparison the mobility curves of homogeneously doped GaAs with donor concentrations $N_D = 4\times10^{13}$ cm^{-3} and $N_D = 10^{17}$ cm^{-3} are also shown (*dashed*). The limiting values of the mobility are determined by the following mechanisms: scattering from ionized impurities (ID) (···); phonon scattering (P) (–·–); piezoelectric scattering (PE) (–··–). (After [12.8])

the z direction (perpendicular to the heterojunction) (Fig. 12.26a). They can move freely only in the direction parallel to the heterostructure. The wavefunction of such an electron thus has Bloch-wave character only parallel to the heterostructure; perpendicular to it (along z), one expects quantum effects like those for electrons in a potential well (Sect. 6.1). One speaks of a two-dimensional electron gas (2DEG).

Similar effects to those in the simple modulation-doped heterostructure (Fig. 12.26a) appear if one epitaxially grows two heterostructures that are mirror images of one another, or even a whole series of layers of semiconductors I and II with different band gaps (Fig. 12.26b). A series of so-called "quantum wells" are then formed in the bandstructure, in which the free electrons of the conduction band accumulate. Such a structure is shown in Fig. 12.26b; it is called a *composition superlattice*, because the crystal lattice has superimposed on it an artificially created structure of potential wells with larger periodicity. One continues to speak of a modulation-doped superlattice when, as in Fig. 12.26b, only semiconductor I with the larger band gap is heavily n-doped, while semiconductor II with the smaller gap is lightly doped. The free electrons in the quantum wells of semiconductor II are again separated from the donor impurities in semiconductor I from which they originate. Impurity scattering is strongly reduced and perpendicular to the superlattice direction (z axis), extremely high mobilities are found at low temperatures as is evident in Fig. 12.27. The band bending shown in Fig. 12.26b, positive in semiconductor I and negative in semiconductor II, corresponds to the sign of the space charge in each region. An unambiguous relationship between these quantities is given by the Poisson equation (12.39). If the quantum wells in Fig. 12.26b are sufficiently narrow, i.e. their extension in the z direction is smaller than or on the order of 100 Å, then quantization effects become evident in the z direction for the thin space-charge zone (Fig. 12.26 a).

This so-called z quantization can be described straightforwardly using the time-independent Schrödinger equation for a crystal electron as in Fig. 12.26 a and b. The electron is "trapped" in one direction (along the z axis), while it is free perpendicular to this direction. The potential V is then a function only of z, and with three effective mass components m_x^*, m_y^*, m_z^*, the following Schrödinger equation applies

$$\left[-\frac{\hbar^2}{2}\left(\frac{1}{m_x^*}\frac{\partial^2}{\partial x^2}+\frac{1}{m_y^*}\frac{\partial^2}{\partial y^2}+\frac{1}{m_z^*}\frac{\partial^2}{\partial z^2}\right)-e\,V(z)\right]\psi(\boldsymbol{r})=E(\psi)(\boldsymbol{r})\ . \tag{12.88}$$

With an ansatz of the form

$$\psi(\boldsymbol{r})=\varphi_j(z)\mathrm{e}^{\mathrm{i}\,k_x x+\mathrm{i}\,k_y y}=\varphi_j(z)\mathrm{e}^{\mathrm{i}\,\boldsymbol{k}_\parallel\cdot\boldsymbol{r}}\ , \tag{12.89}$$

equation (12.88) can be separated into two independent differential equations

$$\left[-\frac{\hbar^2}{2\,m_z^*}\frac{\partial^2}{\partial z^2}-e\,V(z)\right]\varphi_j(z)=\varepsilon_j\varphi_j(z)\ , \tag{12.90}$$

and

$$\left(-\frac{\hbar^2}{2\,m_x^*}\frac{\partial^2}{\partial x^2}-\frac{\hbar^2}{2\,m_y^*}\frac{\partial^2}{\partial y^2}\right)\mathrm{e}^{\mathrm{i}\,k_x x+\mathrm{i}\,k_y y}=E_{xy}\mathrm{e}^{\mathrm{i}\,k_x x+\mathrm{i}\,k_y y}\ . \tag{12.91}$$

The solutions of (12.91) are energy eigenvalues that correspond to the unimpeded motion of an electron perpendicular to z

$$E_{xy}=\frac{\hbar^2}{2\,m_x^*}k_x^2+\frac{\hbar^2}{2\,m_y^*}k_y^2=\frac{\hbar^2}{2\,m_\parallel^*}k_\parallel^2\ . \tag{12.92}$$

For the determination of ε_j in (12.90), the exact form of the potential $V(z)$ must be known. The existence of the space charge means of course that $V(z)$ depends on the density of free electrons and ionized donor cores; i.e., the probability density $|\varphi_j(z)|^2$ is included in the potential via the electron density. Equation (12.90) must thus be solved self-consistently. In a simple approximation, however, $V(z)$ is described by a rigid square-well potential (as in Sect. 6.1) for the case of quantum wells (Fig. 12.26 b), or by a triangular potential for the case of a simple modulation-doped heterostructure (Fig. 12.26 a). If for the square-well potential one assumes in addition infinitely high potential walls, the electron wavefunctions are simple standing waves in the z direction and, as in Sect. 6.1, the eigenvalues ε_j are given by

$$\varepsilon_j\simeq\frac{\hbar^2\pi^2}{2\,m_z^*}\frac{j^2}{d_z^2}\ ,\quad j=1,2,3\ldots\ , \tag{12.93}$$

where d_z is the width of the potential well in the z direction. The total energy eigenvalues for such electron states quantized in the z direction

$$E_j(\boldsymbol{k}_\parallel) = \frac{\hbar^2 k_\parallel^2}{2\,m_\parallel^*} + \varepsilon_j \tag{12.94}$$

are described by a family of discrete energy parabolas along k_x and k_y, so-called sub-bands (Fig. 12.28 b). These two-dimensional (2D) sub-bands have a constant density of states $D(E) = dZ/dE$, as can easily be demonstrated. In analogy to Sect. 6.1 one notes that in the 2D reciprocal space of wave numbers k_x and k_y, the number of states dZ in a ring of thickness dk and radius k is given by

$$dZ = \frac{2\pi k\, dk}{(2\pi)^2} \; . \tag{12.95}$$

Since $dE = \hbar^2 k\, dk/m_\parallel^*$, one obtains a density of states

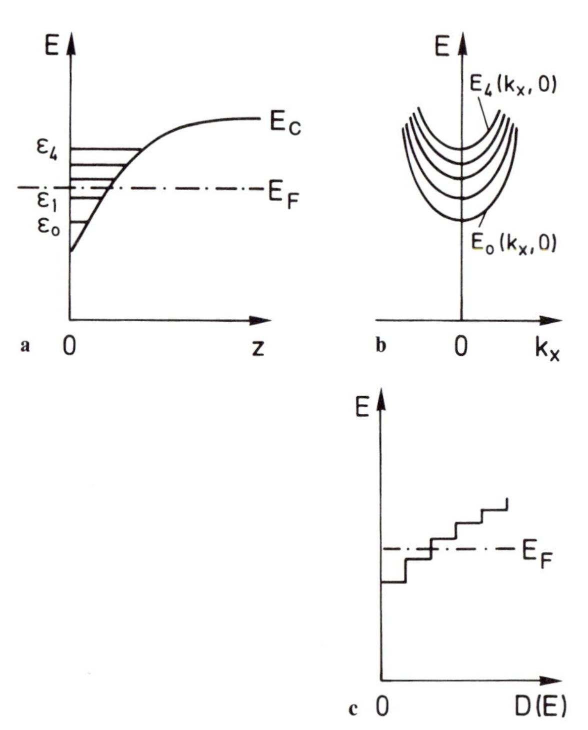

Fig. 12.28 a–c. Quantization of a quasi 2D electron gas in a triangular potential well, as occurs in the strong accumulation zone at a heterojunction ($z = 0$). (**a**) Behavior of the conduction band edge $E_C(z)$; E_F is the Fermi level. $\varepsilon_1, \varepsilon_2, \ldots$ are the energy levels resulting from the quantization of the one-electron states along z; (**b**) energy parabolas of the sub-bands, plotted against the wave vector k_x in the plane of free propagation perpendicular to z; (**c**) density of states of the quasi 2D electron gas reflecting its quantization into sub-bands

$$D = dZ/dE = m_{\|}^{*}/\pi\hbar^2 = \text{const} \,, \tag{12.96}$$

with spin degeneracy accounted for by the factor 2. The total density of states $D(E)$ of all sub-bands is therefore a superposition of constant contributions, or a staircase function as in Fig. 12.28c.

Sharp parabolic sub-bands as in Fig. 12.28 only appear when the eigenvalues ε_j (12.93) are sharp energy levels. This is the case for a single potential well in the conduction band, or when, in a superlattice, the neighboring potential wells are so far apart that the wavefunctions of the individual potential wells do not overlap. If the distance between the potential wells is so small (less than 50–100 Å) that significant overlap between the wavefunctions exists, then this leads to a broadening of the bands. The broadening is completely analogous to that of the individual atomic energy levels of atoms in a crystal (Sect. 7.3). Figure 12.29 shows the theoretically expected broadening for sub-band energies $\varepsilon_1, \varepsilon_2, \ldots$ for the case of a rectangular superlattice, in which the width of the potential wells d_z corresponds to the spatial separation of two heterojunctions. One sees that the energetically lowest sub-band ε_1 is noticeably broadened for a periodicity of less than 50 Å, and splits off as a band. For the higher sub-bands, the broadening begins at even larger distances between the potential wells.

The broadening of the sub-bands and, in particular, the dependence of the sub-band energies on the spatial width of the potential wells is clearly seen in photoluminescence experiments. Photoluminescence spectroscopy is

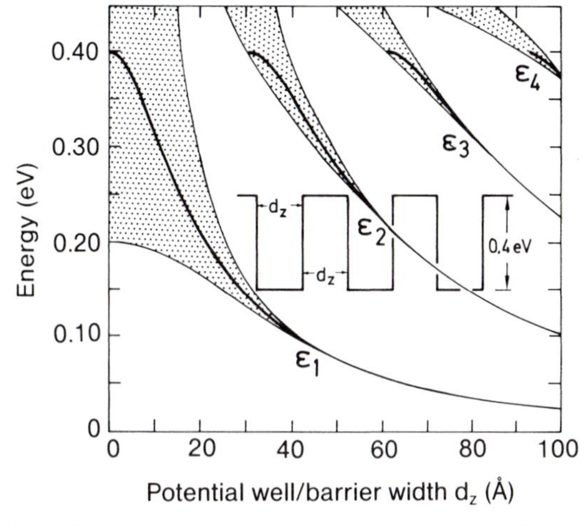

Fig. 12.29. Energy states of electrons confined in the rectangular potential wells (inset) of the conduction bands of a composition superlattice; the potential wells have a width d_z which also corresponds to their distance from one another. For the calculation, an electronic effective mass of $m^* = 0.1\, m_0$ was assumed. The heavy lines in the shaded regions are the results for single potential wells with the corresponding widths d_z; potential wells in a superlattice with sufficiently small separation lead to overlap of wavefunctions and therefore to a broadening into bands (*shaded region*). (After [12.9])

an important optical method for characterizing semiconductor heterostructures and superlattices. It is essentially the reverse of an optical absorption experiment: The semiconductor structure is illuminated with monochromatic laser light of photon energy above the band edge, thus creating electron–hole pairs. These occupy the sub-bands of the conduction or valence bands of the semiconductor or the corresponding excitonic states (Sect. 11.11). In the case of direct-gap semiconductors (Sect. 12.1), the recombination of the free electrons with the holes is associated with strong luminescence, which, for sharply defined sub-bands, is monochromatic and corresponds, apart from the exciton binding energies (Sect. 11.11), to the energetic separation of the sub-bands for electrons and holes. The experiment has to be performed at low temperatures in order to observe the small energy broadening of the sub-band structure $(\varepsilon_1, \varepsilon_2, \dots)$. The low temperature is also the reason why the sub-band excitons are stable and radiative recombination occurs from the excitonic electron–hole states. Figure 12.30 shows a photoluminescence spectrum taken from an AlGaAs/GaAs multilayer structure at a temperature of 2.7 K. The layer structure consists of four GaAs quantum wells of differing thicknesses (20, 30, 60, 100 Å) embedded between barriers of $Al_{0.3}Ga_{0.7}As$. The four different quantum-well widths give rise to four sharp luminescence lines at different photon energies in the range 1.55–1.75 eV. Since the sub-bands are characterized by the fact that an integral number of half electron wavelengths must more-or-less fit into the quantum well, narrower quantum wells yield sub-bands in the valence and conduction bands that lie further apart and thus have higher energy transitions in the photoluminescence. Of course, an exact theoretical analysis of the relationship between luminescence energy and quantum-well width must take into account the Coulomb attraction between the excited electron–hole pairs. This means that the energy difference between electron and hole sub-bands differs from the energy of the emitted photon by the amount of the exciton binding energy. Since local variations in the width of the quantum well lead to a spread in the spectral position of the emitted photoluminescence line, high resolution photoluminescence spectroscopy is

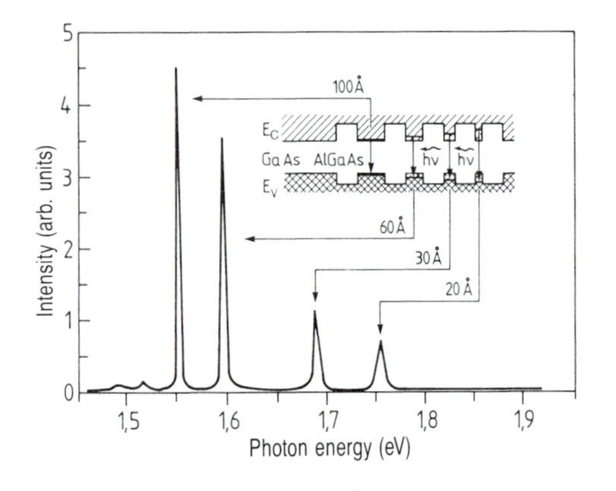

Fig. 12.30. Photoluminescence spectrum of an MBE-grown multiple quantum well structure of $Al_{0.3}Ga_{0.7}As$/GaAs taken at 2.7 K. The various emission lines result from electron–hole recombination processes in quantum wells of different thicknesses (20–100 Å). The bandstructure (E_c, E_v) of the layer structure is indicated showing the correspondence of the quantum wells to the emission lines. (After [12.7])

particularly well suited to measuring spatial variations in the layer thickness of quantum well layers.

Besides the compositional superstructures considered so far, there is a further type of semiconductor superlattice, the so-called *doping superlattice.* Here the superlattice structure consists of one and the same semiconductor, but the material is periodically and alternately n and p doped. In principle it is equivalent to a periodic sequence of p–n junctions (Sect. 12.6). Because quasi intrinsic (i regions) exist between each n and p zone, these structures also have the name "nipi structures". The production of such lattices is also carried out using epitaxy (MOCVD, MBE, etc., Panel XVII). During the growth process, the p and n doping sources are switched on alternately.

The interesting properties of nipi superlattices manifest themselves in the position-dependent electronic bandstructure (Fig. 12.31). Because of the periodic sequence of n- and p-doped regions, the conduction and valence bands must alternately approach the Fermi level. This leads to a periodic modulation of the band edges with position. Excited free electrons (thermal and non-equilibrium) are found in the minima of the conduction band, while excited holes are spatially separated and gather in the maxima of the valence band. This spatial separation of electrons and holes is responsible for the fact that the collision rate between these two particles is drastically reduced. One interesting property which results is an extremely long recombination lifetime for electrons and holes. This is clearly manifest when electron-hole pairs are created by irradiation with light. A photocurrent can exist for considerably longer in such a nipi structure than in a homogeneously doped semiconductor.

Another interesting property of doping superlattices concerns the band gap. In spite of the considerable spatial separation of electron and hole

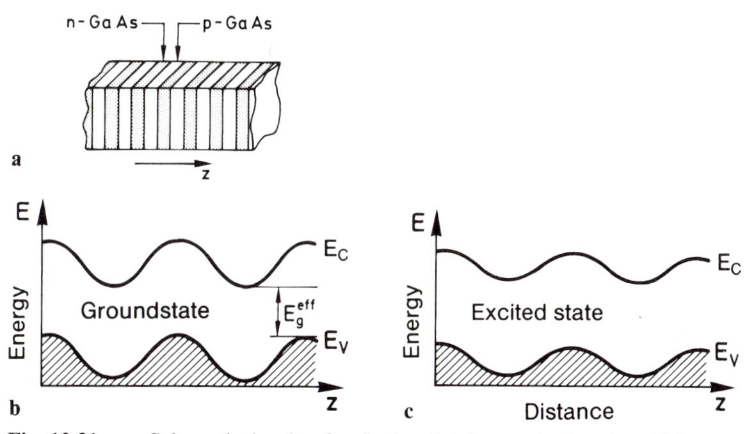

Fig. 12.31 a–c. Schematic bands of a doping (*nipi*) superlattice, in which one and the same semiconductor material (e.g. GaAs) is alternately n- and p-doped. (**a**) Qualitative sketch of the structure; (**b**) band scheme in thermal equilibrium; (**c**) band scheme under strong excitation of electron–hole pairs to above their thermal equilibrium densities

states, there is nonetheless a certain overlap of the wavefunctions in the transition region (*i* region). This results in the possibility of optical transitions in absorption and emission. As a consequence of the band modulation one thus observes optical transitions at quantum energies below the band edge of the homogeneously doped semiconductor (Fig. 12.32). Nipi structures have an effective band gap E_q^{eff}, which can be adjusted in a controlled way by the respective n and p doping levels. In the simplest approximation one assumes, as for a p–n junction (Sect. 12.6), that all impurities are ionized, and applies the Schottky approximation with a rectangular space-charge distribution. Because of the relationship between band bending and space-charge density [described by the Poisson equation (12.39)], it is immediately clear that a reduction of the space charge results in a decrease of the band bending and therefore a flattening out of the band modulation. The effective band gap becomes larger (Fig. 12.31). The space charge consists of ionized donors (positive) and electrons in the n region, and of ionized acceptors (negative) and holes in the p region. Sufficiently energetic excitation of electrons and holes, e.g. by irradiation with light, reduces the space charge and also the band modulation. In nipi superlattices, the effective band gap is dependent on the density of optically excited non-equilibrium carriers. Thus the effective band gap can be optically altered. This may be demonstrated in a photoluminescence experiment (Fig. 12.32), in which the emission due to recombination of optically excited electrons and holes is observed as a function of laser excitation power [12.11]. In a GaAs nipi superlattice with a p and n doping of about 10^{18} cm^{-3} and a translational period of about 800 Å, the band edge luminescence line at low exci-

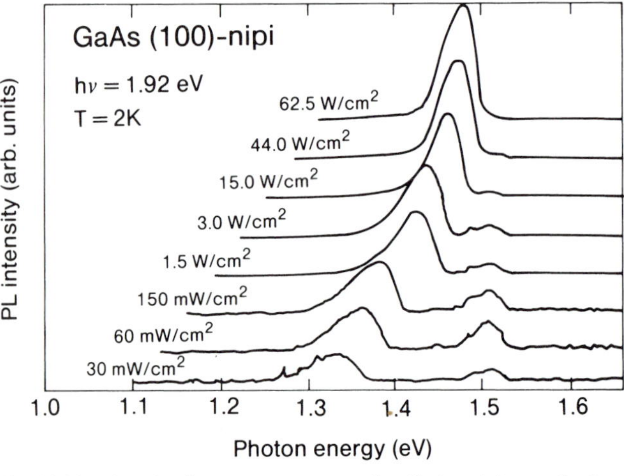

Fig. 12.32. Photoluminescence spectra of a GaAs nipi superlattice, which was deposited by metal organic molecular beam epitaxy (MOMBE) on a GaAs wafer with (100) orientation, p-doping due to carbon, n-doping due to Si. The spectra were taken with different excitation powers and a photon energy of 1.92 eV at 2 K. (After [12.11])

tation power appears slightly above 1.3 eV (Fig. 12.32), whereas the band gap in homogeneously doped GaAs at 2 K is about 1.5 eV. With increasing excitation power the luminescence line shifts to a higher energy, approaching the region of the GaAs band edge, as one would expect for a decrease in the band modulation.

Because of their completely new properties which are not present in homogeneous semiconductors, heterostructures and superlattices have opened up a new dimension in the field of microelectronics. Devices based on new concepts are being realized both in the area of very fast circuits and in optoelectronics. This whole domain is a rapidly developing field with great promises for the future [12.12].

12.8 Important Semiconductor Devices

Modern information technology is based on integrated semiconductor circuits (IC) that are used for data processing (logic), data storage (memory) and data transfer in networks, e.g. glass fiber networks, where optoelectronic devices such as detectors and optical emitters (light emitting diodes, LEDs, and lasers) are the fundamental elements. While more than 90% of the semiconductor ICs are fabricated on silicon wafers, the major material base for optoelectronic devices are the III–V semiconductors, and particularly GaAs. The reason is that the elemental semiconductor silicon permits the highest density of integration, i.e. the smallest size of devices, while the much stronger coupling of electronic transitions to the electromagnetic field in the case of direct band gap III–V semiconductors is essential for optoelectronic devices. Of all active devices, transistors are the most important ones for data processing in logic and memory circuits as well as for power electronics. Transistors are three-port devices, i.e. essentially switches, in which an input signal at one contact controls the resistance between two other contacts. Simultaneously, transistors possess amplifying properties for current and/or voltage.

The essential optoelectronic devices, such as detectors, light emitting diodes and lasers, on the other hand, are two-port devices (diodes) that feature a current flow between two contacts only. Devices can be distinguished further by another criterion: Depending on the transport mechanism, be it by only one type of carriers, electrons or holes, or by both types of carriers, devices are called unipolar or bipolar. According to this criterion two classes of transistors exist, so-called bipolar transistors and the field effect transistors (FETs) which have unipolar character. In this sense laser devices and LEDs where the light emission originates from the recombination of electrons and holes are also bipolar devices.

The Bipolar Transistor

The classical bipolar transistor, invented by Bardeen, Brattain and Shockley at the Bell Laboratories (USA) in 1947, consists of two oppositely biased *pn*-junctions (Fig. 12.33 a). Accordingly there are both, *npn*- and *pnp*-transistors; in *npn* devices the current is essentially carried by electrons, while in the *pnp*

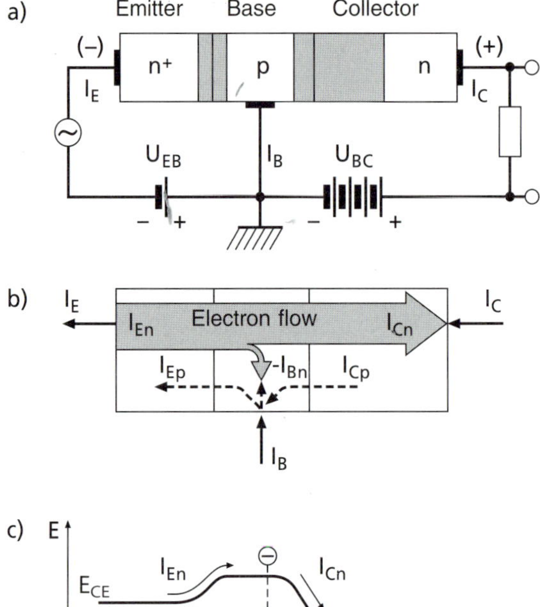

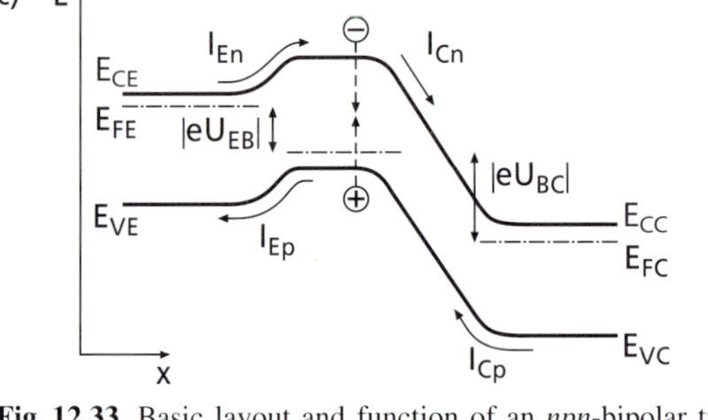

Fig. 12.33. Basic layout and function of an *npn*-bipolar transistor. (**a**) Layout of the transistor together with the external circuitry in the so-called common-base configuration; the space charge regions of the forward-biased emitter (E)-base (B) junction and the reverse-biased base (B)-collector (C) junction are shaded. (**b**) Schematic plot of the currents involved in the function of the transistor. The large shaded arrow depicts the electron current from the emitter to the collector via the base. A small portion of the current is diverted into the base. This current pattern is essential for the performance of the transistor. The directions of the currents (I_E, I_C, I_B, etc.) are drawn according to engineer's convention. (**c**) Electronic band scheme $E(x)$ along the coordinate x from the emitter to the collector, with an emitter-base voltage U_{EB} and a base-collector voltage U_{BC} applied. E_{CE}, E_{VE}, and E_{CC}, E_{VC} are conduction and valence band edges in the emitter (E) and the collector (C) regions, respectively. E_{FE} and E_{FC} are the quasi-Fermi levels in the emitter and collector regions, respectively. The electron-hole recombination process in the base region (which is detrimental to the performance) is indicated by dashed arrows

arrangement holes are the main carrier type. Nevertheless, both carrier types contribute to the device performance, as is expressed by the term "bipolar" transistor. In Fig. 12.33 a the scheme of an *npn*-transistor together with its external circuitry is shown: the first n^+p-junction (high n doping, medium p doping) is biased in the forward direction, such that the p-doped base region is flooded by a high concentration of electrons from the n^+-emitter region through the narrow, low-resistance n^+p space-charge zone. The electrons injected from the emitter into the base region are minority carriers there, and have, therefore, a high tendency to recombine with holes, which represent the majority carriers in the base region. However, in an efficient bipolar transistor the base region is small compared to the recombination length and the major part of the emitted electrons reaches, without recombination the adjacent base-collector junction (biased in the reverse direction). As is seen from the band scheme in Fig. 12.33 c these excess electrons are accelerated by the spatially extended electric space-charge field of the reverse-biased *pn* junction into the collector region. For a sufficiently narrow base region the collector current I_C is smaller than the emitter current I_E but by only a tiny amount, namely by the current I_{Bn} which is lost due to the recombination with holes (majority carriers in the base) (Fig. 12.33 b). The collector current I_C is mainly determined by the forward bias U_{EB} between emitter and base and therefore also by the base current I_B. The collector current I_C is thus essentially controlled by the base current I_B, while the reverse bias at the base-collector junction has only a negligible effect on the current I_C. The external current I_B into the base consists of three contributions: The portion I_{Bn} originates from recombination of holes with electrons being transferred from the emitter region. The hole current (majority carriers in the base) diffuses into the emitter via the valence band barrier. The third contribution is due to holes that are thermally generated within the collector region and are accelerated into the base by the strong space-charge field between base and collector (Fig. 12.33b). For the different current contributions the following relations therefore result:

$$I_E = I_{En} + I_{Ep} , \tag{12.97 a}$$

$$I_C = I_{Cn} + I_{Cp} , \tag{12.97 b}$$

$$I_B = I_E - I_C = I_{Ep} + I_{Bn} - I_{Cp} . \tag{12.97 c}$$

Since the base current I_B is typically below 1% of the collector current I_C the collector current, which flows as an emitter-base current through the base region into the collector, is approximately represented by the relation for a p–n junction (12.74) as

$$I_C \approx I_E \approx A \frac{eD_n n_B}{L_n} [\exp(eU_{EB}/kT) - 1] . \tag{12.98}$$

Here, A is the current-carrying area of the transistor, D_n the diffusion constant of electrons in the base region, n_B their equilibrium concentration therein, and L_n the electronic diffusion length in the p-doped base. Since the diffusion length L_n is related to the diffusion time τ_n according to $L_n = \sqrt{D_n \tau_n}$, the ratio D_n/L_n can be considered as a diffusion velocity of electrons in the base. The collector and emitter current (12.98) can therefore be interpreted as a product of the number of electrons injected into the base and their diffusion velocity. This diffusion velocity thus determines predominantly the speed of the transistor. The exponential increase of the collector current I_c with base-emitter voltage U_{EB} is plotted in Fig. 12.34b. This plot is usually called the *transfer characteristics*. The so-called *output characteristics* $I_C(U_{CE})$ are shown in Fig. 12.34c. They represent essentially the behavior of the reverse-biased base-collector p–n junction, since the base-emitter resistance is small. The output characteristic is therefore the reverse bias currents of a p–n junction, controlled in magnitude by the emitter-base voltage U_{EB}.

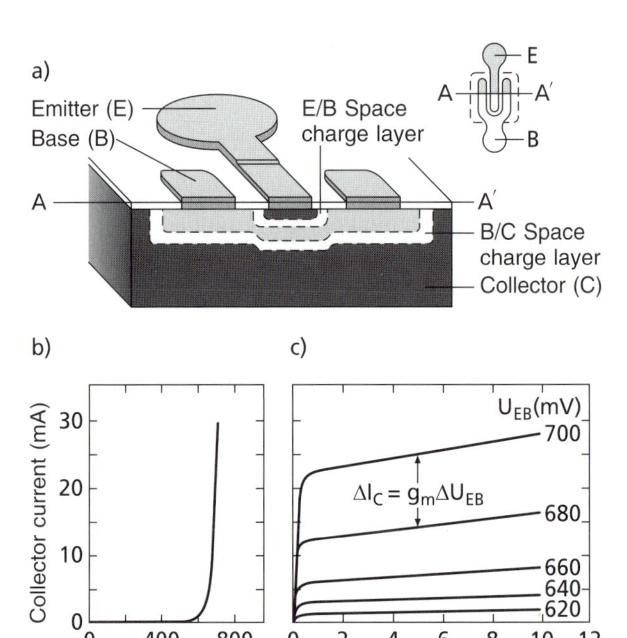

Fig. 12.34. (**a**) Layout of a planar bipolar transistor. The differently doped emitter, base, and collector regions are shaded according to the concentration of dopants, with black denoting the highest concentration. The space-charge layers *E/B* and *B/C* are drawn as unshaded. The top view on the metallic emitter (*E*) and base (*B*) contact in the inset indicates the position of the cross section through the layer structure along *AA′*. (**b**) Input characteristic of a bipolar transistor (after Tietze and Schenk [12.13]). (**c**) Output characteristics of a bipolar transistor with emitter-base voltage U_{EB} as a parameter. An important figure of merit of a transistor is the derivative of the collector current I_C with respect to the emitter-base voltage, the *transconductance* $g_m = \partial I_C / \partial U_{EB}$. (After Tietze and Schenk [12.13])

Depending on the different applications, bipolar transistors can be used in connection with three circuitry configurations [12.13]. The common base configuration in which the base is at ground potential was used in the considerations to Fig. 12.33. There, the collector output current I_C is always smaller than the input current I_E by the (small) amount I_{Bn} ($I_{Bn} \ll I_C$). Nevertheless, significant voltage and power amplification can be achieved since the input circuit of the transistor contains the low-resistance emitter-base p–n junction biased in the forward direction, while the output current I_C ($\approx I_E$) flows through the high resistance of the base-collector p–n junction biased in the reverse direction. This permits the use of a high load resistance in the output circuitry with large voltage drop. More frequently used is the emitter configuration where the emitter contact of the transistor is at ground potential. In this configuration, the base current I_B, small in comparison with I_E and I_C, controls the output collector current I_C. The current amplification of this configuration is obvious.

Apart from current amplification the parameters known as *transconductance* $g_m = \partial I_C / \partial U_{EB}$ and *emitter efficiency* $\gamma = \partial I_{En} / \partial I_E$ are the principle figures of merit of a bipolar transistor. The emitter efficiency indicates what percentage of the total emitter current I_E is carried by the majority carriers rather than by holes from the base region. In order to increase the speed of the transistor one has to enhance the conductance of the base region, i.e. the doping level of the base. This, on the other hand, increases the back current of holes I_{Ep} into the emitter and thus causes a decrease of the emitter efficiency γ. A way out of this problem is provided by modern *hetero-bipolar transistors* (HBTs), where the emitter region is formed by a semiconductor layer of a wide band gap material deposited epitaxially on the base-collector system formed by lower band gap materials (e.g., AlGaAs on GaAs). Given a significant valence band offset (Sect. 12.7) this valence band barrier reduces the back-stream of holes from the base into the emitter, despite a large p-doping of the base. Similar concepts are realized by using low band gap semiconductors as the base material in connection with emitter layers possessing a larger band gap.

Field Effect Transistors (FET)

A field effect transistor (FET) is essentially a resistor that is controlled by an external bias voltage. Its architecture is rather simple. A FET consists of a current channel with two contacts, the source and the drain, and a third contact, the gate, which is separated from the current channel by an insulating barrier. The gate contact can therefore be biased with respect to the conducting channel. Depending on the polarity and magnitude of the gate voltage more or less charge is induced in the channel, and the channel conductance is thereby modified. Since only one type of carriers is controlled by the external bias the FET is a unipolar device. There are three main types of FETs (Fig. 12.35), which are distinguished according to the nature of their channels and gate barriers. The most important FET, which is the

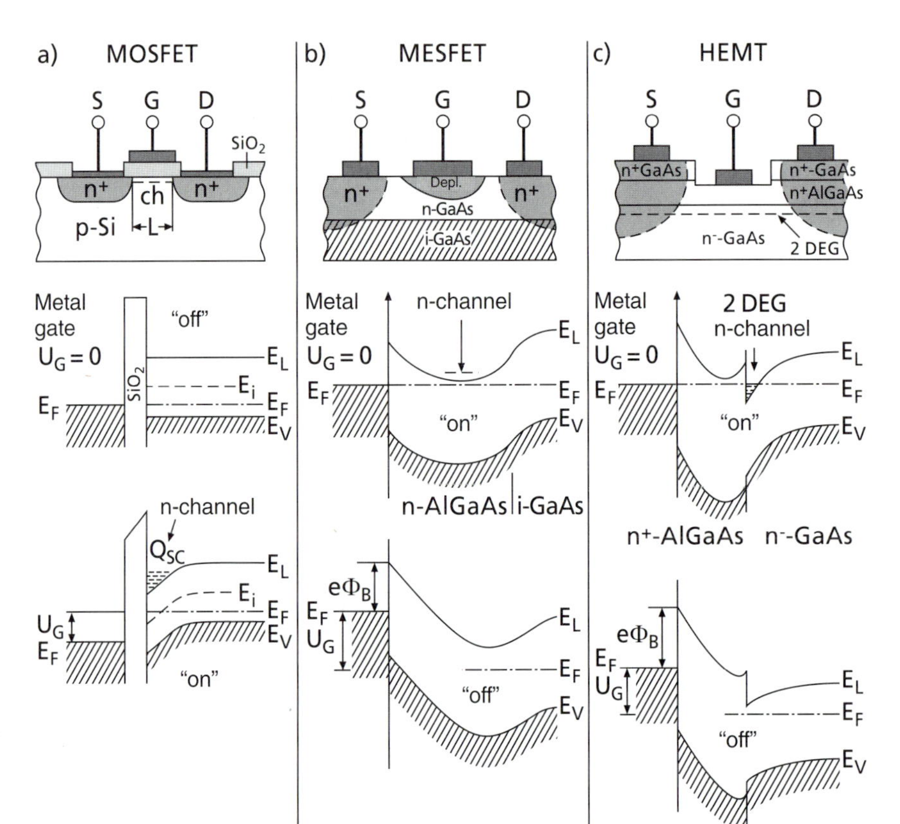

Fig. 12.35. Schematic layout and electronic band schemes for the three most important types of field effect transistors (FET). The figure displays the cross section perpendicular to the wafer surface below the metallic gate electrode. The symbols S, G, and D denote source, gate, and drain, respectively. (**a**) MOSFETs (metal-oxide-semiconductor FET) can only be realized on a Si basis with SiO_2 as gate oxide. The conducting channel and its length are denoted by *ch* and *L*, respectively. For technological reasons the metallic gate electrode is fabricated from highly degenerate doped polycrystalline Si (poly Si). (**b**) MESFETs (metal semiconductor FET) are preferentially realized on III–V semiconductor basis, e.g., GaAs. (**c**) HEMTs (high electron mobility transistor) preferentially realized on III–V semiconductor heterostructures, e.g., AlGaAs/GaAs; the conducting channel is formed by a two-dimensional electron gas (2 DEG) that is connected to the source and drain contacts by highly n^+-doped regions. The band schemes are plotted for the case of zero gate voltage (U_G=0) (upper panel) and for positive and negative gate voltages U_G, for MOSFETs and MESFETs and HEMTs, respectively (lower panel); the (quasi-) Fermi level, the conduction band edge, the valence band edge, and the intrinsic energy level are denoted as E_F, E_C, E_V, and E_i, respectively

basis for all highly integrated silicon circuits, be it logic or memory circuits, is the MOSFET (**m**etal **o**xide **s**emiconductor FET). The schematic cross section in Fig. 12.35 a displays the essential features: Source and drain regions are formed by highly n^+-doped wells that are generated by ion implantation in a moderately p-doped wafer. The two n^+ wells are therefore separated by a p-type region. For any applied voltage between source and drain either one of the two *p*–*n* junctions is biased in the reverse direction allowing only a small current between source and drain. The *p*-type region between the n^+-source and drain is separated from the metallic gate (G) contact by a very thin gate oxide (SiO_2). A bias voltage applied to the gate gives rise to a large electric field across the gate oxide layer that in turn causes a band bending in the underlying semiconductor. For a sufficiently positive gate voltage U_G the initial *p*-region (at $U_G=0$) below the gate is inverted ($E_i<E_F$) to become an *n*-accumulation layer. The two source and drain n^+-wells are thence connected by a conductive *n*-accumulation channel. The transistor has switched from "off" to "on".

The performance of a MOSFET depends essentially on the perfection of the gate SiO_2 layer and its interface to the underlying Si channel. Any residual trap defect states in the oxide or at the Si/SiO_2 interface would be detrimental to the performance since the charge influenced by the gate voltage would merely reside in the defects rather than affecting the inversion channel. Since only Si is able to form thin, but nevertheless tight and stoichiometrically almost ideal oxides, with atomically smooth interfaces to the Si substrate MOSFETs can be fabricated only on Si wafers. III–V semiconductors such as GaAs, while being superior to Si with regard to electron mobility and thus device speed, do not form perfect oxide overlayers. Even if a high performance of the insulator were granted, the interface between the GaAs substrate and the insulating epilayer exhibits such a high density of interface traps (partially due to non-stoichiometry) that MOSFET structures based on GaAs are very ineffective. The MESFET (**m**etal **s**emiconductor **f**ield **e**ffect **t**ransistor) is therefore the most common FET on III–V compounds (Fig. 12.35 b). A highly resistive, intrinsic (i) or *semi-insulating* GaAs (achieved by doping with Cr) with the Fermi level E_F located near the midgap position serves as the substrate on which a thin, crystalline *n*-doped GaAs layer is epitaxially deposited or generated by ion implantation. Due to its high free electron concentration arising from ionized donors the lower conduction band edge in this epilayer lies close to the Fermi level E_F and a conducting *n*-channel is formed that is separated from the metallic gate by the Schottky barrier below the metal/GaAs junction (Sect. 12.6). Since in this metal/GaAs junction the Fermi level is pinned near midgap at the interface, the electronic bands are bent upwards with respect to the conducting *n*-channel. The metallic gate is thus separated from the *n*-channel by a depletion region of high resistance. In contrast to the *p*-channel, source and drain contacts are highly conductive due to a local high n^+ doping below the metallization layer. Thus, a conductive channel between source and drain exists for $U_G=0$ in the arrangement of Fig. 12.35 b. This channel can be pinched off by applying a negative voltage U_G to the gate because the negative bias lifts the Fermi level E_F of the metallic gate with respect to E_F in the i-GaAs. Since on the other hand E_F is pinned near midgap ($\phi_B=$const) at the interface between the gate electrode and the *p*-channel the entire band scheme is lifted by the applied bias voltage U_G, and the channel becomes depleted of electrons and thus highly resistive. Hence, the transistor switches from "on" to "off" when subjected to a sufficiently high negative gate voltage U_G. The performance depends sensitively on the thickness of the n-doped layer as compared with the extension of the Schottky depletion space-charge layer (typically 50 nm) under the gate metal contact.

Complementary to the "normally on"-MESFET described above, the "normally off" MESFET is designed to possess a thin n-GaAs channel with a thickness comparable to the extension of the Schottky depletion layer. A sufficiently high positive gate bias bends the electronic bands below the metal gate downwards and opens the channel; the transistor switches to the "on" state only under this positive gate bias.

Since semiconductor heterostructures of high quality can be produced by modern epitaxy techniques such as MBE and MOCVD (panel XVII) the MESFET concept has been extended to a heterostructure FET, the so-called HEMT (**h**igh **e**lectron **m**obility **t**ransistor), which is extremely fast (20–300 GHz) and exhibits very low noise (Fig. 12.35c). Its major fields of application are therefore radar and satellite communication. As a "heterostructure MESFET" the HEMT is switched by a Schottky barrier, separating the metal gate contact electrically from the conducting channel. The conduction channel, however, is a modulation-doped 2DEG at the interface of an AlGaAs/GaAs heterostructure in this case (Sect. 12.7). The high electron concentration of typically $2\times10^{12}\,cm^{-2}$ in the 2DEG originates from donors in the n^+-AlGaAs barrier that are spatially separated from the 2DEG. The conducting 2DEG channel is thus free of ionized scattering centers which results in low noise and high carrier mobility.

A possible modification that exists for all three types of FETs is to invert the type of doping (from p to n). These "inverted" FETs exhibit the same transistor functions, albeit with inverted external bias (plus to minus). Holes rather than electrons then carry the transistor function. Holes have higher effective masses than electrons, their mobility is lower and therefore p-channel FETs are slower than their n-channel analogues. Nevertheless, nearly all advanced silicon ICs consist of combinations of p- and n-channel MOSFETs (CMOS = complementary MOS). The combination of both transistor types allows the realization of logic gates that carry current only within the switching period between two bit-operations. CMOS circuits, therefore, consume less power, which is of significant advantage in the ever-increasing integration density.

FETs show amplification since very small gate currents induce sufficient charge on the gate such that much larger source-drain currents are controlled. Since the resistance of the SiO_2 layer in the case of the MOSFET is much higher than the resistance of the Schottky depletion space-charge layer under the gate of a MESFET or a HEMT, gate currents in MOSFETs are lower than in MESFETs and HEMTs. Characteristic figures of merit for a FET are its transconductance g (the ratio between drain current and the applied change in the gate voltage) and the gate capacity C_g. The ratio between transconductance and gate capacity determines the cut-off frequency $f_{max} = g/C_g$, up beyond which the power amplification drops below 1. In order to improve the high-frequency performance of a FET, its gate capacity should be decreased and the highest possible transconductance should be achieved. Fast, high-performance HEMTs in the material system GaInAs/InP reach transconductances g of about 600 mS/mm and upper cut-off frequencies $f_{max} \approx 300$ GHz at channel lengths of 0.1 μm [12.14].

Typical source-drain current characteristics (I_{SD} vs. U_{SD}) of FETs, in particular of MOSFETs, are shown in Fig. 12.36, with the gate voltage U_G as a parameter. For small U_{SD}, where the inversion space-charge zone along the channel is essentially not affected by a change in U_{SD}, the current I_{SD} is nearly proportional to the applied voltage U_{SD} (quasi-ohmic behavior). Increasing source-drain voltage U_{SD} causes a significant potential drop along the channel within the semiconductor, while the potential on the metallic gate contact remains constant. The voltage drop between the metallic gate contact and the channel (perpendicular to the direction of the channel) thus increases along the channel. This causes a decrease of the channel width (12.80) along the channel, which results in a saturation of I_{SD} for voltages $U_{SD}>2$ V. The dependence of the saturation current I_{SD} on the gate voltage U_G differs for short- and long-channel transistors. In short channels (Fig. 12.36a), high electric fields of the order of 10^4 to 10^5 V/cm are reached and the carriers are accelerated up to their saturation velocity v_s (Fig. 12.15). Consequently, the current I_{SD} grows proportionally with the

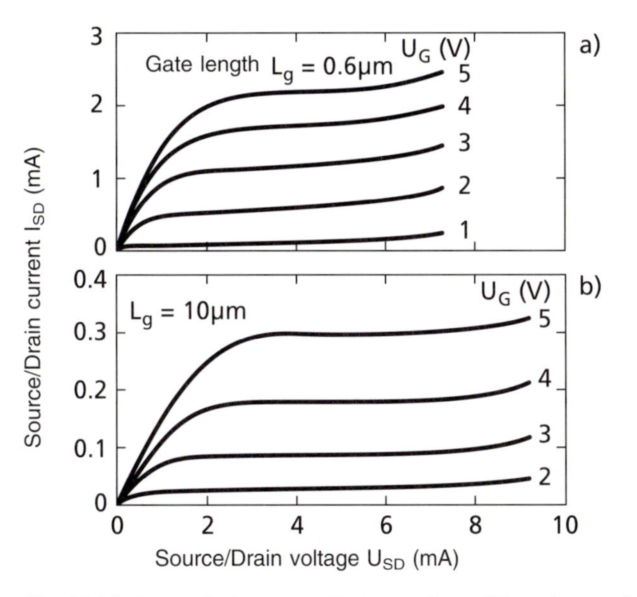

Fig. 12.36. Source-drain current (I_{SD}) vs. voltage (U_{SD}) characteristics of Si MOS-field effect transistors (MOSFETs): $I_{SD}(U_{SD})$ curves are plotted with the gate voltage U_G as a parameter. (**a**) Short-channel MOSFET with gate length L_g=0.6 μm and gate width 5 μm. (**b**) Long-channel MOSFET with gate length L_g=10 μm and gate width 5 μm (after Mühlhoff and McCaughn [12.15])

charge induced on the gate and is thus proportional to U_G (Fig. 12.36a). In the case of long channels, one has to consider the fact that the potential difference between gate and channel varies along the channel. In the saturation limit (high source/drain voltage) the gate may be positively biased with respect to the channel near the source, however, still negatively biased with respect to the channel near the drain. As a consequence, the channel is quenched to a high resistance near the drain. A more positive bias on the gate therefore changes not only the carrier concentration locally but also reduces the length where the channel is effectively quenched. In that case a nearly quadratic dependence of the source-drain current I_{SD} on the gate voltage results (Fig. 12.36b). Similar current-voltage characteristics I_{SD} (U_{SD}) as discussed here for MOSFETs are also found for MESFETs and HEMTs.

Semiconductor Lasers

The performance of optoelectronic devices such as photodetectors, solar cells, light emitting diodes (LED) and lasers is based on the interaction of the electromagnetic light field with electronic excitations in the semiconductor. Three different interaction processes exist: (i) an absorption of light quanta by an excitation of electrons from the valence band into the conduction band, or between electronic defect levels within the forbidden band, (ii) an inverse process, the stimulated emission, in which the incident photon stimulates an electron to be de-excited from a state of higher energy into a state of lower energy, and (iii) the spontaneous emission of light. Both, absorption and stimulated emission depend linearly on the strength of the electric field $\mathscr{E}(\omega)$ of the electromagnetic wave (Sect. 11.10). The stimulated emission is coherently coupled with the light field so that the emitted photon is in phase with the stimulating electromagnetic wave. In contrast, spontaneous emission is not coherently coupled to the light field. All three types of interactions are used in photoelectric devices: For photodetectors and solar cells the absorption process is essential. Light emitting diodes (LED) utilize spontaneous emission, while lasers are based on stimulated coherent emission. In all three processes energy conservation is obeyed

$$\hbar\omega = E_2 - E_1 \ , \tag{12.99}$$

with $\hbar\omega$ the photon energy and E_2 and E_1 the energies of the excited and ground state, respectively. Under stationary illumination with light the transition rates must compensate each other. With A_{21} as the probability for spontaneous emission and B_{21} and B_{12} the probabilities for stimulated emission and absorption, respectively, the stationary state is described by

$$\dot{n}_2 = -B_{21}n_2|\mathscr{E}(\omega)|^2 - A_{21}n_2 + B_{12}n_1|\mathscr{E}(\omega)|^2 = -\dot{n}_1 = 0 \ , \tag{12.100}$$

where n_1 and n_2 are the occupation numbers of the ground state and the excited state, respectively. Apart from a pre-factor the coefficients B_{21} and B_{12}

are equal to the square modulus of the transition matrix element $|x_{ij}|^2$ for stimulated emission and absorption (11.93) and are therefore equal to each other (11.93). The ratio of the stimulated and spontaneous emission rates is

$$\dot{n}_{2\mathrm{stim}}/\dot{n}_{2\mathrm{spon}} = \frac{B_{21}}{A_{21}}|\mathcal{E}(\omega)|^2 \ . \tag{12.101}$$

Since laser action requires an excess of stimulated emission processes, a laser is built as a standing-wave resonator for the wavelength to be emitted. Thereby, a high electric field within the lasing medium is generated. Furthermore, the rate of stimulated emission must exceed that of absorption in a laser. Neglecting spontaneous emission because of (12.101) at high light fields $\mathcal{E}(\omega)$ and using the equality B_{21}=B_{12} one obtains from (12.100) the following laser condition:

$$\dot{n}_2 = -(n_2 - n_1)B_{21}|\mathcal{E}(\omega)|^2 < 0 \quad \text{i.e.} \quad n_2 > n_1 \ . \tag{12.102}$$

This condition describes the *inversion* of the occupation (population) statistics (see Problem 11.8). In thermal equilibrium, just the opposite would hold: the occupation of the excited states is less than the occupation of the ground state. Population inversion is achieved by "pumping" into the excited state. In a semiconductor laser a convenient method for pumping is to bias a p–n junction in the forward direction and to flood the space-charge region with non-equilibrium electrons and holes. In writing (12.100–12.102) it was assumed that the final states of the electronic transitions were unoccupied, i.e. their probability to be empty was "one". While this assumption is correct for an ensemble of single atoms in a gas laser, where the occupation of an excited state results from depopulation of the ground state the laser condition has to be modified for the valence band (E_V) and the conduction band states (E_C) of a semiconductor laser. Here, the transition rates are proportional to the density of *occupied initial* and the density of *empty final* states. The condition for population inversion in a p–n junction biased in the forward direction thus follows as

$$\begin{aligned} &- B_{21}D_\mathrm{C}(E_\mathrm{C})D_\mathrm{V}(E_\mathrm{V})f(E_\mathrm{C})[1 - f(E_\mathrm{V})] \\ &+ B_{12}D_\mathrm{V}(E_\mathrm{V})D_\mathrm{C}(E_\mathrm{C})f(E_\mathrm{V})[1 - f(E_\mathrm{C})] < 0 \ . \end{aligned} \tag{12.103}$$

In (12.103) the occupied and empty energetic regions of the conduction and valence band, respectively, are represented by sharp energy levels E_C and E_V, the integrals over the corresponding energetic ranges ($\sim kT$) are approximated by constant densities of states $D_\mathrm{C}(E_\mathrm{C})$ and $D_\mathrm{V}(E_\mathrm{V})$, respectively. $f(E)$ is the Fermi distribution function, which can not be approximated by the Boltzmann distribution at high doping levels. Because $B_{21} = B_{12}$ (12.103) gives:

$$f(E_\mathrm{C})[1 - f(E_\mathrm{V})] > f(E_\mathrm{V})[1 - f(E_\mathrm{C})] \tag{12.104a}$$

$$f(E_\mathrm{C}) > f(E_\mathrm{V}) \ . \tag{12.104b}$$

The occupation probabilities can be approximated by Fermi functions with E_F^n, E_F^p as quasi-Fermi levels in the n- and p-region, respectively (Fig. 12.37)

$$f(E_C) = [1 + \exp(E_C - E_F^n)/kT]^{-1} , \qquad (12.105\,a)$$

$$f(E_V) = [1 + \exp(E_V - E_F^p)/kT]^{-1} . \qquad (12.105\,b)$$

Hence, the condition for population inversion becomes

$$E_F^n - E_F^p > E_C - E_V = E_g . \qquad (12.106)$$

For laser operation, the quasi-Fermi levels in the n- and p-doped region must energetically be separated from each other by more than the band gap energy E_g. The p- and n-type regions, therefore, have to be doped deep into degeneracy, in order to reach population inversion by an externally applied forward bias $U=(E_F^n - E_F^p)/e$ (Fig. 12.37 a, b). The current through the p–n junction thus defines the onset of laser action. The emitted light power plotted versus current density through the p–n junction increases slowly due to spontaneous emission, until population inversion is reached at a certain *threshold current density* and laser action starts. At this point stimulated emission becomes the determining mechanism (Fig. 12.38). Threshold

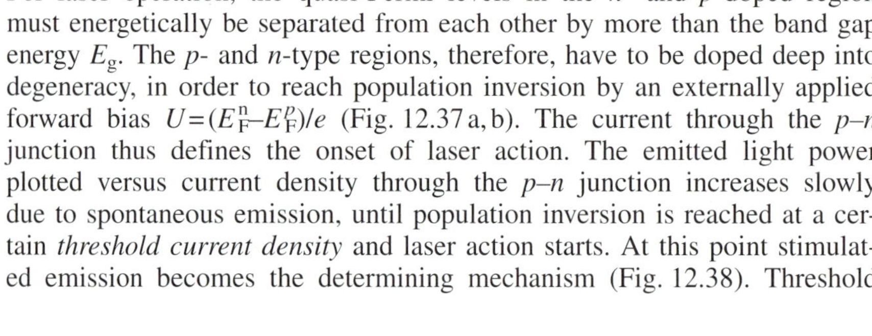
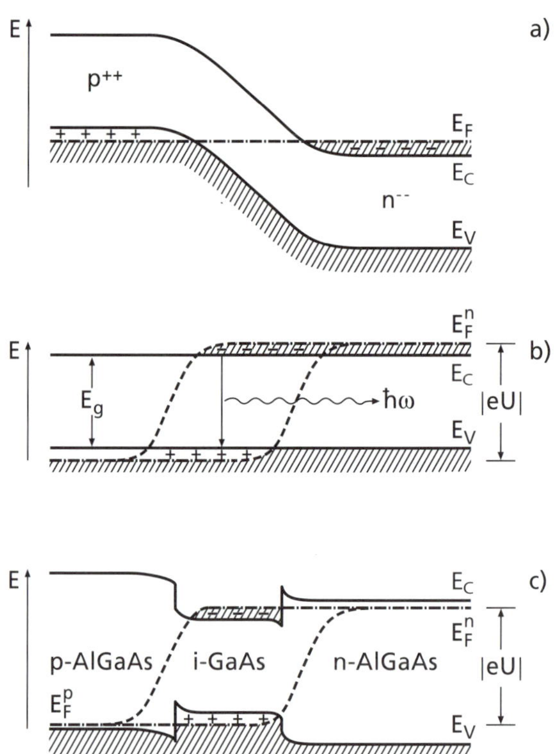
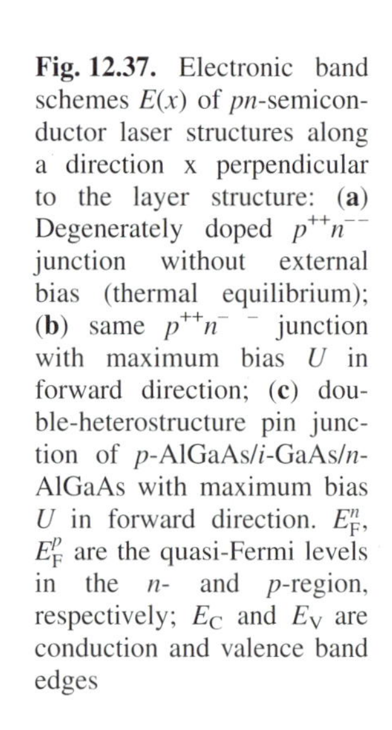

Fig. 12.37. Electronic band schemes $E(x)$ of pn-semiconductor laser structures along a direction x perpendicular to the layer structure: (**a**) Degenerately doped $p^{++}n^{--}$ junction without external bias (thermal equilibrium); (**b**) same $p^{++}n^{--}$ junction with maximum bias U in forward direction; (**c**) double-heterostructure pin junction of p-AlGaAs/i-GaAs/n-AlGaAs with maximum bias U in forward direction. E_F^n, E_F^p are the quasi-Fermi levels in the n- and p-region, respectively; E_C and E_V are conduction and valence band edges

current density is thus the major figure of merit of a semiconductor laser; the lower it is, the more effectively the laser performs. Too high threshold current densities cause too much energy dissipation (Joule heating) which, aside from making the laser less efficient, reduces the lifetime of a laser considerably.

An important breakthrough leading to the commercialization of semiconductor lasers was the invention of the double heterostructure laser. Figure 12.37 c shows the electronic band scheme of such a laser, in this case an AlGaAs/GaAs/AlGaAs heterostructure. A lowly doped or intrinsic (i)-GaAs layer is inserted into a *p–n* junction between two *p*- and *n*-doped wide band gap AlGaAs regions. In spite of the low *p*- and *n*-type doping in the AlGaAs regions on both sides the quasi-Fermi levels in the active (i)-GaAs region are located within the conduction and valence band of the GaAs layer under strong forward bias. Population inversion is thus easily achieved without employing highly doped, degenerate semiconductors and the high current densities that accompany the use of degenerate semiconductors. The active *i*-GaAs region is flooded with non-equilibrium electrons and holes, which are confined to the active region by the conduction and valence band discontinuities, respectively. This effect, which causes enhanced light emission is called "*electrical confinement*". Additionally, this structure provides an "*optical confinement*", since for the wavelength of the emitted light (which corresponds to the GaAs band gap) the refractive index of the active GaAs region exceeds that of the adjacent AlGaAs layers (see Problem 11.9). The light originating from stimulated emission in the GaAs is therefore totally reflected at

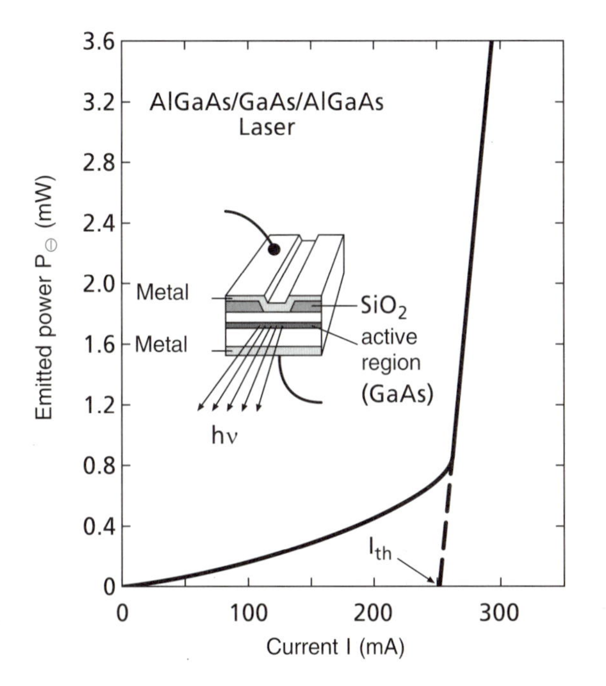

Fig. 12.38. Emission characteristics at room temperature $T=300$ K of an AlGaAs/GaAs/AlGaAs double-heterostructure laser: the emitted light power is plotted as a function of the current I through the laser; I_{th} is the threshold current where laser action begins. Inset: Schematic layer structure of the laser with emitted radiation ($h\nu$). (After Kressel and Ackley [12.16])

the AlGaAs layers; it remains in the active GaAs channel, and is focused and concentrated. The schematic representation of a double heterostructure laser device (inset in Fig. 12.38) shows that the active GaAs channel (resonator) is spatially confined by the two AlGaAs layers on the top and bottom of the GaAs layer, and laterally by a metal stripe deposited into an opening in the SiO_2 mask. The metal contacts the semiconductor layer structure only within this stripe whereby the electric field and the current flow is laterally limited to the width of the stripe.

The optical resonator, i.e. the active GaAs channel (Fabry–Pérot interferometer) requires effective semi-transparent mirrors at its front and back side, through which the light is emitted or reflected. In the simplest case these mirrors are realized by cleavage planes of the GaAs wafer along (110). Typical lengths of these active channels range between 100 and 1000 μm.

The emitted light power as a function of the external current (Fig. 12.38) slowly increases at lower currents due to spontaneous emission. In this current regime the amplification is too low to overcome losses in the resonator. At a certain threshold current I_{th} the amplification of the device matches the resonator losses and laser action starts, and the output light power increases abruptly.

An important figure of merit of a semiconductor laser is thus a threshold current as low as possible (or threshold current density) arising from low losses in the resonator. With increasing temperature the electrical confinement of the laser deteriorates, since carriers populate higher energy states because of a less sharp Fermi edge. The threshold current increases and the laser performs less efficiently. Since Boltzmann statistics are approximately valid, the threshold current I_{th} depends exponentially on the working temperature T:

$$I_{th} \propto \exp(T/T_0) \ . \tag{12.107}$$

A further figure of merit of a semiconductor laser is the critical temperature T_0, which should be as high as possible, so that the laser shows a slow increase of the threshold current with working temperature T. Typical values of T_0 are close to 100°C for double-heterostructure lasers.

In the double-heterostructure lasers (DHL) considered so far, the vertical extension of the active channel, i.e. the thickness of the embedded GaAs layer is much higher than 200 Å. Carriers in the active zone occupy quantum states of a macroscopic 3D crystal. Further improvement of the laser performance is achieved by incorporating into the active zone even thinner layers (thickness $\leq$100 Å) of a semiconductor with a lower band gap than that of GaAs, e.g., InGaAs. Since $In_xGa_{1-x}As$ is not lattice-matched to GaAs these interlayers are highly strained. They can only be grown without dislocations up to a particular critical thickness. Due to their smaller band gap, i.e. their band offsets with respect to GaAs both in the conduction and valence bands, these thin InGaAs interlayers act as quantum wells for electrons as well as for holes. The electronic states in the conduction and in the valence band are quantized perpendicular to the layer sequence. The density of states within these layers for each subband is constant and the superposition for all sub-bands is step-like (Fig. 12.28c). Carriers that occupy the lowest possible energy states are thus even more spatially confined and on even sharper energy levels than in the double heterostructure laser. These so-called *quantum-well lasers* exhibit even better confinement and therefore lower threshold currents and larger critical temperatures T_0.

Problems

12.1 Silicon crystallizes in the diamond structure with a lattice constant $a = 5.43$ Å.

a) How many atoms does the cubic elementary cell contain?
b) Phosphorus atoms incorporated on Si sites act as donors. Calculate the ionization energy of such a donor impurity within the framework of the simple hydrogenic model (dielectric constant $\varepsilon_{Si} = 12$).
c) Calculate the radius r_0 of the first Bohr orbital. What is the probability of finding the electron in its ground state within a sphere of radius $2r_0$? How many Si atoms are located within such a sphere?
d) Discuss the validity of the hydrogenic model and explain its shortcomings.
e) At what impurity concentration does interaction between the donor centers become important? The related impurity band conductivity is assumed to correspond to an average donor separation of $4r_0$.

12.2 The effective mass of electrons at the lower conduction band edge of a semiconductor is three times higher than that of holes at the upper valence band edge. How far is the Fermi energy E_F located from the middle of the forbidden band in the case of intrinsic conduction. The band gap energy E_g must exceed $8kT$; why is this a prerequisite for the calculation?

12.3 A semiconductor with a band gap energy E_g of 1 eV and equal hole and electron effective masses $m_e^* = m_h^* = m_0$ (m_0 is free electron mass) is p-doped with an acceptor concentration of $p = 10^{18}\,\text{cm}^{-3}$. The acceptor energy level is located 0.2 eV above the valence band edge of the material.

a) Show that intrinsic conduction in this material is negligible at 300 K.
b) Calculate the conductivity σ of the material at room temperature (300 K), given a hole mobility of $\mu_p = 100\,\text{cm}^2/\text{Vs}$ at 300 K.
c) Plot the logarithm of the hole concentration, $\ln p$, versus reciprocal temperature $1/T$ for the temperature range 100 to 1000 K.

12.4 Consider the electronic bandstructure of a typical direct-gap III–V semiconductor like GaAs with a high-mobility, sharp conduction band minimum at Γ and a broad side minimum at somewhat higher energy (cf. Fig. 12.4).

a) Discuss qualitatively the dependence of the average drift velocity of free electrons on the strength of an externally applied electric field taking into account scattering into the side valley of the conduction band.
b) Give a similar discussion of an indirect-gap semiconductor such as Si, where only the broad side minimum of the conduction band is occupied (Fig. 7.13).

12.5 Calculate the Hall coefficient R_H for an intrinsic semiconductor in which electric current is carried by both electrons and holes.
a) Discuss the temperature dependence of the Hall coefficient $R_H(T)$ of the intrinsic semiconductor, when hole and electron mobilities μ_p and μ_n are assumed to be equal.
b) In an experimental measurement of the Hall voltage, reversing the sense of the magnetic field yields a different magnitude for this quantity. Explain the origin of this effect.

12.6 Consider the process of photoconductivity in semiconductors. Concentrations of electrons and holes above those of thermal equilibrium are produced from defect sites or from the valence band by means of photoionization.

Write down the rate equations for the direct recombination of electron–hole pairs for the cases of (a) intrinsic conduction and (b) *n*- or *p*-doping.

Calculate the stationary photocurrent at constant illumination and the time dependence of the photocurrent after the illumination is switched off.

12.7 Explain the operation of a solar cell in which a *p*–*n* junction is illuminated. Assume a closed circuit containing an external resistance R. Make qualitative sketches of the current–voltage relationships with and without incident light. What is the minimum frequency which the light must have? What is the theoretical efficiency as a function of the frequency of the light? How can the current–voltage characteristic be used to obtain the photoelectrically generated power? Discuss the various parameters that influence the efficiency. Why is it advantageous to combine several *p*–*n* junctions of semiconductors with different band gaps using thin-film technology?

12.8 Small-gap semiconductors such as InAs (E_g=0.35 eV), and InSb (E_g=0.18 eV) usually exhibit Fermi-level pinning within the conduction band (approximately 100 meV above the lower conduction band edge E_C for InSb). Plot qualitatively the band scheme (band energy versus spatial coordinate) in the vicinity of a metal contact to such a semiconductor that is highly *n*-doped. What is the electrical resistance behavior for both bias directions?

12.9 Calculate the transconductance $g_m = (\partial I_{SD}/\partial U_G)$ as a function of gate capacity C_g and gate length L_g for a field effect transistor (MOSFET, MESFET or HEMT, Sect. 12.8). Note that even without an external gate bias ($U_G = 0$) a "built-in" band bending under the gate with a threshold voltage U_0 is present. For short-channel transistors one can assume in a simple approximation that the carrier velocity in the channel is constant and equal to the saturation velocity v_s for high field strength (Fig. 12.15). The source-drain current I_{SD} consists of the 2-dimensional (2D) carrier density n_{2D} which is induced in the channel by the action of the gate bias U_G.

a) Using Fig. 12.15, estimate the transit time for carriers under the gate for a short channel AlGaAs/GaAs HEMT with gate length $L_g = 100$ nm.
b) What gate capacity can be reached in these transistors for a transconductance $g_m = 150$ mS?

Panel XIV
The Hall Effect

To separately determine the carrier concentration n and the mobility μ appearing in the conductivity $\sigma = n e \mu$, one measures both the conductivity and the Hall effect. A current i is passed through a crystal sample and a magnetic field (magnetic induction $\boldsymbol{B}$) is applied normal to the current. Using two contacts placed opposite to one another (perpendicular to i and $\boldsymbol{B}$), a so-called Hall voltage U_H can be measured (open circuit measurement, $i_H = 0$). This is shown schematically in Fig. XIV.1. Since measured in the absence of current, the Hall voltage U_H which builds up in the sample in the y direction is exactly compensated by the Lorentz force on an electron, which also acts in the y direction. The Lorentz force on an electron moving in the x direction with velocity v_x is:

$$\begin{aligned} F_y &= -e(\boldsymbol{v} \times \boldsymbol{B})_y - e\mathscr{E}_y \\ &= e v_x B - e\mathscr{E}_y = 0 \,. \end{aligned} \quad \text{(XIV.1)}$$

Here $\mathscr{E}_y = U_H/b$ is the so-called Hall field. Assuming that the current is carried exclusively by electrons (n-type semiconductor or metal), we have

$$j_x = I/(b\,d) = -n e v_x \quad \text{(XIV.2)}$$

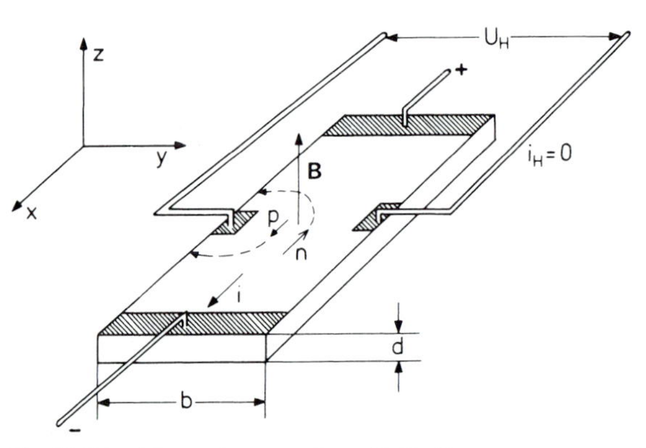

Fig. XIV.1. Schematic set-up for a Hall effect measurement. $\boldsymbol{B}$: magnetic induction; i: current through the sample; U_H: Hall voltage. The dashed lines are the paths which electrons and holes would follow in the absence of a compensating circuit for U_H and with a current flow ($i \neq 0$). For $i_H = 0$, the carriers are forced to move in straight lines (parallel to x) due to the build-up of the Hall voltage U_H

and it thus follows that

$$\mathscr{E}_y = \frac{U_H}{b} = -\frac{1}{ne} j_x B = -\frac{1}{ne}\frac{IB}{bd} . \tag{XIV.3}$$

The quantity $R_H = -(ne)^{-1}$ is called the Hall constant. It can be determined from measurements of U_H, i and B via the relation

$$U_H = R_H I B/d . \tag{XIV.4}$$

The sign of R_H gives the type of carrier (a negative sign corresponding to electrons) and its absolute value gives the carrier concentration n.

If the current in a semiconductor is carried by both electrons (concentration: n, mobility: μ_n) and holes (p, μ_p), then an analogous calculation gives the following expression for the Hall constant (Problem 12.5)

$$R_H = \frac{p\mu_p^2 - n\mu_n^2}{e(p\mu_p + n\mu_n)^2} . \tag{XIV.5}$$

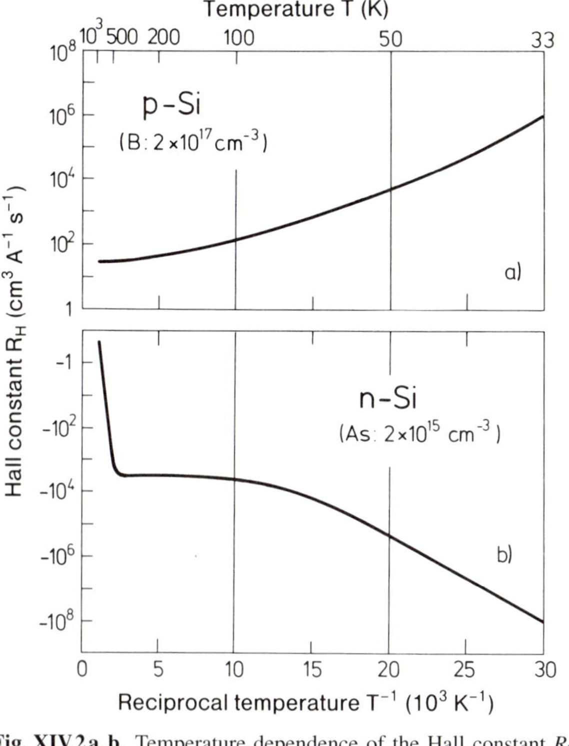

Fig. XIV.2 a, b. Temperature dependence of the Hall constant R_H for p-type (**a**) and n-type silicon (**b**). For p-type with a boron concentration of 2×10^{17} cm^{-3} intrinsic conductivity sets in at temperatures in the vicinity of 1300 K. The curve in part (**a**) would then go through zero and finally convert to the intrinsic branch of (**b**) [XIV.1]

Figure XIV.2 shows experimentally determined values of the Hall constant R_H for boron-doped p-silicon (a), and arsenic-doped n-silicon (b). Because the Hall constant gives the reciprocal carrier concentration in units of the elementary charge e, the shape of the curve in a logarithmic plot is similar (at least in the temperature region 33–500 K) to a typical curve of the carrier density for semiconductors (see e.g. Fig. 12.10). The gradients in the region 33–50 K are governed by the ionization energies of the acceptors and donors (12.27). The steeply rising section of the curve in Fig. XIV.2b at about 10^3 K reflects the intrinsic conductivity due to electron-hole pair creation. The different signs of R_H in Fig. XIV.2a,b correspond to the different kinds of carriers in the p- and n-doped material.

Reference

XIV.1 F.J. Morin, J.P. Maita: Phys. Rev. **96**, 29 (1954)

Panel XV
Cyclotron Resonance in Semiconductors

The effective masses of electrons (m_n^*) and holes (m_p^*) in semiconductors can be determined by cyclotron resonance. A crystal sample is placed in a variable, static magnetic field $\boldsymbol{B}$ and the absorption of a high-frequency alternating field is measured as a function of $\boldsymbol{B}$.

Electrons in the static magnetic field move in $\boldsymbol{k}$-space on surfaces of constant energy in a plane perpendicular to $\boldsymbol{B}$ and around the magnetic field axis (Panel VIII). In real space there is also a closed orbit. For crystal electrons, however, this orbit, like the $\boldsymbol{k}$-space orbit, is not simply a circle (Fig. XV.1). Absorption of the high frequency alternating field occurs when the orbital frequency is exactly equal to the period ω_c ($\omega_c = -e\boldsymbol{B}/m_n^*$). Orbits in $\boldsymbol{k}$-space which enclose an extremal area (maximum or minimum) are especially important since the number of states per frequency interval is particularly high at these points. The high-frequency absorption then has a clear maximum, and for the interpretation of the data one only needs to consider "extremal orbits".

As examples of cyclotron resonance in semiconductors we consider Si and Ge. In a semiconductor, electrons and holes are found in the vicinity of the conduction band minimum and valence band maximum, respectively. As shown in Sect. 12.1, the surfaces of constant energy around the conduction band minimum have the form of ellipsoids with rotational symmetry around either the [100] or [111] directions. For an arbitrary orientation of the magnetic field, there is a different extremal orbit for each ellipsoid pair along a particular [100] or [111] axis. For silicon there are three different values corresponding to the three different [100] axes, and for germanium

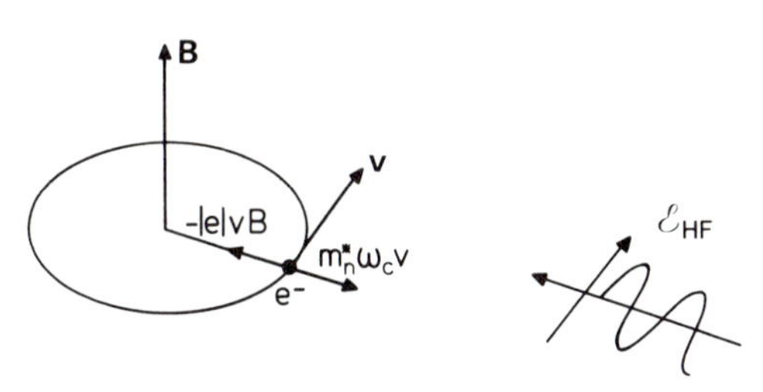

Fig. XV.1. Classical description of cyclotron resonance for electrons with an effective mass m_n^*. On a stable orbit around the magnetic field axis, the centrifugal force and the Lorentz force are in equilibrium. In general, the high frequency field $\mathcal{E}_{HF}$ is maintained at a constant frequency and the magnetic field $\boldsymbol{B}$ is varied

there are four, one for each of the different [111] directions in space (Fig. XV.2 a). If the magnetic field is orientated in a symmetry plane, these numbers are reduced. In Fig. XV.2 b the cyclotron resonance spectra of Dresselhaus et al. [XV.1] are shown. The magnetic field was in the (110) plane. Under these conditions, two of the ellipsoid pairs have the same orientation

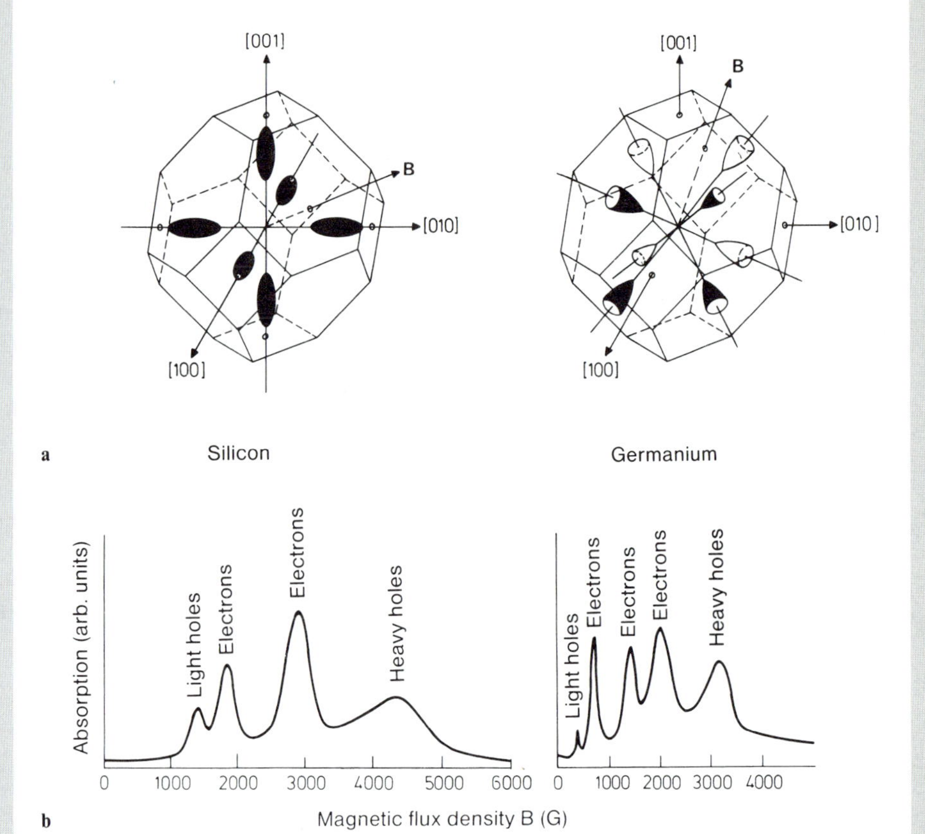

Fig. XV.2. (**a**) Surfaces of constant energy near the minimum of the conduction band for Si (*left*) and Ge (*right*). Electrons move on these surfaces along closed paths perpendicular to the magnetic field. The frequencies of the various orbits for a particular magnetic field are generally different. When the energy surfaces take the special form of ellipsoids, the frequencies are the same for all electrons on *one* ellipsoid. This, however, is unimportant for the observability of cyclotron resonance, since in each case it is the extremal orbit (here the orbit with the largest cross-sectional area) that leads to a clear maximum in the rf absorption. For an arbitrary orientation of the magnetic field there are three different extremal orbits for Si and four for Ge. (**b**) Cyclotron resonance absorption for Si (*left*) and Ge (*right*) with the magnetic field oriented in the (110) plane and angles of 30° and 60° respectively to the [100] direction (magnetic flux density measured in Gauss: $1\ G = 10^{-4}\ T = 10^{-4}\ Vs/m^2$). In both cases, two of the extremal paths are equal for symmetry reasons, such that only two (three) remain distinct. The two valence bands at Γ with their different curvatures show up in the form of the peaks labeled "light" and "heavy" holes. For the observation of cyclotron resonance low temperatures are necessary. The electron-hole pairs were created by irradiation with light. (After [XV.1])

with respect to the magnetic field and the number of absorption maxima for electrons is reduced from three (four) to two (three) for Si (Ge).

Cyclotron resonance is also observed for holes. The direction of their orbit, however, is opposite to that of electrons. It is thus possible to distinguish between electrons and holes by using a circularly polarized high-frequency field incident along the axis of the ***B*** field. Electrons and holes then absorb, according to the direction of their orbit, only right or left circularly polarized radiation. In the absorption spectra (Fig. XV.2b), it can be seen that two maxima are assigned to holes. The interpretation here is different from the case of electrons. Hole states are found at the valence band maximum at the Γ point of the Brillouin zone. In a (simple) treatment, one would expect only spherical energy surfaces on the grounds of the symmetry of the Brillouin zone, that is to say only one cyclotron frequency. As discussed in Sect. 12.1, however (see also Figs. 12.3, 12.4), the band structure at the valence band maximum is rather more complicated: there are both "light" and "heavy" holes.

Cyclotron resonance can only be observed if the carriers are able to complete a large number of closed orbits around the magnetic field prior to suffering collisions with phonons or impurities. This is possible when the relaxation time τ (of the order of the mean time between collisions) is large compared to the inverse cyclotron frequency ω_c^{-1}. Cyclotron resonance is therefore only observed in pure and highly perfect crytals at low (liquid-He) temperature.

Reference

XV.1 G. Dresselhaus, A.F. Kip, C. Kittel: Phys. Rev. **98**, 368 (1955)

Panel XVI
Shubnikov-de Haas Oscillations and Quantum Hall Effect

Precise semiconductor heterostructures and superlattices can be prepared by using atomically controlled epitaxy (Sect. 12.7). Such structures allow the confinement of free conduction electrons in quasi-two-dimensional (2D) quantum-well structures with typical vertical dimensions between 20 and 100 Å. Thereby, quasi-2D electron gases (2DEG) are generated that exhibit highly interesting physical effects, in particular in externally applied strong magnetic fields. An electron gas in a 2D quantum well is confined in one direction z, perpendicular to the hetero-interface or to the layer sequence in a superlattice. Electrons are free to move about only in the x,y-plane perpendicular to the z-direction. Accordingly, the energy eigen-values of an electron within the 2DEG are given as (Sect. 12.7):

$$E_j(\boldsymbol{k}_\parallel) = \frac{\hbar^2}{2m_x^*}k_x^2 + \frac{\hbar^2}{2m_y^*}k_y^2 + \varepsilon_j \, , \quad j = 1, 2, \ldots \tag{XVI.1}$$

where m_x^* and m_y^* are the effective mass components for a free motion in the x,y-plane. The discrete eigen-values that result from the quantization in the z-direction are denoted by ε_j (12.90). For a rectangular quantum well of thickness d_z (Sect. 6.1) with infinitely high barriers, e.g., ε_j correspond to standing electron waves in a potential box (12.93).

$$\varepsilon_j \simeq \frac{\hbar^2\pi^2}{2m_z^*}\frac{j^2}{d_z^2} \, , \quad j = 1, 2, 3, \ldots \, . \tag{XVI.2}$$

Equation (XVI.2) thus describes a set of discrete energy parabolas along k_x and k_y, the so-called sub-bands (Sect. 12.7). Application of a strong magnetic field $\boldsymbol{B}$ perpendicular to the x,y-plane of free-electron movement further decreases the dimensionality of the electron gas. The electrons are forced into cyclotron orbits perpendicular to the $\boldsymbol{B}$-field (Panel XV). The cyclotron frequency (frequency of circulation)

$$\omega_c = -\frac{eB}{m_\parallel^*} \tag{XVI.3}$$

is determined by the balance between the Lorentz force and the centrifugal force. $m_\parallel^*$ is the effective mass of the electrons parallel to the orbital movement (for isotropy in the x,y-plane: $m_x^* = m_y^* = m_\parallel^*$). Since an orbital movement can be decomposed into two linear harmonic oscillations perpendicular to each other, the quantum-mechanical energy eigen-values of the orbits are those of a harmonic oscillator with eigen-frequency ω_c. The magnetic field $\boldsymbol{B}$ therefore causes a further quantization of the sub-bands with single particle energies:

$$E_{j,n,s} = \varepsilon_j + (n + \tfrac{1}{2})\hbar\omega_c + sg\mu_B B . \qquad \text{(XVI.4)}$$

The last term accounts for the two spin orientations in the magnetic field in which $s = \pm 1$ is the spin quantum number, μ_B the Bohr magneton and g the Landé factor of the electron. The magnetic-field-induced quantization (XVI.4) has already been derived for the free electron gas of a metal by a somewhat different method in Panel VIII. In both cases the $\boldsymbol{B}$-field-induced quantization into so-called Landau levels (energetic separation $\hbar\omega_c$) leads to a splitting of the continuous energy parabolas (subbands) into discrete energy eigen-values (Fig. XVI.1 b). Because of the 2D-dimensionality of the bands the density of states at vanishing magnetic field ($\boldsymbol{B} = \boldsymbol{0}$) is a step function (Fig. 12.28 c) for each particular subband. In a finite field $\boldsymbol{B} \neq \boldsymbol{0}$

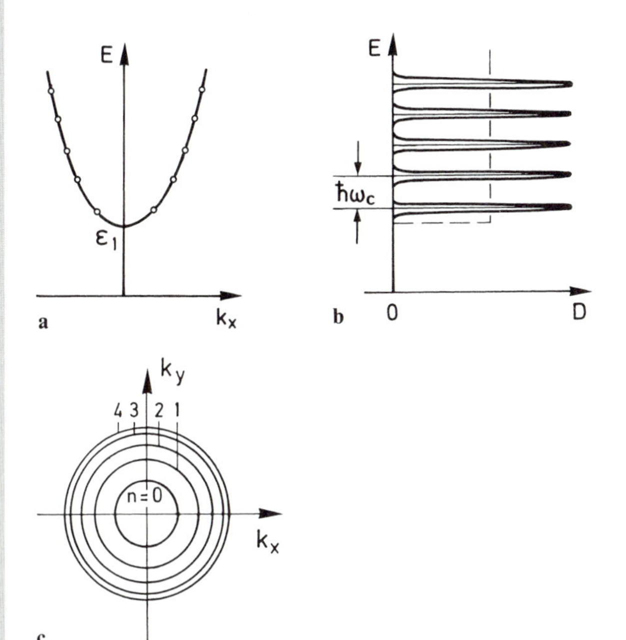

Fig. XVI.1 a–c. Qualitative illustration of the quantization of a 2D electron gas in an external magnetic field $\boldsymbol{B}$ (perpendicular to the (x, y) plane of free-electron propagation). (**a**) Energy parabola of the first sub-band (12.92) of a free 2D electron gas along k_x. A further quantization in the form of Landau states (*points*) appears upon application of an external magnetic field; (**b**) density of states D in the first sub-band of the 2D electron gas, without a magnetic field (*dashed line*) and with an external magnetic field (*solid lines*), $\omega_c = -eB/m_\parallel^*$ is the cyclotron frequency of the electron orbits perpendicular to the magnetic field $\boldsymbol{B}$; (**c**) representation of the Landau splitting $(n + \frac{1}{2})\hbar\omega_c$ in the (k_x, k_y) plane of the reciprocal space

this continuous function splits into a series of δ-like peaks that are separated from each other by the energy $\hbar\omega_c$. The quantum states "condensate" into sharp Landau levels. Since the number of states is conserved in the condensation, a δ-like Landau level must contain exactly the number of states that were originally, at $\boldsymbol{B}=\boldsymbol{0}$, distributed over the energy range between two neighboring Landau-levels, i.e. the degeneracy of a Landau level amounts to

$$N_L = \hbar\omega_c D_0 \, , \tag{XVI.5}$$

where D_0 is the state density of the sub-band at $\boldsymbol{B}=\boldsymbol{0}$. As opposed to (12.96) the spin degeneracy is lifted in the magnetic field and consequently the density of states is a factor of 2 smaller than in (12.96):

$$D_0 = m_{\|}^{*}/2\pi\hbar^2 \, . \tag{XVI.6}$$

Thus the degeneracy of a Landau level is

$$N_L = eB/h = 2.42 \times 10^{10}\,\mathrm{cm}^{-2}\mathrm{T}^{-1} \cdot B \, . \tag{XVI.7}$$

A Landau level with an energy below the Fermi level is occupied with N_L electrons at sufficiently low temperature. A variation of the external magnetic field changes the energetic separation between the Landau levels, as well as their degree of degeneracy (XVI.5). With increasing magnetic field strength, the Landau levels shift to higher energies and finally cross the Fermi energy E_F; the Landau-levels are emptied and the corresponding electrons occupy the next lower Landau level. This becomes possible because of the increased degree of degeneracy (XVI.7). In the case of a sharp Fermi edge at a sufficiently low temperature, the electron system reaches its lowest free energy each time when a Landau level has just crossed the Fermi level. With increasing $\boldsymbol{B}$-field the free energy increases again, until the next Landau level is emptied. As a consequence, the free energy oscillates as a function of the external magnetic field, and so do the material constants, such as the conductivity (Shubnikov-de Haas effect) and the magnetic susceptibility (de Haas-van Alphen effect). The de Haas-van Alphen effect has been of considerable importance in studies of the topology of Fermi surfaces in metals (Panel VIII).

The quantization of states in a magnetic field also gives rise to the quantum Hall effect in semiconductors with 2D-electron gases (2DEG). Since its discovery by Klaus von Klitzing (Nobel prize 1985) the quantum Hall effect [XVI.1] has become an important tool to characterize heterostructures in semiconductor physics. Moreover, the quantum Hall effect had an important impact on the development of the physics of nanostructures in general.

The experimental arrangement for the observation of the quantum Hall effect is similar to that of the classical Hall effect (Panel XIV): a Hall voltage U_H is measured perpendicular to the current through the sample at two opposite contacts. Unlike in the classical Hall-effect, however, the current is carried by a 2DEG in a semiconductor heterostructure, e.g., in a modulation-doped AlGaAs/GaAs heterostructure. The magnetic field $\boldsymbol{B}$ is oriented normal to the current flow and also normal to the plane of the 2DEG (insert in Fig. XVI.2 and Fig. XVI.5c). In order to ensure a sharp Fermi edge in the highly degenerate 2DEG the measurement is performed at low temperature. When the magnetic field $\boldsymbol{B}$ is increased the Hall resistance $r_H = U_H/I$ varies stepwise, with steps at those B values where a sharp Landau level crosses the Fermi energy (Fig. XVI.2a). A simultaneous measurement of the magnetoresistance ϱ_{xx} ($\propto U_L/I$) parallel to the current flow via a pair of contacts placed along the direction of the current yields a sequence of sharp resistance peaks each time when in the Quantum Hall effect r_H jumps from one plateau to the next, i.e. where a Landau level crosses the Fermi energy E_F. For B values in between, ϱ_{xx} is negligible, at least for higher magnetic fields. These resistance oscillations are the Shubnikov-de Haas oscillations of a 2DEG.

The experimentally observed values of the Hall resistance at those magnetic fields where Landau levels cross E_F are derived directly from the relation for the classical Hall effect (Panel XIV). From (XIV.3) one obtains

$$r_H = \frac{U_H}{I} = \frac{-B}{nde} = \frac{-B}{eN_{2D}} , \qquad \text{(XVI.8 a)}$$

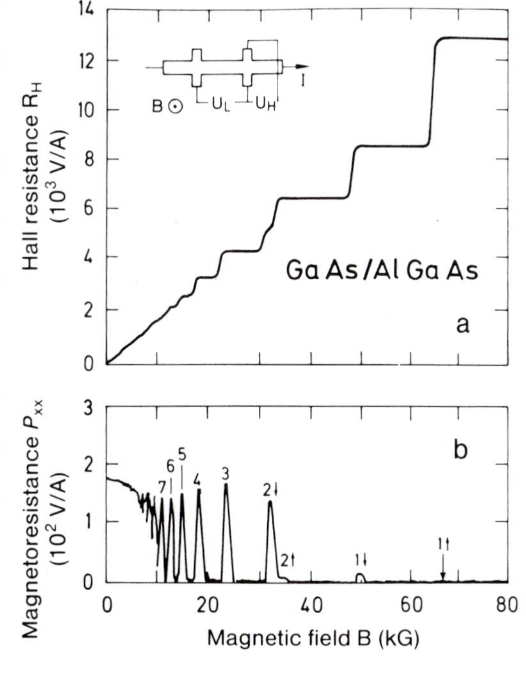

Fig. XVI.2. (**a**) Quantum Hall effect measured at 4 K for the quasi-2D electron gas of a modulation-doped AlGaAs/GaAs heterostructure; 2D electron density $N = 4\times10^{11}$ cm^{-2}, electronic mobility $\mu = 8.6\times10^4$ cm^2/Vs. The Hall resistance $R_H = U_H/I$ was measured as a function of the magnetic field $\boldsymbol{B}$ using the arrangement shown in the inset; (**b**) Shubnikov-de Haas oscillations of the magnetoresistance ϱ_{xx}, measured as indicated in the inset via U_L/I as a function of the external magnetic field $\boldsymbol{B}$. The numbers denote the sub-bands and the arrows indicate the spin direction relative to the $\boldsymbol{B}$ field (after [XVI.2])

where N_{2D} is the two-dimensional carrier density of the 2DEG which, of course, amounts to multiples of the degree of degeneracy N_L (XVI.7) in a magnetic field $\boldsymbol{B}$.

The Hall resistance at Landau level crossing E_F are thus obtained as:

$$r_H = \frac{-B}{\nu N_L e} = -\frac{h}{e^2}\frac{1}{\nu}\,, \quad \nu = 1, 2, 3, \ldots \tag{XVI.8 b}$$

(XVI.8 b) describes correctly single discrete points of the experimental curve in Fig. XVI.2 a, but the characteristic plateaus in the Hall resistance are not at all explained. In order to understand the causes for the existence of plateaus a more profound consideration of the transport in a DEG in the

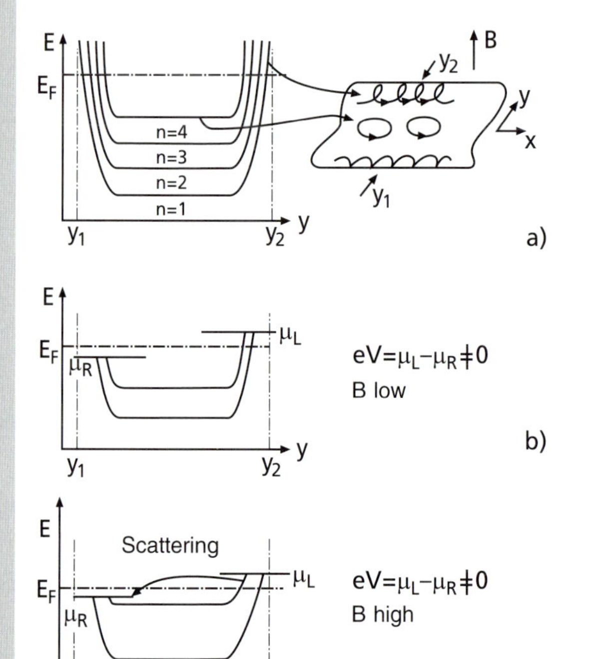

Fig. XVI.3. Explanation of the so-called edge channels for electrons when a magnetic field $\boldsymbol{B}$ is applied perpendicular to the plane of a 2D electron gas (2DEG). (**a**) A strong magnetic field $\boldsymbol{B}$ perpendicular to the 2DEG causes quantization of the electronic states into Landau levels ($n = 1, 2, 3, \ldots$), which correspond to closed cyclotron orbits in the interior of the sample. At the sample boundaries y_1 and y_2 orbiting is interrupted by elastic reflection from the surface. The increased spatial confinement entails a strong enhancement of the energy of the Landau levels, which eventually cross the Fermi energy E_F. (**b**) Under the condition of ballistic current flow the chemical potentials μ_R and μ_L of the right and left edge channels are different and determine the potentials of the corresponding right and left contacts. (**c**) With increasing magnetic field $\boldsymbol{B}$ the splitting between the Landau levels increases, while the upper Landau level approaches the Fermi level E_F. Electronic states near E_F in the interior of the sample become available for scattering processes and the left and right edge channels communicate by these scattering processes

presence of magnetic fields is necessary. In a strong magnetic field, electrons in the 2DEG are forced into cyclotron orbits. Undisturbed, closed orbits, however, are possible only in the interior of the sample (Fig. XVI.3). In that case, the orbital energies are those of the undisturbed Landau levels (XVI.4). Electrons near the boundaries of the sample on the other hand are hindered to complete a full cyclotron orbit. In the simplest case of an elastic mirror reflection at the boundary the electron path becomes a sequence of partial circles (skipping orbits). The combination of cyclotron orbits and surface scattering in the skipping orbits amounts to an additional spatial confinement of the electrons (Fig. XVI.4). By virtue of the uncertainty principle, a stronger confinement is equivalent to a higher kinetic energy. Near the boundaries of the sample the energies of the Landau level are therefore bent upwards and eventually even cross the Fermi level E_F at y_1 and y_2 near the sample surface (Fig. XVI.3 a). Crossing the Fermi level means that one has metallic conduction. Thus, a Landau level that cannot contribute to the electrical conductance in the bulk of the material (since its energy is deep below the Fermi level) does contribute to the conductance at the surface, in the so-called edge channels. The edge channels possess a further important property. Transport in these states is quasi-ballistic, even in macroscopic samples containing impurities. When a carrier in an edge channel is scattered from a defect S (Fig. XVI.4) its trajectory is redirected into the forward direction by the Lorenz force of the strong magnetic field. The total current in the forward direction is therefore not reduced by the presence of impurities and transport in edge channels is therefore quasi-ballistic, i.e. without resistance, regardless of the presence of impurities.

The two edge channels on either side are electrically isolated from each other provided that the Landau levels in the interior of the sample are more than a few k_BT below the Fermi energy E_F. Because of the quasi-ballistic transport in the channels, electrons within one channel are at the same potential. For an electron current flow from left to right this potential is the potential μ_L of the left contact, while electrons in the right "backwards" edge channels are at the potential of the right contact μ_R (Fig. XVI.5). The difference in the potential is the applied voltage multiplied by the electron charge

$$\mu_L - \mu_R = eV \; . \qquad \text{(XVI.9)}$$

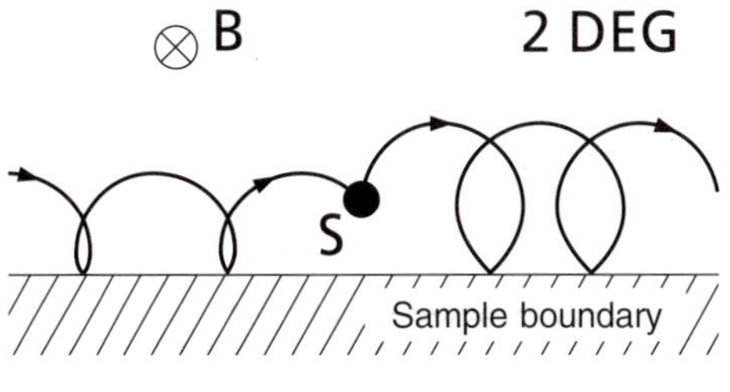

Fig. XVI.4. Schematic trajectory of a carrier in an edge channel that undergoes an elastic scattering process on a defect atom S. The magnetic field is oriented perpendicular to the 2DEG. The 2DEG is assumed to be confined by the sample boundary. The scattering processes at the boundary are assumed to be elastic

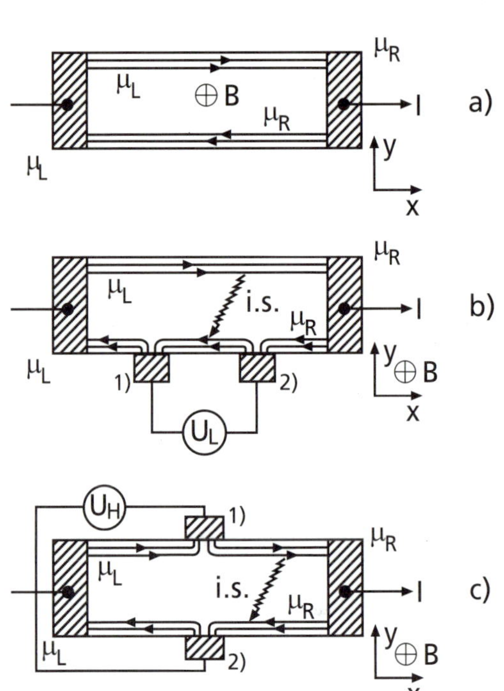

Fig. XVI.5. (**a**) Ballistic electron transport in edge channels that are generated by quantization in a spatially limited 2 DEG due to a strong magnetic field ***B*** perpendicular to the 2DEG. Under the assumption of negligible inelastic scattering between left (L) and right (R) channels the chemical potentials μ_L and μ_R of the left and right channels equal those of the left and right electrical contacts. (**b**) Conduction through edge channels in a 2 DEG in the presence of two electrical contacts 1) and 2) for measuring the Shubnikov-de Haas oscillations in a strong magnetic field. Inelastic scattering (i.s.) between left and right channels causes a potential difference between the contacts. (**c**) Conduction through edge channels in a 2 DEG with two opposite contacts 1) and 2) for measuring the quantum Hall effect

The net electric current carried by the left and right (1D) edge channels is the difference between the contributions from both contacts:

$$I = \sum_{n=1}^{n_c} \int_{\mu_R}^{\mu_L} eD_n^{(1)}(E) v_n(E) dE \ . \qquad \text{(XVI.10)}$$

The sum runs over all occupied Landau levels n (up to a maximum occupied level n_c), which cross E_F as edge channels, $D_n^{(1)} = (2\pi)^{-1}(dE_n/dk_x)^{-1}$ is the 1D density of states of the n-th edge channel, while $v_n = \hbar^{-1}(dE_n/dk_x)$ is the electron velocity within this channel. The n-th edge channel thus contributes to the current an amount

$$I_n = \frac{e}{h}(\mu_L - \mu_R) = \frac{e^2}{h} V \ . \qquad \text{(XVI.11)}$$

The quantum Hall effect is measured as the resistance between left and right edge channels via two contacts perpendicular to the current flow (Fig. XVI.5 c). According to (XVI.11) each channel contributes an amount of e^2/h to the total conductance, i.e. h/e^2 to the Hall resistance. This is exactly the quantum (XVI.8 b) by which the resistance is stepwise enhanced each time a further Landau level crosses the Fermi energy and is thus depleted of electrons. The Shubnikov-de Haas effect is explained in a similar way. In this case, the magneto-resistance is measured along the path of the current via two contacts arranged along the edge channels (Fig. XVI.5 b). As long as the Landau levels in the interior of the sample are sufficiently far away from E_F, no voltage drop occurs, because of the ballistic transport. At a particular magnetic field one of the Landau levels in the interior of the sample reaches the Fermi level E_F and electronic states for scattering between the left and right edge channel become available. Scattering between the forth and back channels induces a resistance along the channel direction. Hence, at these particular magnetic fields a peak-like increase of the resistance is observed (Fig. XVI.2 b). In this interpretation, both quantum Hall effect and Shubnikov-de Haas oscillations are attributed to quasi-one-dimensional carrier transport in edge channels [XVI.4]. The step-like change of the Hall resistance r_H can be determined experimentally with an accuracy of 10^{-7} on samples of different structure and from different groups [XVI.1]. The measurement of quantum Hall effect is therefore presently used for a very precise determination of the fine structure constant.

$$\alpha = \frac{e^2}{h}\frac{\mu_0 c}{2} \cong \frac{1}{137} . \qquad \text{(XVI.12)}$$

If, on the other hand, α is assumed to be known (α is determined by a large number of independent precise experiments) the quantum Hall effect allows the introduction of a standard for the resistance unit "Ohm".

It should finally be mentioned that the quantum Hall effect has the same physical origin as a number of other phenomena related to 1D transport in mesoscopic systems and nanostructures. The fact that edge channels with their 1D conductance (induced by the high magnetic field) each contribute an amount of e^2/h to the total conductivity can directly be transferred to other types of 1D conductance channels. The only requirement is that the width of the channels is comparable with the Fermi wavelength λ_F of the electrons and that the transport is quasi-ballistic, i.e. that the sample dimensions (wire length) are smaller than the mean free path of the carriers. It was, for example, shown that short 1D channels (induced by two metal gates evaporated on the sample) within a 2D electron gas at a modulation-doped AlGaAs/GaAs heterostructure (Sect. 12.7) show a step-like variation of their electrical conductance with varying gate voltage. The channel conductance increases by quanta of $2e^2/h$ when the channel width is increased by a positive bias on the metal gate contacts on top [XVI.5]. The conductance quantum per ballistic channel amounts to $2e^2/h$ in this case, twice the value as in quantum Hall effect. The reason is the spin degeneracy, which is lifted in the quantum Hall effect due to the magnetic field. Even in the breaking of the electrical contact in metallic relays (Au, Ag, Cu) conductivity changes with steps of $2e^2/h$ were observed at room temperature [XVI.6]. The effect is attributed to mesoscopic quasi-1D channels which are produced momentarily by breaking the mechanical contact between the gold-plated relay contacts.

References

XVI.1 K. von Klitzing, G. Dorda, M. Pepper: Phys. Rev. B **28**, 4886 (1983)

XVI.2 M.A. Paalonen, D.C. Tsui, A.C. Gossard: Phys. Rev. B **25**, 5566 (1982)

XVI.3 H. Lüth: *Solid Surfaces, Interfaces and Thin Films* (Springer, Berlin Heidelberg 2001) 4th edition, p. 419

XVI.4 M. Janßen, O. Viehweger, U. Fastenrath, J. Hajdu: *Introduction to the Theory of the Integer Quantum Hall Effect* (VCH, Weinheim 1994)

XVI.5 B.J. van Wees, H. van Houten, C.W.J. Beenakker, J.W. Williamson, L.P. Kouwenhoven, D. van der Marel, C.T. Foxon: Phys. Rev. Let. **60**, 848 (1988)

XVI.6 K. Hansen, E. Laegsgaard, I. Stensgaard, F. Besenbacher: Phys. Rev. B**56**, 2208 (1997)

Panel XVII
Semiconductor Epitaxy

In modern semiconductor physics and device technology, thin crystalline layers are playing an ever increasing role. Of particular interest are multilayer structures, for example, multiple layers of AlGaAs and GaAs, or layers of GaInAs on an InP substrate. Such systems enable one to study novel phenomena such as the quantum Hall effect (Panel XVI), and, at the same time, are the basis for the production of fast transistors and semiconductor lasers. Semiconductor layer structures are almost invariably produced by means of epitaxy. This is a method that nowadays allows extremely precise control of the growth – so precise that single layers of atoms can be reliably deposited under favorable conditions.

Samples for use in fundamental research are most commonly produced by molecular beam epitaxy (MBE), a process in which semiconductor layers are deposited epitaxially on a substrate in an ultra-high-vacuum chamber [XVII.1,2] (Fig. XVII.1). The principle behind the method is simple: a substance such as Ga or Al is evaporated and the vapor deposited on a substrate such as GaAs. Ultra-high-vacuum (UHV) systems with base pressures in the 10^{-8} Pa range are used to prevent contamination and to ensure well-defined conditions, both in the molecular beam and at the surface of the substrate. One can estimate that, at a pressure of 10^{-8} Pa, it will take a few hours for a newly prepared surface to become covered with a monolayer of adsorbates, even if every impinging gas molecule sticks to the surface. Such UHV chambers are made of stainless steel, and the extremely low pressures of less than 10^{-8} Pa (corresponding to a molecular mean free path of some meters) are maintained by ion-getter pumps or turbomolecular pumps. In an ion-getter pump, gas atoms are ionized by high electric fields and then adsorbed onto appropriately charged active metal films (e.g. Ti) – this latter process being known as "gettering". The pumping action of a turbomolecular pump relies on the momentum exchange between gas molecules and the blades of a rapidly rotating turbine.

The UHV chamber that comprises the main part of an MBE machine (Fig. XVII.1) is equipped with an internal cryoshield cooled by liquid nitrogen. This serves to trap stray atoms and molecules, thereby further reducing the background pressure. The substrate material, e.g. a GaAs wafer, onto which an AlGaAs layer is to be deposited, is mounted on a rotating substrate holder. During growth the substrate must be maintained at a temperature of about 500–600 °C in order to ensure a sufficiently high mobility of the impinging atoms or molecules (Ga, As_4, Si, etc.) on the surface. The atoms or

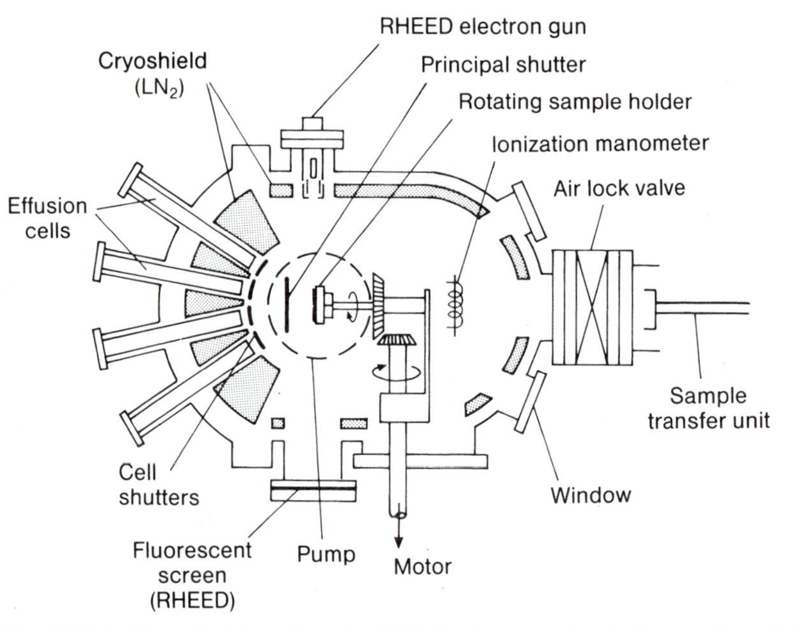

Fig. XVII.1. Schematic illustration of a UHV chamber for molecular beam epitaxy (MBE) of III–V semiconductor layers (viewed in section from above). (After [XVII.3])

molecules to be incorporated into the growing crystal originate from a so-called effusion source in which the starting material, e.g. solid Ga and As for GaAs epitaxy, is vaporized from boron-nitride crucibles by electrical heating. Mechanically driven shutters, controlled from outside, can open and close the individual effusion cells, thereby switching the corresponding molecular beams on and off. The growth rate of the epitaxial layer is determined by the particle flux in the molecular beam, which, in turn, is controlled by varying the temperature of the crucible. To obtain a particular sequence of well-defined layers with atomically sharp boundaries, as is required for a composition superlattice (Sect. 12.7), the crucible temperature and the shutter-opening times must be regulated by a computer program. This is particularly necessary for the epitaxial growth of ternary or quaternary alloys such as $Al_{0.55}Ga_{0.45}As$ with a fixed composition. Homoepitaxy of GaAs layers on a GaAs substrate does not demand any precise regulation of the molecular flux. In this case, nature herself provides a helping hand: stoichiometric growth of GaAs is possible even for a non-stoichiometric beam composition. The growth rate is determined solely by the rate of arrival of Ga atoms. At a substrate temperature of between 500 and 600 °C, arsenic atoms only stick to the surface when sufficient numbers of Ga atoms are present on the growing surface; the sticking coefficient for As approaches one for a Ga excess on the surface and is close to zero when Ga is underrepresented. Thus GaAs epitaxy occurs optimally for an excess of arsenic. For a deeper understanding of the

GaAs growth in MBE, it is also important to know that the arsenic molecular beam produced by evaporation of solid arsenic consists mostly of As_4 molecules, which must be dissociated into As atoms by the thermal energy of the hot substrate surface before they can be incorporated in the growing layer. This dissociation process is sometimes carried out thermally in hot graphite "cracker" cells, which leads to better layer quality. Important dopants in III–V MBE are Si for *n* and Be for *p* doping. These materials are also evaporated from effusion cells which can be switched off and on in a controlled way by the opening and closing of shutters.

As shown in Fig. XVII.1, an MBE system is normally also furnished with an electron gun and a fluorescent screen for the observation of diffraction patterns (reflection high energy electron diffraction, RHEED). Diffraction patterns provide information about the crystallographic structure of the growing surface. An ionization pressure gauge is also included to measure both the chamber pressure and the pressure in the molecular beams.

The MBE system shown in Fig. XVII.1 is of the type frequently employed for the epitaxy of III–V and II–VI semiconductors. Si MBE [XVII.4] is carried out in similar systems, but the Si molecular beam is not produced by thermal evaporation from effusion cells, but rather by electron bombardment of a solid Si target, which thereby becomes so hot that it evaporates.

The decisive advantage of MBE over other epitaxial methods which work at higher presssure is the fast switching time between different sources. This provides a straightforward method of obtaining sharp doping and composition profiles. A typical growth rate is 1 μm/h, which corresponds to about 0.3 nm/s, or one monolayer per second. The time needed to switch between different sources should therefore be well below one second.

Such short switching times are more difficult to achieve using the second important epitaxial method, *metal organic chemical vapor deposition* (MOCVD) [XVII.5]. Here growth takes place in a reactor, whose whole gas volume must be exchanged on switching from one source to another. Compared to MBE, MOCVD has the interesting advantage for industrial applications that the sources can be easily controlled by gas flow regulators. Furthermore the gas phase sources allow almost continuous operation of the system.

We will discuss the principles of MOCVD processes using GaAs epitaxy as an example. We are concerned here with the deposition of solid GaAs from gas phase materials containing Ga and As. The molecule AsH_3 and the metal organic gas trimethylgallium [$TMG = Ga(CH_3)_3$] are frequently used. The overall reaction, which takes place via several complicated intermediate steps, can be written

$$[Ga(CH_3)_3]_{gas} + [AsH_3]_{gas} \rightarrow [GaAs]_{solid} + [3CH_4]_{gas} \, . \qquad (XVII.1)$$

AsH_3 is fed directly from a gas bottle via a gas regulating valve into the quartz reactor (Fig. XVII.2 a). The metal organic component TMG is in a bulb, and its vapor pressure is controlled by a temperature bath. Hydrogen

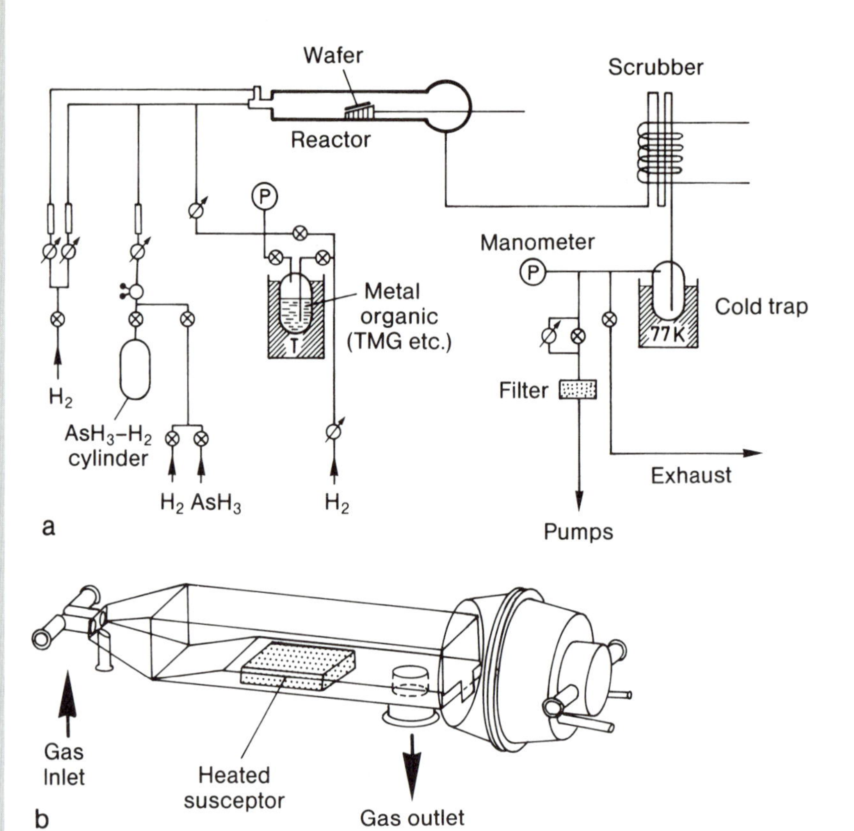

Fig. XVII.2 a, b. Schematic illustration of an apparatus for metal organic chemical vapor deposition (MOCVD) of III–V semiconductor layers. (**a**) General overview; (**b**) quartz glass reactor (typical length 50 cm); the susceptor, which holds the wafer to be epitaxially coated, is heated during growth to 700–1200 K. (After [XVII.6])

(H_2) flows through this bulb and transports the TMG to the reactor. Another gas line allows the whole system to be flushed with H_2. The components which are not consumed in the reaction, and the reaction products, are pumped out at the end of the reactor through a decomposition oven, to eliminate the dangerous excess AsH_3. The pumping system also allows the reactor to be operated at low pressure (so-called low pressure MOCVD). With this method the switching time between different sources can be reduced and most of the advantages of MBE obtained. Besides the sources for the growth of the base material (AsH_3 and TMG) other gas lines are needed for the doping gases, such as SiH_4, $(C_2H_5)_2Te$, $(C_2H_5)_2Mg$ for Si, Te or Mg doping. For the growth of ternary and quaternary III–V alloys further gas lines are also necessary. These supply metal organic materials such as trimethyl-aluminium $(CH_3)_3Al$, trimethyl-antimony $(CH_3)_3Sb$, trimethyl-indium $(CH_3)_3In$ and others. For the growth of phosphorus-containing materials such as InP and GaP, the hydride phosphine PH_3 is used. Figure XVII.2b shows a possible design for a flux reactor. The wafer on which the films are to be deposited, the so-called graphite susceptor, is held at a temperature between 600 and 800 °C during growth. Heating takes place by radiation, direct electrical current or microwave heating.

Compared with MBE, the growth process in MOCVD is considerably more complicated. A stream of gas flows over the growing layer, and from this stream the reaction components diffuse to the surface. Both at the surface and in the gas phase dissociation reactions take place, e.g. AsH_3 is dissociated by collision in the gas phase as well as on the surface itself. After the required surface reactions have occurred, including among others the incorporation of the dissociated As into the growing crystal lattice, the reaction products such as CH_4 are again transported by diffusion away from the surface and into the gas stream. MOCVD growth is therefore largely determined by transport to and from the surface and by surface reaction kinetics. This is seen clearly if one plots the rate of growth in the MOCVD process as a function of the substrate temperature (Fig. XVII.3). The result is a typical curve in which the fall-off at lower temperatures has the form $\exp(-E_a/k T)$, being kinetically limited by surface reactions. Typical activation energies, E_a, in this region are around 1 eV per atom. The temperature range in which this kinetically limited region occurs depends on the source material used. For the relatively stable TMG, the exponential decrease is observed at temperatures below 850 K, but for the more easily dissociated triethyl-gallium [TEG, $(C_2H_5)_3Ga$] it occurs below 700 K. For temperatures above the kinetically limited region, the growth rate shows a

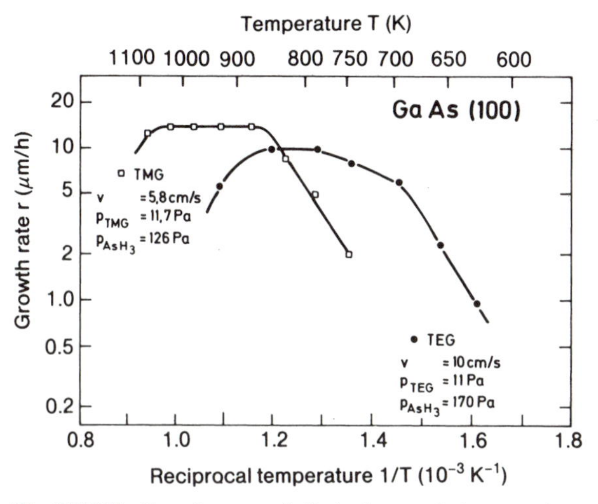

Fig. XVII.3. Growth rates of GaAs layers during metal organic chemical vapor deposition (MOCVD) from the starting materials AsH_3 and trimethyl-gallium (TMG) or triethyl-gallium (TEG); GaAs wafer orientation (100). The gas flow velocities v and the partial pressures P_{TEG}, P_{TMG}, P_{AsH_3} are given. The total pressure in each experiment is 10^4 Pa. (After [XVII.7])

plateau whose height depends on the conditions influencing diffusion to and from the substrate (e.g. the flow velocity in the reactor). Here growth is limited by transport processes in the gas phase. At still higher temperatures, another decrease in the growth rate is observed (Fig. XVII.3). It is probable that the reasons for this are not inherent to the process. A likely cause is the loss of reactants from the gas stream due to deposition on the reactor walls.

By clever process control in low pressure MOCVD, and by using specially designed valves, very short switching times can now be achieved. Thus, even with MOCVD it is possible to produce atomically sharp heterojunctions between two semiconductors.

A third modern epitaxial method, so-called metal organic MBE (MOMBE), also called CBE (chemical beam epitaxy), combines the advantages of MBE and MOCVD [XVII.8]. As in MBE, a UHV system serves as the growth chamber (Fig. XVII.1). The source materials, however, are gases, as in MOCVD, and these are fed through lines with controlled valves into the UHV chamber. Inside the chamber, specially constructed admission systems (capillaries) serve to form the molecular beams, which are directed at the surface of the substrate to be coated. For the growth of GaAs, one might use, for example, AsH_3 and triethyl-gallium [$(C_2H_5)_3Ga$, TEG]. In a MOCVD reactor, collisions in the gas above the hot substrate surface lead to a significant predissociation of the AsH_3, but this does not occur here because of the very low background pressure. In MOMBE, the AsH_3 must therefore be thermally dissociated in the inlet capillary.

All metal organic epitaxy processes suffer from the presence of carbon impurities, which originate in the decomposition of the metal organics. The carbon is most frequently incorporated at As lattice positions and leads to p conduction (although it can also be incorporated as a donor at Ga sites). Carbon inclusion becomes stronger the lower the pressure used for epitaxy. Thus in MOMBE, GaAs layers with low levels of p doping can only be achieved using TEG, whereas in MOCVD it is also possible to use TMG.

References

XVII.1 M.A. Herman, H. Sitter: *Molecular Beam Epitaxy*, 2nd ed., Springer Ser. Mater. Sci., Vol. 7 (Springer, Berlin, Heidelberg 1996)

XVII.2 E.H.C. Parker (Ed.): *The Technology and Physics of Molecular Beam Epitaxy* (Plenum, New York 1985)

XVII.3 A.Y. Cho, K.Y. Cheng: Appl. Phys. Lett. **38**, 360 (1981)

XVII.4 E. Kasper, H.-J. Herzog, H. Dämbkes, Th. Richter: Growth Mode and Interface Structure of MBE Grown SiGe Structures, in *Two-Dimensional Systems: Physics and New Devices*, ed. by G. Bauer, F. Kuchar, H. Heinrich, Springer Ser. Solid-State Sci., Vol. 67 (Springer, Berlin, Heidelberg 1986)

XVII.5 Proc. ICMOVPE I, J. Crystal Growth **55** (1981); Proc. ICMOVPE II, J. Crystal Growth **68** (1984)

W. Richter: Physics of Metal Organic Chemical Vapour Deposition: *Festkörperprobleme* **26**, 335 (Advances in Solid State Physics 26) ed. by P. Grosse (Vieweg, Braunschweig 1986)

XVII.6 H. Heinecke, E. Veuhoff, N. Pütz, M. Heyen, P. Balk: J. Electron. Mater. **13**, 815 (1984)

XVII.7 C. Plass, H. Heinecke, O. Kayser, H. Lüth, P. Balk: J. Crystal Growth **88**, 455 (1988)

XVII.8 H. Lüth: in Surf. Science **299/300**, 867 (1994)

References

Chapter 1

1.1 L. Pauling: *The Nature of the Chemical Bond*, 3rd edn. (Cornell Univ. Press, Ithaca, NY 1960)
1.2 S. Göttlicher: Acta Cryst. B **24**, 122 (1968)
1.3 Y.W. Yang, P. Coppens: Solid State Commun. **15**, 1555 (1974)
1.4 S.P. Walch, W.A. Goddard, III: Surf. Sci. **72**, 645 (1978)

Further Reading

Ballhausen, C.J., Gray, H.B.: *Molecular Orbital Theory* (Benjamin, New York 1964)

Cartmell, E., Fowles, G.W.A.: *Valency and Molecular Structure*, 2nd edn. (Butterworths, London 1961)

Coulson, C.A.: *Valence*, 2nd edn. (Oxford Univ. Press, Oxford 1961)

Hartmann, H.: *Theorie der chemischen Bindung* (Springer, Berlin Heidelberg 1954)

Pauling, L.: *Die Natur der chemischen Bindung* (Chemie-Verlag, Weinheim 1964)

Philips, J.C.: *Covalent Bonding in Crystals, Molecules and Polymers* (The Univ. of Chicago Press, Chicago 1969)

Slater, J.C.: *Quantum Theory of Molecules and Solids* (McGraw-Hill, New York 1963)

Vinogradov, S.N., Linell, R.H.: *Hydrogen Bonding* (Van Nostrand-Reinhold, New York 1971)

Chapter 2

2.1 K. Urban, P. Kramer, M. Wilkens: Phys. Bl. **42**, 373 (1986)
2.2 T.B. Massalski (ed.): *Binary Alloy Phase Diagrams*, 2nd edn. (American Society for Metals, Metals Park, Ohio 44073, 1990)
A database is available under the name TAPP of ES Microwave, 2234 Wade Court, Hamilton, OH 45013, USA

Further Reading

Burzlaff, H., Thiele, G. (eds.): *Kristallographie – Grundlagen und Anwendungen* (Thieme, Stuttgart 1977), insbesondere: Burzlaff, H., Zimmermann, H.: "Symmetrielehre", Bd. I

Hamermesh, M.: *Group Theory and Its Application to Physical Problems* (Addison-Wesley/Pergamon, London Paris 1962)

Heine, V.: *Group Theory in Quantum Mechanics* (Pergamon, London 1960)

Koster, G.F., Dimmock, J.O., Wheeler, R.G., Statz, H.: *Properties of the 42 Point Groups* (MIT Press, Cambridge, MA 1963)

Streitwolf, H.: *Gruppentheorie in der Festkörperphysik* (Akademische Verlagsges., Leipzig 1967)

Tinkham, M.: *Group Theory and Quantum Mechanics* (McGraw-Hill, New York 1964)

Vainshtein, B.K.: *Fundamentals of Crystals: Symmetry and Methods of Structural Crystallography,* Springer Ser. Modern Crystallography, Vol. 1 (Springer, Berlin Heidelberg 1994)

Chapter 3

3.1 W. Marshall, S.W. Lovesey: *Theory of Thermal Neutron Scattering* (Clarendon, Oxford 1971)

3.2 L.D. Landau, E.M. Lifschitz: *Lehrbuch der theoretischen Physik*, Bd. VIII: "Elektrodynamik der Kontinua" (Akademie Verlag, Berlin 1874) p. 436 ff.

3.3 J.B. Pendry: *Low Energy Electron Diffraction* (Academic, London 1974)

3.4 S. Kugler, L. Pusztai, L. Rosta, P. Chieux, R. Bellissent: Phys. Rev. B**48**, 7685 (1993)

3.5 Y. Waseda: *The Structure of Non-Crystalline Materials* (McGraw-Hill, New York 1981)

Further Reading

Bacon, G.F.: *Neutron Diffraction*, 3rd edn. (Oxford Univ. Press, Oxford 1975)

Dachs, H. (ed.): *Neutron Diffraction*, Topics Curr. Phys., Vol. 6 (Springer, Berlin Heidelberg 1978)

Kleber, W.: *Einführung in die Kristallographie*, 16. Aufl. (Verlag Technik, Berlin 1983)

Lovesey, S., Springer, T. (eds.): *Dynamics of Solids and Liquids by Neutron Scattering.* Topics Curr. Phys., Vol. 3 (Springer, Berlin Heidelberg 1977)

Pinsker, Z.G.: *Dynamical Scattering of X-Rays in Crystals*, Springer Ser. Solid-State Sci., Vol. 3 (Springer, Berlin Heidelberg 1978)

Sellin, A. (ed.): *Structure and Collisions of Ions and Atoms*, Topics Curr. Phys., Vol. 5 (Springer, Berlin Heidelberg 1978)

Shull, C.G.: *Early Developments of Neutron Scattering,* Rev. Mod. Phys. **67**, 753 (1995)

Ziman, J.M.: *Models of Disorder* (Cambridge University Press, Cambridge 1979)

Chapter 4

4.1 M. Born, R. Oppenheimer: Ann. Phys. (Leipzig) **84**, 457 (1927)
4.2 G. Leibfried: In *Handbuch der Physik*, Vol. 7/1 (Springer, Berlin Heidelberg 1955) p. 104
4.3 G. Dolling: In *Inelastic Scattering of Neutrons in Solids and Liquids*, Vol. II (Intern. Atomic Energy Agency, Vienna 1963) p. 37
4.4 R.F.S. Hearman: Advances in Physics **5**, 323 (1956)
4.5 W.A. Brantley: J. App. Phys. **44**, 534 (1973)
4.6 G. Simmons, H. Wang: *Single crystal elastic constants and calculated aggregate properties; A Handbook*, 2nd edn., Cambridge, USA (MIT Press, Cambridge 1971)

Further Reading

Bak, T.A.: *Phonons and Phonon Interactions* (Benjamin, New York 1964)

Bilz, H., Kress, W.: *Phonon Dispersion Relations in Insulators*, Springer Ser. Solid-State Sci., Vol. 10 (Springer, Berlin Heidelberg 1979)

Born, M., Huang, K.H.: *Dynamical Theory of Crystal Lattices* (Clarendon, Oxford 1954)

Leibfried, G., Breuer, N.: *Point Defects in Metals I, Introduction to the Theory*, Springer Tracts Mod. Phys. Vol. 81 (Springer, Berlin Heidelberg 1977)

Ludwig, W.: *Recent Developments in Lattice Theory*, Springer Tracts Mod. Phys., Vol. 43 (Springer, Berlin Heidelberg 1967)

Maradudin, A.A., Montroll, E.W., Weiss, G.H.: "Theory of Lattice Dynamics in the Harmonic Approximation" in *Solid State Physics, Advances and Applications*, ed. by H. Ehrenreich, F. Seitz, D. Turnbull (Academic, New York 1971) Suppl. 3

Wallis, R.F. (ed.): *Lattice Dynamics* (Plenum, New York 1965)

Chapter 5

5.1 G. Dolling, R.A. Cowley: Proc. R. Soc. London **88**, 463 (1966)
5.2 *American Institute of Physics Handbook*, 3rd edn. (McGraw-Hill, New York 1972) pp. 4–115
5.3 Wei Chen, D.L., Mills: Phys. Rev. B**36**, 6269 (1987)

5.4 H. Ibach: Phys. Status Solidi **31**, 625 (1969)
5.5 R.W. Powell, C.Y. Ho, P.E. Liley: NSRDS-N13S8 (1966)
5.6 R. Berman, P.G. Klemens, F.E. Simon, T.M. Fry: Nature **166**, 865 (1950)

Further Reading

Rosenberg, H.M.: *Low Temperature Solid-State Physics* (Clarendon, Oxford 1963)

Chapter 6

6.1 A. Sommerfeld, H. Bethe: *Elektronentheorie der Metalle*, Heidelberger Taschenbuch, Bd. 19 (Springer, Berlin Heidelberg 1967)
6.2 Landolt Börnstein, Neue Serie III/6 (Springer, Berlin Heidelberg 1971)
6.3 C.A. Bailey, P.L. Smith: Phys. Rev. **114**, 1010 (1959)
6.4 N.F. Mott: Cdn. J. Phys. **34**, 1356 (1956); Nuovo Cimento **7** (Suppl.), 312 (1958)
6.5 H.B. Michaelson: J. Appl. Phys. **48**, 4729 (1977)

Further Reading

Grosse, P.: *Freie Elektronen in Festkörpern* (Springer, Berlin Heidelberg 1979)

Chapter 7

7.1 W. Shockley: *Electrons and Holes in Semiconductors* (Van Nostrand, New York 1950)
7.2 L.P. Howard: Phys. Rev. **109**, 1927 (1958)
7.3 B. Segall: Phys. Rev. **124**, 1797 (1961)
7.4 R. Courths, S. Hüfner: Phys. Rep. **112**, 55 (1984)
7.5 H. Eckhardt, L. Fritsche, J. Noffke: J. Phys. F**14**, 97 (1984)
7.6 F. Herman, R.L. Kortum, C.D. Kuglin, J.L. Shay: In *II–VI Semiconducting Compounds*, ed. by D.G. Thomas (Benjamin, New York 1967)
7.7 T.H. Upton, W.A. Goddard, C.F. Melius: J. Vac. Sci. Technol. **16**, 531 (1979)
7.8 H. Haberland (ed.): *Clusters of Atoms and Molecules I and II*, Springer Series in Chem. Phys. **52**, 56 (Springer, Berlin Heidelberg 1994)

Further Reading

Ashcroft, N.W., Mermin, N.D.: *Solid State Physics* (Hold, Rinehart and Winston, New York 1976)
Brillouin, L.: *Wave Propagation in Periodic Structures* (Academic, New York 1960)
Callaway, J.: *Energy Band Theory* (Academic, New York 1964)
Harrison, W.A.: *Pseudopotentials in the Theory of Metals* (Benjamin, New York 1966)
Herrmann, R., Preppernau, U.: *Elektronen im Kristall* (Springer, Berlin Heidelberg 1979)
Jones, H.: *The Theory of Brillouin-Zones and Electronic States in Crystals* (North-Holland, Amsterdam 1962)
Loucks, T.L.: *Augmented Plane Wave Method* (Benjamin, New York 1967)
Madelung, O.: *Introduction to Solid-State Theory*, Springer Ser. Solid State Sci., Vol. 2 (Springer, Berlin Heidelberg 1978)
Skriver, H.L.: *The LMTO Method*, Springer Ser. Solid-State Sci., Vol. 41 (Springer, Berlin Heidelberg 1984)
Wilson, A.H.: *The Theory of Metals*, 2nd edn. (Cambridge Univ. Press, London 1965)

Chapter 8

8.1 R.M. White: *Quantum Theory of Magnetism*, Springer Ser. Solid-State Sci., Vol. 32 (Springer, Berlin Heidelberg 1983)
see also D.C. Mattis: *The Theory of Magnetism I and II*, Springer Ser. Solid-State Sci., Vols. 17, 55 (Springer, Berlin Heidelberg 1985, 1988)
T. Moriya: *Spin Fluctuations in Itinerant Electron Magnetism*, Springer Ser. Solid-State Sci., Vol. 56 (Springer, Berlin Heidelberg 1985)
8.2 J.F. Janak: Phys. Rev. B **16**, 255 (1977)
8.3 J. Callaway, C.S. Wang: Phys. Rev. B **7**, 1096 (1973)
8.4 P. Weiss, G. Foex: Arch. Sci. Natl. **31**, 89 (1911)
8.5 P. Weiss, R. Porrer: Ann. Phys. **5**, 153 (1926)
8.6 H.A. Mook, D. McK. Paul: Phys. Rev. Lett. **54**, 227 (1985)

Further Reading

Chakravarty, A.S.: *Introduction to the Magnetic Properties of Solids* (Wiley, New York 1980)
Crangle, J.: *The Magnetic Properties of Solids* (Arnold, London 1977)

Chapter 9

9.1 P. Drude: Ann. Phys. (Leipzig) **1**, 566 (1900)
9.2 E. Grüneisen: Ann. Phys. (Leipzig) (5) **16**, 530 (1933)
9.3 D.K.C. McDonald, K. Mendelssohn: Proc. R. Soc. Edinburgh, Sect. A **202**, 103 (1950)
9.4 J. Linde: Ann. Phys. (Leipzig) **5**, 15 (1932)
9.5 J.M. Ziman: *Principles of the Theory of Solids* (Cambridge Univ. Press, London 1964)
9.6 N.F. Mott: Phil. Mag. B**19**, 835 (1984)
N.F. Mott: in *The Physics of Hydrogenated Amorphous Silicon* Vol. II, ed. by J.D. Joannopoulos and G. Luckowsky, Topics in Applied Physics, Vol. 56 (Springer, Berlin Heidelberg 1984) p. 169
9.7 B. Ruttensperger, G. Müller, G. Krötz: Phil. Mag. B**68**, 203 (1993)

Further Reading

Blatt, F.J.: *Physics of Electronic Conduction in Solids* (McGraw-Hill, New York 1968)
Busch, G., Schade, H.: *Vorlesungen über Festkörperphysik* (Birkhäuser, Basel 1973)
Madelung, O.: *Introduction to Solid-State Theory*, Springer Ser. Solid-State Sci., Vol. 2 (Springer, Berlin Heidelberg 1978)
Nag, B.R.: *Electron Transport in Compound Semiconductors*, Springer Ser. Solid-State Sci., Vol. 11 (Springer, Berlin Heidelberg 1980)
Smith, A.C., Janak, J.F., Adler, R.B.: *Electronic Conduction in Solids* (McGraw-Hill, New York 1967)
Ziman, J.M.: *Electrons and Phonons* (Clarendon, Oxford 1960)

Chapter 10

10.1 H.K. Onnes: Akad. van Wetenschappen (Amsterdam) **14**, 113, 818 (1911)
10.2 J. Bardeen, L.N. Cooper, J.R. Schrieffer: Phys. Rev. **108**, 1175 (1957)
10.3 N.E. Phillips: Phys. Rev. **114**, 676 (1959)
10.4 W. Meissner, R. Ochsenfeld: Naturwissensch. **21**, 787 (1933)
10.5 F. London, H. London: Z. Phys. **96**, 359 (1935)
10.6 L.N. Cooper: Phys. Rev. **104**, 1189 (1956)
10.7 H. Fröhlich: Phys. Rev. **79**, 845 (1950)
10.8 N.N. Bogoliubov: Nuovo Cimento **7**, 794 (1958)
10.9 J. Hinken: *Supraleiter-Elektronik* (Springer, Berlin Heidelberg 1988)
10.10 V.V. Schmidt (eds. P. Müller, A.V. Ustinov): *The Physics of Superconductors* (Springer, Berlin Heidelberg 1997)
10.11 P.L. Richards, M. Tinkham: Phys. Rev. **119**, 575 (1960)

10.12 I. Giaever, K. Megerle: Phys. Rev. **122**, 1101 (1961)
10.13 E. Maxwell: Phys. Rev. **86**, 235 (1952);
B. Serin, C.A. Reynolds, C. Lohman: Phys. Rev. **86**, 162 (1952);
J.M. Lock, A.B. Pippard, D. Schoenberg: Proc. Cambridge Phil. Soc. **47**, 811 (1951)
10.14 V.L. Ginzburg, L.D. Landau: Soviet Phys.-JETP (USSR) **20**, 1064 (1950)
siehe auch V.L. Ginzburg: Nuovo Cimento **2**, 1234 (1955)
10.15 A.C. Rose-Innes, E.H. Rhoderick: *Introduction to Superconductivity* (Pergamon, Oxford 1969) p. 52
10.16 R. Doll, M. Näbauer: Phys. Rev. Lett. **7**, 51 (1961)
10.17 B.S. Deaver Jr., W.M. Fairbank: Phys. Rev. Lett. **7**, 43 (1961)
10.18 T. Kinsel, E.A. Lynton, B. Serin: Rev. Mod. Phys. **36**, 105 (1964)
10.19 J. Schelten, H. Ullmeier, G. Lippmann, W. Schmatz: in *Low Temperature Physics* – LT 13, Vol. 3, ed. by K.D. Timmerhaus, W.J. O'Sullivan, E.F. Hammel (Plenum, New York 1972) p. 54
10.20 J.G. Bednorz, K.A. Müller: Z. Phys. B **64**, 189 (1986)
10.21 M.K. Wu, J.R. Ashburn, C.J. Tornq, P.H. Hor, R.L. Meng, L. Gao, Z.J. Huang, Y.Q. Wang, C.W. Chu: Phys. Rev. Lett. **58**, 908 (1987)
10.22 J.R. Cooper, S.D. Obertelli, A. Carrington, J.W. Loran: Phys. Rev. B **44**, 12 086 (1991)
10.23 R. Wördenweber: Private Mitteilung (ISI, Forschungszentrum Jülich)
10.24 H. Maeda, Y. Tanaka, M. Fukutomi, T. Asano: Jpn. J. Appl. Phys. Lett. **27**, L 209 (1988)
10.25 Z.Z. Sheng, A.M. Hermann: Nature **332**, 55 (1988)
10.26 R. Hott, H. Rietschel, M. Sander: Phys. Bl. **48**, 355 (1992)
10.27 B. Kabius: Private Mitteilung (IFF, Forschungszentrum Jülich)
10.28 S.J. Hagen, T.W. Jing, Z.Z. Wang, J. Horvath, N.P. Ong: Phys. Rev. B **37**, 7928 (1988)
10.29 J. Zaanen, A.T. Paxton, O. Jepsen, O.K. Anderson: Phys. Rev. Lett. **60**, 2685 (1988)
10.30 Ching-ping S. Wang: Electronic Structure, Lattice Dynamics, and Magnetic Interactions, in *High Temperature Superconductivity*, ed. by J.W. Lynn, Graduate Texts Contemp. Phys. (Springer, Berlin Heidelberg 1990) p. 122
10.31 H. Takagi, S. Uchida, Y. Tokura: Phys. Rev. Lett. **62**, 1197 (1989)
10.32 Y. Zhang, Y. Tavrin, M. Mück, A.I. Braginski, C. Heiden, T. Elbert, S. Hampson: Physiol. Meas. **14**, 113 (1993)
10.33 J. Fink, E. Sohmen: Phys. Bl. **48**, 11 (1992)
10.34 C. Politis, V. Buntar, W. Krauss, A. Gurevich: Europhys. Lett. **17**, 175 (1992)

Further Reading

Buckel, W.: *Supraleitung*, 5. Aufl. (VCH, Weinheim 1993)

De Gennes, P.G.: *Superconductivity of Metals and Alloys* (Addison-Wesley, Reading 1992)

Huebener, R.P.: *Magnetic Flux Structures in Superconductors*, Springer Ser. Solid-State Sci., Vol. 6 (Springer, Berlin Heidelberg 1979)

Kuper, Ch.G.: *An Introduction to the Theory of Superconductivity* (Clarendon, Oxford 1968)

Lynn, J.W. (ed.): *High Temperature Superconductivity*, Graduate Texts Contemp. Phys. (Springer, Berlin Heidelberg 1990)

Lynton, E.A.: *Superconductivity* (Methuen's Monographs on Physical Subjects, London 1964). (Deutsche Übersetzung: Bibliographisches Institut, Mannheim 1966)

Rickayzen, G.: *Theory of Superconductivity* (Wiley, New York 1965)

Rose-Innes, A.C., Rhoderick, E.H.: *Introduction to Superconductivity* (Pergamon, Braunschweig 1969)

Schmidt, V.V. (eds. P. Müller and A.V. Ustinov): *The Physics of Superconductors* (Springer, Berlin Heidelberg 1997)

Stolz, H.: *Supraleitung* (Vieweg, Braunschweig 1979)

Tinkham, M.: *Introduction to Superconductivity*, 2nd edn., (McGraw-Hill, New York 1996)

Chapter 11

11.1 B. Schrader, W. Meier: *Raman/IR Atlas organischer Verbindungen*, Bd. 1 (VCH, Weinheim 1974)

11.2 L.H. Henry, J.J. Hopfield: Phys. Rev. Lett. **15**, 964 (1965)

11.3 T.P. Martin: Phys. Rev. B **1**, 3480 (1970)

11.4 M. Hass, B.W. Henri: J. Phys. Chem. Sol. **23**, 1099 (1962)

11.5 G. Frank, E. Kauer, H. Köstlin: Phys. Bl. **34**, 106 (1978)

11.6 J.C. Phillips: in *Solid State Physics*, Vol. 18, ed. by F. Seitz, P. Turnbull (Academic, New York 1966)

11.7 G.W. Gobeli, H.Y. Fan: in *Semiconductor Research*, 2nd Quarterly Report, Purdue Univ. (1956)

11.8 E.J. Johnson: in *Semiconductors and Semimetals*, Vol. 3, ed. by R.K. Willardson, A.C. Beer (Academic, New York 1967) p. 171

11.9 H. Ehrenreich, H.R. Philipp: Phys. Rev. **128**, 1622 (1962)

11.10 M.D. Sturge: Phys. Rev. **127**, 768 (1962)

11.11 H. Lüth: *Solid Surfaces, Interfaces and Thin Films*, 4th edn. (Springer, Berlin Heidelberg New York 2001)

11.12 J. Geiger, K. Wittmaack: Z. Phys. **195**, 44 (1966)

11.13 H. Ibach: Phys. Rev. Lett. **24**, 1416 (1970)

Further Reading

Abeles, F. (ed.): *International Colloquium on Optical Properties and Electronic Structure of Metals and Alloys* (North-Holland, Amsterdam 1966)

Cho, K. (ed.): *Excitons*, Topics Curr. Phys., Vol. 14 (Springer, Berlin Heidelberg 1979)

Fröhlich, H.: *Theory of Dielectrics* (Clarendon, Oxford 1958)

Geiger, J.: *Elektronen und Festkörper* (Vieweg, Braunschweig 1968)

Greenaway, D.L., Harbeke, G.: *Optical Properties and Band Structure of Semiconductors* (Pergamon, Oxford 1968)

Haken, H., Nikitine, S. (eds.): *Excitons at High Densities*, Springer Tracts Mod. Phys., Vol. 73 (Springer, Berlin Heidelberg 1975)

Mitra, S.S., Nudelman, S. (eds.): *Far Infrared Properties of Solids* (Plenum, New York London 1970)

Nakajima, S., Toyozawa, Y., Abe, R.: *The Physics of Elementary Excitations*, Springer Ser. Solid-State Sci., Vol. 12 (Springer, Berlin Heidelberg 1980)

Nudelman, S., Mitra, S.S. (eds.): *Optical Properties of Solids* (Plenum, New York 1969)

Willardson, R.K., Beer, A.C. (eds.): *Semiconductors and Semimetals*, Vol. 3, Optical Properties of III–V-Compounds (Academic, New York 1968)

Chapter 12

12.1 J.R. Chelikowsky, M.L. Cohen: Phys. Rev. B **14**, 556 (1976)

12.2 J.H. Reuszer, P. Fischer: Phys. Rev. **135**, A1125 (1964)

12.3 E.M. Conwell: Proc. IRE **40**, 1327 (1952)

12.4 S.M. Sze: *Physics of Semiconductor Devices*, 2nd edn. (Wiley, New York 1969)

12.5 B. Gelmont, K. Kim, M. Shur: J. Appl. Phys. **74**, 1818 (1993)

12.6 H. Lüth: *Solid Surfaces, Interfaces and Thin Films*, 4th edn. (Springer, Berlin Heidelberg 2001)

12.7 G. Margaritondo, P. Perfetti: The Problem of Heterojunction Band Discontinuities, in *Heterojunction Band Discontinuities, Physics and Device Applications*, ed. by F. Capasso, G. Margaritondo (North-Holland, Amsterdam 1987) p. 59

12.8 H. Morkoç: Modulation-doped $Al_xGa_{1-x}As/GaAs$ Heterostructures, in *The Technology and Physics of Molecular Beam Epitaxy*, ed. by E.H.C. Parker (Plenum, New York 1985) p. 185

12.9 L. Esaki: Compositional Superlattices, in *The Technology and Physics of Molecular Beam Epitaxy*, ed. by E.H.C. Parker (Plenum, New York 1985) p. 185

12.10 A. Förster: Private Mitteilung (ISI, Forschungszentrum Jülich)

12.11 H. Heinecke, K. Werner, M. Weyers, H. Lüth, P. Balk: J. Crystal Growth **81**, 270 (1987)

12.12 F. Capasso (ed.): *Physics of Quantum Electron Devices*, Springer Ser. Electron. Photon., Vol. 28 (Springer, Berlin Heidelberg 1989)
12.13 U. Tietze, Ch. Schenk: *Halbleiterschaltungstechnik* (Springer, Berlin Heidelberg 1983)
12.14 K. Schimpf, M. Sommer, M. Horstmann, M. Hollfelder, A. van der Hart, M. Marso, P. Kordos, H. Lüth: IEEE Electron Device Letters **18**, 144 (1997)
12.15 H.-M. Mühlhoff, D.V. McCaughan: *MOS Transistors and Memories,* in *Handbook of Semiconductors*, Vol. 4, ed. by T.S. Moss, C. Hilsum (North-Holland, Amsterdam 1993) p. 265
12.16 H. Kressel, D.E. Ackley: Semiconductor Lasers, in *Handbook of Semiconductors*, Vol. 4, ed. by T.S. Moss, C. Hilsum (North Holland, Amsterdam 1993) p. 725

Further Reading

Allan, G., Bastard, G., Boccara, N., Lanoo, M., Voos, M. (eds.): *Heterojunctions and Semiconductor Superlattices* (Springer, Berlin Heidelberg 1986)

Bauer, G., Kuchar, F., Heinrich, H. (eds.): *Two-dimensional Systems: Physics and New Devices*, Springer Ser. Solid-State Sci., Vol. 67 (Springer, Berlin Heidelberg 1986)

Brauer, W., Streitwolf, H.W.: *Theoretische Grundlagen der Halbleiterphysik* (Akademie-Verlag, Berlin 1976)

Buchreihe: *Halbleiter-Elektronik*, Bd. 1–20, ed. by W. v. Heywang, R. Müller (Springer, Berlin Heidelberg)

Capasso, F., Margaritondo, G. (eds.): *Heterojunction Band Discontinuities, Physics and Device Applications* (North-Holland, Amsterdam 1987)

Chang, L.L., Ploog, K. (eds.): *Molecular Beam Epitaxy and Heterostructures*, NATO ASI Ser. E, No. 87 (Martinus Nijhoff, Dordrecht 1985)

Kelly, M.J.: *Low-Dimensional Semiconductors*, Materials, Physics, Technology, Devices (Clarendon Press, Oxford 1995)

Lanoo, M., Bourgoin, J.: *Point Defects in Semiconductors I and II*, Springer Ser. Solid-State Sci., Vols. 22, 35 (Springer, Berlin Heidelberg 1981, 1983)

Madelung, O.: *Grundlagen der Halbleiterphysik*, Heidelberger Taschenbuch, Bd. 71 (Springer, Berlin Heidelberg 1970)

Nizzoli, F., Rieder, K.-H., Willis, R.F. (eds.): *Dynamical Phenomena at Surfaces, Interfaces and Superlattices*, Springer Ser. Surf. Sci., Vol. 3 (Springer, Berlin Heidelberg 1985)

Parker, E.H.C. (ed.): *The Technology and Physics of Molecular Beam Epitaxy* (Plenum, New York 1985)

Reggiani, L. (ed.): *Hot-Electron Transport in Semiconductors*, Topics Appl. Phys., Vol. 58 (Springer, Berlin Heidelberg 1985)

Seeger, K.: *Semiconductor Physics*, 4th edn., Springer Ser. Solid-State Sci., Vol. 40 (Springer, Berlin Heidelberg 1988)

Shklovskii, B., Efros, A.L.: *Electronic Properties of Doped Semiconductors*, Springer Ser. Solid-State Sci., Vol. 45 (Springer, Berlin Heidelberg 1984)

Shockley, W.: *Electrons and Holes in Semiconductors* (Van Nostrand, Princeton 1950)

Spenke, E.: *Elektronische Halbleiter* (Springer, Berlin Heidelberg 1965)

Sze, S.M.: *Physics of Semiconductor Devices*, 2nd edn. (Wiley, New York 1969)

Ueta, M., Kanzaki, H., Kobayashi, K., Toyozawa, Y., Hanamura, E.: *Excitonic Processes in Solids*, Springer Ser. Solid-State Sci., Vol. 60 (Springer, Berlin Heidelberg 1986)

Yu, P.Y., Cardona, M.: *Fundamentals of Semiconductors*, 3rd edn. (Springer, Berlin, Heidelberg 2001)

Subject Index

Printing (Computer to Plate): Saladruck Berlin
Binding Lüderitz&Bauer, Berlin